COMPENDIUM of PROBLEMS IN GENETICS

John Kuspira
University of Alberta

Ramesh Bhambhani
University of Alberta

Wm. C. Brown Publishers
Dubuque, Iowa • Melbourne, Australia • Oxford, England

Dedicated to:
George W. R. Walker, a former colleague,
A true friend and an exemplary human being.
In recognition of his excellence as a teacher
And for the inspiration and encouragement
That he so generously provided, throughout
Our many years of association with him.

Book Team

Editor *Megan Johnson*
Developmental Editor *Robin Steffek*
Production Editor *Audrey Reiter*

Wm. C. Brown Publishers
A Division of Wm. C. Brown Communications, Inc.

Vice President and General Manager *Beverly Kolz*
Vice President, Director of Sales and Marketing *John W. Calhoun*
Marketing Manager *Carol J. Mills*
Advertising Manager *Janelle Keeffer*
Director of Production *Colleen A. Yonda*
Publishing Services Manager *Karen J. Slaght*

Wm. C. Brown Communications, Inc.

President and Chief Executive Officer *G. Franklin Lewis*
Corporate Vice President, President of WCB Manufacturing *Roger Meyer*
Vice President and Chief Financial Officer *Robert Chesterman*

Cover photo by CNRI/Science Photo Library/Photo Researchers, Inc.

Copyright © 1994 by Wm. C. Brown Communications, Inc. All rights reserved

A Times Mirror Company

Library of Congress Catalog Card Number: 92–75515

ISBN 0–697–16734–8

No part of this publication may be reproduced, stored in a retrieval system, or transmitted, in any form or by any means, electronic, mechanical, photocopying, recording, or otherwise, without the prior written permission of the publisher.

Printed in the United States of America by Wm. C. Brown Communications, Inc., 2460 Kerper Boulevard, Dubuque, IA 52001

10 9 8 7 6 5 4 3 2

Contents

1. Physical Basis of Heredity: Mitosis, Meiosis and Life (Chromosome) Cycles of Eukaryotes ... 1
2. Monohybrid Inheritance: The Law of Segregation ... 9
3. Dihybrid and Multihybrid Inheritance: The Law of Independent Assortment ... 14
4. Probability and Chi-Square ... 20
5. Gene Interaction and Lethal Genes ... 23
6. Multiple Alleles ... 32
7. Polygenic Inheritance ... 39
8. Sex Determination and Sex Differentiation ... 43
9. Sex Linkage, Sex-Influenced and Sex-Limited Characters ... 52
10. Linkage, Crossing-over and Genetic Mapping ... 63
11. Extranuclear Inheritance and Related Phenomena ... 77
12. Genetics of Bacteria and Viruses ... 86
13. Genotype - Environment Interactions; Expressivity, Penetrance, Phenocopies and Pleiotropism ... 103
14. Euploidy: Haploidy and Polyploidy ... 109
15. Aneuploidy ... 118
16. Chromosome Aberrations ... 126
17. Chemistry, Structure and Replication of Genetic Material and Chromosomes ... 138
18. Mutation and Repair ... 152
19. The Gene: Its Genetics and Interallelic Complementation ... 163
20. Biochemical Genetics ... 174
21. Protein Synthesis: Transcription and Translation ... 185
22. Coding, Colinearity and Suppressors ... 201
23. Development and Regulation ... 214
24. Population Genetics, Inbreeding, Outbreeding and Evolution ... 231
25. Current Approaches to Genetic Analysis: Somatic Cell Hybrids, RFLPs and Recombinant DNA ... 246

Appendix ... 260
    Table 1   Chi-Square Values
    Table 2   Coding Dictionary
    Table 3   Metric Equivalents

Answers to Selected Questions and Problems

Chapter 1	262
Chapter 2	265
Chapter 3	267
Chapter 4	270
Chapter 5	272
Chapter 6	275
Chapter 7	277
Chapter 8	279
Chapter 9	282
Chapter 10	285
Chapter 11	289
Chapter 12	292
Chapter 13	297
Chapter 14	299
Chapter 15	303
Chapter 16	307
Chapter 17	312
Chapter 18	318
Chapter 19	322
Chapter 20	327
Chapter 21	332
Chapter 22	339
Chapter 23	345
Chapter 24	354
Chapter 25	358

Selected texts and their chapters relevant to the questions in this text

Selected Texts

A) *Basic Genetics* - R.F. Weaver and P.W. Hedrick; 1st edition 1991; Wm. C. Brown Publishers

B) *Concepts of Genetics* - W.S. Klug and M.R. Cummings; 3rd edition 1991; Macmillan Publishing Company

C) *Genetics* - P.J. Russell; 2nd edition 1990; Scott, Foresman and Company

D) *Principles of Genetics* - R. Tamarin and R.W. Leavitt; 4th edition 1993; Wm. C. Brown Publishers

E) *An Introduction to Genetic Analysis* - D.T. Suzuki, A.J.F. Griffiths, J.F. Miller and R.C. Lewontin; 4th edition 1989; W.H. Freeman and Company

F) *Genetics* - R.F. Weaver and P.W. Hedrick; 2nd edition 1992; Wm. C. Brown Publishers

G) *Genetics* - M.W. Farnsworth; 2nd edition 1988; Harper & Row, Publishers

H) *Understanding Genetics* - N.V. Rothwell; 4th edition 1988; Oxford University Press

I) *Genetics: Human Aspects* - A.P. Mange and E.J. Mange; 2nd edition 1990; Sinauer Associates, Inc., Publishers

Corresponding chapters in this and selected texts

Chapter in this text	Corresponding chapters in selected texts								
	A	B	C	D	E	F	G	H	I
1	4	2	1	3	3	4	4	2/3	5
2	2	3	2	2	2	2	1	1	2
3	2	3	2	2	2	2	2	1	11
4	2	3		4	5	2	2	6	10
5	3	4	4	2	4	3	3	4	
6	3	4	4	2	4	3	3	4	10/17
7	3	23	23	18	4	3	20	7	23
8	4	5	3	5	3/5	4	5	5	2/8
9	4	5	3	5		4	5	5	2
10	5	6	5/6	6	5/6	5	6	8/9	11
11		22	22	17	20	18	19	20	6/14
12	13	15	7	7	10	14	10/11	15/16	
13	3		2		4	3	2	4	12
14	4	7	17	8	9	4	7	10	
15	4	7	17	8	9	4	7	10	7/8
16	4	7	17	8	8	4	7	10	9
17	12	14	16	16	7/15	6/10	16	14	14
18	6	8/9/10	9/10	14/19	11	12	9	11	4/13
19	10	18	9/10	9	12	14	12	11	13
20		13	14		12	7	8	12	16
21	9/11	12	11/12	10/11	13	9/11	13	12/13	13
22	11/7	11	3	11	13		11	12/14	14
23	14/10	19/20	20/21	15/13	16	10/15/17	17/18	17	6
24	16/17	25/26	24	20	24	19/20	20/22	21	20/21
25	15/5	16	15	12	15	16/17	15	9/18/19	15

Preface

The basis for a well-founded course of instruction in genetics has been eloquently and succinctly described by Sinnot and Dunn[1]:

The principles of genetics have been developed out of arduous study of scores of investigators, and understanding of principles can best be gained by the student through a process similar to that employed in their original discovery. This process begins with, and is continually stimulated by, curiosity as to the methods and the mechanism of inheritance; it proceeds by the collection and study of facts, and by a critical discrimination between those which are true and relevant and those which are untrue and irrelevant; and finally it involves a considerable practice of the reasoning faculty by which deductions are made, and applied or tested on many similar cases. It is only in this way that the process of inheritance can be understood. The learning of facts alone cannot accomplish this.

The authors of *"Problems in Genetics"* have sought to facilitate some of these aspects of instruction that are rarely found in texts, viz. "the collection and study of facts" and "practice of the reasoning faculty by which deductions are made and applied or tested on many similar cases". The authors have attempted to cover all the basic concepts of genetics by including problems on as wide a range of organisms as possible and emphasizing current trends of thought and technique. The problems are intended to assist students in their study and review of lecture topics. A large number of the problems are based on actual experimental data, so as to allow the student to gain a sense of true exploration in the science. To emphasize this feature we have included, within parentheses, the references to the papers and texts which were used or referred to in setting some of the problems. We have tried to include data from several classic and significant papers in genetics. It is hoped that students, especially upper undergraduates and graduates, will use them to explore the science beyond the confines of texts and this book. Instructors are advised to review questions chosen for assignments carefully to make sure that their students are adequately prepared to handle them. *"Problems in Genetics"* should complement any current text in genetics. The authors are confident that instructors will find the text useful for teaching purposes.

In this book taxonomy is concerned only with a clear identification of the organisms used in each experimental study, together with their genetic variants. Some laxity has therefore been permitted in the current binomial nomenclature, so that outdated names are retained (along with their current equivalents) to facilitate reference to original papers. In some instances scientific names have been omitted in favour of their colloquial equivalents.

The text also provides solutions to selected questions and problems from each of the chapters. In most cases the reasoning involved in arriving at the solution to the problem is clearly described.

A list of recently published texts and chapters therein that correspond to chapters in *"Problems in Genetics"* is also provided.

Acknowledgements

We would be remiss if we failed to recognize the contributions of all geneticists, in particular the many scientists including Mendel, whose data we have utilized in formulating the majority of the problems and questions.

We gratefully acknowledge the patience and understanding of our wives, Joanne Marlene and Lael Clarise. The co-operation of all our colleagues in the Department of Genetics at the University of Alberta, in particular that of Drs. Asad Ahmed, John Bell, Ross Hodgetts, Phil Hastings, Heather McDermid, Linda Reha-Krantz, Michael Russell, in providing some excellent questions that have been included in chapters 17, 18 and 25, is sincerely appreciated. We are deeply indebted to James Maclagan and Kym Banks for their expert assistance with the typing and painstaking checking of details of the entire manuscript. We are also grateful to Sophie Ogle, Anil Bhambhani, Darren Bolding and Greg Rairdan for their assistance with various aspects of this undertaking.

---

[1] Sinnot, E.W. and L.C. Dunn *Principles of Genetics*, p xiii, McGraw-Hill Book Company, New York, 1939.

Chapter 1
Physical Basis of Heredity: Mitosis, Meiosis and Life (Chromosome) Cycles of Eukaryotes

1. a) From a geneticist's point of view, what is the major function of mitosis and how is it accomplished?
   b) How does mitosis (understood here to include both nuclear and cytoplasmic division) in animals differ from mitosis in plants?
   c) Is the process of mitosis basically the same or different in diploid and haploid cells of the same organism? In different organisms of (i) the same species, (ii) different species? Explain.
   d) Explain why reductional division cannot occur during mitosis.

2. Rarely do all the somatic cells of a multicellular individual have the same chromosome number; e.g., in humans, although most of the body cells have a 2n=46 chromosome number, at least some of the cells of the liver have 92 (4n) chromosomes. How can a 4n cell arise from a 2n one?

3. In a diploid insect species, there are 10 chromosomes in each of its somatic cells. How many bivalents would you expect to observe in its meiocytes at prophase I of meiosis if:
   (i) pairing was at random between chromosomes within a genome?
   (ii) pairing was at random between chromosomes of the 2 genomes?
   (iii) pairing was specifically between homologues in the 2 genomes?
   Indicate how the meiotic products (gametes) would enable you to distinguish between (ii) and (iii).

   Which of the above situations is characteristic of meiosis as it occurs normally in a diploid species?

4. At interphase chromosomes are not visible with the light microscope. Nevertheless, we know they are there, having maintained their integrity from the previous division. Suggest ways of determining whether a particular interphase nucleus is n, 2n, or polyploid (3n, 4n, etc.).

5. a) Which, if either, of the two cells illustrated is haploid? Why?

   b) Which of the mitotic stages can these cells not be at? Why?

6. a) What in your opinion is the main function of meiosis? How is this accomplished? What makes such a process necessary?
   b) What genetic functions does meiosis perform? Explain.

7. What would be the consequences to future generations of a diploid organism if meiosis was omitted from its life cycle?

8. a) In what main features does the first division of meiosis differ from mitosis?
   b) How does the second division of meiosis differ from mitosis?

9. Explain whether the following statements are true or false:
   (i) A chromosome may synapse with any other in the same cell at zygotene.
   (ii) During mitosis chromosomes divide into chromatids which separate at anaphase to form two daughter nuclei, each with the same number of chromosomes as the mother cell.
   (iii) A secondary meiocyte has half as many chromosomes as a primary meiocyte.

(iv) In a primary spermatocyte containing 18 chromosomes, 15 of these chromosomes may be paternal.
(v) The orientation of one bivalent is random with respect to the poles of a meiocyte and independent of the orientation of other bivalents in the cell.
(vi) A microspore may contain more paternal chromosomes than the somatic cells of the same plant.

10. Is it possible for meiosis to occur in haploid species? Haploid individuals? Explain.

11. If half of the chromosomes of a primary oocyte segregate into the first polar body, why aren't some of the different kinds of chromosomes of the diploid chromosome complement absent from the egg?

12. The somatic cells of humans (*Homo sapiens*) contain 23 pairs of chromosomes (Tjio, J.H. and A. Levan, Hereditas 42: 1, 1956). What is the relationship between the members of each pair of chromosomes in the somatic cells of each individual and those in the somatic cells and meiocytes of the parents of each individual? Explain, using one pair of chromosomes.

13. In *Haplopappus gracilis* (2n=4), one of the Compositae, a zygote receives the chromosomes A and B from the male parent and their homologues A' and B' from the female parent. Which of the following chromosome complements would you expect to find in the somatic cells of the plant arising from this zygote: A'A'BB, AA'B'B', AABB, AA'BB', A'A'B'B'? Why?

14. a) Explain, with the aid of diagrams, how meiosis reduces the chromosome number to one-half the somatic number whereas mitosis produces nuclei with the same chromosome number as the parent nucleus.
b) Is it theoretically possible to simplify meiosis so that only one nuclear division would be required to obtain cells containing the haploid complement of chromosomes and genes? Explain your answer with the aid of diagrams.

15. In *Euschistus variolarius*, a hemipteran insect, there are 14 chromosomes in the somatic cells. Seven bivalent associations were always found at prophase I of meiosis (Montgomery, T.H., Trans. Am. Philos. Soc. 20: 154, 1901).
a) Argue that this is evidence that bivalent associations at prophase I are always made up of one paternal and one homologous maternal chromosome.
b) Show what you would expect if pairing were between chromosomes of a given parent.

16. *Crepis neglecta*, a member of the Compositae, has four pairs of chromosomes (Babcock, E.B., Bot. Rev. 8: 139, 1942). Assume that a plant receives the chromosomes A, B, C, and D from the male parent and the homologues A', B', C', and D' from the female parent.
a) If no crossing-over occurs:
(i) What proportion of the gametes of this plant will be expected to carry all the chromosomes of paternal origin?
(ii) Would the proportion of gametes carrying only chromosomes of the maternal parent be the same? Explain.
(iii) What proportion of the gametes would carry chromosomes of both the male and the female parent?
b) Derive a formula that indicates the number of different kinds of gametic chromosomal combinations an individual with (n) number of pairs of chromosomes can produce given that each chromosome pair is qualitatively different.

17. In sampling cells of a diploid plant species (2n=18) at random and measuring their DNA content, a geneticist obtained the following results which are representative of all other measurements:

Cell #	Amount of DNA per cell nucleus (a.u.)
1	1.2
2	0.6
3	2.4
4	1.3

Answer the following questions giving one reason for each answer:
(i) Which of the cells is representative of the haplont (haploid) generation? Which stage of the cell cycle is it at? Is this cell a gamete?

(ii) How much DNA would the geneticist have detected in a meiocyte nucleus at prophase I?
(iii) How much DNA would you expect to be present in a secondary meiocyte which had just been formed? How much DNA would be present if this cell was at prophase II?
(iv) Which of the cells could be at (a) metaphase II, (b) metaphase of mitosis, (c) prophase of first mitotic division of the male gametophyte?
(v) The geneticist finds a highly sterile but vigorous plant whose somatic cells possess 3.6 amount of DNA per nucleus. What is the chromosome number of this plant? Was DNA content measured at G1 of interphase or later in the cell cycle?

18. In a cytological study of *Brachystola magna*, a grasshopper, Carothers. E.E. (J. Morph. 24: 487, 1913) examined 300 cells showing metaphase I or anaphase I of meiosis in males with 23 chromosomes. She found 11 bivalents, 1 of them heteromorphic (one long, one short), and an unpaired chromosome. This unpaired chromosome (the X) moved in its entirety to one pole or the other at anaphase I. In 154 cells the large chromosome of the heteromorphic pair went to the same pole as the X chromosome, and in 146 cells the shorter one segregated with the X chromosome. What conclusions can you draw regarding the manner in which chromosome pairs orient and segregate in relation to each other?

19. In *Allium cepa* (2n=16), meiosis consists of only one nuclear and cell division in megasporocytes which gives rise to two **haploid** (n) meiotic products in which the genomes possess a C amount of DNA.
(i) Explain, with the aid of diagrams, including pre-meiotic and meiotic events, how the meiocytes give rise to two rather than four haploid meiotic products.
(ii) Would you expect some of the allele pairs in each of the chromosome pairs to show second-division segregation? Explain.

20. Disjunction (segregation) of homologues, chromosome pairing, replication, crossing-over and disjunction of sister chromatids are five functionally important events in the process of meiosis.
(a) In what temporal sequence do the five events occur during a normal meiosis?
(b) Explain briefly which of the above event(s) or features of meiosis account for the orderly segregation of the two alleles at each gene locus in accordance with Mendel's law of segregation.

21. A plant, (2n=8), has a karyotype shown below consisting of rod-shaped and V-shaped chromosomes. What proportion of the progeny will have only rod-shaped chromosomes after four generations of self-fertilization? What proportion of the progeny will have the original karyotype? Explain.

22. In the tomato (*Lycopersicum esculentum*), chromosome 2 has a satellite (S) on the short arm (Brown, S.W., Genetics 34: 437, 1949). Suppose you discover a plant that has a satellite missing from one of the second chromosomes and a satellite attached to one of the third chromosomes, so that the karyotype of the plant with respect to these chromosomes is as shown.

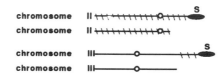

a) What types of meiotic products (with respect to chromosome number and kind) would you expect on the male side and in what proportions would you expect to find them?
b) Would the same results be obtained on the female side? Explain.
c) The plant is self-fertilized. Show the types of chromosome complements you would expect in a large population of offspring and the proportions of each.

23. In armadillos (*Dasypus novemcinctus texanus*) the pattern of armoured bands around the body frequently varies within, as well as between families. However, the pattern is always the **same** in all offspring (usually 4) of a litter (Newman, H.H., Am. Nat. 47: 513, 1913).
a) Explain why the latter is the case?

b) Offer an explanation for the fact that members of different litters from the same parents may differ in their patterns of armoured bands.

c) Would you normally expect members of any litter to be of different sex? Explain.

24. In some coccids, meiocytes and gametes are 2n and n in chromosome number, respectively, as in eukaryotes with the common pattern of meiosis. However, the secondary meiocytes in these insects are 2n, whereas, those in eukaryotes with the common pattern of meiosis are n. Suggest how the events at meiosis in the coccids might differ from those in eukaryotes with the common pattern to account for both the similarities and the difference between the two patterns. (Assume 2n=2; each meiocyte produces four haploid meiotic products).

25. (a) The housefly (*Musca domestica*) has 12 chromosomes in its somatic cells: three long pairs, one metacentric, one acrocentric, and one telocentric; three short pairs, one metacentric, one acrocentric, and one telocentric.
a) What proportion of the spermatids of such a male would you expect to possess:
(i) A long metacentric pair, a long acrocentric pair and a short telocentric pair?
(ii) Two telocentric pairs and one acrocentric pair?
(iii) Two metacentrics, two acrocentrics, and two telocentric chromosomes?
(iv) Three long chromosomes (one metacentric, one acrocentric and one telocentric) and three short chromosomes (one metacentric, one acrocentric and one telocentric)?
b) What kinds of chromosomes should the progeny possess if this male is mated with a female with the same number and kinds of chromosomes?
c) A spermatid is found with the same number and kinds of chromosomes as in the body cells. How could such a cell arise? Illustrate.

26. In all the meiocytes of the grasshopper, *Locusta migratoria*, there is only one chiasma in some bivalents. Moreover, in these bivalents the chiasma is located near the centromere and does not terminalize until metaphase I. In 1978 Tease, C. and G.H. Jones (Chromosoma 69: 163, 1978) using *Brdu*-substitution and fluorescence plus Giemsa (FPG) staining observed that:
(i) Before the occurrence of crossing-over, one chromatid of each chromosome in each of these bivalents was lightly stained and the other was darkly stained.
(ii) In 50 percent of the meiocytes, the bivalents with a single chiasma each revealed a chiasma between a light and a dark non-sister chromatid. In 25 percent of the meiocytes chiasmata were formed between two light non-sister chromatids and in a similar proportion of meiocytes, chiasmata were observed between two dark non-sister chromatids.
(iii) In 50 percent of the meiocytes at anaphase I, one chromatid of each chromosome was composed of light and dark segments. The positions of ligation of light and dark segments in each of these chromatids corresponded to the positions of the chiasmata at prophase I (diplotene).

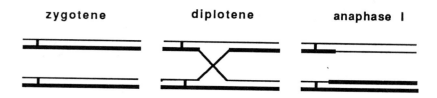

What conclusions can you draw from these data:
(a) regarding the relationship between crossing-over and chiasma formation?
(b) as to the involvement (random or not) of non-sister chromatids in crossing-over and therefore chiasma formation?
(c) with respect to the mechanism of crossing-over?

27. The males in the *Hymenoptera* (bees, wasps), *Icerya* coccids, and a few other groups develop from unfertilized eggs and are haploid (Hughes-Schrader, S., Adv. Genet 2: 127, 1948; White, M.J.D., *Animal Cytology and Evolution,* Macmillan, New York, 1954). Nevertheless they produce haploid gametes that fertilize the haploid gametes of the female to give rise, with few exceptions, to 2n females. In what way might meiosis be modified in these males to produce haploid gametes?

28. In certain organisms the chromosomes have been found to have diffuse centromeres that extend along their entire length, e.g., in homopteran and hemipteran insects (Brown, S.W., and Nelson-Rees, W.A., Genetics 46: 983, 1961; Hughes-Schrader, S., Adv. Genet 2: 127, 1948) and *Luzula*, a member of the plant family *Juncaceae* (Castro, et al., Genet. Iber. 1: 48, 1949). It is probable that certain algae and fungi also have this feature (Godward, M.B.E., Ann. Bot. 18: 144, 1954).
a) In what respects does their mitotic chromosome behaviour differ from that in organisms with localized centromeres?
b) With the aid of diagrams compare and contrast meiosis in organisms with localized centromeres to that in organisms with diffuse centromeres, with respect to:
(i) Synapsis at prophase I.
(ii) Type of orientation (auto- or co-) at metaphase I.
(iii) Type of disjunction and segregation (for a particular locus) at anaphase I.
(iv) The chromosome number of secondary meiocytes.
(v) Second division of meiosis (type of orientation, disjunction, and segregation).
(vi) Chromosome number of the meiotic products.

29. The male and female hybrids from crosses between the horse (2n=64) and the donkey (2n=62) have 63 chromosomes. If there is no pairing between the chromosomes of the two species at meiosis would you expect the hybrids to be fertile or sterile? Why?

30. a) What are the main differences between the life cycles of higher plants and animals?
b) Illustrate with a sketch the salient features of the life cycles of:
(i) *Homo sapiens*, typical of multicellular diploid animal eukaryotes.
(ii) *Neurospora crassa*, a haploid fungus typical of haploid plant eukaryotes.
(iii) Corn (*Zea mays*) or some other plant typical of diploid plant eukaryotes.
c) Compare these life cycles with respect to:
(i) The conspicuous phase.
(ii) The relationship of meiosis and syngamy to the haplontic and diplontic phases.
(iii) Gamete formation (whether direct or indirect).
(iv) Mitosis - its place of occurrence and its function.

31. a) Describe the origin and development of the endosperm. In what group of organisms does it occur, and what is its function?
b) In corn (*Zea mays*), a monoecious species, the 3n endosperm may be *flinty*, F or *floury*, F'. Reciprocal crosses between true-breeding *flinty* and true-breeding *floury* strains give the following results:

Parent		Endosperm type in seeds
♀	♂	
Flinty x	*Floury*	*Flinty*
Floury x	*Flinty*	*Floury*

(i) Sketch as accurately as possible a pollen grain (male gametophyte) and an embryo sac (female gametophyte) from each strain. Identify all nuclei and show the genotypes of each in both structures.
(ii) Explain genetically why reciprocal crosses give different results.

32. In what ways does syngamy in higher animals differ from that in higher plants?

33. Distinguish between cross-fertilization and self-fertilization and state what type or types can occur in:
a) A unisexual (dioecious) species.
b) A bisexual (monoecious) species.

34. Discuss the statement: The human egg is much larger than the human sperm, yet a child inherits equally from both parents.

35. In a number of species, including humans, it is known that some characters are determined by hereditary determiners in the cytoplasm. Would children more often resemble their

mothers or their fathers with respect to characters of this nature? Explain.

36. (a) Which form of reproduction, sexual or asexual, would you expect to generate greater phenotypic variation between parents and offspring and among offspring, and why?
(b) Would you consider genetic variability an advantage or disadvantage to the survival of a species under natural conditions?
(c) What is one possible advantage of sexual reproduction to the maintenance of a species?

37. You find a very small plant species in which some individuals are *green* and others are *bluish* in colour. Only one of the two phases of the life cycle, the diplont or the haplont, is conspicuous (large enough to be seen and studied). After crossing the two, you find that the progeny plants are *green* and *bluish* in equal numbers.
(a) Is the species diploid or haploid? Why?
(b) Outline one type of study you might undertake to verify or refute your answer to (a). Indicate the results expected.

38. In the sea lettuce (*Ulva ulva*), a multicellular alga, both the haplont and the diplont phases of the life cycle are equally large and conspicuous and morphologically the same. A plant with *rough* leaves is crossed with a plant with *smooth* leaves. All the progeny have *smooth* leaves.
(i) Is it possible to determine whether the parents of the *smooth*-leaved progeny were diploid or haploid? Explain with the aid of diagrams and using your own gene symbols.
(ii) Briefly indicate one type of study you might undertake to verify or refute your answer in (i). Indicate the results expected if your answer in (i) is (a) correct, (b) incorrect.

39. Organisms are frequently described as *diploid* or *haploid*. What is meant when these terms are used?

40. A *black* Hydra, isolated from the rest of the members of the species, produces some *black*, some *brown* and some *yellow* offspring all with the same karyotype as the parent. Is the species haploid or diploid? Unisexual or bisexual? Explain?

41. *Crepis capillaris* is a sexually reproducing bisexual plant species. The karyotype of the somatic cells of the diplont phase of the life cycle of all individuals in all generations is identical; all possess a 2n number of six chromosomes with the morphologies sketched below.

After showing the karyotype of a cell of the haplont phase, briefly explain, with the aid of diagrams, why the chromosome number and kind in *Crepis* remains constant (i) within the diplont and haplont phases and (ii) from one diplont and haplont phase to the next.

42. Substantiate with a reason whether each of the following statements is true or false:
a) Syngamy is a process involving two cells; it occurs only in diploid organisms.
b) In dioecious organisms the terms male and female are always applicable.
c) Organisms that are hermaphroditic cannot cross-fertilize.
d) Sex is always necessary for biological multiplication.
e) Cytokinesis and karyokinesis always occur concurrently.
f) In certain isogamous species either gametic type may develop parthenogenetically into an adult, while in oogamous species only the eggs are capable of developing in this manner.
g) Individuals produced by parthenogenesis in diploid organisms are always diploid.

43. If the chromosome carries genetic material, all the body cells derived by mitosis should possess an identical genotype. Describe how you would proceed to test this hypothesis using a plant like the carrot or the geranium.

44. The diplontic sporophyte and the haplontic gametophyte of the brown algae are both large plants. A botanist discovers a new species of this group and identifies two forms A and B, which he suspects represent the two phases of the life cycle. He distinguishes two classes of form B, *elongate* and

ovata, and only one of form A, *ovata*. Experiments show the following:
(i) Isolated plants of form A produce young form B plants of both classes.
(ii) Isolated plants of form B produce no young.
(iii) A pair of plants of form B, one *elongate* and one *ovata*, give rise to young plants of form A.
a) State which form represents the haplont generation and why.
b) Explain the genetic data, showing the genotypes for form A and the two types of form B.
c) Is the species self- or cross-fertilizing?

45. a) In bread wheat (*Triticum aestivum*) the haploid number of chromosomes is 21. How many chromosomes would you expect to find in the following:
(1) The tube nucleus?   (2) A microsporocyte?
(3) One of the polar bodies?   (4) An aleurone cell nucleus?
(5) A root-tip nucleus?   (6) An egg nucleus?
(7) A leaf cell?

b) In humans (*Homo sapiens*) the diploid chromosome number is 46. How many chromosomes would you find in the following:
(1) An egg?   (2) A brain cell?
(3) A red blood cell?   (4) A white blood cell?
(5) A primary spermatocyte? (6) A spermatid?
(7) A polar body?   (8) An oogonium?
(9) A cell in the cornea of the eye?

46. A chromosome mutation in a dog deletes a large segment of heterochromatin from one of the ends of chromosome 9. A viable and fertile male possesses this mutant chromosome 9 as well as the normal homologue; the members of this pair of homologous chromosomes are heteromorphic, one long and one short. At anaphase I the chromosome with the two long chromatids **always** segregates to one pole and therefore the homologue with the two short chromatids **always** moves to the opposite pole.
(i) Where is the centromere located on chromosome 9? Explain with the aid of a simple diagram.
(ii) None of the chromosome pairs in this species has more than two reciprocal exchanges per meiocyte. How many occur in each meiocyte in this heteromorphic chromosome pair? Explain.

47. The common red fox (*Vulpes vulpes*) has 38 chromosomes in its somatic cells; the Arctic fox (*Alopex lagopus*) has 50, some of which are smaller than those in the red fox (Gustavsson I. and C.O. Sundt, Hereditas 54: 249, 1965). Hybrids from crosses between the two species are sterile. Studies of meiosis in these hybrids by Wipf, L. and R.M. Shackleford (P.N.A.S. 35: 468, 1949) have shown that both bivalents and univalents are present at diakinesis and metaphase I.
i) How many chromosomes would you expect to find in the somatic cells of the hybrids?
ii) Account for the observed cytological behaviour.
iii) Is there any relationship between the cytological behaviour and the sterility of the hybrids? Explain.

48. Chromosomes at metaphase I - anaphase I in a hybrid between two phenotypically different plants (one with a known somatic chromosome number of 4) behave as illustrated.

(i) Are the parents of the hybrid likely to be members of the same species or not? Why?
(ii) What proportion of the meiotic products of the hybrid would you expect to have the same chromosome number as its somatic cells?
(iii) What is the somatic chromosome number of the second plant? Explain.

49. A pollen grain (male gametophyte) with nuclei labeled A, B and C fertilized an embryo sac (female gametophyte) with nuclei labeled D, E, F, G, H, I, J and K as shown below:

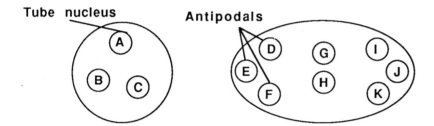

(i) Identify each of the nuclei B, C, G, H, J, and K.
Of the following combinations of nuclei:
(1) ABF (2) CJ (3) BGH (4) BCJ (5) CK
(ii) Which is likely to be found in the zygote?
(iii) Which of these could be found in the aleurone layer of the seed?
(iv) Which combination would be found in the pollen tube?
(v) Which of the nuclei in the male gametophyte would contain genetically identical genomes?
(vi) Which of the nuclei in the embryo sac would be chromosomally and genetically identical?
(vii) Which of the nuclei in these two gametophytes will not be found in zygotes, embryos and endosperms?

Chapter 2
Monohybrid Inheritance:
The Law of Segregation

1. a) Does the phenotype of one generation ever affect the genotype of the next? Explain.
   b) "Individuals of identical phenotype may have different genotypes, and vice versa." Is this statement true or false? Explain.

2. a) Define and explain Mendel's law of segregation.
   b) State, giving evidence, whether or not segregation can occur at either of the meiotic divisions.
   c) Does segregation occur in asexual reproduction? In sexually reproducing homozygotes? Explain.

3. a) What genetic phenomenon that is important in diploids cannot usually be investigated in haploids?
   b) What are some of the advantages of genetic analysis of haploid organisms over that of diploid ones?

4. In guinea pigs *black* vs. *albino* is due to alleles of one gene, with the allele for *black* dominant to that for *albinism*. Castle W.E. and J.C. Phillips (Science, 30: 312, 1909) replaced the ovaries of an *albino* female with those from a homozygous *black* female. The female was then mated with an *albino* male and produced two *black* offspring.
   a) Show whether these results would be expected on the basis of Mendelian inheritance.
   b) What bearing have they on the validity of the Lamarckian doctrine of inheritance of acquired characters?

5. Expain why, in human families, many traits, e.g., *albinism*, *blue eyes*, and *phenylketonuria*, skip generations while traits such as *polydactyly, free earlobes,* and *A* and *B* blood groups do not.

6. a) Does a study of pedigrees always permit a person to determine whether an allele is dominant or recessive? Explain.
   b) Why is it much easier to analyse human pedigrees for autosomal dominant traits than for autosomal recessive ones?
   c) Why are parents of individuals homozygous for rare recessive alleles likely to be related?
   d) Briefly discuss the conditions under which a recessive trait may appear to be inherited as a dominant one and vice versa and the precautions necessary in drawing conclusions from pedigree analysis.

7. A population geneticist collects 400 deer mice (*Peromyscus maniculatus*) from an area near Jasper, Alberta. All these mice have *normal* ears. He raises about 800 animals in each of the following four generations (all matings occur at random) and finds that all except one male in the fourth generation have *normal* ears. The exceptional male is *earless* but otherwise *wild-type* and fertile.
   a) Suggest three possible types of causes that could lead to the appearance of this *earless* male.
   b) Briefly discuss how you would proceed to distinguish among these alternatives and show the results expected with each cause in this experiment.

8. The dorsal pigmentary pattern of the common leopard frog (*Rana pipiens*) is genetically determined (Volpe, E.P., J. Hered. 51: 151, 1960). When *kandiyohi* frogs (mottling between the dorsal spots) are mated reciprocally with *wild-type* frogs (white between the dorsal spots), all the progeny are *kandiyohi*. The F_1 females mated with F_1 males produced 55 *kandiyohi* and 17 *wild-type* frogs. The reciprocal of the above cross produced 49 *kandiyohi* and 16 *wild-type* frogs.
   a) Which of these alternative patterns is due to a dominant allele?
   b) How many of the F_2 of dominant phenotype are expected to be heterozygous?
   c) Approximately how many of the F_2 of recessive phenotype are expected to be homozygous?
   d) How can you determine which of the F_2 individuals with the dominant phenotype are homozygous and which are heterozygous?

e) Why do reciprocal crosses between individuals in the true-breeding *kandiyohi* and *wild-type* lines produce the same F_1 and F_2 results?

9. An observant gardener finds that some of his bean plants have *pubescent* leaves and others have *glabrous* leaves. He crosses plants and obtains the results shown.

Cross	Parents			Progeny	
				Pubescent	*Glabrous*
1	*pubescent*	x	*glabrous*	56	61
2	*pubescent*	x	*pubescent*	63	0
3	*glabrous*	x	*glabrous*	0	44
4	*pubescent*	x	*glabrous*	59	0
5	*pubescent*	x	*pubescent*	122	41

a) Explain these results genetically.
b) Using your own gene symbols, give the genotypes of the parents of each cross.
c) How many of the *pubescent* progeny in crosses 2, 4 and 5 would you expect to produce *glabrous* progeny when self-fertilized?

10. *Peroneal muscular atrophy*, the onset of which occurs between the ages of 10 and 20, consists of a slow progressive wasting of the distal muscles of the limbs. A study of a series of family pedigrees shows that a person never has *peroneal muscular atrophy* unless at least one of the parents has also had it (Macklin, M.T. and J.T. Bowman, J. Am. Med. Assoc. 86: 613, 1926). How is this trait most probably inherited? Explain.

11. *Spastic paraplegia* (a rare trait) vs. *normal* are determined by different alleles of an autosomal gene (Rechtman, A.M. and B.J. Alpers, Arch. Neurol. Psychiat. 32: 248, 1934). An *affected* woman marries a *normal* man, and they have 5 children; 3 *normal* and 2 *affected*. Is the disease determined by a dominant or by a recessive allele? Explain.

12. In rabbits certain *short-haired* individuals when crossed with *long-haired* ones produce only *short-haired* progeny. Other *short-haired* individuals when crossed with *long-haired* ones produce approximately equal numbers of *short-haired* and *long-haired* offspring. When *long-haired* individuals are inter-crossed, they always produce progeny like themselves.
a) Outline an hypothesis to explain these results and show the genotypes of all individuals.
b) How would you proceed to test this hypothesis? Show the results you would expect in the crosses you describe.

13. When 20 purebred *Himalayan* female rabbits are mated with a *gray* male of unknown ancestry, 46 of the offspring are *Himalayan* and 52 are *gray*. A single pair of alleles is involved.
a) Is *Himalayan* determined by a recessive or dominant allele?
b) According to your answer in (a), how many offspring would you expect in each phenotypic class?
c) Test your explanation using the chi-square method and indicate whether you would accept or reject your hypothesis in (a).

14. In sheep, fat colour and ear size are each determined by one gene (Rae, A.L., Adv. Genet. 8: 189, 1956). The allele for *white fat* is dominant to that for *yellow fat*; the allele for *normal ears* is incompletely dominant to that for *earless*, the heterozygous individuals being *short-eared*. Both *yellow fat* and *earless* are undesirable traits.
a) In some Icelandic flocks the proportion of *yellow fat* lambs reaches 25 percent in some years. Breeders attempt to eliminate *yellow fat* animals from their flocks since the meat cannot be sold and the trait cannot be detected until the animals are slaughtered.
(i) If animals that have produced *yellow fat* offspring are not used for further mating, will this eliminate the recessive allele from a flock?
(ii) How could a breeder eliminate the recessive allele from his flock most efficiently?
(iii) A *white fat* ram is mated with two *white fat* ewes. The first ewe has 3 *yellow fat* and 2 *white fat* offspring. The second has 9 offspring, all *white fat*. What are the genotypes of the three parents?
b) (i) Which of these undesirable traits would be easier to eliminate from the flock? Explain.

(ii) In a flock in which all phenotypes occur for both characters how would you establish a true-breeding line for (1) *white fat* and (2) *normal* ears?

(iii) Would it be possible to establish a true-breeding line of *short-eared* sheep? If not, what cross would you make to obtain the maximum number of these individuals?

15. Eriksson, K. (Nord. Vet. Med. 7: 773, 1955) found that 65 of 143 foals from 124 mares mated with the pedigreed Belgian stallion Godvan were afflicted with *aniridia* (complete lack of the iris) and developed cataracts about 2 months after birth. Godvan had *aniridia*, but both his parents were normal. Account for the appearance of the affliction in both the foals and the stallion.

16. Isozymes are multiple biochemical forms of a protein or enzyme which differ in their physical-chemical properties. These variant molecules migrate at different rates in an electrical field because they also differ with respect to their electrical properties. Starch gel electrophoresis permits **all** the molecules of a given isozyme of an enzyme to collect at a specific location and thus be detected as a single band in the gel. Such bands represent the biochemical phenotype of an individual for a given enzyme. Gaur, P.M. and Slinkard, A.E. (J. Hered. 81: 455, 1990), using starch gel electrophoresis, obtained the phenotypes shown below for the enzyme alcohol dehydrogenase (ADH) in P_1, P_2, F_1 and F_2 plants of a cross in chick peas (*Cicer arientinum*).

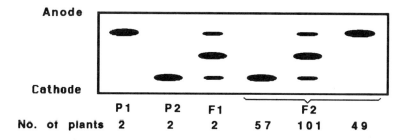

a) Is ADH a monomeric or a dimeric enzyme? Explain.

b) Provide a complete cytogenetic explanation for the results indicating the number of allele pairs, their allelic relationships as well as the genotypes of the P_1, P_2, F_1 and F_2 plants for the ADH enzyme.

c) If the 101 F_2 plants that exhibit a phenotype similar to that of the F_1 plants were self-fertilized, what proportion of these plants would you expect to produce progeny with a phenotype that resembled (i) the P_2 parent? (ii) the F_1 individuals?

17. In Guernsey cattle *normal* vs. *incomplete hairlessness* is due to a single pair of alleles, *H* vs. *h* (Hutt, F.B. and L.Z. Saunders, J. Hered. 44: 97, 1953). The following dispute is referred to you for solution:

Mr. Burch, who never has had any *incompletely hairless* calves in his herd of Guernseys, finds that of 46 calves from the matings of the heifers from his own bull to one bought from Mr. Fraser, 7 are *incompletely hairless*. He insists that Mr. Fraser's bull is responsible. Mr. Fraser states that the bull is only partially responsible. Burch does not agree. Fraser refers Burch to you for verification of his contention.

a) What would you tell Burch?

b) What evidence would you cite to absolve Fraser's bull from full responsibility? What matings would you suggest and what results would you require to confirm your stand?

c) If Fraser's explanation is correct, what numbers of *normal* and *incompletely hairless* calves should be expected among the 46?

d) How many of the 39 *normal* calves would you expect to be heterozygous?

18. A cross between two sweet pea plants produced 41 plants with *pink* flowers, 18 with *white* flowers, and 19 with *red* flowers.

a) What is the phenotype of each parent and why?

b) What phenotypes do you expect and in what proportions among the progeny of the following crosses:

(1) *white* x *pink*? (2) *red* x *red*? (3) *pink* x *pink*?

19. The *palomino* horse has a golden yellow coat with flaxen mane and tail, the *cremello* is almost white, and the *chestnut* is brown. These horses are identical with respect to the basic colour genotype: all are homozygous for a recessive allele (*b*) for *brown*. The table shows the results obtained

upon mating these types in various combinations (Castle, W.E. and F.L. King, J. Hered. 42: 61, 1951).

Parents	Offspring
cremello x cremello	all cremello
chestnut x chestnut	all chestnut
chestnut x cremello	all palimino
palomino x palimino	1 chestnut:1 cremello: 2 palomino
palomino x chestnut	1 palomino : 1 chestnut
cremello x palomino	1 cremello : 1 palomino

a) Describe the genetic control of coat colour as revealed by these results, including the allelic relationships.
b) Diagram the last three matings.
c) If you raised *palominos*, how would you set up your breeding program and why?
d) If a *chestnut* individual is mated with a *cremello* one, what is the chance of getting: (i) a *chestnut* animal? (ii) a *cremello* animal? (iii) a *palomino* animal? Explain.

20. In a certain eukaryotic species one strain is able to synthesize the amino acid arginine and another is not. These strains reproduce themselves true to type. The progeny of a cross between them consists of both types, in the following distributions:

Arginine synthesizers   85
Arginine nonsynthesizers   89

a) Are the two strains diploid or haploid? Explain.
b) What genetic law do these results support?

21. In *Neurospora crassa* a true-breeding strain unable to synthesize the vitamin thiamine is obtained by irradiating spores with X-rays. This strain is crossed with the *wild-type* (thiamine synthesizing) one, and the vegetative mycelia from each of 200 ascospore pairs, isolated in order from 50 asci, are tested for their ability to synthesize thiamine; 100 are synthesizers, and an equal number are not.
a) Diagram this cross as far as the meiotic products of the hybrids, illustrating in your diagram the homologues and alleles concerned.

b) Why is the phenotypic ratio exactly 1:1?
c) Which of the alleles, if any, is dominant?

22. In *Neurospora crassa*, tetrad analysis of a cross between a leucine, leu^+ and a leucineless, leu^-, strain showed a 1:1 segregation for $leu^+ : leu^-$ (Regnery, D.C., J. Biol. Chem. 154: 151, 1944). It was also found that 21 of the 31 tetrads analyzed showed second division segregation.
a) How could these tetrads be recognized?
b) What event during meiosis would lead to their production?

23. Centromere mapping entails the location (in terms of genetic distance) of genes on a chromosome relative to the centromere of that chromosome. It is best accomplished in eukaryotes such as *Neurospora* and *Sordaria*.
a) Discuss the genetic and cytological conditions that a species must possess so as to be amenable to centromere mapping.
b) In *Neurospora crassa*, chromosome 5 carries a gene that determines spore colour. A strain of one mating type (+) carries the allele for *gray* ascospores while a strain of the opposite mating type (-) contains the allele for *white* spores. One hundred and twenty asci, isolated from a cross between the 2 strains, revealed 6 different arrangements of the spore colors *gray* and *white*. Two of the arrangements are shown below of which the first was detected in 40 asci and the second in 10 asci.

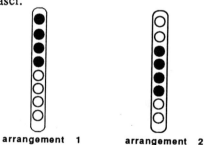

arrangement 1   arrangement 2

i) Account for the occurrence of the two spore colour arrangements that are represented above.
ii) Illustrate the four other arrangements of spore colours that are possible and account for the numbers that you would expect of each amongst the remaining 70 asci.
iii) Calculate the genetic distance between the spore colour locus and the centromere of chromosome 5 in *Neurospora*

crassa. Provide an explanation for the procedure used in your calculation.

24. The Treacher Collins Syndrome (TCS) is a hereditary disorder in humans that affects approximately one in 50,000 individuals. Some of its clinical features include abnormalities of the head, face and ears, resulting in hypoplasia of facial bones, cleft palate, deafness and other symptoms. The pedigree shown below was published by Dixon, M.J. et. al. (Am. J. Hum. Genet. 49: 17, 1991) and is typical of the disorder's mode of inheritance.

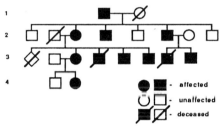

a) Provide a plausible explanation for the genetic basis of the disorder, indicating the number of genes, their chromosomal location (autosomal or X-linked) and the allelic relationship at each locus.
b) Using appropriate symbols for the gene(s), indicate the genotypes of all the individuals in the pedigree.

25. The figure below is part of a pedigree for *alkaptonuria* (from Khachadurian, A. and K.A. Feisal, J. Chronic Dis. 7: 455, 1958), an inborn error of metabolism that is simply inherited. State whether this portion alone unequivocally shows that the allele for *alkaptonuria* can be (1) dominant only, (2) recessive only, (3) either dominant or recessive.

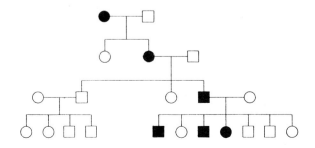

b) The trait is a very rare one. State how this observation will change your answer to (a) and why.
c) The complete pedigree of the family reported on by Khachadurian and Feisal is shown below. Does this confirm your answer to (b)? Why?

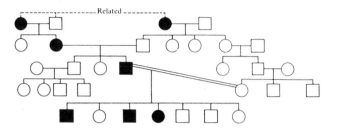

26. In the human pedigrees illustrated the rare traits (represented by solid squares and circles) are determined by alleles of single genes. For each pedigree:
a) State whether the trait is due to a dominant or recessive allele, giving your reasons.
b) Determine the genotype, or alternative genotypes, of each individual.

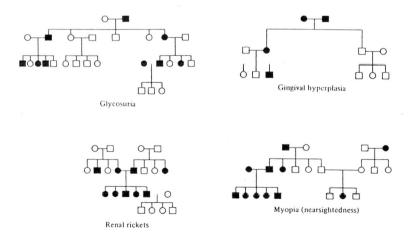

Chapter 3
Dihybrid and Multihybrid Inheritance:
The Law of Independent Assortment

1. What proportion of the progeny of the following crosses, in which the different allele pairs are segregating independently, will be homozygous?

   (a) *AaBb* x *AaBb*   (b) *AaBbCC* x *AaBbcc*   (c) *aabbcc* x *AAbbCC*

2. What evolutionary advantages do segregation, independent assortment, and random union of gametes confer upon a sexually reproducing species over an asexually reproducing one?

3. In silkworms (*Bombyx mori*), the hemolymph ('blood') may be *deep-yellow* or *white*. Fully developed larvae may be *plain* or *moricaud* (heavily lined and dotted) (Tanaka, Y., Adv. Genet. 5: 240, 1953). A cross is made between the two true-breeding strains *deep-yellow, plain* and *white, moricaud*. When 10 F_1 females are crossed with 10 F_1 males, the following progeny result:

Deep-yellow, moricaud	293
Deep-yellow, plain	96
White, moricaud	104
White, plain	38

   a) State how many allele pairs are involved and which alleles are dominant; give reasons for your answer.
   b) Would all F_1's have the same phenotype? Why?
   c) Using your own gene symbols, give the genotype of:
   (i) Each parent.
   (ii) The F_1's.
   (iii) The four F_2 phenotypes.
   d) Explain why the F_2 ratio is approximately 9:3:3:1. How many silkworms would you expect in each F_2 phenotypic class? Using the chi-square method, show that the actual results do not deviate significantly from the 9:3:3:1 ratio.
   e) What phenotypic ratio do you expect from crosses of F_1 males with females from the *white, moricaud* strain? Of F_1 males with females of the true-breeding *deep-yellow, plain* strain? Would you expect results different from these if F_1 females were crossed with males from the two strains? Why?
   f) Suppose you made a cross between a true-breeding *deep-yellow, moricaud* and a true-breeding *white, plain* strain? Would you expect the F_2 results to closely approximate those given? Explain.

4. In sesame both the number of seed pods per leaf axil and the shape of the leaf are monogenetically determined. The allele for *one-pod* condition is dominant to *three-pod*, and that for *normal* leaf is dominant to *wrinkled* (Langham, D.G., J. Hered. 36: 245, 1945). The two characters are inherited independently. The results of five crosses, each between a single pair of plants, gave the results shown. Determine the genotypes of the parents of each cross.

Parents	Progeny			
	1-*pod* normal	1-*pod* wrinkled	3-*pod* normal	3-*pod* wrinkled
1-*pod, normal* x 3 *pod, normal*	318	98	323	104
1-*pod, normal* x 1-*pod, wrinkled*	110	113	33	38
1-*pod, normal* x 3-*pod, normal*	362	118	0	0
1-*pod, normal* x 3-*pod, wrinkled*	211	0	205	0
1-*pod, wrinkled* x 3-*pod, normal*	78	90	84	88

5. In corn, the endosperm (3n) may be *sugary* or *starchy* and *floury* or *flinty*. Reciprocal crosses are made between true-breeding, *sugary, flinty* and *starchy, floury* strains. The phenotypes of the F_1's are shown. Why do reciprocal crosses give the same results for the first but not the second pair of traits? Illustrate your answer diagrammatically.

Parents		
♀	♂	F_1
sugary, flinty × starchy, floury		starchy, flinty
starchy, floury × sugary, flinty		starchy, floury

6. In corn, each of the characters kernel colour, height, and reaction to rust is determined by a single gene. The allele P for *purple* is dominant to p for *white*; the allele T for *tall* is dominant to t for *dwarf*; and the allele R for *resistance* is dominant to r for *susceptibility*. The allele pairs are on different (nonhomologous) chromosome pairs. A $PPTTRR$ strain is crossed with a $ppttrr$ one, and the F_1 is testcrossed.
a) With the aid of the branching method where needed, give:
(1) Genotypes of F_1's and the testcross progeny.
(2) The F_1 gametic ratio.
(3) The testcross genotypic and phenotypic ratios.
b) Explain *cytogenetically* (viz. by illustrating the number and kinds of metaphase I orientations and anaphase I segregations expected in a large population of meiocytes and their consequent products) why the genotypic ratio in the F_1 gametes is as given in (2).
c) How many genotypes and phenotypes do you expect in the F_2 and in what proportions?

7. A poultry geneticist made reciprocal crosses between two true-breeding strains of fowl, one with *rose-combs* and *blue-splashed white* plumage, the other with *single combs* and *black* feathers. After the cross was made, the investigator left for a year, instructing his technician to record F_1 phenotypes and to cross some of the F_1 birds with the *rose-comb, blue-splashed white* strain and other F_1's with the *single-comb, black* strain. The technician omitted the recording of the F_1 phenotypes but did make the crosses. The records of the offspring of the crosses are shown. Analyze the results and state your conclusions concerning:

a) The number of allele pairs segregating and the traits determined by the alleles of each gene.
b) The allelic relationship for the alleles of each gene.
c) Whether inheritance is independent or interdependent.

F_1 × *single-comb, black*		F_1 × *rose-comb, blue-splashed white*	
rose-comb, blue	64	rose-comb, blue-splashed white	108
rose-comb, black	57	rose-comb, blue	113
single-comb, blue	61		
single-comb, black	59		

8. The pedigree shows the pattern of transmission of two rare human traits, *cataracts* and *pituitary dwarfism*. Individuals with *cataracts* are indicated by a solid upper half of the symbol; those with *pituitary dwarfism* are indicated by a solid lower half.
a) What is the mode of inheritance of each of these traits? Explain.
b) IV-1 marries IV-6, and they have 5 children, 3 *dwarfs* with *no cataracts* and 2 *dwarfs* with *cataracts*. Does this verify the hypothesis formulated in (a)? Give the genotypes of the parents of this marriage.

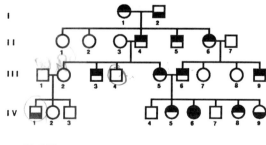

c) What phenotypes would you expect among progeny from the following marriages:
(i) III-5 × IV-1?
(ii) III-4 × IV-5?
d) What is the probability of:
(i) II-6 being heterozygous for the alleles for both traits?
(ii) II-2 being homozygous for the allele for *dwarfism*?

(iii) The first child of a mating of IV-2 and IV-4 being a *dwarf*?

9. Ability to *accept* or *reject* tissue transplants is a genetic property determined by a large number of histocompatibility genes in humans and mice. Skin transplants are accepted from one part of the body to another and from one individual to another of an identical genotype. When individuals are genotypically different, acceptance or rejection will depend on the actual genotypes of the donor and recipient. An "immune" (antigen-antibody) reaction appears to be responsible for acceptance or rejection of grafts. **For the graft to be accepted the donor must not contain any antigen for which the recipient can produce antibodies.** Typical results of crosses between different inbred (homozygous) lines in mice are as follows:

Donor	Recipient	Reaction
A	A	Accept
B	B	Accept
A	B (or reverse)	Reject
A or B	F_1 hybrid from cross of A x B	Accept
F_1	A or B	Reject
All F_2's	F_1	Accept
F_1	32 of 128 F_2's	Accept
A	72 of 128 F_2's	Accept
B	72 of 128 F_2's	Accept

The alleles at each histocompatibility locus are codominant. Analyse the above data and give:
a) the number of allele pairs involved and
b) the genotypes of inbreds A and B, F_1's and the F_2's that can accept skin from inbred A, inbred B and F_1's.

10. In his stocks of Mexican swordtail fish (*Xiphophorus helleri*) a pet-shop owner has two *montezumas* (bright orange-red), one with a *crescent* spot at the base of the caudal fin and the other with *twin* spots at this position. When crossed, they gave 12 *montezumas* and 9 *wild-type* (olive-green) progeny all with *crescent* markings. The *montezuma* F_1's when intercrossed produced:

montezuma, crescent	66	*wild-type, crescent*	31
montezuma, twin spot	24	*wild-type, twin spot*	12

The *wild-type* F_1's when intercrossed produced 122 progeny, all of which were *wild-type*; 88 of them had *crescent* markings and 34 had *twin* spots.
a) Explain the manner of genetic control of the two pairs of traits.
b) *Montezumas* with *twin* spots are in great demand. Outline the breeding procedure a breeder should use to reap the maximum profit from these types.

11. In the soybean (*Glycine max*), the cultivar Bragg is normally *nodulating* (i.e., it produces nitrogen-fixing nodules) and produces *white* hypocotyls. Mathews, A., et. al. (J. Hered. 80: 357, 1989) crossed a *nonnodulating* line of Bragg with the cultivar Clark which is *nodulating* and produces *purple* hypocotyls. The F_1's were *purple* and *nodulating*. The phenotypes of the F_3 progenies derived from self-fertilizing 133 F_2's are shown below:

white, nodulating	10
white, nodulating and *white, nonnodulating* in a 3:1 ratio	17
purple, nonnodulating	9
purple, nodulating and *white, nodulating* in a 3:1 ratio	13
white, nodulating	8
purple, nodulating; purple nonnodulating; white, nodulating and *white, nonnodulating*	33
purple, nonnodulating and *white, nonnodulating* in a 3:1 ratio	16
purple, nodulating	8
purple, nodulating and *purple, nonnodulating* in a 3:1 ratio	19

How many allele pairs are involved? Are they assorting independently? Explain by giving F_2 genotypes and their proportions.

12. A testcross of an F_1 results in a 1:1:1:1 phenotypic ratio.
   a) How many genes are involved?
   b) Is it possible to determine allelic relationships from these results? Explain.
   c) What other information is given by these results?

13. Mendel, G. (Verh. Naturforsch. Ver. Bruenn. 4: 3, 1866) crossed a pure-breeding strain which had *round, yellow* seeds with a pure-breeding strain that had *wrinkled, green* seeds. The F_1 seeds were all *round, yellow*. When he backcrossed the F_1 plants to the *wrinkled, green* parent, the progeny consisted of 55 *round, yellow*, 51 *round, green*, 49 *wrinkled, yellow*, and 53 *wrinkled, green* seeds. The cross F_1 x *round, yellow* parent gave rise to *round, yellow* progeny only. Explain his results and show the results you would expect in the F_2 generation of the original cross.

14. Radishes (*Raphanus sativus*) may be *long, round,* or *oval* in shape. The colour may be *red, white* or *purple* (Uphof, J.C., Genetics 9: 292, 1924). A *long, white* variety crossed with a *round, red* one produced *oval, purple* F_1's. The F_2 progeny segregated into nine phenotypic classes in the following numbers:

long, red	9	*long, white*	8
long, purple	15	*round, purple*	16
oval, red	19	*round, white*	8
oval, purple	32	*oval, white*	16
round, red	9		

   a) How many allele pairs are involved? Do they assort independently? What phenotypes would you expect in crosses between the F_1's and each of the parental strains?
   b) Give the genotypic and phenotypic ratios expected among the progeny of a cross between:
   (1) A *long, purple* and an *oval, purple* plant.
   (2) An *oval, purple* and a *round, white* plant.
   c) How many true-breeding varieties of radishes could be established?
   d) If *oval, purple* radishes were commercially preferred, state what lines should be maintained to produce them most profitably and why.

15. A man whose hobby is the rearing of Mexican swordtail fish (*Xiphophorus helleri*) crosses a strain true-breeding for *stippled* and *patternless* (lacking spots at the base of the caudal fin) with a strain true-breeding for *nonstippled* and caudal *twin spot*. He finds that all the offspring are *stippled* and possess a caudal spotting pattern unlike that of either parent, called *crescent spot*. When the offspring are permitted to mate at random, he finds that the progeny fall into six categories in the following actual numbers:

Crescent, stippled	83	*Twin-spot, nonstippled*	9
Crescent, nonstippled	27	*Patternless, stippled*	35
Twin-spot, stippled	29	*Patternless, nonstippled*	12

   He comes to you for a genetic explanation of these results, being particularly interested in raising true-breeding lines for the two *crescent* phenotypes. Using your own symbols, carry out the genetic analysis required, showing genotypes of parents and of F_1 and F_2 offspring and suggest how he might develop the true-breeding lines in question.

16. Determine the probable mode of inheritance of tongue colour in dogs (*Canis familiaris*) from the pedigree chart. State how it could be verified.

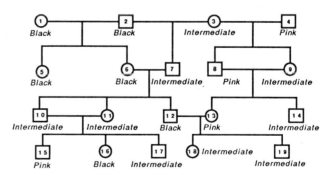

17. A corn plant homozygous for dominant alleles at four loci on four different (nonhomologous) chromosomes ($AABBCCDD$) is crossed with one homozygous for the corresponding recessive alleles ($aabbccdd$) to produce a tetrahybrid ($AaBbCcDd$) which is self-fertilized. The allele pairs control the expression of different pairs of traits.
   a) Give:

(i) The number of gametic genotypes produced by the tetrahybrid and their proportions.
(ii) The number of F_2 genotypes and their proportions.
(iii) The number of F_2 phenotypes and their proportions.
(iv) The proportions of the F_2 that would resemble each of the parents of the F_1 respectively.
b) What proportion of the tetrahybrid gametes will carry the dominant alleles at all four loci? What proportion will carry dominant and recessive alleles?
c) What is the probability that an F_2 individual will get all the dominant alleles? All the recessive alleles? All eight alleles?
d) What proportion of the F_2 genotypes are expected: to show the recessive phenotype for all four loci? To be homozygous for all dominant alleles?
e) Would your answers be different if the cross was $AAbbccDD \times aaBBCCdd$? Explain.

18. *Crepis capillaris* is a diploid bisexual plant species with 2n=6 chromosomes. A plant receives chromosome 1 with segments C and d and chromosome 2 with segment E from the male parent and their homologues with segments c and D and e respectively from the female parent. This plant's karyotype and genotype, including the correct positions of the c, d, and e segments, are shown below:

Five meiocytes in this plant produce the following kinds of tetrads with respect to these three pairs of segments:

Meiocyte	Tetrad
1	2 CDE and 2 cde
2	2 Cde and 2 cDE
3	1 Cde, 1 cDE, 1 CDe and 1 cdE
4	2 CdE and 2 cDe
5	1 CdE, 1 Cde, 1 cDE and 1 cDe

Classify the meiocytes and tetrads as PD (parental ditype), NPD (non-parental ditype) or TT (tetratype) and diagram the essential features of chromosome behaviour during pachytene and metaphase I in each of the five types of meiocytes that will account for its tetrad type.
In each of the first three meiocytes indicate whether each pair of corresponding segments shows first-division segregation or second-division segregation.

19. In *Chlamydomonas reinhardii*, arg^+ (non-arginine requiring) vs. arg^- (arginine-requiring) and thi^+ (non-thiamine requiring) vs. thi^- (thiamine-requiring) are each determined by a single pair of alleles (Levine, R.P. and W.T. Ebersold, Annu. Rev. Microbiol. 14: 197, 1960). A cross between arg^+thi^+ and arg^-thi^- gives 100 ordered (linear) tetrads of six different classes:

Tetrad Class					
1	2	3	4	5	6
arg^+thi^+	arg^+thi^-	arg^+thi^+	arg^+thi^-	arg^-thi^-	arg^-thi^+
arg^+thi^+	arg^+thi^-	arg^+thi^-	arg^+thi^+	arg^-thi^+	arg^+thi^+
arg^-thi^-	arg^-thi^+	arg^-thi^+	arg^-thi^-	arg^+thi^-	arg^-thi^-
arg^-thi^-	arg^-thi^+	arg^-thi^-	arg^-thi^+	arg^+thi^+	arg^+thi^-
38	41	7	6	4	4

a) Classify the 6 different tetrad classes for whether they are PD, TT, or NPD.
b) Are the two pairs of alleles assorting independently? Explain.

20. Tetrads of *Neurospora crassa* are ordered. Lindegren, C.C. (Bull. Torrey Bot. Club 60: 133, 1933) obtained the results shown from tetrad analysis of a cross between the strains *nonfluffy*, f^+ "plus" mating type, mt^+ and *fluffy*, f^-, "minus" mating type, mt^-.

Tetrad Class						
1	2	3	4	5	6	7
f^+mt^+	f^+mt^-	f^+mt^+	f^+mt^+	f^+mt^-	f^+mt^-	f^+mt^+
f^+mt^+	f^+mt^-	f^+mt^-	f^+mt^+	f^-mt^-	f^-mt^+	f^-mt^+
f^-mt^-	f^-mt^+	f^-mt^+	f^+mt^-	f^+mt^+	f^+mt^-	f^+mt^-
f^-mt^-	f^-mt^+	f^-mt^-	f^+mt^-	f^-mt^-	f^-mt^+	f^-mt^+
16	20	6	60	3	1	3

a) Explain how the data reveal that the allele pairs segregate independently.

b) Make a diagram to illustrate how each class originated and state whether segregation occurs at the first or second division for each locus.

c) Suppose that 200 ascospores from the cross, isolated at random, were analyzed. Show the results you would expect.

21. In *Neurospora crassa* the *thiamine-4* (thiamine requirement) gene is close to the centromere on chromosome 4 and the *adenine-1* (adenine requirement) gene is close to the centromere on chromosome 6. If only first-division segregation occurs at both loci, what kinds and proportions of tetrads are expected in the cross thi^+ad^+ x thi^-ad^-? Would the same results be obtained in the cross thi^+ad^- x thi^-ad^+? Explain, using illustrations.

22. In yeast, *large* vs. *petite* colonies is determined by the allele pair Ll; mating type (+ or -) by the allele pair mt^+mt^-. A *large, mt^+* strain crossed with a *petite, mt^-* strain produced the tetrads shown. Are the gene loci on the same chromosome or not? Explain.

Kind of Tetrad				Number
L mt^-	L mt^-	l mt^+	l mt^+	40
L mt^+	L mt^+	l mt^-	l mt^-	36
L mt^+	l mt^+	L mt^-	l mt^-	18

Chapter 4
Probability and Chi-Square

1. In the domestic fowl the allele pairs *Rr* and *Bb*, which are located on different chromosome pairs, carry the potential for the expression of *rose* vs. *single* comb shape and *black* vs. *white* feather colour, respectively. What is the probability of obtaining:
   a) A *rose, black* bird from a cross between birds of the genotypes *Rrbb* and *rrBb*?
   b) An *RB* gamete from an *RrBB* bird?
   c) An *rrbb* zygote from the cross *RRBb* x *RrBB*?
   d) A *rose, white* bird from the cross *RrBB* x *RrBb*?
   e) An *rB* gamete from *RrBb*?
   f) An *RrBB* zygote from the cross *RrBb* x *RrBb*?

2. In rabbits the autosomal allele *S* for *short* hair is dominant to the allele *s* for *long* hair. Assuming that II-2 and II-5 are *SS*, what is the probability that any offspring of a mating between III-1 and III-2 will have (i) *long* hair, (ii) *short* hair? Show how you arrive at your answer.

   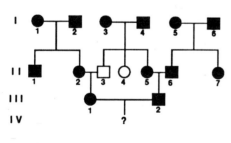

3. At an autosomal locus in sheep, the allele *B* for *black* wool is dominant to the allele *b* for *white* wool. A heterozygous ram is mated with 44 *black* ewes each of which produces 3 lambs; 10 of the ewes produce 1 lamb with *white* and 2 with *black* wool; 5 ewes produce 2 *white* lambs and 1 *black* lamb. All 3 lambs from each of the other 29 ewes have *black* wool.
   a) How many of the ewes are definitely heterozygous?
   b) How many more heterozygous ewes probably exist in the flock?
   c) If each of the 29 ewes had produced 6 lambs, what would be the chance of any heterozygous ewe producing only *black* offspring?

4. *Galactosemia* is a simply inherited autosomal recessive trait. A *normal* couple have an *affected* child. Find the probability that:
   a) The next two children will be *galactosemic*.
   b) Of the next 4 children, 1 will be *galactosemic*.
   c) The father of the *galactosemic* child is heterozygous for the recessive allele.
   d) The paternal grandmother is heterozygous for the recessive allele.
   e) The next child will be a heterozygote.
   f) A child of a *normal* sister of the *affected* child will be heterozygous.
   g) Any sibling of the *galactosemic* will be *affected*.
   h) There will be no *affected* individuals among the next 3 siblings of the *galactosemic*.
   i) Of the next 3 children, 2 will be heterozygous and the other homozygous for the dominant allele.
   j) Any child will be (1) a *galactosemic* girl, (2) a *normal* boy.
   k) The first child will be a *normal* boy and the second a *galactosemic* girl.
   l) In a family of 4: (1) at least 2 of the children will be *normal*, (2) 2 of the children will be *galactosemic* girls and 2 will be *galactosemic* boys.
   m) Of the first 2 of these children either the 2 will be *galactosemic* or the 2 will be *normal*.

5. A plant is heterozygous *AaBb*. The genes are inherited independently.
   a) Find the probability that a pollen-grain nucleus will contain:
   (i) An *A* allele.
   (ii) An *A* allele and a *b* allele.
   (iii) The *B* allele or the *b* allele.
   b) If pollen grains of this plant are used to fertilize eggs of a plant of the same genotype, find the probability that a seed embryo will contain:
   (i) Two *A* alleles.

(ii) One *A* allele and one *a* allele.
(iii) Two *A* alleles and two *b* alleles.
(iv) The genotype *AaBb*.

6. The synthesis of *normal* vs. *sickle-shaped* red blood cells in human beings is determined by alleles Hb^A and Hb^S respectively, of an autosomal gene. A fairly large sample of black children in Africa of genotypes Hb^AHb^A and Hb^AHb^S were tested for degree of infection by the malarial parasite *Plasmodium falciparum*, with the results shown.

	Hb^AHb^A	Hb^AHb^S
Heavily infected	206	340
Not infected or lightly infected	48	133

Are the heterozygotes more resistant to infection by the parasite than the normal homozygotes? Explain.

7. In Toronto, Canada in 1940, of 18,762 deaths, 424 were from *tuberculosis* and 622 were from *diabetes*. From these data derive the probability that the next death will be from:
a) *tuberculosis*.
b) either *tuberculosis* or *diabetes*.

8. Assume that the sex ratio at birth in humans is 1:1.
a) Determine the probability that a sibship of 6 will comprise:
(i) 4 boys and 2 girls in any order.
(ii) All children of the same sex.
(iii) At least 3 girls.
(iv) No fewer than 2 girls and 2 boys.
(v) 3 or more girls.
(vi) 4 boys and 2 girls or 2 boys and 4 girls.
(vii) What is the chance of the eldest child being a boy and the youngest a girl?
b) Determine:
(i) The most frequently expected number of males and females in a sibship of six.
(ii) The percentage of all families of 6 expected to have 3 boys and 3 girls.

9. When a *wire-haired* rabbit is mated to a *normal-haired* one, all F_1 progeny are *wire-haired*. When these F_1 rabbits are mated among themselves, they produce 22 rabbits, 14 *wire-haired* and 8 *normal-haired* ones.
a) How many of the F_2 *wire-haired* rabbits would you expect to be homozygous?
b) What results do you expect in the F_2? Determine whether the actual results deviate significantly from the expected. If they do, does this mean your hypothesis is wrong? Explain.

10. In humans, *hypotrichosis* (sparse body hair) is recessive to *normal*.
a) Two *normal* parents have 5 children, the first 2 with *hypotrichosis* and the others *normal*.
(i) What are the genotypes of the parents? Explain.
(ii) What is the chance of all the *normals* being heterozygous?
b) A *normal* man and *hypotrichotic* woman have 3 children, 1 *hypotrichotic* and 2 *normal*.
(i) What is the man's genotype?
(ii) What is the probability of the *normal* children being heterozygous?
(iii) If one of these *normals* marries a *normal* from the marriage described in (a), what is the probability of their first child being *hypotrichotic*? If they have 4 children, what is the chance of all being *normal*?

11. A self-fertilized F_1 tomato plant from a cross between a plant with a *simple* inflorescence and a plant with a *compound* inflorescence produced 360 progeny, 75 of which developed a *compound* inflorescence. Test the hypothesis that a single autosomal pair of alleles is involved with dominance of the allele *S* for *simple* inflorescence over the allele *s* for *compound* inflorescence in these crosses.

12. In the silkworm, a true-breeding line in which females lay *white* eggs is mated with a true-breeding line in which females lay *pink* eggs. The F_1 females, all of which lay *white* eggs, were crossed with their brothers and produced

208 F_2 females of which 141 laid *white* eggs, 49 laid *black* eggs and 18 laid *pink* eggs.

a) Present a hypothesis to account for these results.

b) Test your explanation using the chi-square method and indicate whether your hypothesis is consistent with the data.

c) Crosses between F_1 males and females laying *pink* eggs produced 19 females that laid *pink*, 27 that laid *black*, and 54 that laid *white* eggs. Are these results consistent with the hypothesis in (a)? Explain.

13. Guinea pigs of a strain with *short, yellow* hair mated with those of a strain having *long, white* hair produced F_1's with *short, cream* hair, which, when crossed with *long, white* animals, produced 40 *short, cream*, 11 *long, cream*, 9 *short, white*, and 38 *long, white* progeny.

a) Propose a hypothesis to explain these results and indicate the results expected if:

(i) F_1's were crossed among themselves.

(ii) F_1's were crossed with pigs from the *short, yellow* strain.

b) On the basis of your hypothesis indicate whether:

(i) Alleles of each pair are transmitted by F_1's in equal (1:1) proportions.

(ii) The testcross results are consistent with the hypothesis that the different allele pairs are inherited independently.

14. In humans, *gingival hyperplasia* (overgrowth of the gums) is dominant to the *normal* condition (Garn, S.M. and C.E. Hatch, J. Hered. 41: 41, 1950). *Cataracts* is also dominant to the *normal* condition (Danforth, C.H., Am. J. Opthalmol. 31: 161, 1912). A woman with *cataracts* whose father suffered from it marries a man with *gingival hyperplasia* whose mother also had this latter condition.

a) What kinds of children could this couple have? State the probability for each kind.

b) A son with both *gingival hyperplasia* and *cataracts* marries a *normal* woman, and their first child is *normal*. What is the genotype of the son? What is the chance of a child having *cataracts* and *normal* gums?

15. Two breeds of dogs, one with *straight, black* fur and the other with *curly, brown* fur, were crossbred. The hybrids were intercrossed and produced the following progeny:

Brown, straight	35	*Black, wavy*	179
Brown, curly	29	*Black, straight*	95
Brown, wavy	62	*Black, curly*	89

a) Propose a hypothesis to explain these results. What was the phenotype of the F_1 animals?

b) Test your hypothesis using the chi-square method and indicate whether you accept it or reject it.

c) What fundamental principles or laws are illustrated by these data?

16. When fowl from two true-breeding lines with *white* feathers were crossed, all the F_1's were phenotypically identical to the parents and the F_2 consisted of 124 *white* and 36 *coloured* birds. These results may be explained by either a single autosomal pair of alleles with dominance of the allele for *white* over that for *coloured* or by the interaction of two independently inherited pairs of alleles involving dominant and recessive epistasis.

a) Using the chi-square test, determine whether each hypothesis is consistent with the data.

b) Outline a genetic test that would clearly indicate which of the two hypotheses is the correct one.

17. The distribution of girls in 240 families, each with 4 children, is shown below:

No. of boys	No. of girls	No. of families
0	4	12
1	3	69
2	2	84
3	1	57
4	0	18

Do the data indicate that the sexes occur in a 1:1 ratio? Provide a statistical explanation.

11. In breeding experiments designed to study fruit colour in summer squash (*Cucurbita pepo*), Sinnott, E.W. and G.H. Durham (J. Hered. 13: 177, 1922) found that strains with *white* fruits occasionally produced plants with *green* and plants with *yellow* fruits. Never did strains with *green* or *yellow* fruits produce plants with *white* ones. Moreover, strains with *green* never produced plants with *yellow*. Crosses between true-breeding *green* and *yellow* strains produced *yellow* only. These *yellow* plants when intercrossed produced 81 *yellow* and 29 *green*. Homozygous *white* x homozygous *yellow* produced plants with *white* only. These when intercrossed produced 155 *white*, 40 *yellow* and 10 *green*. Give a complete genetic explanation of these results.

12. In the Mexican swordtail fish (*Xiphophorus helleri*), the *wild-type* colour is *olive-green*. Many other colour variants have been developed since 1909, including a *golden* and an *albino*. The *golden* owes its distinctive colour to the fact that the skin has many yellow cells but few black ones. The eyes, in which pigment cells are more frequent, are black. The *albino* has a few yellow cells but no black ones, and the eyes are pink. Matings involving these three phenotypes give the tabulated results shown below (Gordon, M. J., Hered. 32: 221, 1941).

Parents	Progeny
golden x *golden*	*golden*
albino x *albino*	*albino*
golden x *albino*	*wild-type*
wild-type x *golden*	68 *wild-type*: 52 *golden*
wild-type x *albino*	33 *wild-type*: 23 *albino*
wild-type* x *wild-type**	202 *wild-type*: 65 *golden*: 72 *albino*

*Obtained from *golden* x *albino* matings.

a) Explain the reversion or atavism to *wild-type* in matings between *golden* and *albino*, giving the probable number of allele pairs and the form of gene interaction involved.
b) Is it possible for the *wild-types* to breed true? Explain.
c) When a *golden* is mated with a *montezuma*, the progeny consist of *montezuma* and *wild-type* in the ratio of 1:1, and when the *montezuma* offspring of *montezuma* x *golden* cross are mated back to the *golden*, about one-quarter of the young are *wild-type*. Explain how this reversion can occur, giving genotypes of parents and offspring.

13. When *red* cattle are crossed with *brindle* (irregular narrow stripes of black on a red background), the F_1 are *brindle* and the F_2 consists of *brindles*, *black-and-reds* and *reds* in a 9:3:4 ratio. In certain crosses between the Angus (*black*) and Jersey (*black-and-red*) breeds, the F_1 are all *black*, and the F_2's segregate 12 *black*: 3 *black-and-red*: 1 *red*. In other crosses between these breeds the F_1's are also *black*, but in the F_2 four phenotypes appear: *black*, *brindle*, *black-and-red* and *red* in a 48:9:3:4 ratio (Ibsen, H.L., Genetics 18: 44, 1933). Describe the genetic basis for coat colour in these breeds, stating the number of allele pairs and the type of interaction and showing the genotypes of the Angus and Jersey parents in each of the three crosses described.

14. The self-fertilized F_1 from a cross of two true-breeding durum wheat varieties with *spring* growth habit produced 91 plants: 6 with *winter* growth habit and the remainder with *spring* growth habit. Indicate:
a) The number of allele pairs involved.
b) The phenotype of the F_1's.
c) The genotypes of the F_2 plants with *winter* growth habit.

15. In garden stock (*Matthiola incana*), the results shown were obtained in crosses between single- and double-flowered varieties and among different double-flowered varieties.

Parents	F_1	F_2 or backcross
single x *double*	*single*	193 *single* : 152 *double*
double x *double*	*single*	281 *single* : 213 *double*
F_1 (*double* x *double*) x *double*		1 *double* : 1 *single*
F_1 (*double* x *double*) x *double*		3 *single* : 5 *double*

a) From an examination of the F_1 results of the first cross, which of these traits appears to be recessive? Why does it not breed true in crosses 2, 3 and 4?

b) Describe the type of gene interaction involved and give the genotypes of the parents and the F_1 and F_2 phenotypic classes for the four crosses.

16. When certain true-breeding strains of *platinum* mink (*blue-grey* fur) are brought together and single-pair matings occur at random within the combined population, it is found that some of the matings produce only *blue-grey* offspring while others produce only *wild-type* (Shackleford, R.M., "Genetics of the Ranch Mink", Pilsbury New York, 1950).
a) On the basis of this information only, state the simplest hypothesis that will explain the inheritance of coat colour in these animals.
b) When *wild-type* individuals are intermated, they produce 192 *wild-type* and 148 *blue-grey*; when mated with *blue-greys*, they produce 125 *wild-type* and 131 *blue-grey*. Does this alter or confirm your previous explanation?
c) A mink breeder finds that *wild-types* occasionally appear in his *platinum* stock. Since these throwbacks reduce his profits, he wishes to eliminate them entirely. How should he proceed?

17. The seed capsule of shepherd's purse (*Capsella bursa-pastoris*) may be either *triangular* or *top-shaped* (Shull, G.H, Z. Indukt. Abstamm. -Vererbungsl. 12: 97, 1914). A cross between *triangular* and *top-shaped* gave a *triangular* F_1. In the F_2, 146 plants were *triangular* and 12 *top-shaped*.
a) How many allele pairs determine *triangular* vs. *top-shaped* in this cross, and how do they interact?
b) Assigning your own symbols to the genes, state the genotypes of the parents that would produce the following proportions of offspring:
(i) 3 *triangular* : 1 *top-shaped*
(ii) All *triangular*
(iii) 7 *triangular* : 1 *top-shaped*
(iv) 1 *triangular* : 1 *top-shaped*

18. Two true-breeding *white*-flowered lines of sweet peas (*Lathyrus odoratus*) when crossed produced *purple*-flowered F_1's and these plants produce 100 *purple*-flowered and 72 *white*-flowered F_2's (Bateson, W., et al., Rep. Evol. Comm. R. Soc. 2: 80, 1905).

a) Show statistically that this could not be considered a random deviation from a 1:1 ratio.
b) Describe the form of inheritance involved, indicating the most likely theoretical ratio. Diagram the cross, indicating the genotypes for each phenotype.
c) Suppose that all the seed you had of the two true-breeding *white*-flowered lines is lost. How could you establish two lines from these F_2's that possess genotypes identical to these two parental lines?

19. The A, B and O blood groups (antigens) are determined by a multiple allelic series. The codominant alleles I^A and I^B specify the synthesis of the A and B antigens respectively; the recessive allele i fails to determine the synthesis of either A or B antigen. An unusual result was observed by Bhende, Y.M., et. al. (Lancet 1: 903, 1952) in a family whose pedigree is shown below:

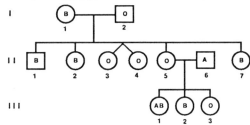

a) Which of the matings in this pedigree produces offspring with unexpected phenotypes? Offer a simple genetic explanation to account for the unusual results.
b) Biochemical studies of individuals of the different blood types reveal the following with respect to the synthesis of the H substance and the A and B antigens.

Blood type	H substance	A antigen	B antigen
O	+	−	−
A	+	+	−
B	+	−	+
AB	+	+	+
O*	−	−	−

*Exceptional individual
(i) How does this information necessitate a modification of your explanation in (a)?

(ii) Offer a plausible genetic explanation to account for the synthesis of the various antigens.

20. Most pedigree studies reveal that *albinos* are homozygous for a recessive autosomal allele for *albinism*. In some rare pedigrees, however, *albino* couples have been found to have both *albino* and *nonalbino* children or *nonalbinos* only. Explain genetically.

21. a) In a true-breeding *purple-eye, pr* strain of *Drosophila melanogaster* a *wild-type* individual appears. What could be a possible cause of this reversion to *wild-type*?
b) A true-breeding strain of the reverted *wild-type* is produced and crossed with an unrelated true-breeding *wild-type* strain. Explain the F_1 and F_2 results genetically.

F_1	F_2	
wild-type	wild-type	80
	purple	6

22. If "fancy" breeds of animals or plants are allowed to mate at random, the population will in time contain many "wild" (or "ancestral") types. This can be perceived in flocks of pigeons whose ancestors escaped from pigeon fanciers' stocks and whose matings are not controlled by humans. In New York City's Battery Park, among the thousands of pigeons that come fluttering down to pick up crumbs along the paths, a few birds with *black* and *white* or *red* and *white* mottling and some that are predominantly *red* or *deep blue* are found. The majority, however, resemble the ancestral, wild blue rock pigeon (*Columbia livia*) (Gordon, M. J., Hered. 28: 221, 1937).
a) Propose a possible explanation for this reversion to *wild-type* in the New York pigeons.
b) How would you test your hypothesis?

23. A *prolineless Neurospora crassa* (haploid) mutant (having a nutritional requirement for proline) reverts to a *wild-type* and then breeds true for the reversion. This reverted *wild-type* is mated with an unrelated *wild-type*, and the progeny segregate 1 *prolineless* : 3 *wild-type*. Explain.

24. In *Neurospora crassa* the difference between *wild-type* and each of the arginine-requiring (*arginineless*) mutants *arg-1* and *arg-3* is due to alleles of different genes (Perkins, D.D., Genetics 25: 178, 1959). A cross was made between *arg-1* and *arg-3* strains, and 400 ascospores analyzed at random showed a ratio of 1 *wild-type* : 3 *arginineless*. Explain these results.
b) Suppose the ratio had been 3 *wild-type* : 1 *arginineless*. How would you explain these results?

25. In corn (*Zea mays*) a cross is made between a homozygous *red*-kernel and a homozygous *white*-kernel strain. The F_1 individuals have *red*-kernels only, and the F_2 express both phenotypes in a distribution of 285 *red*-kernel : 378 *white*-kernel. Explain stating the number of allele pairs involved and the type of gene interaction.

26. Different kinds of *white* phenotype are known in poultry. White silkies and rose comb bantams are *white* because of a recessive allele, *c*, which causes the lack of a chromogen necessary for colour. White Wyandottes, Minorcas, Dorkings, and other breeds are *white* because of a different recessive plumage allele, *o*, which causes the lack of an enzyme converting the chromogen into pigment. White Leghorns carry a dominant allele, *I*, which inhibits pigment formation in the homozygous state (Hutt, F.B., "*Genetics of the Fowl*", McGraw-Hill, New York, 1949).
a) What phenotypes would you expect in the F_1 and F_2 of a cross between a Rose-comb Bantam and a Minorca? Describe the type of interaction involved.
b) What phenotypes would appear in the F_2 of a cross between a White Leghorn and a Minorca or Rose-comb Bantam? Describe the type of interaction involved.
c) What genotype must coloured breeds possess?

27. A wheat plant with *red*-kernels produces 320 seeds upon self-fertilization: 314 give rise to plants with *red*-kernels and 6 to plants with *white*-kernels. The latter group upon self-fertilization breed true for *white*-kernels. Give a genetic explanation to account for these results.

28. Emerson, R.A. (Cornell Univ. Agric. Exp. Stan. Mem. 39: 1, 1921) worked out the genetic basis for stem and foliage colour in corn. The F_1 of a cross between a plant with *green* stems and one with *purple* stems was phenotypically like the second parent. The F_2's segregated as follows:

Purple	265	*Brown*	91
Sun-red	85	*Dilute-sun-red*	32
Dilute-purple	93	*Green*	70

a) How many allele pairs are probably involved, and what is the allelic relationship of each?
b) Is a modification of an F_2 ratio involved? If so, what is the method of interaction?

29. a) Three *white*-kernel strains of corn were crossed in all possible combinations. The F_1's were always *red* with each F_2 exhibiting a segregation ratio of approximately 9 *red* : 7 *white*. How many loci are involved in the determination of kernel colour in these strains? Answer by giving the genotypes of all three strains.
b) A fourth *white* strain when crossed with a *purple* one produces *purple* F_1's, which upon self-fertilization produced approximately 81 *purple* : 27 *red* : 148 *white*. Moreover, when the *purple* strain is crossed with a fifth *white* strain, the F_1 are all *white* and one-fourth of the F_2 are *purple* (Emerson, R.A., Cornell Univ. Agric. Exp. Stn Mem. 16: 231, 1918). Determine the genetic basis of kernel colour and give the complete genotype of each of the parental strains used in these crosses.

30. The self-fertilized *purple-pod* F_1 from a cross between true-breeding *green-pod* and true-breeding *purple-pod* strains of garden peas produce *purple-pod, green-pod* and *yellow-pod* plants in close approximation to a 36:21:7 ratio.
a) Why are 36 out of every 64 plants *purple-pod*?
b) What are the genotypes of plants with *green* and *yellow* pods respectively?

31. A *black* coat strain of cattle is crossed with an *albino* strain. The F_1 are *brindle* and the F_2 consist of 29 *black* : 62 *brindle* : 31 *red* : 41 *albino*. What type of intergenic relationship can explain these results? Outline the crosses you would make to test your explanation and show the results expected in each case.

32. In the sweet pea, two *white*-flowered strains when crossed produced *purple*-flowered F_1's which upon self-fertilization gave 265 *purple*, 92 *red* and 282 *white*-flowered plants (Bateson, W. and R.C. Punnett, Rep. Evol. Comm. R. Soc. 2: 80, 1906). What is the genetic basis for the observed F_2 segregation ratio?

33. This fairly extensive pedigree of *deaf-mutism* in humans is from a family observed in Northern Ireland by Stevenson, A.C. and E.A. Cheeseman (Ann. Hum. Genet. 20: 177, 1956). State the most probable mode of inheritance of *deaf-mutism*. Give reasons for your conclusions.

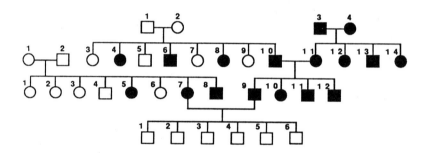

34. The pedigree illustrates the results obtained when horses with *black, chestnut, bay* and *liver* phenotypes are mated (Castle, W.E. and W.R. Singleton, J. Hered. 51: 127, 1960).

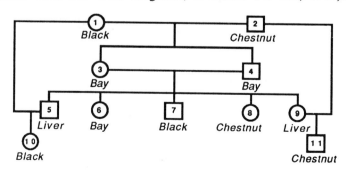

a) Determine the genetic basis for coat colour.

b) When true-breeding *chestnut* horses are mated with *albinos* that appear within the *chestnut* strain, the F₁'s are all *palomino*. The *palominos* when mated, produce *palominos*, *chestnuts* and *albinos* in a 2:1:1 ratio. These *albinos* when mated with *livers* produce *palomino* progeny only and when mated with *bays* produce *buckskin* F₁'s which, mated inter se in sufficient numbers, produce *albinos*, *buckskins*, *bays*, *palominos* and *chestnuts* in a 4:6:3:2:1 ratio. No *blacks* or *livers* appear. A novice horse breeder, eager to establish a *palomino* herd, is somewhat baffled by all this. Outline the essential components of the *palomino* genotype and the matings he should make to ensure the greatest return for his efforts.

35. a) Distinguish between incompletely dominant lethal genes and recessive lethal genes. Which are easier to detect and why?
b) Are incompletely dominant lethals easier or harder than recessive lethals to eliminate from a population? Give reasons to support your answer.
c) Why are lethals much less common in haploid organisms than in diploid ones?

36. *Albinism* is lethal in plants, yet many species produce *albinos* among their progeny. If *albinos* always die before reproducing, why is the trait not eliminated?

37. Most lethals that have been discovered in humans, regardless whether they are dominant, incompletely dominant or recessive, produce their effects shortly before or after birth. On the other hand, many lethals in organisms such as mice and *Drosophila* have been found that produce their effects in early embryonic stages. Why should it be difficult to detect early acting lethals in humans but not in organisms like mice?

38. In the swordtail fish (*Xiphophorus hellcri*), a very colourful variety called the *montezuma* occurs. When two *montezumas* are mated, about one-third of the progeny are always throwbacks, or reversions, to *wild-type* (an olive-green colour) and the remainder are *montezumas*. Explain the reversion to *wild-type* and why the *montezumas* fail to breed true.

39. In cattle, as well as in sheep, certain matings between *normal* parents produce only *normal* progeny, but when the daughters are mated with their fathers, the results are as shown.

Animal	Classification of the progeny of father x daughter matings
Cattle	98 *normal* : 12 *hairless* (lethal)
Cattle	102 *normal* : 13 *amputated* (lethal)
Sheep	29 *normal fetal muscle development*: 4 *fetal muscular degeneration* (lethal)

For each character state:
a) (i) The genotypes of the father and mother, the daughters and the progeny of the father x daughter matings. Diagram each cross.
(ii) The type of lethal allele involved.
b) If the allele produced its effects during early embryonic development, would it have been detected? What conditions are necessary to detect this type of lethal?

40. In crosses between two *crested* ducks approximately three-quarters of the eggs hatch. The embryos of the remaining quarter develop nearly to hatching and then die. Of the ducks that do hatch about two-thirds are *crested* and one-third are *crestless*.
a) Explain these results genetically.
b) What would you expect from the cross of a *crested* with a *crestless* duck?
c) Is it possible to establish a true-breeding strain of *crested* duck? Explain.

41. In bees, the wings of worker females may be *droopy* or *normal*; those of the males (drones) are never *droopy*. In hives where *droopy-winged* workers occur, a number of the eggs fail to hatch (within a hive all individuals are the offspring of a single mating between a queen and a male). Why do *droopy-winged* individuals occur among females only?

42. *Blufrost* mink have light-blue underfur and silver guard hairs; *normal* mink have dark fur. In 1947 Moore, L. and C.E. Keeler (J. Hered. 25: 341) obtained the results shown.

Parents	Offspring	Average litter size
blufrost x blufrost	78 blufrost : 80 normal	5.27
normal x blufrost	345 blufrost : 325 normal	5.11
blufrost x blufrost	19 blufrost : 10 normal	3.65

Propose a hypothesis to explain these data and outline how you would test your explanation.

43. The F_1's of a cross between two breeds of turkey were mated with each other. The females laid 642 eggs, of which 603 produced *normal* birds and 39 did not hatch. The embryos in these 39 eggs terminated development on about the seventh day of incubation. How would you explain these results?

44. *Multiple telangiectasia* (Snyder, L.H. and C.A. Doan, J. Lab. Clin. Med. 29: 1211, 1944) and *hereditary sebaceous cysts* (Munro, T.A., J. Genet. 35: 61, 1937) are rare traits which never appear in a child unless also present in at least one of the parents. In extremely rare cases two *affected* individuals have married. In none of these marriages were the children all phenotypically like the parents. In some such marriages one or more of the offspring regularly died at about one year of age.
a) What is the most likely mode of inheritance of each of these traits?
b) Describe the critical mating that would definitely substantiate your hypothesis and indicate the kinds of progeny expected from this mating.

45. Cole, R.K. (J. Hered. 52: 47, 1961) reported on genetic studies involving *paroxysm*, a new mutant in fowl, characterized by seizures, poor growth, and stilted gait. Individuals with the condition hatch normally and do not show any of the symptoms until about the age of two weeks or more. Death occurs in 14 weeks or less after birth. A *normal* rooster mated with 10 *normal* unrelated hens produced 58 male and 52 female offspring; 3 males and 4 females died within a week of hatching, and by the fourteenth week a further 28 birds, all females, had died from *paroxysm*.
a) Explain these results genetically.
b) What proportion of the progeny would you expect to be heterozygous for *paroxysm*? Explain.
c) Outline a breeding program to eliminate the condition in one generation.

46. In *Drosophila melanogaster* flies may have *red* (*wild-type*) or *plum* eye colour and *normal* or *stubble* bristles (less than half normal length). The progeny of a mating between two *plum, stubble* flies gave the following results:

Plum, stubble	162	Red, stubble	84
Plum, normal	80	Red, normal	42

a) Suggest a hypothesis to account for these results.
b) Outline how you would test your explanation and indicate the results expected in your experiment.

47. A poultryman buys a registered tom turkey and turns him out with the hens of his flock. One-fourth of the eggs fail to hatch. Later it is found that about two-thirds of the progeny are males. Explain these results genetically; illustrate your explanation with a diagram of the cross.

48. In bread wheat (*Triticum aestivum*) the varieties Trumbull, Marquillo, and F.H. 27 are *normal* (in colour and development). The seedlings from crosses between Trumbull and Marquillo are *inviable*; death of the plant occurs about the sixth to eighth week after seeding. However, when F.H. 27 is crossed with these two varieties, the F_1's of both crosses are *normal* (Caldwell, R.M. and L.E. Compton, J. Hered. 34: 65, 1943).
a) Suggest an explanation for these results.
b) Would it be possible to test your hypothesis? If so, explain how and outline the results expected within each cross.

49. The pedigree shown is a modification of one presented by Eldridge, F.E. and F.W. Atkeson (J. Hered. 44: 265, 1953) of

streaked hairlessness in Holstein-Friesian cattle, a condition characterized by an abnormality of the hair coat in which narrow, irregular hairless streaks, running transversely around the trunk, appear.

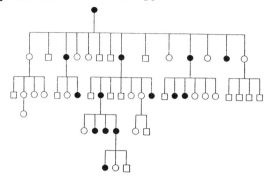

What is the most likely mode of inheritance? Give reasons for your decision.

50. The pedigree shows the inheritance of *brachydactyly* (short fingers and toes). All individuals examined were ten years of age or older.

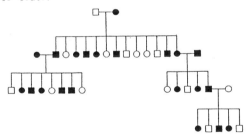

a) Is it due to an allele of an autosomal or sex-linked gene? Is the allele dominant or recessive?
b) Compare the progeny of marriages between *affected* individuals and those between *affected* and *unaffected* individuals. What distributions of phenotypes do you obtain? Explain how this could possibly occur.
c) What is the probable genotype of each *affected* individual? Why?

Chapter 6
Multiple Alleles

1. State whether the following are true or false and why:
   a) It is possible to prove the existence of multiple allelism in an organism that reproduces by asexual means only.
   b) Blood groups can be used to exclude, but not to prove, paternity.
   c) Complementation as a test for allelism can be applied only to recessive mutations.

2. In rabbits, the blood agglutinogens H_1 and H_2 are genetically determined: H_1 is produced by individuals in blood group H_1 and H_2 by those in group H_2. Both agglutinogens are synthesized by H_1H_2 individuals and neither is present in animals belonging to group O. Tests with anti-H_1 and anti-H_2 sera (containing antibodies that agglutinate H_1 and H_2 antigens respectively) gave the results shown below (Castle, W.E. and C.E. Keeler, P.N.A.S. 19: 92, 1933):

		Progeny			
Mating	Parents	H_1H_2	H_1	H_2	O
1	$O \times O$				7
2	$O \times H_1$		9		12
3	$O \times H_2$			10	9
4	$H_1 \times H_2$	17	13	15	20

   a) Present two hypotheses to explain these data, showing the genotypes for each of the phenotypes in each case.
   b) When H_1H_2 individuals were mated with those of group O, 140 H_1 and 162 H_2 individuals were obtained. Which hypothesis do these results support and why?

3. The *striped, moricaud* and *plain* types of larval colour patterns in the silkworm are genetically determined. The following table shows the results of crosses between *striped, plain,* and *moricaud* silkworms in all possible combinations:

Parents	F_1	F_2
striped x *plain*	*striped*	998 *striped* : 314 *plain*
striped x *moricaud*	*striped*	1,300 *striped* : 429 *moricaud*
moricaud x *plain*	*moricaud*	763 *moricaud* : 243 *plain*

State, with reasons, the number of genes and alleles involved and their allelic relationships.

4. Johansson, I. (Hereditas 33: 152, 1947) presented data on the inheritance of *silver*, *platinum*, and *white face*, three coat colour phenotypes in the fox. The average litter size from each of the matings of *silver* x *silver, silver* x *platinum,* and *silver* x *white face* was 4.48; that from all other matings was reduced to 3.56. Give a plausible genetic explanation of these results accounting for the lethality that occurs in the latter three matings.

	Number of progeny		
Parents	*Silver*	*Platinum*	*White face*
silver x *platinum*	4,157	3,842	
silver x *white face*	3,038		2,986
platinum x *platinum*	58	127	
white face x *white face*	267		483
platinum x *white face*	167	182	188

5. *Fucosidosis* is a debilitating disease that is caused by a deficiency for the lysosomal enzyme α-fucosidase which occurs in different molecular (isozymic) forms. This deficiency leads to the accumulation of fucose-containing glycolipids and glycoproteins in various tissues. The disease's clinical manifestations include susceptibility to respiratory infections, severe neurological dysfunction, progressive psychomotor regression and physical and

mental retardation. A diagrammatic representation of the isozyme composition of each of the enzyme's three common phenotypes (1, 2, and 2-1) is shown below:

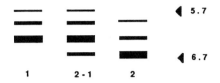

In the pedigree shown below, obtained from Turner, B.M. et al. (Nature 257: 391, 1975), the two brothers III-1 and III-2 are both affected with *fucosidosis* and deficient for α-fucosidase:

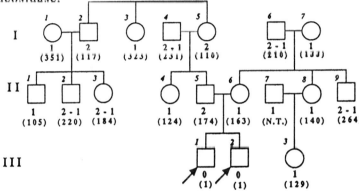

*α-fucosidase activity in leucocytes is shown in parentheses.

On the basis of the above pedigree:
a) Deduce the mode of inheritance of *fucosidosis* and the α-fucosidase phenotypes 1, 2 and 2-1.
b) Indicate whether the alleles are amorphs, hypomorphs or hypermorphs.
c) What is the functional relationship of the alleles?

6. In the nasturtium (*Tropaeolum majus*) the flowers may be *single, double* or *superdouble*. *Superdoubles* are female sterile; they originated in a *double-flowered* variety. In a series of crosses designed to determine the genetic relationships of the three traits, Eyster, E.H. and D. Burpee (J. Hered. 27: 51, 1936) obtained the following results:

T.B*. *single* x T.B. *double* → F_1 *single* → F_2 78 *single* : 27 *double*
T.B. *double* x *superdouble* → 112 *superdouble* : 108 *double*
T.B. *single* x *superdouble* → 8 *superdouble* : 7 *single*
T.B. *double* x the 8 *superdouble* → 18 *superdouble* : 19 *single*
T.B. *double* x the 7 *single* → 14 *double* : 16 *single*
*T.B. = true-breeding

a) These results indicate that the phenotypes *single*, *double* and *superdouble* flowers are determined by a series of three alleles of one gene. Why?
b) What is the order of dominance of the three alleles?

7. Werret, W.F. et al. (Nature 184: 480, 1959) obtained the tabulated results in the fowl starting with a mutant *brown* male, which appeared in the F_1 of a cross between a *brown* Leghorn male and a *silver* (white) Light Sussex female.

Parents		Sex	Progeny		
♂	♀		Silver	Brown	Semialbino
mutant x *silver*		♂	27	0	0
(from T.B. line)		♀	0	12	8
mutant x *semialbino*		♂	0	18	6
(from previous mating)		♀	0	13	12
		?*		5	2
semialbino x *semialbino*		♂	0	0	4
		♀	0	0	8
		?*	0	0	12
semialbino x *brown*		♂	0	18	0
(from T.B. line)		♀	0	0	23
		?*	0	0	3
silver x *semialbino*		♂	1	0	0
		?*	0	1	0

*Sex not determined.   T.B. = true breeding.

Give a complete genetic explanation of these results, indicating the number of genes involved, their mode of interaction if any, allelic relationships, location of gene or genes, and the genotypes and sex of the unsexed progeny. Give reasons for your answers.

8. In mink (*Mustela vison*), animals in the wild may have *black*, *platinum* (blue-grey) or *sapphire* (very light blue) coats. One of many domesticated breeds has a *sapphire* coat. Three *black* animals (A,B,C), a *sapphire* (D), and one *platinum* one (E) from the wild mated with animals from the domesticated breed produced the results shown. A x C matings produced progeny which were *apparently* true-breeding.

Parents			
Wild		Domesticated	F_1
A *black*	x	*sapphire*	26; all *black*
B *black*	x	*sapphire*	14 *black* : 15 *sapphire*
C *black*	x	*sapphire*	21; all *black*
D *sapphire*	x	*sapphire*	21; all *sapphire*
E *platinum*	x	*sapphire*	16 *platinum* : 14 *sapphire*

a) Outline the simplest hypothesis possible to account for these results. Using your own symbols, give the genotypes of the domesticated animals and the five wild ones.
b) If your hypothesis is correct, what types of progeny and in what proportions would you expect them in the following matings: A x E? B x E? C x D?

9. The pedigree typifies the transmission characteristics of the *black*, *tortoise-shell* and *red* coat phenotypes in the guinea pig known to be determined by alleles of one gene (Wright, S., Genetics 12: 530, 1927). State whether two or three alleles are involved.

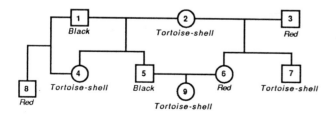

10. In *Drosophila melanogaster wild-type* (w.t.) flies have *dull-red* eyes because of the presence of both red and brown pigments, which are produced via specific but different pathways:

→ A → B → C →   red
                              dull-red (*wild-type*)
→ D → E → F →   brown

Five recessive true-breeding *brown-eyed* mutant strains, designated *l*, *m*, *n*, *o*, and *p* arose idependently after treating *wild-type* flies with the mutagen ethyl methanesulfonate (EMS). Phenotypes of F_1's, from crosses between the mutants in all possible combinations, were as shown:

	♀ parent				
♂ parent	l	m	n	o	p
l	all brown	all w.t.	all w.t.	all brown	♀ w.t.* ♂ brown
m	all w.t.	all brown	all brown	all w.t.	♀ w.t. ♂ brown
n	all w.t.	all brown	all brown	all w.t.	♀ w.t. ♂ brown
o	all brown	all w.t.	all w.t.	all brown	♀ w.t. ♂ brown
p	all w.t.	all w.t.	all w.t.	all w.t.	♀ brown ♂ brown

*w.t. = *wild-type*

a) In how many different genes did the five mutations occur? Explain.
b) What is the chromosomal location (autosome or X) of each of the mutant genes? Explain.
c) If F_1's from mutant l x mutant m were intercrossed, what results would you expect in the F_2?

11. In the house mouse (*Mus musculus*) coat colour is genetically determined (Gruneberg, H., Bibliog. Genet. 15:

1, 1952). The results of matings involving individuals with *yellow*, *tan*, and *black* phenotypes are shown below. Give a complete genetic explanation of these results.

Cross	Parents	Offspring
1	black x yellow	391 black : 402 yellow
2	tan x yellow	214 tan : 210 yellow
3	yellow* x tan	184 yellow : 188 blackish-tan
4	yellow# x black	106 yellow : 110 blackish-tan
5	yellow* x yellow#	218 yellow : 113 blackish-tan
6	blackish-tan x tan	132 blackish-tan : 138 tan
7	black x blackish-tan	141 black : 144 blackish-tan

*from cross 1   #from cross 2

12. Kuspira, J. et al. (Can. J. Genet. Cytol. 28: 88, 1986) obtained the following results in the Einkorn wheat, *Triticum monococcum*, in crosses between a true-breeding line 68 with *spring* growth habit and six different true-breeding lines (155, 198, 258, 278, 282, and 287) all of which expressed the *winter* growth habit.

Cross No.	Parents spring x winter	# of F$_1$ spring	F$_2$ spring	F$_2$ winter	Backcross spring	Backcross winter
1	68 x 198	7	103	26		
2	68 x 258	7	96	39		
3	68 x 155	7	144	42		
4	68 x 287	7	184	56		
5	68 x 282	7	140	52		
6	68 x 278	7	328	112		
7	F$_1$(68x258) x 258				40	41
8	F$_1$(68x155) x 155				51	59
9	F$_1$(68x287) x 287				30	34
10	F$_1$(68x278) x 278				44	38

None of the F$_1$ plants of diallel crosses among the six *winter* lines showed *spring* growth habit.

Offer a complete genetic explanation for these results indicating the number of genes involved and the number of alleles as well as the allelic relationships at each locus.

13. *Phenylketonuria* (PKU) and *mild hyperphenylalaninemia* (MHP) are two different phenotypes that result from variant forms of the enzyme phenylalanine hydroxylase. Four alleles ph^a, ph^b, ph^c, and ph^d have been identified on the bases of their different restriction-fragment-length-polymorphisms (RFLPs) using seven different restriction enzymes. With respect to the pedigree from Ledley et al. (New Engl. J. Med. 314: 1276, 1986), which is similar to other pedigrees for the same condition, answer the following questions:

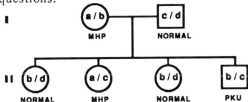

a) Why does the pedigree permit the conclusion that *normal* vs. *PKU* vs *MHP* are determined by different alleles of a single autosomal gene?
b) Which of the possible genotypes does not occur among the children in the pedigree? What phenotype would it most likely express? Explain.
c) Which of the alleles determines the *normal*, *MHP* and *PKU* phenotypes? Explain.

14. Mathews, A. et al. (J. Hered. 80: 357, 1989) obtained the following results in the normally *nodulating* (produces nitrogen-fixing nodules) soybean (*Glycine max*) cultivar 'Bragg':
(i) Three nonnodulating mutants *nod 49*, *nod 139* and *nod 772* were produced by treating cv. 'Bragg' with ethyl methanesulphonate.
(ii) Crosses between each of the mutants and the cv. 'Bragg' produced basically identical results. For example, the cross *nod 49* x 'Bragg' produced *nodulating* F$_1$'s and 203 *nodulating* and 76 *nonnodulating* F$_2$'s.
(iii) All F$_1$'s from each of the crosses (a) *nod 49* x *nod 772* (b) *nod 49* x *y'* (a spontaneously occurring nonnodulating mutant) and (c) *nod 772* x *y'* were *nonnodulating*. F$_1$'s from crosses between *nod 139* and each of the other three mutants produced *nodulating* F$_1$'s.

What is the minimum number of genes involved in nodule formation? What are the allelic relationships at each locus? Explain.

15. *A, B* and *O* blood group differences in humans are determined by different alleles of one gene. The allele *i* (no antigen) is recessive to the two other alleles I^A (antigen A) and I^B (antigen B) which are codominant. Natural antibodies occur that agglutinate these antigens so that an individual never has antibodies that would agglutinate his or her antigens. In 1901 Landsteiner, K. (Ulien. Klin. Wscher. 14: 1132) tested the blood serum from 6 individuals against red blood cells (antigens) from the same individuals. The results he obtained are shown below:

		Red blood cells of the six individuals					
		1	2	3	4	5	6
Serum of the six individuals	1	-	+	+	+	+	-
	2	-	-	+	+	-	-
	3	-	+	-	-	+	-
	4	-	+	-	-	+	-
	5	-	-	+	+	-	-
	6	-	+	+	+	+	-

+ = agglutination   - = no agglutination

a) Do any of the 6 individuals belong to blood group *O*? Explain.
b) Which, if any, belong to blood group *AB*? Explain.
c) If the blood type of individual 2 is *B*, what are the blood types of 3, 4, and 5?
d) If individual 2 was of blood group *A*, how would your answer to (c) be modified?

16. How many different diploid genotypes are possible when a gene exists in five allelic forms?

17. A person of blood type *B*, Rh^- is injured and must be given a blood transfusion. The attending physician is provided with blood from four different individuals whose blood group phenotypes are *O*, Rh^+; *A*, Rh^-; *B*, Rh^-; and *AB*, Rh^+. Which individual's blood would be accepted and whose would be rejected? Why?

18. The dorsal pigmentary pattern of the common leopard frog (*Rana pipiens*) is genetically determined. It may be *kandiyohi*, *burnsi*, *wild type*, or *mottled burnsi*. The following table shows the results of matings among these phenotypes (Volpe, E.P., J. Hered. 51: 151, 1960):

parents	F_1	F_2
kandiyohi x *wild type*	*kandiyohi*	3 *kandiyohi* : 1 *wild type*
burnsi x *wild type*	*burnsi*	3 *burnsi* : 1 *wild-type*
F_1 *kandiyohi* x F_1 *burnsi* (from cross 1)(from cross 2)		1 *mottled burnsi* : 1 *kandiyohi* 1 *burnsi* : 1 *wild-type*

a) What conclusions can you draw from the results of the first two crosses regarding the inheritance of pigmentation patterns and dominance relationships of the alleles involved?
b) The results of the third cross do not permit a distinction between multiple allelism and independent assortment of two allele pairs. Using your own symbols, show how these results can be explained by each of these mechanisms.
c) The table below shows the results obtained by Volpe in reciprocal matings of *mottled burnsi* with *wild-type* frogs. Which hypothesis do these results support? Perform a chi-square test to substantiate your answer.

Cross	Mottled burnsi	Kandiyohi	Burnsi	Wild-type
1	14	18	24	20
2	16	19	22	25

19. In mink, the *wild-type* is dark in colour; two mutant colour patterns called *black cross* and *royal silver* have originated in recent times. The patterns are similar but not identical. The following breeding data were recorded by Shackelford ("*Genetics of the Ranch Mink*", Pilsbury, New York, 1950):

1. *Wild-type* (true-breeding) x *black cross* gives 1 *black cross* : 1 *wild-type*.
2. *Wild-type* (true-breeding) x *royal silver* gives 1 *royal silver* : 1 *wild-type*.
3. When *black cross* mink from the first cross were crossed with *royal silver* from the second cross, the kits appeared in the ratio 2 *black cross* : 1 *royal silver* : 1 *wild-type*.

a) These results do not permit a distinction between (1) multiple allelism and (2) independent assortment of two allele pairs. Using your own symbols, show how these results can be explained by each of these two mechanisms. Some of the *black cross* progeny of the third cross, mated with *wild-type* produced *black cross* and *wild-type* kits only, whereas two other *black cross* from the same cross when intermated produced *royal silver* and *black cross* offspring only.
b) Do these results support either of the hypotheses outlined above? Explain.
c) The two *black cross* offspring that produced *royal silver* kits when intermated were also mated to *wild-type* mink. A small number of offspring were produced, consisting of *royal silver* and *black cross* kits only.
(1) Which hypothesis do these results support?
(2) What would constitute critical evidence that *royal silver* and *black cross* are determined by different alleles of the same gene?

20. Proxy motherhood is frequently used in cattle breeding. The table below shows the blood types of a bull, a cow bred only to this bull, and the calf she gave birth to. Explain whether or not the calf resulted from the transplantation of a fertilized egg from another animal into this cow.

	Blood type														
Individual	A	B	O_2	Y_1	A_1	P	Q	E_3^1	D	F	E_1^1	J	W_3	S	R
Bull	-	+	+	+	+	-	-	+	-	-	+	+	-	-	-
Cow	-	-	-	-	+	+	-	+	+	+	-	-	-	-	+
Offspring	+	+	+	+	+	-	-	+	-	+	+	+	+	+	-

21. In studies of human twins a reliable means of classifying twin pairs as monozygotic or dizygotic is essential. In many studies, this is accomplished by comparing each pair of twins for many genetic characters, among which the blood groups play a relatively important role.
a) Two pairs of twins have the blood-types shown below. Classify each pair of twins for monozygosity vs. dizygosity.

	Anti-A	Anti-B	Anti-N^S	Anti-M^s	Anti-M^S	Anti-N^s	Anti-Rh^+
Twin 1	+	+	+	-	+	-	+
Twin 2	+	-	+	-	-	-	+
Twin 1	-	-	-	+	+	-	-
Twin 2	-	-	-	+	+	-	-

b) Most geneticists would insist on blood-typing the parents as well.
(i) Why?
(ii) Which of the pairs of twins referred to above would necessitate this? Would the parents have to be tested for the MNS blood types?
(iii) Both sets of parents were tested and found to be of the following blood types:
Parents of pair 1   A M^S N^S Rh^+; B N^S Rh^+
Parents of pair 2   O M^S Rh^-; O M^S Rh^-
Explain why your answer for one pair of twins in (a) might have been in error and list the ways in which the validity of assessment can be increased.

22. The father of a certain family belongs to blood group AB, and the mother to blood group O. They have four children, one belonging to group AB, one to A, one to B, and one to O. One of these children is adopted and another is a child from an earlier marriage of the mother. State, with reasons, which of the children is adopted and which is the child from an earlier marriage.

23. a) A woman sues a man for the support of her child. She is of blood type B, her child has type O blood, and the man has type A blood. Could the man be the father of her child? Explain.

b) Further tests show that both the man and woman are *Rh-negative* while the child is *Rh-positive*. What bearing does this information have on the case?

24. The alleles C, c^k, c^d, c^r, c^a in the *albino* series of the guinea pig have the effects shown in the table on the relative amounts of melanin pigmentation in the coat (Wright, S., Genetics 44: 1001, 1959).

Genotype	Percent melanin	Phenotype
$C\_$	100	*Full* colour
$c^k c^k$	88	
$c^k c^d$	65	
$c^k c^r$	54	
$c^k c^a$	36	*Intermediate*
$c^d c^d$	31	colour
$c^d c^r$	19	
$c^d c^a$	14	
$c^r c^r$	12	
$c^r c^a$	3	
$c^a c^a$	0	*White*

Classify each allele as normal (fully morphic), hypomorphic, or amorphic. Explain your classification and arrange the alleles in order of decreasing activity with respect to phenotypic effect.

25. The human hemoglobin molecule consists of two identical halves, each containing an α and a β chain. In addition to normal hemoglobin (*Hb-A*) abnormal ones occur: *Hb-S* (found in sickle-cell anemics), *Hb-C* (in sickle-cell C disease), *Hb HO-2* (found in Hopkins-2 hemoglobinemics), and others. The first two abnormal hemoglobins are defective in the α chain, the latter in the β chain. From the pedigrees, which are typical for these traits, determine whether the defects in the different chains, and therefore *Hb-S* and *Hb-A* on the one hand and *HO-2* on the other, are determined by alleles of the same or different genes.

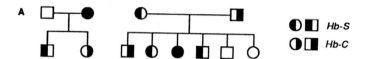

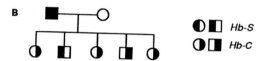

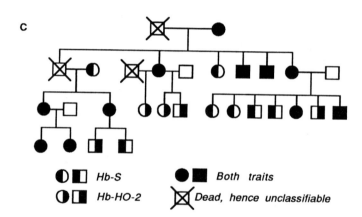

Chapter 7
Polygenic Inheritance

Note: Arithmetic means additive and geometric multiplicative; multiple genes are the same as polygenes. At polygene loci, alleles are of two kinds: contributors, which may act additively or multiplicatively, and neutrals, which have no effect.

1. a) Describe the kinds of observations that would lead you to suspect that a certain human character was determined by polygenes.
   b) Discuss the statement: No new principles of genetics have originated from the study of quantitative characters.
   c) Why is it more difficult to study the inheritance of quantitative characters such as size, weight, and intelligence than qualitative ones such as ABO and Rh blood antigens?
   d) Distinguish between the terms *polygene* and *modifying gene*. What do the two kinds of genes have in common?

2. In a cross between a *large* and a *small* strain of rabbits the F_1's are phenotypically uniform. On the average, their size is *intermediate* (midway) between that of the two parental strains. Among 2,025 F_2 individuals, 8 are of the same size as the *small* parent and 7 the same size as the *large* strain.
   a) How many polygenes are probably involved in the cross?
   b) What is the effect of each contributing allele to rabbit size?

3. In humans *normal* vs. *brachydactyly* (short index fingers) is genetically determined (Farabee, W.C., Pap. Peabody Mus. Harv. Univ. 3: 6, 1905; Mohr, O.L. and C. Wriedt, Carnegie Inst. Wash., D.C., Publ. 295, 1919). If individuals are classified for *normal* vs. *brachydactylous*, we find that the character is monogenically determined, with the allele for *brachydactyly* incompletely dominant to the one for *normal*. *Brachydactylous* individuals, however, show a range in the length of the index finger from extremely short to only slightly short. What might explain this variation in the expression of *brachydactyly*?

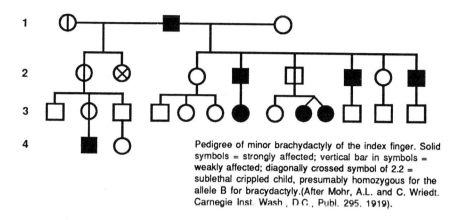

Pedigree of minor brachydactyly of the index finger. Solid symbols = strongly affected; vertical bar in symbols = weakly affected; diagonally crossed symbol of 2.2 = sublethal crippled child, presumably homozygous for the allele B for bracydactyly.(After Mohr, A.L. and C. Wriedt. Carnegie Inst. Wash., D.C., Publ. 295, 1919).

4. According to Davenport, C.B. (Carnegie Inst. Wash., D.C., Publ. 188, 1913) the difference in skin colour between *blacks* and *whites* is determined by two pairs of alleles independently inherited, with the contributors at the two loci having equal and additive effects. On this basis five grades of pigment concentration are recognized as phenotypes in the progenies of marriages between *blacks* and *whites*: *white, lightly pigmented, moderately pigmented, darkly pigmented,* and *black*.
   a) Why should the offspring of some *moderately pigmented* parents differ in skin colour while those of other such parents do not?
   b) (i) If those with various degrees of skin pigmentation married *white* individuals could they have *black* children? Explain.
   (ii) Could nonpigmented couples with *black* ancestry have a *black* child? Explain.
   c) Can matings among the first-generation progeny of different *white-black* matings produce *black* offspring? *White* offspring?
   d) If the number of genes determining skin pigmentation is actually four (or even five or six), as suggested by Harrison, G.A. and J.J.T. Owen (Ann. Hum. Genet. 28: 27, 1964) and Nicholls, B. (Hum. Hered. 23: 1, 1973):

(i) How would the expected results of crosses between *whites* and *blacks* differ from those considered above?

(ii) Why has there been so much uncertainty concerning the number of genes determining skin pigmentation?

e) The first-generation progeny of *black-white* matings, called *mulattoes*, are not uniform in pigmentation, some being considerably lighter than others. Besides the effect of varying environment, what other cause would you suggest for this variation?

f) Do you think all *whites* are identical in genotype with respect to pigment genes? Explain.

5. In 1913 Emerson, R.A. and E.M. East (Nebr. Agric. Exp. Res. Bull. 2) recorded the data shown for ear length in corn.

Parent and generation	\multicolumn{17}{c}{Length of ear (cm)}																
	5	6	7	8	9	10	11	12	13	14	15	16	17	18	19	20	21
P60		4	21	24	8												
P54									3	11	12	15	26	15	10	7	2
F_1				1	12	12	14	17	9	4							
F_2		1	10	19	26	47	73	68	68	39	25	15	9	1			

a) For each of the parents and for the F_1 and F_2 generations calculate the mean and the standard deviation for ear length.

b) Compare the standard deviation: (i) of the two large parents and (ii) of the F_1 and F_2. Offer an explanation for any large differences.

c) Calculate the theoretical and geometric mean of the F_1. Compare these values with the observed F_1 mean. What type of gene action is probably involved?

d) Note that neither of the parental extremes was recovered in the above F_2 population of 401 plants. What conclusion can be drawn from this about the minimum number of genes determining ear length?

e) Determine the approximate number of genes involved with the use of the formula:

$$N = \frac{(\text{difference between parental means})^2}{(F_2 \text{ variance} - F_1 \text{ variance})} = \frac{D^2}{8}$$

f) Delineate the experimental conditions required for the valid use of this formula.

6. An experiment carried out under uniform environmental conditions to determine whether two true-breeding strains of dogs were genotypically identical with respect to weight gave the data shown.

Strain A, 97 lb.		Strain B, 97 lb.
	F_1, 97 lb.	
Testcross progeny	Weight (lb.)	No. of animals
F_1 × 16 lb. true-breeding strain	97	32
	43	124
	25	190
	19	134
	16	32

Note: The segregating generation is a testcross, not an F_2.
The 16 lb. strain is the smallest strain known.

a) Determine the number of polygene loci at which the parents carry different alleles.

b) Using your own symbols, give the genotypes of the parents of the F_1.

c) What type of cumulative gene action (additive or multiplicative) appears to be expressed in this cross? Why?

d) How much of the weight above that produced by the residual genotype (16 lb.) was each contributing allele responsible for?

e) Had the F_1's been intercrossed, what proportion of the F_2 would be expected to express each of the extreme phenotypes and what would these phenotypes have been? Briefly indicate how you arrive at your answer.

7. In bread wheat (*Triticum aestivum*) kernel colour varies from *red* to *white*. The contributors at the different loci for kernel colour act additively. Phenotypic classification of segregating progenies between *red* and *white* kernel varieties is possible (Nilsson-Ehle, H., Lunds Univ. Aarskr., NF. Afd., (2) 3(2): 1, 1909).

a) A cross is made between a *red* kernel, $R_1R_1R_2R_2$, and a *white* kernel, $r_1r_1r_2r_2$, variety. Give the genotype and phenotype of the F_1 and the F_2 genotypes and the proportions of the phenotypes *red*, *dark*, *medium*, *light* and *white*.

b) From a second cross between *red* and *white* varieties, 1/64 of the F_2 progeny have *red* and 1/64 *white* kernels. How many allele pairs determine kernel colour in this cross? What is the formula for determining the proportion of parental types in the F_2?

c) Nilsson-Ehle observed that only plants with *red* kernels were produced in both the F_1 and 78 F_2 progeny of a cross between the *red* kernel wheat variety Swedish Velvet and a *white* variety. The 78 F_2 plants were self-fertilized. Eight gave a 3:1 F_3 ratio (307 *red*:97 *white*), 15 gave a 15:1 ratio (727 *red*:53 *white*), 5 gave a 63:1 ratio (324 *red*:6 *white*), and the remaining 50 F_2 plants bred true (2,317 *red*).
(i) Offer a genetic explanation to account for the F_2 results.
(ii) On the basis of your hypothesis give the genotypes of the parents of the cross.
(iii) Do the F_3 results agree with those expected on the basis of your hypothesis?

8. The data shown are from a series of crosses between two true-breeding strains, under uniform environmental conditions, to study fruit weight (in grams) in oranges.

Strain A, 24 g	Strain B, 32 g
F_1	28 g

F_2 Weight, g	No. of plants
36	2
34	14
32	60
30	108
28	140
26	114
24	52
22	18
20	2

Assuming that the contributing alleles of all genes have equal effects:
a) Determine the number of polygene loci at which the parents carry different alleles.

b) Using your own symbols, give the genotypes of the parents of the F_1.
c) What type of cumulative gene action (additive or multiplicative) appears to be expressed in this cross? Why?
d) How much of the weight above that produced by the residual genotype (20g) is each contributing allele responsible for?

9. Three watermelon plants bear 4 lb. fruits. Plants A and B when self-fertilized breed true, but C when self-fertilized produces progeny with fruits ranging in weight from 3 to 5 lb. A cross of plant A with plant B produces F_1's with 3 lb. fruits and an F_2 with fruits ranging from 3 to 5 lb.. Selection in this and succeeding generations cannot increase fruit weight above 5 lb.. A cross of plant A with plant C gives F_1's with fruits ranging from 3.5 lb. to 4.5 lb. Selection among the F_2 can raise the fruit weight to 6 lb. A cross of plant B with plant C produces F_1's with fruits from 3.5 to 4.5 lb. By selection among the F_2 it is possible to raise the weight of fruit to 5.0 lb. Explain these results, giving the genotypes of the three parents.

10. Each human being has a specific pattern and number of finger print ridges that is differentiated by the end of the fourth week of fetal development. The number of ridges varies from 0 to about 300. As a consequence of many years of work Holt, S. (Brit. Med. Bull. 17(3): 247, 1961) obtained the following correlations among relatives for total ridge-counts:

Type of relationships	Correlation coefficient	Number of pairs used
Parent-child	0.48	810 (200 families)
Mother-child	0.48 ± 0.04	405 (200 families)
Father-child	0.48 ± 0.04	405 (200 families)
Parent-Parent	0.05 ± 0.07	200 (200 families)
Mid-parent-child	0.66 ± 0.03	405 (200 families)
Sib-Sib	0.50 ± 0.04	642 (290 sibships)
Monozygotic twin-twin	0.95 ± 0.01	80
Dizygotic twin-twin	0.49 ± 0.08	92

a) Is total ridge-count a genetically determined character? Explain referring to specific correlations as a basis for your answer. If your answer is an affirmative one, do the contributing alleles at the various loci function additively or multiplicatively? Explain.
b) Does the environment play a role in the phenotypic expression of this character? Is the role a large or small one?
c) Why is it not possible from the information provided to at least estimate the number of genes affecting the expression of this metrical character?

11. A corn breeder has 10 corn plants 70 in. tall. He crosses them in pairs and self-fertilizes the offspring in each of the five crosses for several generations, selecting the taller plants in each generation. His results for the five crosses are:
Cross 1 produces only 70 in. offspring; selection fails to raise their height.
Cross 2 produces offspring varying from 50 to 90 in.; selection among these fails to raise the height above 90 in.
Cross 3 produces progeny varying from 60 to 80 in., and by selection it is possible to raise the height to 110 in.
Cross 4 produces progeny varying from 50 to 90 in.; selection raises the height to 110 in.
Cross 5 produces offspring varying from 50 to 90 in.; selection raises the height to 130 in.
Explain these results giving the genotypes of the parents.

12. Burton, E.W. (Agric. J. 43: 409, 1951) presented extensive data on the inheritance of quantitative characters in pearl millet (*Pennisetum glaucum*). The data for one of these characters, number of leaves per stem, from one of six crosses, are given below:

Parent and generation	8	9	10	11	12	13	14	15	16	17	18	19	20	21	22	23	N
P19	7	36	62	37	23	3											168
P782					2	4	4	11	24	37	26	21	26	13	9	2	179
F_1				2	1	3	17	52	73	52	17	3					220
F_2			4	16	49	163	282	379	375	161	72	19	3				1523

a) Calculate the mean and standard deviation for each parent and for the F_1 and F_2.

b) Why would you expect the standard deviation for the F_2 to be greater than each of the other standard deviations?
c) Compare the mean of the F_1 and that of the F_2 with the two parental means. Do your findings agree with the results expected according to the polygene hypothesis?
d) Is it possible to determine whether the action of the genes is arithmetic or geometric? Explain.
e) Suggest reasons why it is not possible to determine the actual number of genes involved from these data. Indicate how one would proceed to estimate the number of loci involved.

13. a) A breeder in China has a strain of rice (*Oryza sativa*) that has been self-fertilized for 20 generations. He has repeatedly tried to increase the average seed yield per acre of this strain by selection, without success. Explain why.
b) He crosses this strain with another that yields the same. The F_1's yield the same as their parents, but by selection in the F_2 and F_3 he is able in a few generations to increase the yield considerably. Explain how this is possible.

14. Breeds of cattle like the Holstein-Friesian have a *white-spotted* coat. There is, however, considerable variation in the amount of *white* from one individual to another. Some animals are *solid-coloured* (black), others have varying amounts of *white spots*, and still others are almost completely *white* with few *black* spots. In certain crosses between *solid-coloured* and *spotted* individuals all the F_1 are *solid-coloured*, and their F_2 progeny consist of *solid-coloured* and spotted individuals in a 3:1 ratio. Selection within the *spotted* group of animals can increase or decrease the amount of *white*. Selection has no effect in the parents of the F_1 and F_2's that give a 3:1 ratio. Why does the polygene hypothesis fail to account for these observations? Suggest an alternative one that suits them better.

Chapter 8
Sex Determination and Sex Differentiation

1. Distinguish between:
   a) Primary and secondary sex characters.
   b) Sex determination and sex differentiation.
   c) Autonomous and nonautonomous sex differentiation.
   d) Intersexes and gynandromorphs.
   e) Hermaphrodites and monoecious individuals.
   f) Sex mosaics and gynandromorphs.
   g) Heterogametic and homogametic sex.

2. Klinefelter's syndrome (XXY, XXXY, XXYY, etc.) occurs approximately once in every 400 to 600 male births, whereas Turner's syndrome (XO) occurs only about once in every 5,000 female births. Discuss with the aid of diagrams the mechanism by which each of these conditions could arise and suggest what relation the discrepant frequency of occurrence has to your explanation.

3. a) Cite evidence for the fact that the gonads of vertebrates are potentially dual in function.
   b) Explain why YO human beings have not been found.
   c) Is it possible for the members of a pair of monozygotic twins to differ in sex? Explain.
   d) Why do XO monosomics develop as males in *Drosophila melanogaster* but as females (with Turner's syndrome) in *Homo sapiens*? Why are XXY trisomic individuals females in *D. melanogaster* but males in humans?
   e) In humans, males are more vulnerable to mishaps in sexual development than are females. Why?

4. In *Drosophila melanogaster*, flies with two sets of autosomes and two X chromosomes (AAXX) are female, while those with two sets of autosomes plus one X and one Y chromosome (AAXY) are males.
   a) Why is this information alone insufficient to determine the location of the sex-determining genes? Bridges, C.B. (Science 54: 252, 1921 and Am. Nat. 59: 127, 1925) observed the following phenotypes of different aneuploids and polyploids in this species:

   AAXO- males, sterile   AAAAXXXX- female
   AAXXY- females, fertile   AAAAXX- male
   AAAXX- intersex   AAXXX- metafemale
   AAAXXX- female   AAAXY- metamale
   AAAAXXX- intersex

   b) Which chromosomes carry female determining genes and which carry male determiners? Explain.
   c) What is the primary mechanism (signal) that determines the direction of sexual differentiation? Discuss.

   The following results with two (*da* and *Sxl*) of many genes involved in sex determination in this species are based on the findings of Bell, A.E. and T.W. Cline and others (See Cline, T.W. Genetics 96: 903, 1980):
   (i) $da^+da^-Sxl^+Sxl^+$ ♀ x $da^-da^-Sxl^+/Y$ ♂ → 1/2 AAXX, (1/2 of these die as embryos the other 1/2 are viable normal females); 1/2 AAXY, all viable, normal males.
   (ii) $da^+da^+Sxl^+Sxl^+$ ♀ x $da^+da^+Sxl^+/Y$ ♂ → 1/2 AAXX, all viable, normal females; 1/2 AAXY, all viable, normal males.
   (iii) $da^+da^+Sxl^+Sxl^-$ ♀ x $da^+da^+Sxl^-/Y$ ♂ → 1/2 AAXY, all viable, normal males; 1/2 AAXX, 1/2 of these die as early embryos, remaining 1/2 are viable, normal females.

   Note: da^+ acts maternally; it produces a da^+ protein which is stored in the egg cytoplasm; da^- either does not produce da^+ protein or it produces a non-functional one.

   d) Is the da^+ product required for *Sxl* function in sex determination in either or both sexes? Explain.
   e) Is *Sxl* required for male determination? Female determination? Explain.
   f) What is the sequence in which these two loci and the primary sex determining signal function in sex determination? Discuss.

5. In *Drosophila melanogaster*, the recessive allele *tra*, of a gene on chromosome 3, when homozygous transforms diploid

females into sterile males (Sturtevant, A.H., Genetics 30: 297, 1945).
a) What sex ratio would you expect in the progeny of a cross between a *Tratra* female and a *tratra* male?
b) Briefly discuss the significance of the *tra* gene.
c) Suggest how *tra* may act on the sex phenotype.

6. The recessive (null) alleles at the sex-lethal (*Sxl*), transformer (*tra*), transformer-2 (*tra-2*), doublesex (*dsx*) and intersex (*ix*) loci in *Drosophila melanogaster* have the following effects when homozygous (Steinmann-Zwicky, M. et al., Adv. Genet. 27: 189, 1990):
Sxl⁻ - lethal to XX flies; no effect in XY flies
tra - transforms XX flies into males (sterile); no effect in XY flies
tra-2 - transforms XX flies into males (sterile); no effect in XY flies
ix - transforms XX flies into intersexes; no effect in XY flies
dsx - transforms XX and XY flies into intersexes

ix and *dsx* produce very similar intersexual phenotypes while *tra* and *tra-2* produce identical phenotypes in XX flies.
a) The *dsx* gene is unusual in its function. Explain.
b) Does this information permit determination of the sequence in which these five genes function in sexual development? Explain.

Baker, B.S. and K.A. Ridge, (Genetics 94: 383, 1980) observed the following phenotypes of AAXX individuals homozygous for all the double mutant combinations of *ix*, *tra*, *tra-2*, and *dsx*:
(i) *dsx dsx, tra tra*; *dsx dsx, tra-2 tra-2* and *dsx dsx, ix ix* all exhibit the *dsx* intersex phenotype
(ii) *tra tra, tra-2 tra-2* show the same male phenotype as either of the single mutants
(iii) *tra tra, ix ix* and *tra-2 tra-2, ix ix* both express the same male phenotype

c) Do these genes, which regulate sex determination, function in the same or different pathways? Explain.
d) What is the epistatic hierarchy among these 4 loci?

e) Does this epistatic hierarchy permit determination of the sequence in which these genes function in the pathway? If not, what other information is necessary?

7. *Melandrium dioicum* (a member of the pink family) is a dioecious species in which the female is homogametic and the male heterogametic. Many polyploid and aneuploid plants have been obtained and used by Warmke, H.E. (Am. J. Bot. 33: 648, 1946) and Westergaard, M. (Adv. Genet. 9: 217, 1958) to study the roles of the X chromosome, the Y and the autosomes in sex determination. Some of these chromosome constitutions, with their corresponding sexes are shown in the table. Determine the locations and relative strengths of the male- and female-determining genes.

Chromosome Constitution	Sex	Chromosome Constitution	Sex
AAXX	♀	AAAAXXXYY	♂
AAXXX	♀	AAXXY	♂
AAAXX	♀	AAAXXY	♂
AAAAXXX	♀	AAAXXXY	♂
AAXY	♂	AAAAXXY	♂
AAXYY	♂	AAAAXXXYY	♂
AAAXY	♂	AAAAXXXY	♂
AAAAXY	♂	AAAAXXXXY	♂ or ♀

8. A variety of sex mosaics have been found in humans, for some of which the somatic cell chromosome numbers and sex-chromosome constitutions are as shown.

Chromosome number	Sex-chromosome constitution	Phenotype
45-47	XO-XXY	Turner's syndrome
47-48	XXX-XXXY	Klinefelter's syndrome
46-46-45	XX-XY-XO	Male pseudohermaphrodite
46-47	XX-XXY	Klinefelter's syndrome
45-46	XO-XY	True hermaphrodite

Suggest a common mechanism by which these individuals may have originated. Use any two of these mosaics to illustrate the mechanism.

9. In 1962, Gartler, S.M. (PNAS 48: 332) reported that a XX/XY hermaphrodite had two populations of red blood cells for the Rh blood antigen system: CDe/cdE and CDe/cde. Her left eye was hazel like her mother's; her right eye was brown like her father's. The mother's Rh genotype was CDe/cdE; the father's genotype was cdE/cde. What is the most likely method of origin of this hermaphrodite?

10. a) Humans (2n=46) have an XX-XY sex chromosome constitution (Tjio, J.H. and A. Levan, Hereditas 42: 1, 1956; Ford, C.E. and J.L. Hamerton, Nature 178: 1020, 1956). Individuals with Turner's syndrome (female external genitalia, short stature, webbed neck, small uterus, ovaries represented by fibrous streaks, etc.) are classed as being essentially female, and have the chromosome constitution AAXO (Ford, C.E. et al., Lancet 1: 711, 1959). Individuals with Klinefelter's syndrome (male external genitalia, small testes, sparse body hair, female-like breast development, etc.) are classed as males, and usually have the chromosome constitution AAXXY (Jacobs, P.A. and J.A. Strong, Nature 183: 302, 1959). From this information determine the roles, if any, of the X, the Y and the autosomes in sex determination in humans.

b) The table below shows other types of sex-chromosme variants found by various workers (data from McKusick, V.A. *"On the X-chromosome of Man"*, American Institute of Biological Sciences, Washington, 1964; Mittwoch, U. *"Sex chromosomes"*, Academic Press, New York, 1967)

Chromosome number	Chromosome constitution	Sex phenotype	Fecundity
47	AAXXX	Female	Many fertile
48	AAXXXX	Female	Sterile
47	AAXYY	Some normal male, others Klinefelter's	Some fertile
48	AAXXYY	Klinefelter's	Sterile
48	AAXXXY	Klinefelter's	Sterile
49	AAXXXXY	Klinefelter's	Sterile

(i) Do these findings give additional information on the location of sex-determining genes? Explain.
(ii) The extent to which the autosomes determine femaleness is not clear from these findings. What types of individuals with respect to chromosome number and constitution should provide information bearing on the problem?

11. a) In humans, the Y-chromosome carries the male-determining gene(s). Using molecular techniques, Sinclair, A.H. et al. (Nature 346: 240, 1990) identified a gene (*SRY*) near the pseudoautosomal region of the Y that codes for a protein 233 amino acids long. The gene is transcribed and produces its protein only in testes but not in the ovaries or in lung, kidney, and other tissues of males. This gene is present on the Y-chromosome of all other mammals tested as well as in the rare XX human males.
(i) Could *SRY* have a male determining function? Explain.
(ii) Do the above facts provide evidence for identity of the *SRY* and *TDF* (testis determining factor) genes? Explain.

b) The *Sry* gene in mice is homologous to the *SRY* gene in humans. Koopman, P. et al. (Nature 351: 117, 1991) injected fertilized eggs of mice with *Sry* genes and Y-linked *Zfg-1* (zinc-finger protein) genes and isolated three transgenic XX embryos with many copies of the *Sry* but none of the *Zfg-1* genes. Two of these embryos had testes indistinguishable from those of normal XY sib embryos. The third transgenic XX embryo was allowed to develop to adulthood. It possessed normal testes and a normal male phenotype with respect to both internal and external genitalia.
(i) Is *Sry* the only Y-linked gene required for male development? Is *Sry* the testes-determining gene (*Tdf*) in mice? Explain.
(ii) Why is the *Zfg-1* gene not the *TDF* gene?

12. Human males afflicted with the *testicular feminization* syndrome exhibit female external genitalia, breast development and a blind vagina but lack a uterus and the internal female reproductive tract. They have a normal male karyotype AAXY, very little pubic hair, and fail to menstruate. The pedigree, from Schreiner, W.E. (Geburtsch. Frauenkeilk 19: 1110, 1959) is typical of the mode of inheritance of this trait.

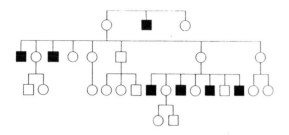

a) What are the possible modes of inheritance of *testicular feminization*? Explain. What modes are definitely eliminated? Explain.
b) By what means could you distinguish between the possible modes of inheritance?
c) Is the causative allele for *testicular feminization* at the male (*testis-determining = TD*) locus? Explain.
d) Is it at the locus that codes for the Mullerian inhibiting hormone? Explain.

13. Normally the ontogeny of AAXX embryos leads to the formation of ovaries and a normal female phenotype accompanied by a regression of Wolffian ducts and the differentiation of Müllerian ducts. On the other hand AAXY embryos develop testes concomitant with development of Wolffian ducts and the disappearance of Müllerian ducts. Jost, A. (The Harvey Lectures, Ser. 55, Academic Press, New York, 1961) reported the following findings:
(i) Castration of XY and XX embryos before the onset of sexual differentiation results in a development of Müllerian ducts and a regression of Wolffian ducts thus resulting in a manifestation of the female phenotype.
(ii) Administration of testosterone to castrated embryos sustains development of Wolffian ducts which develop into the vas deferens, epididymus, etc., and the urogenital sinus which in turn differentiates into the prostrate and penis glands. The Müllerian ducts, however, are not destroyed thus leading to a hermaphroditic phenotype.

a) Do you agree with Ohno, S. et al. (Hereditas 69: 107, 1973) that "... retention of Müllerian ducts and regression of Wolffian ducts are automatic processes which do not require the presence of ovaries"? Explain.
b) Is testosterone itself sufficient for differentiation of XY embryos into normal males? Discuss.

14. In humans the sex chromosome constitution of males is usually XY and that of females XX. The X-Y bivalent in normal males has a single reciprocal exchange usually in the position shown in the diagram below:

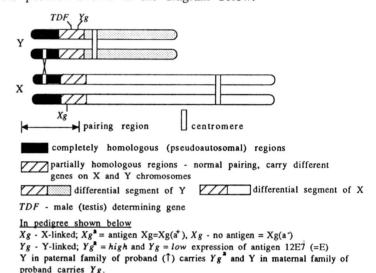

Xg - X-linked; Xg^a = antigen $Xg=Xg(a^+)$, Xg - no antigen = $Xg(a^-)$
Yg - Y-linked; Yg^a = *high* and Yg = *low* expression of antigen 12E7 (=E)
Y in paternal family of proband (↑) carries Yg^a and Y in maternal family of proband carries Yg.

Exceptional males (XX) and females (XY) have each been found to occur in a frequency of about 1 in 25,000.
a) Propose a plausible explanation to account for the origin of the exceptional males and females.

In 1984 de la Chapelle, A. et al. (Nature 307: 170) presented a pedigree shown below which has had a significant bearing on our comprehension of the mechanism of origin of these exceptional males and females.

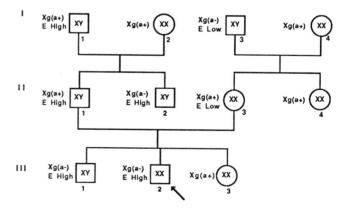

b) Does this pedigree substantiate your explanation in (a)? Explain.

15. In silkworms (*Bombyx mori*) the larval epidermis is *opaque* or *translucent*. In a series of reciprocal crosses involving these traits, Tanaka, Y. (J. Genet. 12: 163, 1922) obtained the results shown:

Parents			
♀	♂	F$_1$	F$_2$
translucent	x *opaque*	♀ 1,374 *opaque*	♀ 1,362 *opaque*
			♀ 1,301 *translucent*
		♂ 1,354 *opaque*	♂ 2,510 *opaque*
			♂ 0 *translucent*
opaque	x *translucent*	♀ 774 *translucent*	♀ 610 *opaque*
			♀ 456 *translucent*
		♂ 782 *opaque*	♂ 524 *opaque*
		1 *translucent*	♂ 453 *translucent*

a) Which sex is heterogametic? Give reasons for your answer.
b) Account for the *translucent* male in the F$_1$ of the second cross.

16. a) *Asparagus officinalis*, the garden asparagus, is normally dioecious with *staminate* (male) and *pistillate* (female) plants occurring in approximately equal numbers. Occasionally, stamens develop in *pistillate* flowers and pistils in *staminate* flowers. Although usually nonfunctional, pistils of this nature sometimes produce viable seed by self-fertilization. From 61 such plants Rick, C.M. and G.C. Hanna (Am. J. Bot. 30: 711, 1943) obtained 198 seeds in this way; when planted and grown to maturity, 155 proved to be male and 43 female.
(i) What do these results suggest regarding the genetic control of sex in this species?
(ii) Which is the heterogametic sex?

b) When 25 of the 155 male plants were crossed with *pistillate* plants, 8 gave male progeny only and 17 gave progeny with males and females in a 1:1 ratio.
(i) Do these results support your explanation to (a)?
(ii) Is the allele for maleness dominant or recessive?
(iii) Show the kinds of progeny and their expected proportions among seed (obtained by self-fertilization) of rare *pistillate* flowers with functional stamens.
(iv) Designating the chromosome pair carrying the sex-determining locus XX in the homogametic sex and XY in the heterogametic sex, show how Rick and Hanna's results can be explained by an XY mechanism.

c) According to certain investigators, *staminate* plants outyield *pistillate* ones by about 25 percent. Outline a method by which seed could be obtained that would give *staminate* plants only.

17. In the fish, *Lesbistes reticulatus*, there is a *black* spot on the dorsal fin. In crosses of *black-spot* males with *normal* females half the progeny have the *black spot* and the other half do not. *Black spot* occurs only in males. Moreover, *black spot* does not occur in males unless the father expressed the trait (Winge, O., J. Genet. 12: 145, 1922). Present a diagrammatic explanation of these results showing which is the heterogametic and which the homogametic sex. Show what would be expected if the gene determining *black-spot* was located on the alternative sex chromosome to that you have postulated.

18. In *Drosophila* species, the male is heterogametic (XY) and the female homogametic (XX). In birds the mechanism is reversed. In *Drosophila* species such as *parthenogenetica*, individuals that arise parthenogenetically are almost always females, but birds that arise by parthenogenesis are, with few exceptions, males.
a) Offer a genetic explanation to account for this difference.
b) Explain the occurrence of rare parthenogenetic males in *Drosophila*.

19. Two varieties of *Ecballium elaterium* (a cucurbit) are found in Spain. Northern populations consist of monoecious types only (var. *monoicum*); those of southern Spain (var. *dioicum*) are dioecious. The results of extensive intra- and

intervarietal crosses by Galan, F. (Acta Salmanticensia, Cliencias, Secc. Biol. 1: 7, 1951) are given below:

Parents		Progeny		
		Dioecious		Monoecious
♀	♂	♀	♂	
dioicum x dioicum		1	1	0
monoicum x monoicum		0	0	all
dioicum x monoicum		0	0	881
monoicum x dioicum		0	423	382
dm x md		16	0	34
md x md		0	105	109
md x d		35	48	30
m x md		0	483	550
dm x m		0	0	59

Key: d = dioecious; m = monoecious; dm = F_1 from dioecious x monoecious cross; md = F_1 from monoecious x dioecious cross.

Note: The results are given as proportions in the first two crosses only.

a) Outline a genetic hypothesis to explain these results, stating which is the heterogametic sex in *E. elaterium* var. *dioicum*.
b) Designating the sex chromosomes of the homogametic sex XX and those of the heterogametic one XY, give the sex genotypes of male, female, and monoecious plants.
c) Contrast the roles of the Y chromosomes in *Ecballium* and *Drosophila* with regard to sex and viability.

20. a) Ford, E.B. ("*Moths*", Collins, London, 1955) illustrated a bilateral gynandromorph of the waved umber moth (*Hemerophila abruptaria*) in which the right side was male and the left female. In addition, a colour difference was involved, the left side being *normal* and the right a *deep-chocolate* shade. This colour difference is autosomally determined. How would you account for the occurrence of this moth?

b) Ford also showed a different colouring on the two sides of a normal female currant moth (*Abraxas grossulariata*). Suggest an explanation for the occurrence of this bicoloured female.
c) What do gynanders in this and other invertebrate species tell us regarding the role of hormones in sex determination and the type of sex differentiation in insects?

21. The following data were obtained from studies on sex determination in a certain diploid animal species in which males and females occur in a 1:1 ratio:
(i) Normal females crossed with neomales (genetic females converted into phenotypic and functional males by high temperature in early embryogenesis) produce progeny of both sexes.
(ii) As a consequence of nondisjunction in the homogametic sex, in addition to normal males and females, individuals are obtained with the chromosome constitutions and sex phenotypes tabulated.

No. of autosomal sets	No. of sex chromosomes of homogametic sex	Sex phenotype
2	1	Female (sterile)
3	2	Intersex
2	3	Supermale

(iii) All males have a *chocolate-coloured* body; females are *chocolate* or *gray*. Body colour is determined by a single pair of autosomal alleles, the allele for *chocolate* being dominant to that for *gray*. An occasional individual appears that is *chocolate* on one side of the body and *gray* on the other.
State, giving reasons for your statements:
a) Which sex is heterogametic.
b) The location of male- and female-determining genes.
c) The mechanism of sex determination.
d) The mechanism of sex differentiation.
e) How a *chocolate gray* individual could arise, and whether the individual started out as a female or male, indicating one possible genotype of the zygote.

22. a) In the creeping vole (*Microtus oregoni*), both sexes are gonosomic mosaics. Eight pairs of autosomes are a constant

feature of all somatic and germinal cells in both sexes. The males have an XY pair in somatic cells but only a Y in primary spermatocytes; they are described as OY/XY. The females have one X in somatic cells and two X's in the primary oocytes; they are described as XX/XO (Ohno, S. et al., Cytogenetics 2: 232, 1963). Account for this unusual sex determining mechanism by diagrammatically illustrating meiosis and its consequences in both sexes, the sex-chromosome constitution of the zygotes for both sexes and the probable origin of the OY and XX meiocytes.

b) A pair of X-linked alleles *Aa* is present in a population of voles. What genetic results would you expect in a three-generation cross that would help you support the cytological findings described above if:
(i) female was *a*, male was *A*. (ii) female was *A*, male was *a*. If cytological studies had not been conducted with this organism, what other interpretation might you give such results? Explain.

c) Considering that mammalian females appear to require only one functional X chromosome, would you consider this form of sex determination more advanced or less advanced than that in other mammals? Explain.

23. Corn (*Zea mays*) is normally monoecious. A recessive allele, *sk* (*silkless*), eliminates functional female flowers, making *sksk* plants phenotypically and functionally male. On another chromosome several recessive alleles of a multiple allelic locus, *ts* (*tassel seed*) cause pistillate flowers to replace staminate ones, making the plant effectively female (Jones, D.F., Genetics 19: 552, 1934). The ts_2 allele is epistatic to *sk* and *Sk*. Plants ts_2ts_2 are female and fertile.
a) What is the genotype of *normal* corn? Of *silkless* plants? Of female plants?
b) A monoecious corn plant of genotype $SkskTs_2ts_2$ is self-fertilized. Show the types of progeny that can occur and their proportions. What type of gene interaction is involved?
c) Using these two genes, show how a stable sex-determining system can be established in which the male is the heterogametic sex.

d) What conclusions can be drawn from such results regarding the complexity of the differences between unisexual and bisexual individuals or species?
e) What further mutational steps are needed for the conversion of such a dioecious system into one resembling the XY mechanism in *Melandrium* and humans?
f) Ts_3, a dominant allele at the *tassel seed* (*ts*) locus, converts staminate flowers into functional female flowers. Using Ts_3 and *sk*, illustrate whether it is possible to establish a stable sex-determining system with males and females in a 1:1 ratio. If so, which sex would be heterogametic?

24. Naito, T. and H. Suzuki (J. Hered. 82: 101, 1991) obtained the following results in single-pair matings in the turnip sawfly *Athalia rosae ruficornis*:

		Progeny					
		Male			Female		
Mating		n	2n	3n	n	2n	3n
Female x male	F_1	789	0	0	0	885	0
F_1 females x F_1 males	F_2 group A	506	0	0	0	943	0
	F_2 group B	551	402	0	0	641	0
F_2 females x F_2 B*n males	F_3	296	239	0	0	293	0
F_2B females x F_2B 2n males	F_3 group A	142	0	0	0	0	0
	F_3 group B	624	0	0	0	0	39
	F_3 group C	222	0	56	0	0	87
2n females x 3n males		561	0	0	0	0	0

*B = F_2 group B
a) Offer a plausible genetic explanation for sex determination in this species. Give the genotypes of the parents and their offspring for the first four matings.
b) Account for the triploid progeny from crosses between diploid females and diploid males.
c) Why did the last mating fail to produce diploid and triploid progeny?

25. Administration of female hormone (estradiol) to male larvae of the African water frog (*Xenopus laevis*) causes them to develop as functional females (called neofemales). Thirteen such females mated with normal males produced 1,624 males and no females (Gallien, L., Bull. Biol Fr. Belg. 90: 163, 1956; Chang, C.Y. and E. Witschi, Proc. Soc. Exp. Biol. Med.

89: 150, 1955). State what conclusions can be drawn regarding:
a) The chromosomal mechanism of sex determination and the heterogametic sex.
b) The process of sex-differentiation.

26. Sex reversed individuals are such that their sex phenotype is either opposite to that normally associated with the genotype or they are hermaphroditic. The results of experiments with sex reversed and hermaphroditic amphibians, are described below:
(1) Witschi, E. (Biol. Zentralbl. 43: 83, 1923) intercrossed hermaphrodites of the frog *Rana temporaria*. He also crossed these with normal males and females. The sex distributions found in the progenies are tabulated.

Parents	Progeny		
	♀	♂	♂
normal ♀ x ♂	182	0	0
♂ x ♂	45	0	0
♂ x *normal* ♂	132	135	0
normal ♀ x *normal* ♂	128	127	0

2) Humphrey, R.R. (Am. J. Anat. 76: 33, 1945) working with the Mexican axolotl (*Ambystoma mexicanum*) succeeded in masculinizing genetic females by heterologous embryonic grafts (graft of testis on a genetic female). These *neomales* were crossed with *normal* females and gave a sex ratio of 3 female : 1 male. Seventeen F_1 females, chosen at random and mated with *normal* males, produced the following offspring:
6 F_1 ♀ x *normal* ♂ → all (833) ♀ progeny
11 F_1 ♀ x *normal* ♂ → 370 ♂ : 378 ♀
Some F_1 females that gave only female progeny were similarly masculinized. These *neomales* mated with *normal* females gave only female progeny (Humphrey, R.R. J. Exp. Zool. 109: 171, 1948).

a) For either or both result (1) and result (2), verify that one sex is homogametic and the other heterogametic. State which is the heterogametic sex and whether the hermaphrodites (sex-reversed) individuals are heterogametic or homogametic.
b) What do the results of these experiments tell us regarding the role of hormones in sex determination and sex differentiation in these organisms?

27. In the silkworm (*Bombyx mori*), with the ZZ-ZW type of sex determination both sexes have a 2n=56 chromosome number. Shown below are the sex phenotypes of aneuploid and polyploid types (Yokoyama, T., "*Silkworm Genetics Illustrated*", Japanese Society for the Promotion of Science, Tokyo, 1959):

AAZO ♂	AAAZZ ♂	AAAAZZZW ♀
AAZZW ♀	AAAZZZ ♂	AAAAZZWW ♀
AAZZ ♂	AAAZZW ♀	
AAAZZZW ♀	AAAZWW dies	

a) Determine the location of female-determining genes. Is it possible to do the same for male determiners? Explain.
b) What parallel do you see between silkworms and humans in terms of sex determination?

28. In the axolotl (*Ambystoma mexicanum*), the female is heterogametic (ZW) and the male homogametic (ZZ). Humphrey, R.R. (Am. J. Anat. 76: 33, 1945) reported that in a cross between the *sex-reversed* females and *normal* females, 76 percent of the progeny were female and 24 percent male.
a) What do these results suggest regarding the viability and sex of WW animals in this genus?
b) How could you test your hypothesis further?

29. The early embryo of mammals and of amphibians such as the salamander is sexually neutral. The rudimentary gonad consists of a *cortex* (an outer layer of tissue which in AAXX individuals dominates and develops into an ovary) and a *medulla* (an inner mass which in AAXY individuals develops into testes). Witschi, E. (Biol. Zentralbl. 43: 83, 1934; Witschi, E. and H.M. McCurdy, Proc. Soc. Exp. Biol. Med. 26: 655, 1929) carried out elegant grafting experiments with

salamanders and found that when male and female embryos in the same early stage of development are grafted together, the male dominates in development, the testes are normal, and ovary development in the female graft partner is often entirely suppressed. If, however, the male graft is small in comparison with the female graft partner, the female influence predominates and the male gonads are often converted into ovaries. Explain how these results illustrate that individuals are bipotential with respect to sex and that differentiation is under the control of hormones.

30. In many organisms, e.g., the bedbug (*Cimex lectularius*) and the praying mantis (*Sphodromantis viridis*), sex is determined by "multiple sex chromosomes". In the praying mantis the male's sex-chromosome complement is X_1X_2Y, and the female's $X_1X_1X_2X_2$; while in hops the male and female complements are XY and XX, respectively.
a) Show how the sex chromosomes would probably behave during male and female meiosis in these two organisms to give a stable sex-determining system.
b) Suggest how the sex-chromosome complement in the praying mantis could have originated (see White, M.J.D., *"Animal Cytology and Evolution"*, 2d, ed., Cambridge University Press, New York, 1954).
c) Certain mantid species, e.g., *Mantis mantis*, with the XX-XO method of sex determination have one chromosome pair more than related species such as *S. viridis* in which males possess the X_1X_2Y sex-chromosome complement and the females have a $X_1X_1X_2X_2$ sex-chromosome constitution. Indicate, with the aid of illustrations, how the latter could have arisen from the former.

31. Assuming the bisexual state to be ancestral:
a) Outline a hypothesis that would explain the evolution of a monoecious species into a dioecious one like the cucurbit *Ecballium elaterium* with the XX-XY method of sex determination in which the Y is morphologically the same as the X and in which YY individuals are viable.
b) What are the possible stages in the evolution of the XX-XY mechanism described in (a) to one in which the X and Y differ morphologically and genetically to such a degree that YY individuals are inviable but the Y still plays a decisive role in sex determination (as in humans and *Melandrium*)?

c) How might the XX-XY mechanism of humans and mouse might change to one in which the Y chromosome is present but nonfunctional in sex determination and possibly completely inert, as in *Drosophila* and *Lygaeus*?
d) Outline a possible mode of origin for the XX-XO system, which occurs in many grasshoppers and in other organisms, from an XX-XY system.

Chapter 9
Sex Linkage, Sex-Influenced and Sex-Limited Characters

1. a) Explain why most of the sex-linked traits thus far discovered in humans are recessive whereas most of the autosomal ones are dominant.
   b) Would you expect sex-linked recessive traits in organisms with an XY sex-determining mechanism to be more frequent in males than in females? Why?
   c) What evidence would be required to prove that a character in humans is due to a gene on the segment of the Y chromosome that is not homologous to any portion of the X chromosome?

2. Discuss the various lines of evidence that suggest that a Barr body is derived from one X chromosome.

3. State, with reasons, whether the following statements are true or false:
   a) For a rare X-linked recessive trait, one-fourth of the sons of all daughters of carrier females are expected to be *affected*.
   b) The critical feature of the pedigree pattern of a rare sex-linked trait is the absence of male-to-male transmission.
   c) Father-to-son transmission of X-linked traits can occur.

4. A virgin *Drosophila* female whose thorax bristles are very *short* is mated with a male having normal (*long*) bristles. The F_1 progeny are 1/3 *short*-bristle females, 1/3 *long*-bristle females, 1/3 *long*-bristle males. A cross of the F_1 *long*-bristle females with their brothers gives only *long*-bristle F_2 progeny. A cross of *short*-bristle females with their brothers gives 1/3 *long*-bristle females, 1/3 *short*-bristle females, and 1/3 *long*-bristle males. Explain genetically.

5. In the moth *Abraxas grossulariata*, the *wild-type* has fairly large spots on the wings; in the *lacticolor* mutant these spots are greatly reduced in size. The table shows the results of reciprocal crosses made by Doncaster, L. and G.H. Raynor (Proc. Zool. Soc. (Lond.) 1: 125, 1906).

Parents			
♀	♂	F_1	F_2
lacticolor	x wild-type	♀ *wild-type* ♂ *wild-type*	♀ 1/2 *lacticolor* : 1/2 *wild-type* ♂ all *wild-type* 1 *lacticolor* : 3 *wild-type*
wild-type	x lacticolor	♀ *lacticolor* ♂ *wild-type*	♀ 1/2 *wild-type* : 1/2 *lacticolor* ♂ 1/2 *wild-type* : 1/2 *lacticolor* 1 *wild-type* : 1 *lacticolor*

Explain cytogenetically why the reciprocal crosses give different results?

6. The cinnamon variety of canary has *pink* eyes. The green variety has *black* eyes. Eye colour is monogenically determined. When two *pink*-eyed birds are mated, the progeny are always *pink*-eyed. When *pink*-eyed hens are mated with *black*-eyed cocks, all the offspring of both sexes are *black*-eyed; and when *black*-eyed hens and *pink*-eyed cocks are bred together, all male offspring are *black*-eyed and females, with a few exceptions, are *pink*-eyed (Durham, F.M., J. Genet. 17: 19, 1927).
   a) Did all the eggs of the latter mating hatch? Explain.
   b) *Black* eyes is due to a dominant Z-linked allele. Why?
   c) What results would you expect in the F_1 and F_2 generations of the cross *black* male x *pink* female?
   d) Describe the breeding procedure you would use with the F_2 progeny from the cross in (c) to establish a true-breeding variety of *pink*-eyed canaries. Would it be easier to establish a true-breeding variety of *black*-eyed canaries? Explain.

7. a) Mothers *afflicted* with the rare trait *ocular albinism* always have *afflicted* sons, but the sons of *afflicted* fathers

are hardly ever *afflicted* (Gillespie, F.D., Arch. Opthalmol. 66: 774, 1961). Explain.
b) Occasionally, however, in a marriage between a woman with *ocular albinism* and a *normal* man, a son is *normal* and a daughter has *ocular albinism*. Explain how such results may occur and how you might confirm your hypothesis cytologically.

8. In the fish *Lesbistes reticulatus*, certain individuals possess a *black spot* on the dorsal fin as a result of the presence of the *maculatus, Ma* gene. The trait has the following transmission characteristics: A male having the trait mated with a female lacking it transmits its determiner to all the male offspring but to none of the females. These females, moreover, do not transmit the determiner to their offspring. The offspring of *unspotted* parents never show the trait. The trait does not appear in sons unless the father also expresses it (Wingo, O. J., Genet. 12: 145, 1922).
a) Describe and illustrate diagrammatically the mode of inheritance of this trait.
b) State which is the heterogametic sex. Why?
c) Is the *maculatus* gene dominant or recessive? Explain.

9. In the guinea pig (males XY, females XX), a series of matings between *albino* females (all same genotype) and *brown* males (all same genotype) produced the following results: 96 *albino* males and 99 females (25 *black*, 50 *brown* and 24 *cream* which died at birth). With reasons for your answers, indicate the number of allele pairs involved, their location, types of allelic and genic interactions, and any other features you may detect.

10. In *Melandrium*, a flowering plant belonging to the pink family, the males are XY, the females XX. In 1931 Winge, O. (Hereditas 15: 127) reported studies involving the heritable trait *aurea* (foliage is a blotchy yellowish-green). *Aurea* males crossed wtih *normal* females produced either all *normal* progeny or 1 *normal* female : 1 *normal* male : 1 *aurea* male. *Normal* males in crosses with *normal* females always produced *normal* female progeny, but in some such matings the males, instead of being all *normal* consisted of *normals* and *aureas* in a 1 : 1 ratio. No *aurea* females were ever found. What is the genetic basis for this condition?

11. Four common varieties of platyfish are *ruber* (grayish-red with spots), *golden ruber* (gold with spots), *gray* (grayish red without spots), and *gold* (gold without spots). All wild platyfish are *ruber*. From these, the other three types have originated. The series of crosses tells the story of the origin of the *gold* variety (Gordon, M. J., Hered. 26: 97, 1935). Discuss the underlying genetic control of these phenotypes, diagramming the path that leads to the formation of the *gold* variety.

Parents		Offspring
♀	♂	
ruber	x *ruber*	all *ruber*
ruber[1]	x *ruber* F_1	all *ruber*
ruber[2]	x *ruber* F_1	3/4 *ruber* : 1/4 *golden ruber*
golden ruber	x *golden ruber*	all *golden ruber*
golden ruber	x *gray*	♀ *gray*, ♂ *ruber*
gray[3]	x *ruber*[3]	3 *ruber* : 1 *golden ruber*: 3 *gray* : 1 *gold*
gold	x *gold*	*gold* only

1 Certain F_1 females; 2 Other F_1 females; 3 Offspring of previous cross

12. Beginning with crosses between true-breeding *red* and *white* strains, Aida, T. (Genetics 6: 554, 1921) obtained the following results in the freshwater fish *Aplochelius latipes*:

Parents		F_1 cross		Progeny	
♀	♂	♀	♂	♀	♂
white x *red*		*red* x *red*		F_2 38 *red*; 36 *white*	75 *red* ; 0 *white*
		*white** x *red*		Testcross	
				2 *red* ; 177 *white*	183 *red* ; 1 *white*

*homozygous

Note: All fish were diploid.
Which of the three modes of inheritance (i) autosomal, (ii) completely sex-linked, or (iii) incompletely sex-linked can account for these results? Explain with the aid of diagrams, and give reasons for rejecting the others.

13. Glucose-6-phosphate dehydrogenase (G6PD), an enzyme that is specific to erythrocytes was studied by Ohno, S. et al. (Science 150: 1737, 1965) in each of the two European species of wild hares, *Lepus europaeus* and *L. timidus*, and in the reciprocal hybrids between them. Starch-gel electrophoresis revealed the following:
1. The single band of the enzyme of *L. europaeus* was faster than that of *L. timidus*.
2. Each male hybrid had a single band of enzyme identical to that of its mother. Both parental types of G6PD band were detected in female hybrids.
What do these data indicate about the inheritance of G6PD? Explain.

14. tudies of *resistant rickets* (patients do not respond to normal doses of vitamin D) indicate that the trait is determined by an X-linked dominant allele (Winters, R.W. et al., Medicine 37: 97, 1958).
a) If this is true, what phenotypes would you expect among the male and female children of the following marriages:
(i) An *affected* male and a *normal* female?
(ii) An *affected* female from marriage in (1) and a *normal* male?
b) Explain why more women than men show the trait.

15. In cats, a sex (X)-linked pair of alleles, *Bb*, determines the colour of the fur. The allele *B* for *yellow* is incompletely dominant over *b* for *black* so that *Bb* individuals are *tortoise-shell*, a splotchy mixture of yellow and black hairs.
a) You and a friend are strolling down the street and see a *tortoise-shell* cat and you tell your friend the cat is female. Your friend doubts your statement. How would you explain to him that you are correct?
b) A *yellow* male is crossed with a *tortoise-shell* female. If the female has a litter of 6 males, what phenotypes might they express?
c) A *yellow* cat has a litter of 2 *tortoise-shell* and 1 *yellow*. What is the probable sex of the *yellow* kitten?
d) A *tortoise-shell* female has a litter of 7; 2 *yellow* females, 2 *tortoise-shell* females, 1 *black* and 2 *yellow* males. What was the probable genotype and phenotype of the father?
e) A *black* female has a litter of 6; 3 *black* males, 2 *tortoise-shell* females, and 1 *black* female. What were the genotypes and phenotypes of the parents?

f) A *tortoise-shell* cat brings home her litter of *black*, *yellow* and *tortoise-shell* kittens. By what criteria might it be possible to incriminate the *yellow* tom living next door as the probable father if there are not any other *yellow* toms in the neighbourhood? What criteria would exonerate him?
g) Occasionally sterile *tortoise-shell* males occur with 39 chromosomes instead of the normal number of 38 chromosomes (Thuline, H.C. and D.E. Norby, Science 134: 554, 1961). Advance a hypothesis to explain these findings. Show the genotype of the parents.

16. The pedigree shown is typical for *incontinentia pigmenti*, a congenital human skin abnormality usually associated with skeletal and other malformations. In the fully developed disease, the skin shows swirling patterns of melanin pigmentation, especially in the trunk, giving a "marble cake" appearance. The pigmentation fades gradually and usually disappears completely by age 20 (Haber, H., Br. J. Dermatol. 64: 129, 1952).

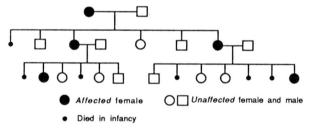

a) What is the most likely mode of inheritance and why?
b) What other modes are possible, and what criteria would be needed to eliminate them as possibilities?

17. *Hemophilia B* (Christmas disease) in humans is determined by a recessive sex (X)-linked allele (Whittaker, D.L. et al., Am. J. Hum. Genet. 14: 149, 1962).
a) Could a father and son both be *hemophilic*?
b) Explain why *hemophilic* mothers always have *hemophilic* sons yet this is rarely true when fathers are *hemophilic*.
c) What are the chances that a *normal* daughter of a *hemophilic* father and *normal* mother who marries a *normal* man will have a *hemophilic* child? Would it be male or female?

18. A *normal* man, an X-ray technician in an industrial plant, and his *normal* wife, have a son who is *affected* with the X-

linked recessive trait, *nystagmus*. There are no known cases of the disease in his ancestry or that of his wife's for the past two generations. He sues the industrial plant, claiming that it failed to provide him with adequate protection from the radiation and that his son is affected as a result of an induced mutation. Explain what your testimony would be in a court of law.

19. In *Drosophila melanogaster* the allele pairs *Ww*, *Yy* and *Mm*, which determine *red* vs. *white* eyes, *gray* vs. *yellow* body and *normal* vs. *miniature* wing, respectively, are X-linked. A bilateral gynandromorph is found with one half showing the traits *white*, *yellow* and *miniature* and the other half exhibiting all the dominant traits.
a) How could such an individual arise?
b) What is the genotype of the X chromosome in the portion of the fly exhibiting the recessive traits? Show diagrammatically how this fly may have been produced.

20. In the insect, *Pseudococcus nepae*, whose genome has 5 chromosomes, many genes have been identified and assigned to all 5 linkage groups. All these genes have an identical transmission pattern. Specifically, reciprocal crosses between homozygous lines with different traits of the same character produce the transmission patterns shown for *slow* (*S*) vs. *rapid* (*s*) metamorphosis.

P$_1$ slow ♀ x rapid ♂   rapid ♀ x slow ♂

F$_1$ 1/2 males - *slow*   1/2 males - *rapid*
        1/2 females - *slow*   1/2 females - *slow*

Explain these results cytogenetically.

21. a) A person with Klinefelter's syndrome (XXY) whose parents are *normal* is red-green *colour-blind* (a condition known to be caused by a recessive allele at an X-linked locus). Explain this anomaly of sex-linked inheritance.
b) A second Klinefelter individual whose parents are also *normal* is red-green *colour-blind* in one eye and *normal* in the other. How does the probable course of events leading to this condition differ from that in (a)?

22. In *Drosophila* the eye colours *red*, *blood* and *white* are determined by alleles of the same gene. From the pedigree determine their order of dominance. Do the results indicate whether the locus is autosomal or sex-linked? Give reasons for your answers.

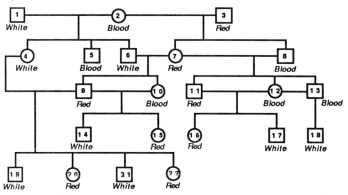

23. In papaya (*Carica papaya*), a unisexual species, a female from a line true-breeding for *glabrous* (*glab.*) stems, *resistance* (*resist.*) to root rot and *lack* of spots on petals was crossed with a male with *pubescent* (*pub.*) stems, *susceptible* (*suscept.*) to root rot and *spots* on petals. The F$_1$ and F$_2$ results were as follows:

P$_1$ ♀ Glab., resis., no spots   x   ♂ pub., suscept., spots

F$_1$ 1/2 ♀ Pub., resist., no spots;   1/2 ♂ Glab., resist., spots

F$_2$			
400 ♀♀		400 ♂♂	
194	Pub., resist., no spots	2	Pub., resist, spots
		203	Pub., suscept., spots
206	Glab., resist., no spots	192	Glab., resist., spots
		3	Glab., suscept., spots

Give a complete cytogenetic explanation of these results. In your explanation indicate the number of allele pairs determining each pair of traits, allelic relationships at each locus, the locations of these genes within the genome (autosome or sex chromosome) and whether the sex chromosomes in the heterogametic sex are completely homologous.

24. The incompletely sex-linked pair of alleles *Bb* in *Drosophila* (males XY, females XX) determines bristle size; the dominant allele B causes *wild-type* bristle whereas the recessive allele b causes *bobbed* bristles.
a) Is it possible for this pair of traits to show crisscross inheritance? Illustrate.
b) How would you prove that both the X and Y chromosomes carry the locus for this pair of alleles?

25. In the mouse, which has an X-Y mechanism of sex determination, *Ta ta* is an X-linked pair of alleles; *Ta* is incompletely dominant to *ta*. In 1959 Welshons, W.J. and L.B. Russell (PNAS 45: 560) reported the occurrence of unexpected phenotypes among the female progeny of several crosses, two of which are presented below:

Cross	♀ progeny Expected	Exceptional
tata x *Ta/Y*	152 heterozygous *tabby*	2 *wild-type*
Tata x *ta/Y*	501 *wild-type* and *tabby* heterozygous	2 *tabby*

a) Outline two hypotheses (one implicating events in the male parent and the other in the female parent as being responsible for the condition) to explain the occurrence of the exceptional females and give the genotypes of these females according to each hypothesis. In this question it is assumed that the roles of the X and Y in sex determination are not known as they are in man.
b) Outline genetic tests required to distinguish between these hypotheses and show the expectations on the basis of each.
c) In a cross of exceptional *tabby* females derived from the second cross with *wild-type* males the following offspring were obtained; 7 heterozygous *tabby* females, 7 *wild-type* females and 12 *tabby* males. Which hypothesis do these results support? Are the results of this cross evidence that your hypothesis is correct? Explain.
d) What chromosome counts are possible in the exceptional females if your hypothesis is correct? What is the most probable mechanism causing these exceptional females? On the basis of the data given, does it occur primarily in the male or female? Why?
e) Is it possible on the basis of the above genetic data and cytological observations to determine the location of male-determining genes? Female-determining genes? Explain.

26. The following typical pedigrees, based on studies by Wirth, B. et al. (Genomics 2: 263, 1988) show the pattern of transmission of *retinitis pigmentosum* (*RP*), a condition that results in gradual loss of peripheral and night vision:

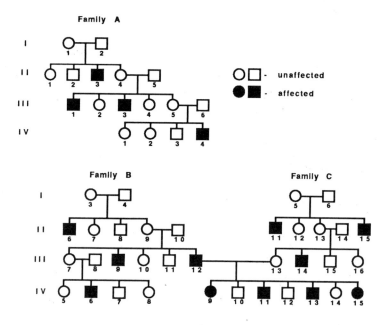

a) Account genetically for the inheritance of *RP* in each of the 3 families.
A mating between IV-4 and IV-9 produces 6 children with phenotypes as shown:

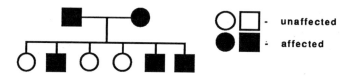

b) If your explanation in (a) is correct why are some of the daughters of this mating *unaffected*?

c) Would you expect the females IV-9 and IV-15 to be as severely *affected* as the *affected* males? Explain.

27. A holandric gene is known in man which causes *hypertrichosis* (long hair growth) of the ears. If men with *hypertrichosis* marry *normal* women:
a) What proportion of the sons would you expect to express *hypertrichosis*?
b) What proportion of the daughters would you expect to express the trait?

28. In the domestic fowl, feathering is determined by a Z-linked gene; the dominant allele *S* determines *slow-feathering* and the recessive allele *s* determines *fast-feathering*. The phenotypes can be distinguished on the first day after hatching. In the egg-producing business it is highly desirable that chicks be females. Outline a method of breeding which would make sex determination possible on the basis of this character.

29. The following pedigree is based on a genetic and molecular study of a family by Drummond-Borg, M. et al. (Am. J. Hum. Genet. 43: 675, 1988).

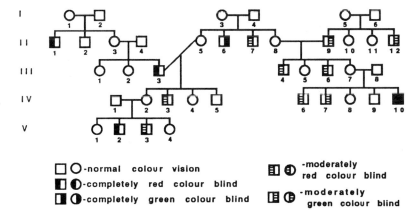

a) Are the genes for *red* and *green* colour vision allelic? Explain.
b) Are the genes autosomal or X-linked? Explain.
c) What kinds of mutant alleles could produce complete and partial defects in *red* and *green* colour vision? Discuss.
d) The males with *moderate red* colour blindness and *moderate green* colour blindness possess fusion genes (each with one segment from the gene for *red* pigment and the remaining region from the gene for *green* pigment - see question 31). What bearing, if any, does this have on your answer to (c)?
e) Postulate two plausible modes of origin for the completely *red-green* blind male IV-10.

30. In humans, *normal* colour vision vs. *red-green* colour blindness is determined by alleles *C* (dominant) and *c* (recessive) at an X-linked locus.
a) Can a *normal* daughter have a *colour-blind* father? A *normal* father? A *colour-blind* mother? A *normal* mother? Two *colour-blind* parents?
b) Answer the same questions for a *normal* son.
c) Can two *normal* parents have a *colour-blind* son? A *colour-blind* daughter?
d) A brother and sister are both *colour-blind*. Is it possible for them to have: (1) A *normal* brother? (2) A *normal* sister? (3) One parent *normal*, one *colour-blind*?

31. Human colour (*red, green blue*) vision is based on light-sensitive pigments (apoproteins) each of which is coded for by a single specific gene. Using molecular techniques Nathans, et al. (Science 232: 193, 1986; Science 232: 203, 1986) concluded that: (i) the amino acid sequences of the red and green pigments are about 96% identical and therefore the genes coding for these apoproteins must be highly homologous. (ii) *Colour-normal* individuals possess only one gene for *red* pigment and one or more genes for *green* pigment. (iii) these genes reside in a head-to-tail tandem array on the X-chromosome as illustrated below:

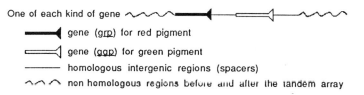

a) What is a likely mechanism for the origin of these highly homologous genes? Explain and illustrate.
b) Some X-chromosomes can have more than one gene for *green* pigment. What is the most likely mode of origin of these tandem arrays? Illustrate.

c) In tandem arrays of three or more of these genes why is the gene specifying *red* pigment always at one end of the array?

32. The following pedigrees show typical transmission patterns of rare or fairly rare hereditary traits. If the mate of a parent is not shown, assume the mate is of normal phenotype and genotype. In each case assume the gene is completely penetrant.

I) Ectodermal dysplasia (■)

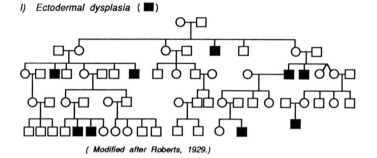

(Modified after Roberts, 1929.)

II) Deficiency for glucose-6-phosphate dehydrogenase.

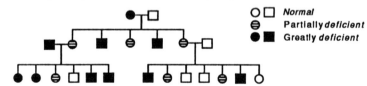

III) Hypophosphatemia (●■)

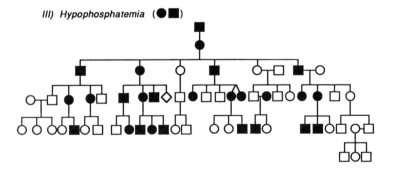

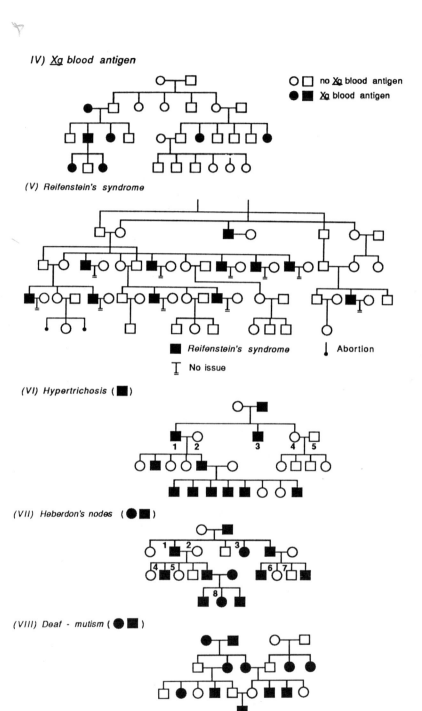

IX) *Congenital cataracts* (■)

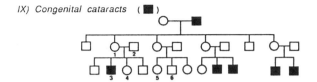

X) *Achondroplasia* (●■)

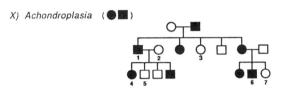

XI) *Oxalic urinary calculi* (■)

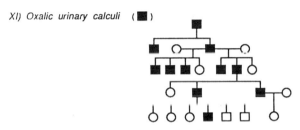

XII) *Amelogenesis imperfecta* (●■)

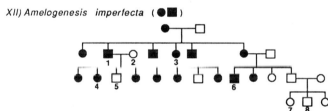

a) State which of the following kinds of genes could produce each trait and why:
(1) An autosomal dominant gene.
(2) An autosomal recessive gene.
(3) A sex-linked dominant gene.
(4) A sex-linked recessive gene.
(5) An incompletely dominant lethal gene.
(6) A holandric gene.
(7) A sex-influenced gene (dominant in males, recessive in females).
(8) A sex-influenced gene (dominant in females, recessive in males).
b) In the pedigrees chosen, derive the genotypes of the original ancestors and of the individuals designated numerically.

33. a) Distinguish between the following, giving the locations of the genes, the mode of transmission and relationship to sex:
(i) Sex-influenced and holandric characters.
(ii) Sex-limited and sex-influenced characters.
(iii) Sex-linked and sex-influenced characters.
(iv) Sex-limited and holandric characters.
b) The male bison differs from the female in having a well-developed mane. Which of the four genetic explanations might be advanced to explain this dimorphism?

34. Reciprocal crosses between true-breeding *bearded* and *beardless* breeds of goats give the results shown (Asdell A.S. and A.D.B. Smith, J. Hered. 19: 425, 1928). State, giving reasons for your answers:
a) The number of pairs of alleles involved and whether they are autosomal or sex-linked.
b) The type of association with sex.

Parents		F_1	F_2
♂	♀		
bearded x *beardless*		♂ *bearded*	♂ 3 *bearded* : 1 *beardless*
		♀ *beardless*	♀ 3 *beardless* : 1 *bearded*
beardless x *bearded*		♂ *bearded*	♂ 3 *bearded* : 1 *beardless*
		♀ *beardless*	♀ 3 *beardless* : 1 *bearded*

35. In Ayrshire cattle, the allele M^m for *mahogany-and-white* is dominant in males recessive in females. The reverse is true for the allele M^f for *red-and-white*.
a) A *red-and-white* male is crossed with a *mahogany-and-white* female. Show the expected F_1 and F_2 genotypic and phenotypic proportions.
b) A dairy farmer who raises this breed of cattle would like to be able to distinguish the male from the female calves in his herd by a difference in colour. How should he proceed?
c) An Ayrshire breeder has used a *red-and-white* bull to sire his herd. A geneticist visiting the farm comments that all the *mahogany-and-white* calves are males. The farmer is impressed to find that this is true in all cases. Explain why the geneticist was able to predict the sex of the *mahogany-and-white* calves.

d) A *mahogany-and-white* cow has a *red-and-white* calf. What is the calf's sex?

36. In 1937 Petterson, G. and G. Bonnier (Hereditas 23: 49) reported genetic studies of a rare human heritable type of intersex known as *testicular feminization*. Progenies in which the trait appeared consisted of 8 *normal* males, 28 *normal* females and 22 *affected*. Outwardly and physiologically the *affected* individuals are females and consider themselves *normal* females.
a) What chromosome constitution might you expect in the *affected* individuals?
b) What are the possible modes of inheritance of *testicular feminization*?

37. In humans, *pattern baldness* shows the following features of inheritance:
1. The trait can occur in both sexes.
2. *Affected* persons are usually of one sex.
3. A pair of *normal* parents may have an *affected* child.

a) State which of the following modes of inheritance are eliminated by the above information, and specify one feature by which you eliminate each:
(1) Sex-limited   (2) Autosomal recessive
(3) Sex-linked recessive (4) Sex-linked dominant
(5) Autosomal dominant   (6) Holandric
(7) Sex-influenced, dominant in males
(8) Sex-influenced, dominant in females

b) The pedigree shown below for this trait (solid symbols) should enable you to decide among the remaining alternatives. State what the mode of inheritance is and why.

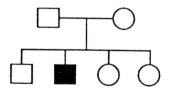

38. In the clover butterfly (*Colias philodice*) the wing colour of males is always *yellow*, while that of females may be *yellow* or *white*. The data shown are summarized from studies by Gerould, J.H. (Genetics 26: 495, 1941) and Remington, C.L. (Adv. Genet. 6: 403, 1954). Answer the following questions, giving reasons for your answers:
a) Is *white* a sex-limited or a hologynic trait?
b) How many genes control its expression?

Parents		F_1	
♀	♂	♀	♂
yellow x yellow		all yellow	yellow
yellow x yellow		1/2 yellow : 1/2 white	yellow
yellow x yellow		all white	yellow
white x yellow		all white	yellow
white x yellow		1/4 yellow : 3/4 white	yellow

c) Is *white* determined by a dominant or recessive allele?
d) Is the gene (or genes) on (1) the W chromosome, (2) the Z chromosome, or (3) an autosome?

39. A typical pedigree for the rare sex-limited trait *cryptorchidism* (the failure of one or both testes to descend into the scrotum) in dogs represented by solid symbols is shown below. Since the condition is rare it is highly unlikely that the normal males and females marked with an asterisk carry the gene for this trait.

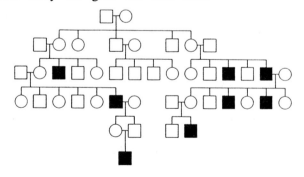

a) Why can't this trait be determined by an allele of a holandric gene?
b) A single gene is involved. Is it on an autosome or on the differential segment of the X-chromosome? Is the causal allele dominant or recessive? Explain.

c) In the following pedigree the shaded symbols represent girls with *transverse vaginal septum*. The trait is due to an allele of an autosomal gene rather than the result of a recessive allele of an X-linked gene. Explain why this is so.

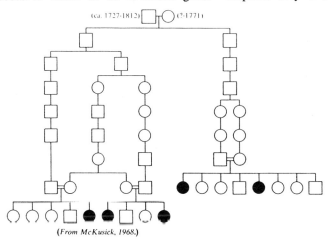
(From McKusick, 1968.)

40. In the damselfly *Ischnura damula*, the males all have the same (*andromorphic*) colour pattern on the dorsal synthorax; the females have two patterns, one like the male and another termed *heteromorphic*. Johnson, C. (Genetics 49: 513, 1964) reported studies on the genetic basis of this female dimorphism. In all crosses a virgin female was mated with a single male. The results were as follows:
(1) From 14 crosses the ratio of male to female progeny was approximately 45:55.
(2) Certain females of either phenotype produced female progeny that included both *andromorphs* and *heteromorphs*. When both types of females occurred in the progeny, the ratios were 1:1 from *andromorphic* and 1:1 or 3:1 from *heteromorphic* females, respectively.
(3) Other females of either phenotype produced only one type of female progeny. The female progeny of *andromorphic* females were either all *andromorphic* or all *heteromorphic*; the female progeny of *heteromorphic* females were all *heteromorphic*.
(4) No *heteromorphic* males were encountered.
Pairs of traits having unequal frequencies of expression in the two sexes could result from a variety of mechanisms, including:
(i) Differential lethality between the forms in one sex before classification.
(ii) Higher frequencies of a recessive sex-linked trait in the heterogametic males than in females, which could be homozygous for the recessive allele.
(iii) Expression of a sex-linked gene in only one sex.
(iv) Expression of an autosomal gene in only one sex.
Which of these mechanisms is the most likely explanation for the female dimorphism present in *Ischnura*?

41. Stehr, G. (J. Hered. 46: 263, 1955) in studies of intraspecific variability in eastern North American populations of the spruce budworm (*Choristoneura fumigerana* (Clem.)) obtained the tabulated results in crosses between *brown* and *gray* moths from true-breeding strains. Discuss, with the aid of diagrams, the genetic basis for colour dimorphism in this organism.

	Parents			Progeny	
Generation	♀	♂	Generation	♀	♂
Cross 1:					
P_1	brown x	gray	F_1	brown 0 : gray 76	gray
F_1	gray x	gray	F_2	brown 43 : gray 44	gray
F_2	brown x	gray	F_3	brown 0 : gray 128	gray
	gray x	gray			
F_2	brown x	gray	F_3	brown 50 : gray 50	gray
	gray x	gray			
Cross 2:					
P_1	gray x	gray	F_1	brown 39 : gray 0	gray
F_1	brown x	gray	F_2	brown 118 : gray 123	gray
F_2	brown x	gray	F_3	brown 81 : gray 70	gray
	gray x	gray			
F_2	brown x	gray	F_3	brown 126 : gray 0	gray
	gray x	gray			

42. In the bug *Euchistus variolarius*, which has the X-Y mechanism of sex determination, the male has a *black spot* on the abdomen; the female does not. In the related species *E. servis* neither sex has this spot. A female of the *variolarius* species was crossed with a male of the *servis* species. The F_1 females did not have an abdominal spot, but all the males did. In the F_2 the females did not have spots, but approximately three-fourths of the males did (Foot, K. and E.C. Strobell, Arch. Zellforsch. 12: 485, 1914).

a) What kind of trait is this?

b) Explain why the Y chromosome cannot carry the gene.

c) Is it correct to conclude that the gene is not X-linked? Why?

d) An *unspotted* female was mated to a male of the same phenotype, and all male progeny were *spotted*. What were the genotypes of the parents?

43. In the silver-washed fritillary butterfly (*Argynnis paphia*) all males have *rich-brown* wings; the females have either *rich-brown* wings like the males or *dark olive-green* ones (Ford, E.B., Adv. Genet. 5: 43, 1955).

a) The *dark olive-green* trait may appear in the progeny of matings between two *rich-brown* individuals. Can it be a holandric trait? Explain.

b) In certain crosses the female progeny show a 3:1 ratio of *dark olive-green: rich brown*.

(i) Is the gene sex-linked or autosomal?

(ii) What is the genotype and phenotype of the female parent?

c) A gynandromorph shows a *rich-brown* colour on one side and a *dark olive-green* one on the other.

(i) How could such an individual arise?

(ii) Did it start out as a male or as a female?

(iii) What are the possible genotypes of each of its two sides?

(iv) If the female parent of the gynandromorph was *dark olive-green*, what were the genotypes of the male parent and the zygote?

Chapter 10
Linkage, Crossing-over and Genetic Mapping

1. a) Distinguish between linkage and sex linkage.
   b) Why is linkage an exception to Mendel's second law?
   c) In certain organisms the linkage relationships between sex-linked genes are much better known than those between autosomal genes. Moreover, linkage maps of sex-linked genes can often be made before those of autosomal genes are established. Why is this to be expected?

2. State, with reasons, whether the following statements are true or false:
   a) Crossing-over always results in genetic recombination.
   b) Crossing-over occurs at the two-strand stage of meiosis.
   c) Segregation of alleles can occur only during the first meiotic division.
   d) Centromeres always separate reductionally at anaphase I.
   e) The percent recombination between two linked genes is always half that of the percent of crossing-over.
   f) The percent recombination between any two syntenic genes is always a true estimate of the physical distance between them.
   g) Recombination between linked genes is the consequence of a reciprocal exchange of homologous segments between non-sister chromatids.
   h) It is possible to establish linkage relationships of genes in organisms that reproduce exclusively by asexual means.
   i) Regardless of environmental conditions, the percent recombination between any two genes always remains the same.

3. In 1911, Morgan, T.H. (Science 34: 384), crossed a *white*-eyed, *yellow*-bodied *Drosophila* female with a *red*-eyed, *gray*-bodied male. In the F_1 the females were phenotypically like the father and the males like the mother. The F_1 flies were intercrossed, and produced the following F_2 progeny:

Phenotype		Sex	
Eyes	Body	♀	♂
White	Yellow	543	474
Red	Gray	647	512
White	Gray	6	11
Red	Yellow	7	5

   Outline an hypothesis to explain these results.

4. In tomatoes (*Lycopersicon esculentum*), the allele *P* for *smooth* fruit pubescence is dominant over *p* for *peach* and the allele *R* for *round* fruit is dominant over *r* for *long*. MacArthur, J.W. (Genetics 13: 410, 1928) obtained the following results when he testcrossed the F_1 generation of a cross between the homozygous strains *smooth, long* and *peach, round*.

Smooth, round	12	*Peach, round*	133
Smooth, long	123	*Peach, long*	12

   a) Classify each of the phenotypes as either parental or recombinant.
   b) Give the percentage crossing-over and the map distance between the genes.

5. In the house mouse (*Mus musculus*) *trembling* vs. *normal* and *rex* (short) vs. *long* hair are each determined by an autosomal pair of alleles. The alleles for *trembling* and *rex* are both dominant. In crosses of heterozygous *trembling, rex* females with *normal, long* males the following results were obtained by Falconer, D.S. and W.R. Subey (J. Hered. 44: 159, 1953):

Trembling, rex	21	*Normal, rex*	54
Trembling, long	52	*Normal, long*	22

a) Explain whether or not these two genes are located on the same chromosome.
b) Were the *trembling, rex* females heterozygous in coupling or repulsion? Show how you determine this.
c) Determine the percentage recombination.
d) Would the results have differed if these workers had made the reciprocal cross? Explain.
e) If heterozygous *trembling, rex* males and females were crossed, what phenotypic ratio would be expected in the progeny?

6. In corn, *starchy* vs. *sugary* endosperm and *susceptibility* vs. *resistance* to *Helminthosporium* are monogenically determined. The F_1 of a cross between *starchy, susceptible* and *sugary, resistant* are testcrossed and the progeny phenotypes and their distributions are as follows:

Starchy, susceptible	92	Sugary, susceptible	91
Starchy, resistant	86	Sugary, resistant	88

a) What are the possible explanations for these results?
b) Many genes have been allocated to each of the ten linkage groups in corn. Knowing this, how would you determine whether these two genes are members of the same linkage group or not?

7. The allele pairs Ee (*gray* vs. *ebony* body) and Cc (*normal* vs. *curled* wings) in *Drosophila melanogaster* are located on chromosome 3. Males of a true-breeding *gray, normal* strain were crossed with females of an *ebony, curled* one. The resulting F_1's produced the following F_2 progeny:

Gray, normal	288	Ebony, curled	88
Gray, curled	14	Ebony, normal	10

a) Are the E and C genes linked? Explain.
b) Show how you derive the map distance between E and C.

8. In mice, the allele pairs $Sh\text{-}2, sh\text{-}2$ for *nonshaker* vs. *shaker* and $Wa\text{-}2, wa\text{-}2$ for *straight* vs. *wavy* hair are on the same chromosome, 25 map units apart (Green, M.C. and M.M. Dickie, J. Hered. 50: 3, 1959). True-breeding *nonshaker, straight* mice are crossed with homozygous *shaker, wavy* ones, and the F_1's are testcrossed. Diagram the parental, F_1 and testcross generations of this cross, showing the genes on the chromosome. Show the expected frequencies of F_1 meiocytes with and without crossing-over and demonstrate that the expected frequency of recombinant gametes is half that of crossover meiocytes.

9. In the fowl, *silver* vs. *gold* feathering and *slow* vs. *rapid* feather development are each determined by a single pair of alleles. In crosses between males of a *silver, slow* strain and females of a *gold, rapid* strain the F_1's were phenotypically like the male parents. The F_2 phenotypes were distributed as shown:

Phenotype	♀	♂
Silver, slow	300	120
Silver, rapid	0	35
Gold, rapid	0	123
Gold, slow	0	32

a) Explain the male-female discrepancy of the results, and determine whether the 2 pairs of traits are determined by independently segregating or linked pairs of alleles.
b) Are the genes located on an autosome or a sex chromosome?

10. In *Drosophila melanogaster*, *arc* wings and *black* body are determined by recessive alleles of different genes. Crosses between F_1's derived from matings of *arc, gray* and *normal, black* flies produced the following F_2 progeny (Bridges, C.B. and T.H. Morgan, Carnegie Inst. Wash., D.C., Publ. 278, p. 123, 1919):

Normal, gray (wild-type)	923	arc, gray	387
Normal, black	401	arc, black	0

a) Are these genes syntenic? Explain.
b) Is it possible to determine the percent recombination between the two loci? Explain.

11. In *Drosophila melanogaster*, the allele pairs *Cncn* (*dull red* vs. *cinnabar* eyes) and *Roro* (*smooth* vs. *rough* eyes) are on chromosomes 2 and 3 respectively. A mutant fly with *bent* wings is discovered in the homozygous double-recessive *cinnabar, rough* strain. A true-breeding *cinnabar, rough, bent* strain is established, and females from this strain are crossed with males from a true-breeding *dull-red, smooth, straight* stock. A series of crosses between F_1 males and *cinnabar, rough, bent* females gives the progeny shown:

Phenotype	♀	♂
Dull-red, smooth, straight	38	36
Cinnabar, smooth, straight	34	38
Dull-red, rough, straight	35	36
Dull-red, smooth, bent	39	34
Cinnabar, rough, straight	40	35
Cinnabar, smooth, bent	34	33
Dull-red, rough, bent	37	34
Cinnabar, rough, bent	38	36

a) Is the mutant gene for *bent* wings located on either of the chromosomes 2 or 3? Explain.
b) Would the results of the reciprocal cross allow the same unequivocal answer? Explain.

12. In meiocytes of the grasshopper, *Romalea microptera*, regardless of their number per bivalent, the chiasmata do not terminalize until metaphase I. In a classic ^3H-thymidine labelling experiment in 1965, Taylor, J.H. (J. Cell Biol. 25: 57), observed the following:
(i) Before the occurrence of crossing-over only one complete chromatid per chromosome in each bivalent was labelled.
(ii) The following results were obtained with chromosome pairs 4, 5 and 6. At diplotene, in 74 bivalents, the average number of chiasmata per bivalent was 2.62 and the frequency of chiasmata per chromosome was 1.31. Of 76 chromosomes at anaphase I (38 at each pole), 49 chromatids had labelled and unlabelled segments.
In recombinant chromatids the positions of association of labelled and unlabelled segments corresponded to the positions of chiasmata at diplotene. For example:

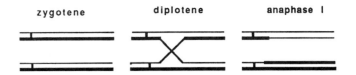

a) What is the expected frequency of crossing-over between labelled and unlabelled non-sister chromatids? Explain and illustrate using the above chromosome pair, assuming one such event occurs per bivalent per meiocyte.
b) Are Taylor's data evidence that the mechanism of crossing-over is by copy-choice? By breakage-and-reunion? Explain.
c) What conclusions can you draw from these data regarding the relationships between crossing-over and chiasma formation?

13. In the housefly, *Musca domestica* (2n=12; XY sex determination) the results shown were obtained with the second-chromosome pair of alleles *Bb* (*gray* vs. *brown* body) in crosses involving a mutant, *bb* and three *BB* strains: the *normal*, the *furen* and the *ND* (Hiroyoshi, T., Genetics 50: 373, 1964).

Cross		F_1	F_2 or testcross			
♀	♂		Gray		Brown	
			♀	♂	♀	♂
mutant x	normal	gray	2590	2290	773	778
normal x	mutant	gray	1999	1967	587	600
mutant x	furen	gray	921	4079	748	0
furen x	mutant	gray	2120	1990	699	689
mutant x	F_1 (mutant ♀ x ND ♂)	gray	0	1224	889	0
mutant x	F_1 (ND ♀ x mutant ♂)	gray	788	640	464	360
mutant x	F_1 (mutant ♀ x furen ♂)	gray	0	2856	839	0
mutant x	F_1 (furen ♀ x mutant ♂)	gray	530	495	448	459

Outline a plausible genetical and cytological hypothesis to account for these results.

14. a) In *Drosophila melanogaster*, *Pr* (*red* eye) is dominant to *pr* (*purple* eye), and *B* (*gray* body) is dominant to *b* (*black*

body). Bridges, C.B. and T.H. Morgan (Carnegie Inst. Wash., D.C., Publ. 278, p.123, 1919) crossed a *red, black* female with a *purple, gray* male. The F_1 males when testcrossed produced the following progeny:

Red, black	74	*Purple, gray*	71
Red, gray	0	*Purple, black*	0

When the F_1 females were testcrossed, the results were:

Red, black	383	*Purple, gray*	382
Red, gray	22	*Purple, black*	16

i) Do these reciprocal testcrosses give evidence of linkage? Explain how you determine this. What is the percent recombination for each of the crosses?
ii) Compare the recombination values from reciprocal crosses and give a possible reason for the discrepancies.
iii) Was the cross made in the repulsion or coupling phase?

b) When the F_1's were intercrossed, the surprising F_2 phenotypic distribution shown below was observed:

Red, black	300	*Purple, gray*	371
Red, gray	684	*Purple, black*	0

Are these results consistent with your hypothesis? Diagram the cross in answering this question.

15. In *Drosophila melanogaster*, alcohol dehydrogenase (*ADH*) is an enzyme composed of two polypeptide chains (a dimer) which exists in three electrophoretically different forms (isozymes) determined by two alleles of one gene:

$Adh^F Adh^F$	*Fast* isozyme; both chains identical
$Adh^S Adh^S$	*Slow* isozyme; both chains identical
$Adh^F Adh^S$	*Intermediate* isozyme; both chains different as well as *fast* and *slow* isozymes

Grell, E.H. et al. (Science 149: 80, 1965) crossed true-breeding *cy bl / cy bl, ubx vno / ubx vno* (*wild-type*) males that were $Adh^S Adh^S$ with females of the balanced lethal stock *Cy bl / cy Bl, Ubx vno / ubx Vno* that were true-breeding for $Adh^F Adh^F$.
F_1 *Cy bl / cy bl, Ubx vno / ubx vno*, $Adh^S Adh^F$ males were crossed with *wild-type* females that were $Adh^S Adh^S$. All F_2 *Cy/cy* flies produced the *intermediate* as well as the *fast* and *slow* isozymes. The *cy/cy* flies produced the *slow* isozyme only. Half the *Ubx* flies produced the same isozymes as the *Cy* flies; the other half produced the *slow* isozyme only. Which chromosome carries the *Adh* gene and why? Note: *Cy* (*curly* wings) and *Bl* (*short* bristles) are incompletely dominant lethal genes on the second chromosome; *Ubx* (*enlarged* halteres) and *Vno* (*veins missing*) are incompletely dominant lethal genes on the third chromosome.

16. In corn, the aleurone layer of the kernel may be *coloured* or *colourless*, and the endosperm may be *starchy* or *waxy*. From crosses between true-breeding lines A and B and C with D, Bregger, T. (Am. Nat. 52: 57, 1918) obtained F_1 seed with *coloured* aleurone and *starchy* endosperm. He testcrossed one F_1 plant from AxB and one from CxD with a true-breeding *colourless, waxy* line. The single ear produced by the F_1 plant from AxB possessed 403 kernels and that produced by the F_1 plant from CxD had 292. The table below shows the distribution of phenotypes on each of these ears:

F_1 Plant	*Coloured starchy*	*Colourless waxy*	*Coloured waxy*	*Colourless starchy*
From A x B	147	133	65	58
From C x D	46	32	103	111

a) Explain why the two ears give different distributions of phenotypes.
b) State the genotypes of the two F_1's and the true-breeding parental lines from which they were derived.
c) Combine the two sets of data to derive the percentage recombination.

17. In corn, *tunicate* vs. *nontunicate*, *glossy* vs. *nonglossy* seedling and *liguled* vs. *liguleless* are each determined by a single pair of alleles. A trihybrid *tunicate, glossy, liguled* plant crossed with a *nontunicate, nonglossy, liguleless* one produces the following offspring:

Tunicate, liguleless, glossy	53
Tunicate, liguleless, nonglossy	10
Tunicate, liguled, glossy	50
Tunicate, liguled, nonglossy	8
Nontunicate, liguled, glossy	11
Nontunicate, liguled, nonglossy	48
Nontunicate, liguleless, glossy	9
Nontunicate, liguleless, nonglossy	54

a) Which, if any, of these allele pairs are linked and which are segregating independently? Show how you arrive at your decision(s).
b) Are the allele pairs on the same or different chromosome pair(s)? Explain.
c) Using your own symbols give the genotypes of the parents of the F_1 and of the F_2.
d) If linkage exists, state whether the recombination value is an accurate measure of the distance between the genes and why.
e) If all 3 genes are on the same chromosome what is their sequence? Explain.
f) Were all F_1's heterozygous in the coupling or repulsion phase? Explain.

18. In rabbits, *C (coloured* coat) vs. *c (albino*) and *B (black)* vs. *b (brown)* are autosomal pairs of alleles and show dominance as indicated (Robinson, R., Bibliog. Genet. 17: 229, 1958). Rabbits from a true-breeding *brown* strain are crossed with *albinos* of genotype *ccBB*; the F_1's are crossed with *albinos* of genotype *ccbb*, with the following results:

Black   102   Brown   198   Albino   300

a) What type of gene interaction is involved?
b) Explain why these data provide evidence of linkage between genes *B* and *C*. Calculate the percent recombination.
c) If the *brown* rabbits were intercrossed, what phenotypes would appear in the progeny and in what proportions? Explain.

19. In rabbits, three *rex* mutants numbered 1, 2 and 3, each characterized by short, soft, plushlike fur, have been found. The first was discovered in 1919 by Abbé Gillet in France, the second in 1926 by a breeder in Hamburg, Germany, and the third in 1927 by Mme. Du Barry in France. *Normal* is dominant to *rex*. Castle, W.E. and H. Nachtsheim (PNAS 19: 1006, 1933) crossed *rex*-1 with *rex*-2 and obtained F_1's with *normal* fur and 391 F_2's, 195 of which were *normal* and 196 *rex*.
(i) What type of gene interaction is involved and why?
(ii) State whether the genes are linked or not and give reasons for your answer.
(iii) If the genes are linked, is it possible, from these data, to determine the distance between them?

The genotypic distribution of 51 F_2 *rex* individuals as determined by backcrossing to one or the other or both of the pure races *rex*-1 and *rex*-2 was:

18	$r_1r_1R_2R_2$	5	$r_1r_1R_2r_2$
21	$R_1R_1r_2r_2$	7	$R_1r_1r_2r_2$

What is the percentage of recombination? (Hint: *rex*-1 is r_1R_2/r_1R_2, *rex*-2 is R_1r_2/R_1r_2 and the F_1's are r_1R_2/R_1r_2).

20. In poultry, *white* is due to the genotypes $cc\_\_$ or $\_\_I\_$. Individuals are coloured only when they have the genotype C_ii. True-breeding *coloured* birds are crossed with *white* fowl of genotype *ccII*. The F_1's are testcrossed, and 65 *coloured* and 206 *white* birds are obtained. Explain these results and show that the wrong cross was made if the purpose was to determine whether linkage exists between the loci.

21. The brine shrimp *Artemia salina* has *black (wild-type)* eyes. Bowen, S.T. (Biol. Bull. 124: 17, 1963) found one *white* eyed male, which he mated with a *wild-type* female. Pedigree A shows the distribution of *white* eyes for six generations, including the initial cross. Shaded symbols

represent the *white* phenotype, and open symbols represent *black*. The number under a symbol is the number of progeny in that class. Note: Males are homogametic, females heterogametic.

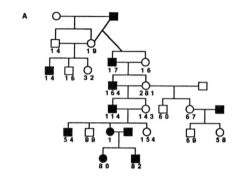

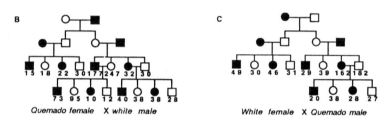

a) Outline a hypothesis to account for:
i) The mode of transmission of *black* vs. *white* eyes.
ii) The origin of the *white*-eyed female indicated by an asterisk.
iii) The reasons why the cross *white* female x *white* male produces only *white*-eyed progeny.

b) Bowen, S.T. (Genetics 52: 695, 1965) further pursued the study of *black* vs. *white* in reciprocal crosses between selected *white*-eyed strains and *wild-type* strains from seven American races. Reciprocal crosses between a *wild-type* from the salt lake near Quemado, New Mexico and inbred *white* stock number 11 gave the results illustrated in B and C. Are these results in agreement with the hypothesis proposed above? Explain by giving the genotypes of all phenotypes in the Quemado female x *white* male cross. Also sketch the essential features of the chromosomes involved.

22. In *Triticum monococcum*, the allele *B* (*black* glume) is dominant to *b* (*white* glume) and allele *P* (*pubescent* glumes) is dominant to *p* (*glabrous* glumes). Data obtained by Kuspira, J. and R.N. Bhambhani (Theor. Appl. Genet. 68: 61, 1984) involved 4 different true-breeding strains and are shown below:

Parents	F_1	F_2
Black, pubescent x *White, glabrous*	*Black, pubescent*	66 % *Black, pubescent* 9 % *Black, glabrous* 9 % *White, pubescent* 16 % *White, glabrous*
Black, glabrous x *White, pubescent*	*Black, pubescent*	51 % *Black, pubescent* 24 % *Black, glabrous* 24 % *White, pubescent* 1 % *White, glabrous*

a) Calculate the percent recombination between the linked allele pairs Bb and Pp from both crosses. Show your calculation and explain its logic.
b) Were the F_1's heterozygous in the coupling or the repulsion phase? Give the genotypes of the parents of the F_1's and F_2's.
c) What results would you expect among the progenies if you testcrossed the F_1's of both crosses?

23. In *Drosophila melanogaster*, the allele pairs Yy (*black* vs. *yellow* body) and $Snsn$ (*long* vs. *singed* bristles) are X-linked. Both loci are on the same side of the centromere; *sn* is closer to the centromere than *y*. Females of genotype *ySn/Ysn* usually express the *wild-type* (*black, long*) phenotype throughout the entire abdomen and thorax. Stern, C. (Genetics 21: 625, 1936) found that a few females exhibited patches (*twin spots*) of relatively equal size, of *yellow* and *singed* tissue adjacent to each other on the thorax or abdomen. Less frequently single *yellow* patches (spots) were found, and still less frequently single *singed* spots, without the adjacent patch of opposite phenotype.
a) Show:
(i) Using appropriate diagrams, how the "twin spots" and the "single spots" could arise by crossing-over.
(ii) Why the spots were infrequent.
(iii) Why they occurred only in the females.

b) Why can mutation be ruled out as an explanation for these observations?

c) Using diagrams, show whether similar phenotypes could be detected in females carrying these genes in coupling, e.g. YSn/ysn.

24. *Hemophilia A* and *red-green* colour blindness in humans are determined by recessive alleles at different X-chromosome loci (McKusick, V.A., *On the X-Chromosome of Man*, American Insitute of Biological Sciences, Washington., 1964).

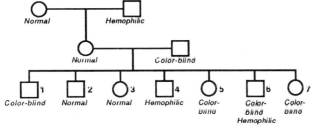

Normal = normal color vision, no hemophilia; *Hemophilic* = normal color vision, has hemophilia A; *Color-blind* = red-green color-blind, no hemophilia.

Which of the third-generation offspring in the pedigree are: recombinants? non-recombinants? cannot be classified? Illustrate or explain your answers.

25. The *ABO* blood groups and the *nail-patella* syndrome are determined by different genes. A small part of one of the pedigrees studied by Renwick, J.H. and S.D. Lawler (Ann. Hum. Genet. 19: 312, 1955) is given below in which the *ABO* genotypes appear below the symbols and the individuals with the syndrome are represented by solid symbols.

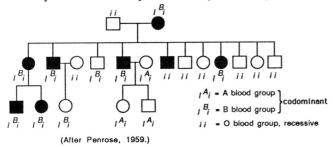

(After Penrose, 1959.)

a) Is the *nail-patella* syndrome determined by a mutant allele that is dominant or recessive? Is the gene autosomal or X-linked?

b) Do the above data indicate there is linkage between the two genes? Explain, using appropriate illustrations of the relevant chromosomes and their homologues.

26. The genes for haptoglobin ($Hp=H$), tyrosine aminotransferase ($TAT=T$) and chymotrypsinogen B($CRTB=C$) in humans reside at the q22 region of chromosome 16. Westphal, E.M. et al. (Genomics 1: 313, 1987) using modern molecular techniques have identified 3 alleles at each locus: $H^1, H^2, H^{2v}, T^a, T^b, T^c$, and C^1, C^2, C^3. Two pedigrees they studied are presented below:

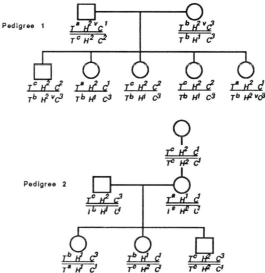

Note: Indicated beneath each individual are their genotypes at the 3 loci, not necessarily in the correct gene sequence. From these 2 pedigrees it is possible to determine the sequence of these 3 genes on chromosome 16. What is the sequence? Explain how you arrive at your answer.

27. The blood of adult human beings contains two kinds of hemoglobin molecules: hemoglobin *A* contains two α and two β polypeptide chains; hemoglobin A_2 has two α and two δ polypeptide chains. The β and δ polypeptides are specified by different genes. $β^a$ (in normal hemoglobin *A*) vs. $β^s$ (in the *S* hemoglobin of sickle-cell anemics) are specified by the allele pair $β^aβ^s$. $δ^{a2}$ (in normal hemoglobin A_2) vs. $δ^{a21}$ (a rare variant of hemoglobin A_2) are specified by the allele pair $δ^{a2}δ^{a21}$. The following modified pedigree is from

Horton, B.F. and T.H.J. Huisman (Am. J. Hum. Genet. 15: 394, 1963):

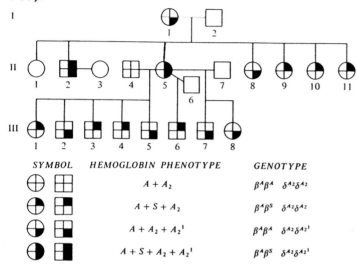

SYMBOL	HEMOGLOBIN PHENOTYPE	GENOTYPE
⊕ ⊞	$A + A_2$	$\beta^A\beta^A \; \delta^{A2}\delta^{A2}$
⊕ ⊞	$A + S + A_2$	$\beta^A\beta^S \; \delta^{A2}\delta^{A2}$
⊕ ⊞	$A + A_2 + A_2^1$	$\beta^A\beta^A \; \delta^{A2}\delta^{A2^1}$
⊕ ⊞	$A + S + A_2 + A_2^1$	$\beta^A\beta^S \; \delta^{A2}\delta^{A2^1}$

a) Does this pedigree indicate linkage of the genes determining the formation of the β and δ polypeptide chains? Justify your answer.

b) Horton and Huisman also reported the inhertaince patterns of the two genes in 21 offspring from parents of which one was $\beta^a\beta^s\delta^{a2}\delta^{a21}$, and the other $\beta^A\beta^A\delta^{A2}\delta^{A2}$. Their phenotypes were as follows:

$A + A_2$	0	$A + A_2 + A_2^1$	8
$A + S + A_2$	13	$A + S + A_2 + A_2^1$	0

Is the inheritance pattern, beyond reasonable doubt, of the type the pedigree in (a) suggests? Explain.

28. a) With the aid of diagrams explain how copy-choice differs from chromatid exchange (breakage-and-reunion) as an explanation of crossing-over.

b) Moses, M.J. and J.H. Taylor (Exp. Cell Res. 9: 474, 1955) and Rossen, J. and M. Westergaard (C.R. Trav. Lab. Carlsberg 35: 233, 1966) have shown that the major period of DNA replication of genes (at pre-meiotic interphase) occurs prior to the onset of chromosome synapsis. Explain what bearing this has on the validity of the copy-choice hypothesis.

c) In *Chlamydomonas reinhardii* a cross involving three linked genes in the sequence A B C produced the following results:

Cross	Tetrad
ABC x abc	ABc
	AbC
	aBC
	abc

Why is this data evidence against the copy-choice hypothesis?

29. a) In 1933 Lindegren, C.C. (Bull. Torrey Bot. Club, 59: 119) crossed a *Neurospora crassa* strain with *pale* conidia and *normal* growth habit to one with *orange* conidia and *fluffy* growth habit. The numbers showing first- and second-division segregation, respectively, for each of these two pairs of traits and for the mating-type traits were as follows:

Pairs of traits	First division	Second division
Mt^+ vs. mt^- mating type	97	12
Pale vs. *orange* conidia	73	36
Fluffy vs. *normal* growth	42	67

Lindegren explained these results on the basis of the hypothesis that:
i) Crossing-over occurred at the four-strand stage.
ii) Each crossing-over event involved only two of the four chromatids.
iii) The homologous centromeres of each chromosome pair segregated at the first division of meiosis.

1) Do you agree with Lindegren's explanation?
2) Calculate each of the gene-centromere distances.

b) In a cross between an *arginineless* and an *histidineless* strain of *Neurospora crassa*, the spore arrangements shown below are found.

1	arg⁺his⁺	arg⁺his⁺	arg⁻his⁻	arg⁻his⁻	78
2	arg⁺his⁻	arg⁺his⁻	arg⁻his⁺	arg⁻his⁺	83

c) How are the allele pairs located on the chromosome(s) with respect to their centromeres and to each other?

30. In 1982 Bird, T.D. et al. (Am. J. Hum. Genet. 34: 388) presented 2 pedigrees (one of which is modified) showing the mode of inheritance of *normal* vs. *Charcot-Marie-Tooth* (*CMT*) disease (a rare hereditary neuropathy) and the codominant alleles Fy^a (=F^a) and Fy^b (=F^b) of the Duffy blood group locus located on chromosome 1.

Pedigree 1

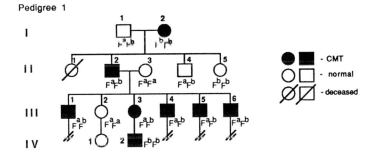

Pedigree 2

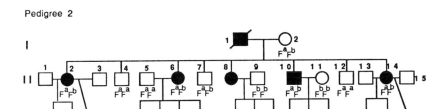

a) What is the mode of inheritance of *CMT*? Explain.
b) Which progeny in each of the two pedigrees are recombinants?
c) Is the *CMT* locus on chromosome 1? Explain.

31. Houlahan, M.B. et al. (Genetics 34: 493, 1949) observed the following ordered tetrads in *Neurospora crassa* which were derived from a cross between a strain with *yellow* conidia (ylo^-) which was unable to synthesize *cystine* (cys^-) and a *wild-type* strain.

ylo⁺cys⁺	ylo⁺cys⁻	ylo⁻cys⁻	ylo⁻cys⁻	ylo⁺cys⁺	ylo⁻cys⁺	ylo⁻cys⁺
ylo⁺cys⁺	ylo⁺cys⁺	ylo⁻cys⁺	ylo⁺cys⁻	ylo⁻cys⁻	ylo⁺cys⁻	ylo⁺cys⁻
ylo⁻cys⁻	ylo⁻cys⁺	ylo⁺cys⁻	ylo⁺cys⁺	ylo⁺cys⁺	ylo⁻cys⁺	ylo⁺cys⁺
ylo⁻cys⁻	ylo⁻cys⁺	ylo⁺cys⁺	ylo⁺cys⁺	ylo⁻cys⁻	ylo⁺cys⁻	ylo⁻cys⁻
74	13	45	5	12	2	4

a) Explain why these 2 genes are linked.
b) Determine the genetic distance between them.
c) Are the genes on the same or different arms of the chromosome? Explain.

32. The results of Lindegren, C.C. (Bull. Torey Bot. Club 60: 133, 1933) involving the 4 loci *p* (*pale*), *f* (*fluffy*), *c* (*crisp*) and *mt* (*mating type*) in *Neurospora crassa* were presented by Catcheside, D.G. ("*Genetics of Micro-organisms*", Sir Isaac Pitman & Sons, London, 1951).

	Classes of ordered ascospore arrangements						
	a⁺b⁺	a⁺b⁻	a⁺b⁺	a⁺b⁺	a⁺b⁺	a⁺b⁻	a⁺b⁺
	a⁺b⁺	a⁺b⁻	a⁺b⁻	a⁺b⁺	a⁻b⁻	a⁻b⁺	a⁻b⁻
	a⁻b⁻	a⁻b⁺	a⁻b⁺	a⁺b⁻	a⁺b⁺	a⁺b⁻	a⁺b⁻
	a⁻b⁻	a⁻b⁺	a⁻b⁻	a⁻b⁻	a⁻b⁻	a⁻b⁺	a⁻b⁺
f⁺mt⁺ × f mt⁻	16	20	6	60	3	1	3
p⁺mt⁺ × p⁻mt⁻	62	1	10	34	0	1	1
c⁺p⁻ × c⁻p+	2	52	12	3	0	9	0
p⁺mt⁻ × p⁻mt⁺	3	46	8	17	0	3	1
c⁺mt⁺ × c⁻mt⁻	55	2	9	9	2	0	1

Note: (i) + = *wild-type* allele; − = *mutant* allele
(ii) *a* and *b* refer to the first and second loci in the genotypes of the parents in each cross.
a) Determine the distance of each gene from the centromere.
b) Determine the percent recombination between each pair of loci.
c) With respect to syntenic genes, determine for each cross whether the genes are on the same or opposite arms of the chromosome and construct a linkage map. Explain your answer in each case.

d) Do these results provide evidence for crossing-over at the four-chromatid stage? Explain, using the results of any one cross as an example.

33. In *Aspergillus nidulans* the mutant genes *w* (*white*) and *acr* (*acriflavine*) are linked and on chromosome 2. *White* diploid mitotic segregants were selected from the diploid strains Y,(*acr w / Acr W*) and Z, (*Acr w / acr W*). All the *white* segregants from Y were homozygous *acr/acr* and from Z were homozygous *Acr/Acr*. On the other hand, of the *acr/acr* segregants from diploid Y, 87% were *w/w*, and 13% were *W/w*. And of those from diploid Z, 82% were *W/W*, and 18% were *W/w* (Pontecorvo, G. and E. Kafer, Proc. R. Phys. Soc. (Edinb.), 25: 16, 1956). What is the linkage relationship of the two genes and the centromere? Explain.

34. Ebersold, W.T. et al. (Genetics 47: 531, 1962) studied 5 different pairs of alleles in *Chlamydomonas reinhardii*. The following results were obtained in 2-point crosses between single mutant strains:

Cross	Observed number of tetrads		
	PD	NPD	TT
ac-14 x *ac-14b*	53	0	0
pf-15 x *pf-17*	22	23	46
ac-12 x *pf-12*	74	0	22
pab-1 x *thi-2*	36	2	47
ac-17 x *pf-17*	54	53	9
ac-51 x *pf-16*	65	0	41
pf-13 x *nic-11*	21	22	44

a) Why are only PD tetrads produced in the first cross?
b) Show which allele pairs, if any, are linked and the genetic distance between them.

35. Analysis of diploid mitotic recombinants in *Aspergillus nidulans* indicated the linked sequence *ad-23* (*adenineless*)-*w*(*white*)-centromere. Other such studies revealed the linked order *pu* (*putrescine*)-*thi-4* (*thiamineless*)-centromere. Diploid mitotic recombinants did not reveal linkage between these two groups. A study of 85 *white* haploids from heterozygous diploids, one of whose parents was *wild-type* and the other true-breeding for all four mutant genes, showed that all were genotypically identical, *ad-23 w pu thi-4*.
a) Are the genes on the same or different chromosomes? Explain.
b) If on the same chromosome, where is the centromere located relative to the four genes? Why?

36. The genes *ad-14* (*adenineless*), *pro-1* (*prolineless*), *pab-1* (*p-aminobenzoic acidless*), *y* (*yellow*) are in linkage group 1 in *Aspergillus nidulans*.
a) Pontecorvo, G. and E. Kafer (Adv. Genet. 9: 71, 1958) selected 371 *yellow* mitotic recombinants (homozygotes) from a diploid strain heterozygous for the four markers and of genotype *ad-14 Pab-1 Y Pro-1 / Ad-14 pab-1 y pro-1* (not necessarily in this order). Of these, 96 were prototrophs (*Pab-1 - Ad-14 - Pro-1*), 245 required p-aminobenzoate, and 30 required proline as well as p-aminobenzoate. What is the sequence of these genes with respect to each other and the centromere? Show your reasoning.
b) None of the 371 *yellow* homozygotes was homozygous for *ad-14*. Another diploid strain, *ad-14 y / Ad-14 Y*, produced *ad-14, ad-14* homozygous mitotic recombinants, all of which were *Yy*. Is *ad-14* on the same or different arm of chromosome 1 as the other three genes? Explain.

37. A diploid strain of *Aspergillus nidulans* heterozygous for the mutant genes *y* (*yellow*), *sm* (*small*), *pan⁻* (*pantothenicless*), *nic-8* (*nicotineless*), *phe-2* (*phenylalanineless*) and *cho⁻* (*cholineless*) produced a large number of haploid segregants of which 113 *yellow* ones were selected and tested for the presence or absence of the other mutant genes. The genotypes of these *yellow* haploids and their numbers were as follows:

y nic-8 cho⁻ phe-2 pan⁻ sm	31
y Nic-8 Cho phe-2 pan⁻ sm	26
y nic-8 cho⁻ Phe-2 Pan Sm	24
y Nic-8 Cho Phe-2 Pan Sm	32

What are the linkage relationships of these genes? Explain with illustrations.

38. In 1964, Holliday, R. (Genet. Res. 5: 282) presented a molecular model to explain crossing-over (reciprocal exchange). Evidence for its essential correctness, although not necessarily the details, has been provided by Dressler, Meselson, Radding, Holliday himself and others (see Watson, J.D. et al., *Molecular Biology of the Gene*, Volume 1, 4th edition, Chapter 11, The Benjamin/Cummings Publ. Co. Inc., Menlo Park, CA 1987 for an excellent discussion of the subject). For discussion purposes assume a chromosome pair is heterozygous for two pairs of alleles as shown below:

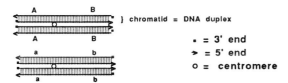

Knowing that crossing-over is a complex process involving the actions of endonucleases, DNA polymerases and ligases, describe with the aid of appropriate illustrations how cross-over chromosomes with recombinant genotypes can be generated according to the Holliday model.

39. a) Why are extremely short regions used in establishing genetic maps?
    b) For how many allele pairs is heterozygosity required so as to detect double crossovers? Explain.
    c) Would you expect estimates of map distances to be increased or decreased as a result of undetected double crossovers? Why?
    d) Explain why, regardless of the number of double crossovers, the theoretical percentage recombination between any two syntenic genes can never be more than 50.

40. You have two testcross populations, each numbering 150; one for two closely linked genes, the other for two more distantly linked genes. Which of the populations would you expect to permit a more reliable estimate of the map distance and why?

41. a) In three-point testcrosses why are the parental types most frequent and the double-crossover types least frequent?
    b) Explain how the information in (a) can be used to determine the relative positions of three genes on the linkage map.

42. In 1955 Brown, S.W. and D. Zohary (Genetics 40: 850) studied two plants of *Lilium formosanum* in each of which the homologues of chromosome pair A were heteromorphic: one homologue, having lost 2/3 of its short arm, was shorter than the other. At least one crossing-over event occurs in the long arm of this chromosome pair in every meiocyte. A reciprocal exchange also occurs in the short arm in some meiocytes. The consequences of crossing-over and its absence in the short arm are as follows:

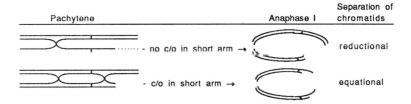

The meiotic data for the two plants under different temperature conditions were as follows:

Plant	Diplotene-Metaphase I		Anaphase I	
	no. chiasmata	%	no. equational segregations	%
1 normal temp.	452	70	172	71
2 reduced temp.	120	51	60	50

Do these data support Janssens, F.A. (La Cellule 25: 287, 1909) chiasmatype theory that each chiasma is a consequence of crossing-over or the classical theory (McClung, C.E., Qt. Rev. Biol. 2: 344, 1927) that chiasmata are simply the regions where non-sister chromatids lie across each other, likely break and rejoin, giving rise to cross-over chromatids?

43. Barring chromosomal aberrations, the order of genes on a genetic map of a chromosome corresponds completely with the order on a cytological map. Why do the distances on the two maps not always correspond?

44. In the mouse, the genes studied to date fall into 20 linkage groups. How many chromosomes would you normally expect to find in a somatic cell of a mouse? Explain.

45. In rye (2n=14), approximately 400 genes have been located in each linkage group. All linkage groups are of the same length with the total map length of all the linkage groups being 350 map units.
a) What is the average length of each linkage group?
b) On the average, how many reciprocal exchange events occur per meiocyte in each chromosome pair?
c) Is each bivalent likely to possess an open or closed appearance at late diakinesis-metaphase-I? Illustrate.

46. The following is a linkage (chromosome) map for the genes *y*, *sh* and *c* in corn:

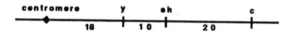

Assume that double crossing-over does not occur in either the *y-sh* or the *sh-c* regions and that it occurs without interference in the *y-c* region.
a) What proportion of the meiocytes in the trihybrid *Y Sh C / y sh c* would you expect to undergo:
(i) Double crossing-over in the *y-c* region?
(ii) Single crossing-over in the *y-sh* region?
(iii) Single crossing-over in the *sh-c* region?
(iv) No crossing-over in the *y-c* region?

b) Answer the questions in (a) assuming the coefficient of coincidence is 0.5.

47. In *Drosophila melanogaster*, the allele pairs *Vv* (*red* vs. *vermilion* eyes), *Mm* (*normal* vs. *miniature* wings) and *Ff* (*normal* vs. *forked* bristles) are on the X chromosome at positions 33.0, 36.1 and 56.7 respectively, from the left end. A *vermilion, miniature, forked* female is crossed with a *red, normal, normal* male and the F_1's are intercrossed.

a) Determine the kinds of gametes and their proportions that are likely to be produced by the F_1 progeny.
b) What phenotypic ratio would be expected in the F_2?
c) In the reciprocal of this cross would you expect results similar to those shown? Explain.

48. In corn, the following allele pairs are on chromosome 9: *Cc* (*aleurone colour* vs. *no aleurone colour*), *Shsh* (*full* vs. *shrunken* endosperm), *Wxwx* (*starchy* vs. *waxy* endosperm). F_1 plants heterozygous for all three pairs of alleles were testcrossed and the tabulated phenotypes were obtained by Emerson, R.R. et al. (Cornell Univ., Agric. Exp. Stn. Mem. 180, 1935):

Phenotype	Number of Progeny
White, shrunken, starchy	116
Coloured, full, starchy	4
Coloured, shrunken, starchy	2,538
Coloured, shrunken, waxy	601
White, full, starchy	626
White, full, waxy	2,708
White, shrunken, waxy	2
Coloured, full, waxy	113

a) Determine the sequence of the genes on the chromosome, the map distances and the genotypes and phenotypes of the homozygous parents of the F_1.
b) Derive the coefficient of coincidence for interference.
c) How many double-crossover phenotypes would you expect in this cross per 1,000 progeny? How many double-crossover chromosomes?
d) Derive the amount of underestimation of the map distance between the end genes if the phenotypes of the progeny for the middle one had not been scored.

49. In the guinea pig *black* vs. *white* coat, *short* vs. *long* hair and *wavy* vs. *straight* hair are each determined by a single pair of alleles. Crosses are made between homozygous *white, short, wavy* and homozygous *black, long, straight* animals. The *black, short, wavy* F_1's are crossed with animals from a

true-breeding *white, long, straight* strain and produce the offspring shown:

Phenotype	Number of Animals
White, short, wavy	46
White, short, straight	20
White, long, wavy	28
White, long, straight	6
Black, short, wavy	6
Black, short, straight	32
Black, long, wavy	20
Black, long, straight	42

Are the allele pairs on the same chromosome pair? Explain, using your own symbols.

50. In the mink (*Mustela vison*) a true-breeding line with *black, wavy, short* hair was crossed with a true-breeding line with *white, straight, long* hair. The *black, wavy, long* haired F$_1$ animals produced the following results when crossed with a true-breeding *white, straight, short* haired strain:

Black, straight, long	42	Black, wavy, short	348
White, wavy, short	38	White, wavy, long	9
Black, straight, short	7	White, straight, long	356

a) How many allele pairs are involved in causing these phenotypic differences? Explain.
b) All genes involved in this cross are syntenic. Are they all linked? Why?
c) Were the F$_1$'s heterozygous in the coupling or repulsion phase? Give the genotypes of the parental strains and the F$_1$'s.
d) How many phenotypic classes are missing? Which ones are they and why are they absent?
e) Derive the sequence of the genes and their map distances.
f) What is the coefficient of coincidence? Why?

g) Is the recombination value for each of the 3 intergenic regions an accurate measure of the genetic distance between the genes? Why?

51. In *Neurospora crassa* the allele pairs *Lys lys* and *A d a d* determine the expression of *no lysine* vs. *lysine requirement* and *no adenine* vs. *adenine requirement*. From the data below for the cross *Lys Ad* x *lys ad* determine the positions of these genes with respect to each other and their centromeres.

	Tetrad type			
	lys Ad	Lys Ad	Lys ad	lys ad
	lys Ad	Lys Ad	Lys ad	lys ad
	Lys ad	lys ad	lys Ad	Lys Ad
	Lys ad	lys ad	lys Ad	Lys Ad
Number of asci	39	36	40	35

52. In *Drosophila simulans*, the following three allele pairs are on the X chromosome: *Yy* (*gray* vs. *yellow* body), *Cm cm* (*red* vs. *carmine* eyes), and *Ff* (*normal* vs. *forked* bristles). F$_1$'s heterozygous for these allele pairs were testcrossed by Sturtevant, A.H. (PNAS 7: 235, 1921). Parental and double-crossover-progeny classes were:

cm y F	725	Cm y F	34
Cm Y f	719	cm Y f	32

a) Show diagrammatically why the order of the genes cannot be *Cm Y F*.
b) What is the correct gene order? Explain how you determine it.
c) What were the genotypes of the true-breeding parents and of the F$_1$?

53. In *Drosophila melanogaster*, the *Adh* gene specifying the synthesis of alcohol dehydrogenase (ADH) is located on the second chromosome. The co-dominant alleles *Adh^F* and *Adh^S* specify the *fast* and *slow* electrophoretic variants (isozymes) of the enzyme. Grell, E.H. et al. (Science 149: 80,

1965) crossed males from a true-breeding *wild-type* stock homozygous for Adh^S with females homozygous for Adh^F and the recessive alleles *b* (*black* body; map position 48.5), *el* (*elbow* wings; map position 50.0), *rd* (*scraggly* bristles; map position 51.0), and *pr* (*purple* eyes; map position 54.5) located on chromosome 4. The F_1 females were backcrossed to males homozygous for Adh^F, *b el, rd* and *pr*. From among the backcross progeny individuals with single crossovers in one of the three regions *b-el, el-rd* and *rd-pr* were selected and tested for their ADH isozymes. The results are tabulated below:

Genotype of gamete received from F_1	Number of progeny in each class
b El Rd Pr Adh^S	10
b El Rd Pr Adh^F	0
B el rd pr Adh^S	0
B el rd pr Adh^F	6
b el Rd Pr Adh^S	3
b el Rd Pr Adh^F	25
B El rd pr Adh^S	17
B El rd pr Adh^F	2
b el rd Pr Adh^S	0
b el rd Pr Adh^F	5
B El Rd pr Adh^S	5
B El Rd pr Adh^F	0

a) In which of the three regions of chromosome 4 is *Adh* located?

b) What is its position on the genetic map?

54. Genes at loci *K, L, M* are linked, but their order is unknown. F_1's from the cross *KK LL MM* x *kk ll mm* are testcrossed. The most frequent phenotypes in the testcross progeny will be *KLM* and *klm* regardless of gene order. What phenotypic classes will be least frequent if locus *K* is in the middle? If locus *L* is in the middle?

Chapter 11
Extranuclear Inheritance and Related Phenomena

1. a) What are the basic features by which cytoplasmic inheritance is distinguishable from nuclear inheritance in almost all eukaryotic organisms? Suggest three forms of inheritance which behave in a manner similar to cytoplasmic inheritance and show how they can be distinguished from one another.
   b) (i) What specific properties do chromosomal genes possess?
   (ii) Explain how cytoplasmic genes show these properties, citing examples where possible.

2. In the mouse, *Mus musculus*, *Mta* is a cell surface antigen which serves as a target for specific T killer lymphocytes. The antigen occurs in two forms Mta^+ and Mta^-. Some strains are Mta^+, others are Mta^-. All F_1 progeny of the cross Mta^+ female x Mta^- male are Mta^+; those from the reciprocal cross are Mta^- (Lindahl, K.F. and B. Hausmann, Genetics 103: 483, 1983). Among possible explanations of the reciprocal-cross differences are:
   (1) Sex Linkage
   (2) Dauermodification
   (3) Maternal inheritance (influence)
   (4) Cytoplasmic inheritance
   (5) Maternal influence via placenta, milk or egg

   Outline the series of crosses you would make to determine which of these explanations is the correct one and show the results expected in each cross with each form of inheritance.

3. Van Wisselingh, C. (Z. Indukt. Abstamm.- Vererbungsl. 22: 65, 1920) working with *Spirogyra*, found cells with two types of plastids, one with *normal* pyrenoids and the other lacking this structure. After many cell divisions both plastid types were still present in each cell. Does this constitute evidence that plastids possess their own hereditary determiners? Explain.

4. Extensive use of a certain drug can lead to addiction; children from matings between *addicted* mothers and *nonaddicted* fathers are *addicted* while those from reciprocal matings are *nonaddicted*. If the reciprocal cross differences are not due to cytoplasmic genes, offer an explanation to account for these results.

5. A particular phenotype in the progeny of mammals is either due to the transmission of cytoplasmic substances through the female gamete or to the transfer of a substance(s) from mother to the fetus through the placenta. In rabbits and mice it is possible to remove ovaries and replace them with ovaries from other females. Many of the transplanted ovaries become established and produce and shed eggs normally. How might this technique be used to distinguish between the two alternatives mentioned above?

6. In the four-o'clock (*Mirabilis jalapa*) variegated plants occur in which certain branches have normal *green* leaves, other branches have leaves entirely *white* or *pale*, and still others have leaves that are *green* with *pale* or *white* areas. Correns, C. (Z. Indukt. Abstamm.-Vererbungsl. 1: 291, 1909) found that irrespective of the pollen parent, seed borne on flowers on *green* branches produce *green* plants, those on *white* branches produce *white* seedlings, and those on *variegated* branches gave *green*, *white* and *variegated* seedlings in widely differing ratios.
   a) Explain these results.
   b) What would be the critical observation for the acceptance of your explanation?

7. a) Through inbreeding and selection, lines of mice have been developed that differ markedly in the incidence of mammary cancer. The level of incidence in a given line is constant from generation to generation. The inheritance of this character was studied by Bittner, J.J. (Ann. N.Y. Acad. Sci. 71: 943, 1958) who found that when lines with *high* incidence were crossed as females with lines possessing a

low incidence of the disease, almost all the F_1 and F_2 females developed the cancer. In reciprocal crosses only a few F_1 and F_2 females were *afflicted*. Suggest possible explanations for this reciprocal cross difference.

b) Newborn mice of the *high*-incidence strain were transferred, before they could be suckled by their mother, to foster mothers of the *low*-incidence strain; among them *mammary-cancer* development was no more frequent than in the *low*-incidence line. When the reciprocal transfer was performed, many of the females developed *mammary cancer*, and the majority of their subsequent female progeny were also *afflicted*. Based on this information what is the most plausible explanation for the mode of transmission of this disease?

8. In *Chlamydomonas reinhardii*, a unicellular haploid alga with two mating types mt^+ and mt^-, the zygote contains all the contents, both nuclear and cytoplasmic, of both gametes. *Wild-type* cells are *streptomycin-sensitive, ss*. Sager, R. (Science 132: 1459, 1960) treated large populations of these with 500 ug/ml of this antibiotic and isolated a *resistant* strain, *sr*. The cross $sr\ mt^+ \times ss\ mt^-$ produced haploid progeny that segregated 1:1 for mating type. The *sr* clones, after years of vegetative propagation under a variety of environmental conditions in the absence of streptomycin, showed the same degree of resistance as the *resistant* parental strain. The $sr\ mt^+$ F_1 clones backcrossed to $ss\ mt^-$ produced *sr* offspring only (4:0), but mating type still segregated in a 1:1 ratio. The same results were obtained in three subsequent backcross generations. The reciprocal backcross ($F_1\ sr\ mt^- \times ss\ mt^+$), produced only *streptomycin-sensitive* (0:4) progeny. In subsequent backcrosses resistance did not appear.

What mode of transmission appears to be the most plausible explanation for the 4:0 segregation? Explain. For the 0:4 segregation? Explain.

9. *Chlamydomonas reinhardii* is a one-celled alga, with each cell possessing one chloroplast (and its DNA genome, *cp* DNA) in the cytoplasm. Despite each of the mt^+ and mt^- parents contributing a chloroplast and *cp* genome to the zygote, usually only the *cp* genome of the mt^+ parent is transmitted to the progeny (Sager, R. and Z. Ramanis, PNAS 65: 593, 1970).

Offer a plausible mechanism for the retention and transmission of only the mt^+ *cp* genome to the progeny. (See R. Sager and Z. Ramanis, Theor. Appl. Genet. 43: 101, 1973 for the mechanism involved).

10. In the diploid protozoan *Paramecium aurelia*, two conjugants undergo a reciprocal exchange of gamete nuclei during short sexual fusion; during long fusion they exchange cytoplasm as well. Exconjugants that are heterozygous for a chromosomal gene can undergo an autogamous fission, producing individuals homozygous for each allele. The following data regarding the *killer* trait are from crosses made by Sonneborn, T.M., PNAS 29: 329, 1934 and Beale, G.H., "*The genetics of* Paramecium aurelia", Cambridge University Press, Cambridge, 1954. *Killer* individuals contain 200 to 300 DNA-containing particles called kappa in their cytoplasm; *sensitive* animals lack these entities.

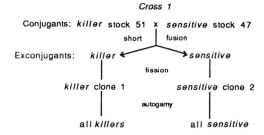

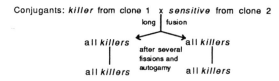

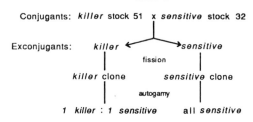

Do you agree with Sonneborn's explanation that kappa is responsible for the *killer* trait, but a nuclear gene *K* is necessary for the multiplication and maintenance of kappa? Explain.

11. In humans, *myoclonic epilepsy* is a rare disorder characterized by deafness, seizures, dementia, other clinical symptoms, ragged red muscle fibers and abnormal appearance of mitochondria in cells of muscles. In 1985, Rosing, H.S. et al. (Ann. Neurol. 17: 228) published the following extensive pedigree indicative of the mode of inheritance of this syndrome:

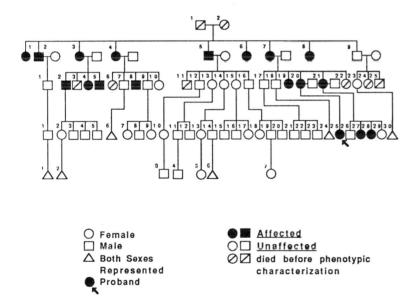

a) There is only one pattern of transmission that will account for this mode of inheritance. What is it and why?

Using molecular techniques, Shoffner, J.M. et al. (Cell 61: 931, 1990) identified a mutant allele in the mitochondrial DNA (*mt*DNA) gene that codes for tRNA^{LYS}. Depending on the *affected* individual, between 73% and 98% of the *mt*DNA molecules are *mutant* and the remainder are *wild-type*. 75 individuals (controls) from *unaffected* families possessed only *wild-type mt*DNAs.

b) Do these additional findings confirm the mode of inheritance of this condition that was postulated in (a)?

c) The degree of expression of the disease among *affected* individuals varies in that some are *mildly affected*, others *moderately* and still others *severely affected*. Moreover, all the progeny of *affected* individuals do not necessarily express the syndrome. In light of the current understanding of number of mitochondria per cell, their reproduction and distribution, provide a plausible explanation for the disease's variable expressivity.

12. In the grass, *Poa annua*, *resistance* to the herbicide atrazine is maternally inherited (Darmency, H. and J. Gasquez, New Phytol. 89: 487, 1981). Barros, M.D.C. and T.A. Dyer (Theor. Appl. Genet. 75: 610, 1988), showed that this trait is due to a mutant allele of the gene coding for the *D1* protein of the photosynthetic system II.

Is this information sufficient evidence that the *D1* gene resides in the chloroplast genome? Explain. How would you proceed to provide unequivocal evidence for its location in the chloroplast genome and not one of the chromosomes in the nucleus? Show the results expected in each case.

13. In corn, true-breeding mutants with *green* and *white* striping of the foliage are fairly common. One such mutant is called *iojap* (Jenkins, M.T., J. Hered. 15: 467, 1924). Reciprocal crosses between the true-breeding *iojap* line and true-breeding *green* lines gave the following results. *Green* females crossed with *iojap* males produced F$_1$'s that were all *green*. When these were self-fertilized they produced 2,498 *green* and 782 *iojap*. Twenty-six of the *green* F$_2$'s were self-fertilized; 9 of which bred true for *green* and 17 segregated for *iojap*.

Rhoades, M.M. (PNAS 29: 327, 1943) crossed *iojap* females and *green* males. The F$_1$'s were either all completely *white* (these die), all wholly *green*, or progenies comprised *green*, *white* and *striped* plants in varying proportions. A backcross of these *striped* plants as females with *green* ones as males produced the following:

(i) Some ears gave progenies consisting entirely of *green* plants.
(ii) Others produced progenies with varying percentages of *green*, *white* and *striped* seedlings.
(iii) Occasionally all the progeny consisted of *white* seedlings only.
(iv) Many generations of backcrossing *striped* females to *green* males resulted in *green*, *white* and *striped* seedlings in varying proportions.

a) Show why it can be concluded that the *iojap* trait is determined by both a nuclear gene and a cytoplasmic determinant.
b) How might the chromosomal allele pair be associated with a non-chromosomal determinant in the expression of this form of striping?
c) Do these results indicate whether or not the cytoplasmic determinant is actually located in the chloroplast? Explain.
d) Are the cytoplasmic genes autonomous in:
(i) determining the phenotype?
(ii) their duplication?

14. Corn (*Zea mays*) is a monoecious species which has male flowers at the top of the plant and female flowers along the side of the plant. A strain is developed in which almost all the plants are *male-sterile*; the few *male-fertile* ones produce very little pollen. The F_1's between the *male-sterile* line and different *male-fertile* ones are *male-sterile*. Male fertility is not restored by backcrossing the F_1 and offspring of subsequent backcross generations to the *male-fertile* line. When female flowers on plants in the *male-fertile* line are fertilized by the rare pollen from plants in the *male-sterile* line, all the progeny are *male-fertile*. These plants when self-fertilized breed-true. Offer a genetic explanation of these results.

b) When the above *male-sterile* line is crossed with *male-fertile* strain A, all the progeny are *male-fertile*. These F_1's backcrossed as females to the *male-sterile* line produced 52 *male-sterile* and 55 *male-fertile* plants. The reciprocal of this cross also produces the same results. When the F_1's are self-fertilized, they produce 123 *male-fertile* and 42 *male-sterile* plants.

(1) Explain why these results are different from those in (a) despite the fact that the same *male-sterile* line is used.
(2) On the basis of your hypothesis would you expect *male-sterile* plants in the following crosses and, if so, in what proportions:
(i) F_1 (*male-sterile* x *male-fertile* strain A) x *male fertile* strain A?
(ii) F_1 (*male-sterile* x *male-fertile* strain A) x *male-fertile* strain discussed in (a)?

15. In peanuts (*Arachis hypogaea* L.) commercial varieties possess either a *runner* or *bunch* growth habit. Ashri, A. (Genetics, 50: 363, 1964) obtained the tabulated results in reciprocal crosses between true-breeding lines with these two traits.

♀	♂	F_1	F_2
V4 x	NC2	runner	648 runner : 499 bunch
V4 x	123	runner	581 runner : 456 bunch
V4 x	G2	runner	441 runner : 368 bunch
NC2 x	V4	bunch	290 runner : 624 bunch
123 x	V4	bunch	28 runner : 47 bunch
G2 x	V4	bunch	89 runner : 180 bunch

Propose a model based on genic and gene-cytoplasm interaction to account for these results.

16. A number of investigators have shown that treatment of plants and animals with various environmental agents (e.g. chemicals, high temperature) can induce the expression of abnormal phenotypes which are transmitted maternally for a few generations with slowly diminishing intensity of expression until they gradually disappear. For example, Hofmann, J. in 1927 (see Jollos, V., Grundbegriffe der Vererbungslehre, "Handbuch der Vererbungswissenschaft", Borntrager, Berlin, 1939) found that treatment of bean (*Phaseolus vulgaris*) plants with chloral hydrate induced the formation of leaf aberrations which were transmitted by the female only and persisted for six generations before disappearing. Offer an explanation for the origin of these

modifications and the fact that they last for only a few generations.

17. In some perfect-flowered, cross-fertilizing species, e.g., onions and sugar beets, hybrid F_1 seed is commercially desirable but not economically feasible because of the high cost of emasculation (removal of anthers). Male sterility obviates this necessity. In onions, the inheritance of male sterility was studied by Jones, H.A. and A.F. Clarke (Proc. Am. Soc. Hort. 43: 189, 1943) using the male-sterile line 13-53 and a large number of commercial varieties. The results of the crosses are tabulated.

Parents		Progeny
♀	♂	
Male-Sterile 13-53 x Male-Fertile 1		all male-fertile
Male-Sterile 13-53 x Male-Fertile 2		all male-sterile
Male-Sterile 13-53 x Male-Fertile 3		1 male-fertile : 1 male-sterile
F_1 male-fertile (Self-Fertilized)		3 male-fertile : 1 male-sterile
Male-Sterile 13-53 x F_1 Male-Fertile		1 male-fertile : 1 male-sterile
Male-Fertile F_1 from cross 3 x Male fertile P_1		3 male-fertile : 1 male-sterile
Male-Fertile parent x Male-Fertile F_1 from cross 3		all fertile

a) State with reasons whether *male sterility* is determined by nuclear genes or not.
b) Demonstrate that *male sterility* is determined by cytoplasmic genes and show whether there is evidence of their nonautonomous duplication.

18. a) Mitchell, M.B. and H.K. Mitchell (PNAS 38: 442, 1952) found an erratically slow-growing *Neurospora crassa* mutant called *poky, po,* which showed an unusual pattern of transmission:

protoperithecial parent		conidial parent	progeny (ascospores)
wild-type	x	*poky*	all *wild-type*
poky	x	*wild-type*	all *poky*

(i) How is the mutant allele inherited?
(ii) What further crosses would you make to confirm your conclusion and to ascertain whether the determiner is acting autonomously?

b) In 1953 Mitchell, M.B. et al. (PNAS 39: 606) studied three other *slow-growing* mutant strains, *mi-3*, *c115* and *c117*. All were found to carry an abnormal system of respiratory enzymes. Their transmission patterns are shown in the following table:

protoperithecial parent		conidial parent	progeny (ascospores)	
			wild-type	*slow*
mi-3	x	*wild-type*	0	2,071
wild type	x	*mi-3*	1,113	3*
wild-type	x	*c115***	596	590
wild-type	x	*c117***	1,050	1,035

*Due to slow growth or other factors.   ** *Female-sterile*

(i) Give a genetic interpretation of these results.
(ii) What results would you expect in crosses between *c115* or *c117* and *mi-3* or *po*?

c) The same workers in 1956 showed that certain strains of *poky* (*fast-poky*) had reverted to nearly *wild-type* growth, although they still retained the abnormal metabolism (and cytochromes).
(1) A cross with *fast-poky* as protoperithecial parent and *wild-type*, a conidial parent, gives 1 *poky* : 1 *fast-poky* ascospores in all asci.
(i) Interpret these results genetically, showing what they indicate about the autonomy of the cytoplasmic determiner in *poky*.
(ii) Show the results expected for the reciprocal of the cross described. Illustrate and explain your answer (assume no cytoplasm enters with the male nucleus).

(2) Reciprocal crosses were made between a *poky* and a *fast-poky* colony, both derived from ascospores of a single ascus of the cross described in (a). The two crosses gave the same

results, a 1:1 ratio of *poky* : *fast-poky*. *Fast-poky* x *wild-type* female produced only *wild-type* progeny.
(i) Is your interpretation borne out?
(ii) What other conclusions can you draw from these results?

19. In *Chlamydomonas reinhardii*, cytoplasmic genes usually show uniparental inheritance. Specifically, only the cytoplasmic genes from the mt^+ (female) parent in zygotes are transmitted to all the progeny. Low doses of ultraviolet (UV) irradiation of the mt^+ parent result in zygotes retaining the chloroplast genome (*cp*DNA) of both mating types and this leads to biparental inheritance of genes in the *cp* genome. If parents carry different alleles of one or more cytoplasmic genes in the *cp* genome, the biparental zygotes will be cytohets (heterozygous for cytoplasmic genes). In 1965 Gillham, N.W. (PNAS 54: 1560) crossed UV-treated streptomycin-*resistant* mt^+ cells with streptomycin-*sensitive* mt^- ones and found that all four products of each zygote (meiocyte) were *resistant* and each of the cells resulting from postmeiotic mitoses expressed either the *resistant* or the *susceptible* phenotype.

Acetate requirement in this species is determined by mutant alleles of different genes. In 1970, Sager, R. and Z. Ramanis (PNAS 65: 593) UV-irradiated cells of the ac_2^+ ac_1^- mt^+ parent and crossed them with cells of the ac_2^- ac_1^+ mt^- strain. Each of the meiotic products of each zygote (meiocyte) was allowed to go through mitosis as was each of the 8 resulting daughter cells to form 16 spores. The colonies arising from each of the spores were classified phenotypically and genotypically. The results were as follows:

Number of colonies	Genotypes of colonies			
	$ac_2^+ac_1^-$ / $ac_2^+ac_1^-$	$ac_2^-ac_1^+$ / $ac_2^-ac_1^+$	$ac_2^+ac_1^+$ / $ac_2^+ac_1^+$	$ac_2^+ac_1^-$ / $ac_2^-ac_1^+$
559	16.5%	17.5%	2.2%	64.0%

a) Do these data provide evidence for the segregation of alleles of cytoplasmic genes? Explain. Do they indicate that alleles of different cytoplasmic genes recombine? Explain.
b) Do these 2 phenomena occur during meiosis or is their occurrence pre- or post-meiotic?
c) What is the linkage relationship of the 2 genes for acetate-requirement? Explain.

20. The mutant *grandchildless* in *Drosophila subobscura* is caused by a recessive allele *gs* of an autosomal gene (Spurway, H.S., J. Genet. 49: 126, 1948). Although *gsgs* males are viable and fertile and have grandchildren, *gsgs* females, regardless of their fertile male mate, produce only sterile offspring: females with rudimentary ovaries and males which have no testes. Offer an explanation to account for the delayed phenotypic effect of *gs* which occurs only in females.

21. The freshwater snail *Limnaea peregra* is a hermaphroditic species, capable of cross- or self-fertilization. The species is polymorphic, some individuals having right-handed (*dextral*) coiling of the shell and others having left-handed (*sinistral*) coiling. Boycott, A.E. and C. Diver (Proc. R. Soc. (Lond.) B95: 207, 1923) noted that all the progeny from single individuals by self-fertilization are of the same phenotype (viz. *dextral* or *sinistral*). In 1930 Boycott, et al. (Philos. Trans. R. Soc. Lond. B219: 51) in crosses between true-breeding *dextral* and true-breeding *sinistral* lines found that the direction of coiling of the F_1 progeny was the same as that of their mother; the F_2 snails were all *dextral*, and in the F_3 1,192 *dextral* and 401 *sinistral* broods were obtained, each brood being the result of self-fertilization of an F_2 snail. Offer an explanation to account for these results and see whether it agrees with the ingenious hypothesis of Sturtevant, A.H. (Science 58: 269, 1923).

22. In the meal moth *Ephestia kuhniella* the dominant allele *A* is necessary for the conversion of tryptophan to the diffusible hormonelike substance kynurenine, a precursor of brown pigment. As a consequence, *A*_ larvae have a *pigmented* skin and as adults have *dark-brown* eyes; *aa* larvae are *non-pigmented*, and as adults have *red* eyes (no brown pigment). Reciprocal crosses between true-breeding *pigmented*, *AA* and *non-pigmented aa*, moths produce

pigmented F_1's. Reciprocal crosses between *Aa* and *aa* animals give the following results (Kuhn, A., Naturwiss 24: 1., 1936):

Cross	Result
Aa ♀ × *aa* ♂	All larvae are *pigmented* 1/2 the adults have *red* eyes 1/2 have *brown* eyes
aa ♀ × *Aa* ♂	Half of the larvae are *pigmented* and, as adults, have *brown* eyes

a) Can these results be explained on the basis of sex-linked inheritance? Explain with the aid of appropriate diagrams.
b) Why do reciprocal crosses give different results? What is this phenomenon called?
c) A male and female, both *pigmented* as larvae, are crossed. About half their adult progeny have *dark-brown* eyes and half have *red* eyes. What eye colour did the male and female parents have as adults?

23. Hagberg, A. (Hereditas 36: 228, 1950) reported on the association between genotype and phenotype for *bitter* vs. *sweet* in the lupine, *Lupinus augustifolius*. Reciprocal crosses were made between the *sweet* variety Borre and the *bitter* variety Blue.
a) All F_1 plants from *sweet* female × *bitter* male were *sweet* when tested within 5 to 6 weeks after the seed was sown. F_1 plants from the reciprocal cross tested within the same period were *bitter*. What forms of inheritance may be responsible for these F_1 results?
b) The leaves of F_1's after 5 to 6 weeks' growth and the ripe seeds of the F_1's were *bitter* regardless of which variety was used as the female parent. Which of the forms of inheritance suggested in (a) are eliminated by this information? Explain.
c) Forty F_2 seeds, all *bitter*, from *sweet* female × *bitter* male gave rise to plants whose leaves tested *bitter* until approximately a month after seeding. The mature F_2 plants segregated 28 *bitter* : 12 *sweet* for both their leaves and ripe seed (either *bitter* for both or *sweet* for both).
(i) At what stages in the ontogeny of lupines are the traits *bitter* and *sweet* affected by the genotype of the parental generation? By their own genotype?
(ii) Diagram the reciprocal crosses to illustrate your answer to (i).

24. In the Mexican axolotl (*Ambystoma mexicanum*) *fluid imbalance* is an embryonic condition in which localized deposits of jellylike fluid are found. In 1948 Humphrey, R.R. (J. Hered. 39: 255) showed that *fluid imbalance* is caused by an autosomal recessive allele, *f*. In 1960 he reported detailed studies of the time and mode of action of this gene, as summarized in the table. The ovaries of *ff* donors transplanted to *FF* or *Ff* hosts behave autonomously.

Mother	Offspring Genotype	Phenotype
Ff	*FF, Ff, ff*	Develop normally; exhibit *fluid imbalance* in its usual form; excess fluid disappears after circulation is established; most hatch and survive to adult stage.
ff	*Ff*	Fluid deposits accumulate during cleavage, eventually escapes; about 85% of individuals survive.
	ff	Fluid deposits accumulated during cleavage do not escape; circulation is not established; all die.

Propose an explanation of these results.

25. a) In *Drosophila melanogaster*, females from a true-breeding *normal*-winged strain crossed with *fused*-wing males produce F_1's with *normal* wings. The F_1's when intercrossed produced 3 *normal* : 1 *fused*; all the *fused* offspring were males. When the reciprocal cross was made, only females

were produced in the F_1, the adult progeny being about half as numerous as in the first cross. Analyze these results and state with reasons (i) the number of allele pairs involved and (ii) whether they are located on an autosome or the X chromosome.

b) Lynch, C.J. (Genetics 4: 501, 1919) made other crosses involving the allele *fused*. The results of these are as follows:

Normal female (from *normal* ♀ x *fused* ♂ or reciprocal cross) x *fused* ♂ produced 1 *fused* ♂ : 1 *fused* ♀ : 1 *normal* ♂ : 1 *normal* ♀

Fused ♀ x *fused* ♂ produced no adult progeny

Fused ♀ x *normal* ♂ produced *normal* ♀, no ♂

Propose an explanation to account for the action of the *fused* and *normal* alleles on viability.

26. a) Mitochondria and chloroplasts contain DNA (Granick, S. and A. Gibor, Prog. Nucleic Acid Res. Mol. Biol. 6: 143, 1967; Luck, D.J.L. and E. Reich, PNAS 52: 931, 1964). Is this proof that these DNAs carry genes responsible for the expression of cytoplasmically inherited traits? Explain.

b) In 1962 Gibor, A. and S. Granick (J. Cell Biol. 15: 599) irradiated the cytoplasm of *Euglena* with ultraviolet light at wavelengths close to 2600 Å while shielding the nucleus and *vice versa*. Irradiation of the nucleus had no effect on the chloroplasts, but irradiation of the cytoplasm caused a hereditary bleached condition, in which bleached plastids remained tiny and colourless and reproduced as such from cell generation to cell generation. Do these results indicate that DNA of plastids is capable of replication? Explain.

27. Densitometer measurements of whole cells and isolated chloroplasts may be used to identify different classes of DNA according to their molecular weights. In the following figure curve (a) represents DNA from whole leaves. A major band α is present. M represents marker DNA from *Micrococcus lysodeikticus*. Curve (b) represents DNA isolated from presumably purified chloroplasts. In addition to the α band, two heavier minor (satellite) bands (β and γ) are present. In curve (c), the β band, has been further purified by repeated centrifugation in cesium chloride.

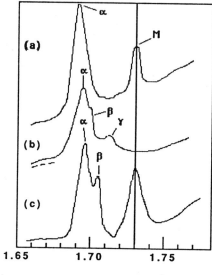

Densitometer tracings of DNA from beet leaves. [Redrawn from E.H. L. Chun, et al., J.Mol Biol., 7:133 (Fig.2)(1963).]

Similar studies with *Euglena* show only one band, that corresponding to α, to be present in mutant strains lacking chloroplasts (Leff, J. et al., Biochem. Biophys. Res. Commun. 13: 126, 1963) while both this and the satellite band β are present in strains carrying chloroplasts (Edelman, M. et al., PNAS 52: 1214, 1964). The α and β-band DNAs differ in base ratios as shown in the table (Edelman, M. et al., PNAS 52: 1214, 1964).

	A	T	G	C
α	0.52	0.45	0.48	0.55
β	0.74	0.70	0.26	0.36

Do chloroplasts contain their own DNA? What is the evidence? (See Gibor, A. and S. Granick, Science 145: 890, 1964).

28. Studies by Wallace, D.C. et al. (Science 242: 1428, 1988) have revealed that *Leber's hereditary optic neuropathy* (*LHON*) a form of central optic nerve death associated with acute bilateral blindness in young adults is due to a mutant allele in the *ND4* gene that codes for NADH dehydrogenase subunit 4. Furthermore, the degree of expression of the disease varies among *affected* individuals, even those of the same sibship. The following pedigree is based on studies and pedigrees presented by Seedorff, T. (Acta Opthal. 47: 23, 1969), Wallace, D.C. et al. and others and is typical of its mode of inheritance:

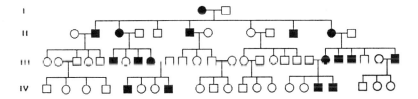

a) Is the *ND4* gene located: (i) in the chloroplast genome? (ii) on a chromosome in the nuclear genome? (iii) the mitochondrial genome? Explain your decision.
b) Offer an explanation to account for the variable expressivity of the disease.

Huoponen, K. et al. (Am. J. Hum. Genet. 48: 1147, 1991) found that a mutant allele of gene *ND1*, which codes for subunit 1 of the 7 subunits of NADH dehydrogenase, also results in *LHON*.
c) If an *affected* female mutant for *ND4* married an *affected* male mutant for *ND1* would you expect their children to be *affected*? *Unaffected*? Explain.

29. In the fungus *Neurospora crassa* the heteroplasmon (fused cytoplasm from both parents) between the *poky* and *mi-4* mutants has the growth rate of *wild-type*, unlike the heteroplasmon between *poky* and *mi-3*, which has a *mutant* phenotype (Pittenger, T.H., PNAS 42: 747, 1956).
a) What do these observations tell us about the probable relationships between the *mutant* plasmagenes?
b) Suggest experiments for clarifying this relationship.

Chapter 12
Genetics of Bacteria and Viruses

Genetics of Bacteria

1. In a U-tube experiment involving two auxotrophic strains of *Escherichia coli* some of the offspring of the recipient strain showed certain donor traits.
   a) Which mechanism(s) of gene transfer can be eliminated as a cause of the hereditary changes in the recipient strain?
   b) How would you proceed to determine the mechanism(s) causing the change?
   c) What results would you expect with each mechanism?

2. To detect bacterial recombinants in conjugation crosses, cells of the donor strain must be prevented from growing on selective media. This counterselection is necessary so that only recombinants are selected. For example, in the cross Hfr($met^+thr^+str^s$) x F⁻($met^-thr^-str^r$) the donor carries the marker str^s (*streptomycin sensitive*). By growing the conjugants on a streptomycin medium all the donors are eliminated. Should the gene controlling the counterselection trait, e.g., streptomycin sensitivity, be located near the O (origin) or the Hfr end of the chromosome? Explain.

3. A *proline-requiring* Salmonella mutant is infected with phage from another *proline-requiring* mutant. Minute colonies never arise, but large colonies are formed with very low frequency. Account for these observations.

4. The chromosome of *E. coli* is circular, yet during conjugation it opens up and is transferred to the female as a linear structure with part of the sex factor at its trailing end. Explain how this change from a circular to a linear structure apparently occurs.

5. Crosses in *E. coli* give the following results (Lederberg, J. et al., Genetics 37: 720, 1952; Cavalli, L. et al., J. Gen. Microbiol. 8: 89, 1953; Hayes, W., CSHSQB 18: 75, 1953):
   (i) F⁺ x F⁺ : fertile (recombinants produced).
   (ii) F⁻ x F⁻ : sterile (no recombinants produced).
   (iii) F⁺ x F⁻ : fertile (recombinants produced).
   (iv) One F⁺ cell x many F⁻ cells : all F⁻'s rapidly converted to F⁺'s that transmit F⁺ (by fission) to all their progeny.
   (v) Hfr x F⁻ : fertile, produce 100 to 20,000 times as many recombinants as F⁺ x F⁻ crosses. The recombinants remain F⁻ with rare exceptions.
   (vi) Hfr can arise from F⁺ and revert to F⁺. Crosses between these reverted F⁺ and F⁻ are fertile and rapidly convert F⁻ to F⁺.

   The ability of Hfr cells to produce high frequencies of recombinants is unaffected by pretreatment with streptomycin. What properties of the F factor do these data reveal?

6. a) Outline the experimental evidence which indicates that F:
   (i) Can integrate at many different places on the *E. coli* chromosome and can also deintegrate.
   (ii) Is composed of genetic material.
   (iii) Is carried extrachromosomally in F⁺ strains.

   b) Like the *E. coli* chromosome the F factor is DNA and appears to be circular. Illustrate how F can integrate into the chromosome and deintegrate.

   c) What properties does F possess when it is integrated in a bacterial chromosome?

7. a) The F factor upon deintegration can occasionally carry a chromosome segment from one *E. coli* strain to another, where it may be integrated into the recipient chromosome (sexduction or F-duction). Beginning with an Hfr gal^+ strain and an F⁻ gal^- one, diagram the process of sexduction.
   b) How is sexduction similar to and different from transduction?
   c) Why are *E. coli* harbouring F' factors called intermediate males?

d) F' carries a portion of the bacterial genome. In what ways do you expect its breeding behaviour to differ from that of F, which carries no bacterial chromosome DNA?

8. a) What features distinguish the bacterial chromosome from the chromosomes of eukaryotes?
b) Explain how sex is determined in bacteria.
c) Outline the events that occur from the time a portion of a donor chromosome enters an F⁻ recipient to the appearance of a recombinant haploid segregant carrying it.
d) What properties do proviruses confer on lysogenic bacteria?
e) Both the sex factor F and the genetic material of temperate phages are episomes. Discuss.
f) Why does simultaneous transduction of two loci indicate their close linkage?

9. The *wild-type* strain K12 of *E. coli* is a prototroph. By treating this strain with X-rays and ultraviolet light Lederberg, J. and E.L. Tatum (Nature 158: 558, 1946; CSHSQB 11: 113, 1946) obtained single, double, and triple auxotrophic mutants which grow on minimal medium only if they are supplemented with appropriate growth factors. In pure cultures of the different mutant strains prototrophs were sometimes found. These proved to be back mutations for a single growth-factor allele only e.g., *threonine-requiring* (thr^-) to *threonine-independent* (thr^+). The two triple mutant strains, Y10, requiring threonine, leucine, and thiamine, $thr^-leu^-thi^-bio^+phe^+cys^+$, and Y24, requiring biotin, phenylalanine, and cysteine, $thr^+leu^+thi^+bio^-phe^-cys^-$, were grown in pure and mixed cultures on appropriate selective media. The only new types found in pure cultures of each of the triple mutants were those which reverted to prototrophy for a single factor. In mixed cultures a variety of types were detected: many *wild-types* (complete prototrophs) and some single- and double-mutant strains (requiring one and two nutrients, respectively). The prototrophs bred true on minimal medium.
a) What are the possible explanations for the occurrence of the prototrophs in the mixture of the two triple mutant strains? Which is the most probable explanation and why?
b) Outline tests that would rule out all but the most probable explanation.

10. Two triple auxotrophic strains of *E. coli*, $thr^-leu^-his^-pro^+pan^+bio^+$ and $thr^+leu^+his^+pro^-pan^-bio^-$, are allowed to conjugate in a liquid medium for 30 minutes. After dilution of the broth, the bacteria are plated onto a complete agar medium in a petri dish which serves as a master plate from which six replicas are made onto plates containing minimal medium and an additional nutrient(s) as indicated below.

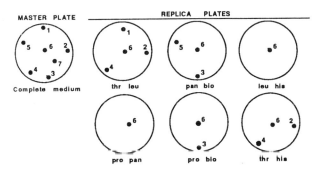

a) Determine the genotype of each of the first six clones. Explain how you arrive at the genotype of each.
b) What is the most likely genotype of clone 7? Why? How would you determine and verify this?

11. a) If the DNA in a bacterial nucleus is approximately 1,200 µm long and the entire chromosome is transferred from Hfr males in 90 minutes, how much of this molecule is transferred to a female, F⁻, strain after 35 minutes of conjugation?
b) The *E. coli* chromosome is 10^7 base pairs long. The DNA is continuous along the chromosome; if 1 time unit (1 minute) is equal to 20 recombination (map) units and the chromosome is 90 time units long, how many base pairs would there be per recombinational unit?

12. In crosses between F⁺ and F⁻ strains the majority of the unselected markers among recombinants are derived from the F⁻ parent. To explain this Hayes, W. (CSHSQB 18: 75, 1953) suggested that the male usually transfers only a portion of its genetic material to the female to form a merozygote, whereas Lederberg, J. (PNAS 35: 178, 1949) and Nelson, T.C. and J. Lederberg (PNAS 40: 415, 1954) suggested that although the entire genome of the male was

always transferred, only a randomly selected portion was integrated into the recipient chromosome. To determine the true nature of gene transfer, Wollman, E.L. and F. Jacob (C. R. Acad. Sci. 240: 2449, 1955) in one set of experiments mated an Hfr H strain carrying $thr^+leu^+azi^sT1^Slac^+gal^+\lambda$ with an F⁻ strain carrying the alternative markers. Conjugation was interrupted at different times by subjecting samples of the mixture to a shearing force in a blender, and the kinds of recombinants were as shown below:

Time after mixing, minutes	Recombinants with Hfr markers
0.0	0
8.0	thr^+
8.5	thr^+leu^+
9.0	$thr^+leu^+azi^S$
11.0	$thr^+leu^+azi^ST1^S$
18.0	$thr^+leu^+azi^ST1^Slac^+$
25.0	$thr^+leu^+azi^ST1^Slac^+gal^+$
26.0	$thr^+leu^+azi^ST1^Slac^+gal^+\lambda$

a) Which hypothesis do these results support and why?
b) What other conclusions can you draw, e.g., with respect to the probable number of linkage groups?

13. a) Although by 1951 it had been shown that *E. coli* can reproduce sexually and that its genes are arranged on linear structures analogous to the chromosomes of higher organisms, there was no answer to the question: Are the members of a pair of conjugating bacteria sexually equivalent; i.e., can either conjugant act as donor or recipient of DNA? The following results obtained by Hayes, W. (Nature 169: 118, 1952) will enable you to answer this question: Mixed cultures of the two *streptomycin-sensitive* auxotrophs, 58-161, met^-, and W677, $thr^-leu^-thi^-$, produced prototrophs, $met^+thr^+leu^+thi^+$, under normal conditions. When both strains were pretreated with streptomycin (initially prevents fission and later kills cells) before mixing, no recombinants were formed. When only W677 was pretreated with streptomycin, no recombinants were formed. When 58-161 was pretreated, prototrophs occurred. Provide an explanation for the above question.
b) Pedigree studies of cells isolated following conjugation reveal that the vegetative progeny of one of the exconjugants never show recombination of genes that marked the conjugants. Cells derived by fission of the other exconjugant show recombinant types for marker genes of the original conjugants (Anderson, T.F. CSHSQB 23: 47, 1958). Do these results confirm the conclusions drawn in (a)? Explain.

14. The *E. coli* data shown are modified from Taylor, A.L. and E.A. Adelberg (Genetics 45: 1233, 1960). Each of three different Hfr strains (A, B, and C) which arose independently and possess the same markers, his^+, gal^+, pro^+, met^+, mtl^+, xyl^+, mal^+, ser^+, tyr^+, and str^S, were separately mixed with the same F⁻ strain carrying the alternative markers. Conjugation was interrupted at various times after mixing over a period of 95 minutes. The bacteria were then plated on appropriate media and recipients scored for various Hfr markers. The times (in minutes after mixing) at which recombinants carrying particular donor markers first appear, and therefore the times at which these first enter the F⁻ cell and then integrate into the recipient genome, are given below:

Donor markers	Time of appearance, min.		
	Strain A	Strain B	Strain C
his^+	6.5	68.5	29.5
gal^+	30.0	46.0	52.0
pro^+	34.5	41.5	56.5
met^+	58.0	20.0	76.0
mtl^+	67.0	11.0	85.0
xyl^+	68.5	9.5	86.5
mal^+	73.0	5.0	3.0
ser^+	81.5	86.5	11.5
tyr^+	88.0	78.5	19.5

What do these results suggest regarding:

a) The type of linkage group (rodlike, branched, circular) in *E. coli*?
b) The location of F in the different strains?

15. In a conjugation-interruption experiment in which samples were removed after mixing at 2- to 5-minute intervals Taylor, A.L. and M.S. Thoman (Genetics 50: 659, 1964) obtained the following results in crosses between Hfr strain AB259 carrying all *wild-type* markers and six different recipient strains carrying the alternate markers:

recipient strain	Time in minutes of appearance of selected markers of the donor parent									
	thr^+	$pyrA^+$	leu^+	$proA^+$	$proB^+$	$purE^+$	lys^+	gal^+	$aroA$	$purB^+$ met^+
AT2213	7.25	7.5	7.75							
AT2217			8.0	16.0						
AT2270			15.0		19.5			24.0		
AT2036			14.5			21.5	23.5			
AB1321			15.0					23.5	29.0	
AB1325			15.0					24.0		32.0

a) Map the chromosomal segment transferred by the Hfr strain AB259.
b) Would you expect any of these markers to be cotransducible? Why?
c) None of the recombinants were Hfr. What does this indicate regarding the location of the Hfr locus?

16. Certain mutants, uv^r, of *E. coli* are more *resistant* to ultraviolet irradiation than the parent strains, uv^s. To determine the genetic basis for *resistance* vs. *sensitivity* to ultraviolet light, Greenberg, J. (Genetics 49: 771, 1964) used the strains whose characteristics are shown in the following Table.

Strain	Sex	met	thi	lac	ara	mal	xyl	gal	T6	T1	val	str	uv
K12W1895	Hfr1	-	+	+	+	+	+	+	s	s	s	s	r
K12/W4531	Hfr2	+	-	+	+	+	+	+	s	s	s	s	r
BPam5	F	+	+	-	-	+	+	+	r	r	r	s	s
BPAM7	F	+	+	-	-	-	-	-	r	r	r	r	s

Hfr1 Order of transmission of markers is T6lacT1ara...
Hfr2 order of transmission of markers is araT1lacT6...

The frequency with which *radio-resistance* appeared as an unselected marker trait in crosses between W1895 and BPAM7 and in crosses between W4531 and various derivatives of strain B when progeny were selected for various other markers is summarized in the following Table:

Frequency of occurrence of donor markers from the cross W1895 X BPAM7

Recomb. selected	Number examined	uv^r	$T6^S$	lac^+	$T1^S$	ara^+	mal^+	met^-	val	xyl^+	str^S
$lac^+ str^r$	100	86	87	100	72	63	22	8	4	3	-
$ara^+ str^r$	100	50	50	57	65	100	15	12	6	1	-
$mal^+ str^r$	100	57	51	62	61	76	100	62	46	29	-
$xyl^+ str^r$	100	26	27	27	20	25	23	38	53	100	-

Frequency of occurrence of donor markers among recombinants between W4531 and BPAM5

Recombinants selected	ara^+	$T1^S$	lac^+	$T6^S$	uv^r	gal^+	xyl^+	mal^+
$ara^+ thi^+$	100	75	60	53	53	-	-	0
$lac^+ thi^+$	69	62	100	78	73	-	-	-

a) How many genes determine *resistance* vs. *sensitivity* to ultraviolet irradiation, and where are they located relative to the other markers?
b) Is the point of origin the same in the two donor strains? Explain.

17. An *E. coli* strain unable to synthesize *leucine*, leu^-, is transduced by a strain unable to synthesize *threonine*, thr^-. An aliquot of the diluted culture medium is plated on minimal medium, and another on minimal supplemented with threonine; 25 colonies grew on the minimal plates and 125 on the supplemented plates. Calculate the amount of recombination between *leu* and *thr*.

18. The sequence of entry of donor genes as determined by conjugation-interruption experiments for the Hfr strain H and the mutant strains HCR1 and HCR2, derived from it by treatment with nitrogen mustard, were as shown below (Jacob, F. and E.L. Wollman, *Sexuality and the Genetics of Bacteria*, Academic Press, New York, 1961). What were the

effects of the mutagenic treatment in each of the mutant strains?

Strain	Sequence
H	O thr leu azi T1 pro lac ad gal H S-G Sm mal xyl mtl ile met arg thi
HCR1	O thr leu azi T1 pro xyl mtl ile met arg ad gal H S-G Sm mal thi
HCR2	O lac ad gal H S-G thr leu azi T1 pro Sm mal xyl mtl ile met arg thi

19. When F is transferred by conjugation from *E. coli* to a different genus of bacteria, e.g., *Serratia* or *Proteus*, a DNA band appears in the cesium chloride density gradient that is not found when the host DNA alone is centrifuged (Marmur, J.S. et al., PNAS 47: 972, 1961; Falkow, S. et al., J. Bacteriol. 87: 209, 1964).
a) What do these results indicate regarding the chemical characteristics of F?
b) Why was it necessary to transfer F to a different genus to determine its chemical and physical properties by density-gradient centrifugation?

20. Jacob, F. and E.L. Wollman (*Sexuality and the Genetics of Bacteria*, Academic Press, New York, 1961) mixed F-P678 ($thr^-leu^-azi^rT1^rlac^-gal^-mal^-xyl^-man^-str^r$) with HfrH carrying the alternative alleles of these genes. After 60 minutes $thr^+leu^+str^r$ recombinants were selected and tested for the presence of other donor traits, with the results shown:

Types of recombinants selected	Frequency of Hfr traits, %							
	thr^+leu^+	azi^s	$T1^s$	lac^+	gal^+	mal^+	xyl^+	man
$thr^+leu^+str^r$	100	91	72	48	27	0	0	0
$gal^+ str^r$	83	78	79	81	100	0	0	0

a) With respect to $thr^+ leu^+$ are the unselected markers distal or proximal to O? Explain.
b) Which of the selected recombinants, $thr^+leu^+str^r$ or gal^+str^r, permit determination of gene sequence and why? What is the gene order?

21. When F⁻ *E. coli* are lysogenic for λ and Hfr are nonlysogenic, the lysogenic trait segregates among the recombinants and can be located accurately on the chromosome.
a) Assuming any location for λ you wish, show the results you would expect in a conjugation-interruption experiment between Hfr H (λ⁻) $thr^+leu^+azi^+lac^+gal^+$ and F⁻ (λ⁺) $thr^-leu^-azi^-lac^-gal^-$.
b) What results would you expect if Hfr was λ⁺ and F⁻, λ⁻? Why?

22. Zinder, N.D. and J. Lederberg (J. Bacteriol. 64: 679, 1952) performed a series of experiments with the mouse-typhoid bacterium (*Salmonella typhimurium*) to determine whether this organism undergoes conjugation. On the basis of their observations, presented below:
a) State which of the mechanisms of genetic transfer (conjugation, transduction, or transformation) can be eliminated and why.
b) Describe and illustrate the mechanism accounting for genetic transfer and explain all the facts stated.

Observations
1. Mixed cultures of met^-thr^+ and met^+thr^- strains in appropriate medium produced numerous prototrophs (met^+thr^+) that resembled the recipient parent (met^-thr^+) except for the trait met^+ from the donor parent met^+thr^-.
2. These prototrophs were much more frequent in mixed than in unmixed cultures.
3. In a U-tube experiment the strains were grown in opposite arms separated by a filter through which no bacteria could pass. After allowing the nutrient to pass from the donor end of the tube to the recipient end, some cells of the recipient strain on transfer to selective medium showed traits of the donor.
4. Treatment of the recipient culture with pure DNA from the donor strain produced no recombinants (prototrophs).
5. Adding deoxyribonuclease to the solution in the U-tube did not prevent the appearance of cells with donor traits in the recipient strain.
6. Broth from a culture of donor cells added to a culture of recipient cells resulted in no heritable changes.

7. Filtered broth from a culture containing both strains produced hereditary changes in fresh cultures of recipient cells.
8. After a culture of donor cells was exposed to fluids from the recipient strain and the fluid from this donor culture was mixed with other donor strains, it was found that all donors could cause hereditary changes in the recipient.
9. Viruses were found in donor cultures treated with fluids from the recipient strain.

23. a) Distinguish between complete and abortive transduction.
b) In *Salmonella*, *flagellated* strains can swim and spread to produce a cloudy (spreading) swarm throughout the culture medium; *nonflagellated* strains cannot move and form new colonies. Stocker, B.A.D. (J. Gen. Microbiol. 15: 575, 1953) studied transduction of a certain strain from *nonmotility* to *motility*. He found that after transduction some of the transduced recipient cells produced spreading swarms whereas others produced trails, e.g., linear trail of colonies, each colony consisting entirely of *nonmotile* cells. Assume *motile* are F and *nonmotile* ones f.
c) In *Salmonella*, *motility* vs. *nonmotility* depends on two genes, F and P. The alleles Ff determine *presence* vs. *absence* of flagella, whereas Pp determine the kind of protein b vs. i, of which the tail is composed. *Motile* (flagellated) strains swim and spread, producing a cloudy swarm throughout the culture medium. *Nonmotile* (nonflagellated) strains cannot move and hence form restricted colonies rather than swarms. When Stocker, B.A.D. et al. (J. Gen. Microbiol. 9: 410, 1953) treated the *nonmotile* strains SW543 and SL13, both of which carry protein b, with lysates of *motile* strain TM2 carrying protein i, (1) a number of swarms were observed; (2) most of these swarms had the protein b characteristic of the recipient, and only a few possessed the flagella protein i.
a) State which of these types of swarms represents linked transduction and why.
b) What kind of gene interaction is this an example of? Explain.

24. When each of three *ath* (adenine-thymine-requiring) mutants of *Salmonella typhimurium* were infected with phages from each of the other two mutants and the *wild-type* strain, the results shown were obtained (based on data of Ozeki, M., Carnegie Inst. Wash., D.C., Publ. 612, p. 97, 1956).

Recipient	Colonies	Wild-type	ath-1	ath-4	ath-6
ath-1	large	many	none	very few	many
	minute	+	-	-	+
ath-4	large	many	very few	none	many
	minute	+	-	-	+
ath-6	large	many	many	many	none
	minute	+	+	+	-

Key: + = minute colonies formed; - = no minute colonies

a) Explain how the *large* colonies originate and why there are many in some transductions and very few or none in others.
b) Why do *minute* colonies arise, and what is the significance of their occurrence?
c) What is the minimum number of genes involved in adenine-thymine formation?

25. In *Salmonella typhimurium* four genes determine tryptophan synthesis. The sequence of biochemical steps in tryptophan synthesis and the steps mediated by each of the four genes are as follows:

$$\cdots \xrightarrow{A} \text{anthranilic acid} \xrightarrow{B} \text{indoleglycerol phosphate} \xrightarrow{C} \text{indole} \xrightarrow{D} \text{tryptophan}$$

Three-gene reciprocal crosses were made by Demerec, M. and Z. Hartman (Carnegie Inst. Wash., D.C., Publ. 612, p. 5, 1956) by transduction. Each parent (strain) was alternatively used as a donor and recipient, by growing the transducing phage on each mutant strain (donor) and then using it to transuce the other strain (recipient). The results of the crosses are tabulated. Assume that the *cysB* locus is to one side of the *trp* loci.

Reciprocal Experiment	Genotype of recipient strain	Genotype of donor strain	Number of *wild-type* colonies on minimal medium
1	trpD⁻cysB⁻trpB⁺	trpD⁺cysB⁺trpB⁻	15
	trpD⁺cysB⁺trpB⁻	trpD⁻cysB⁻trpB⁺	292
2	trpD⁻cysB⁻trpA⁺	trpD⁺cysB⁺trpA⁻	16
	trpD⁺cysB⁺trpA⁻	trpD⁻cysB⁻trpA⁺	456
3	trpA⁻cysB⁻trpB⁺	trpA⁺cysB⁺trpB⁻	59
	trpA⁺cysB⁺trpB⁻	trpA⁻cysB⁻trpB⁺	43
4	trpD⁻cysB⁻trpC⁺	trpD⁺cysB⁺trpC⁻	26
	trpD⁺cysB⁺trpC⁻	trpD⁻cysB⁻trpC⁺	183
5	trpA⁻cysB⁻trpC⁺	trpA⁺cysB⁺trpC⁻	260
	trpA⁺cysB⁺trpC⁻	trpA⁻cysB⁻trpC⁺	286
6	trpB⁻cysB⁻trpC⁺	trpB⁺cysB⁺trpC⁻	24
	trpB⁺cysB⁺trpC⁻	trpB⁻cysB⁻trpC⁺	40

a) Are the tryptophan loci linked to each other? To the *cysB* locus? What is the sequence of the *trp* loci on the genetic map with respect to each other and the *cysB* locus?

b) Compare the sequence of the tryptophan genes with the steps they control in the tryptophan-biosynthesis pathway. What might your results suggest?

26. In *E. coli*, z^- mutants are unable to synthesize β-galactosidase and therefore cannot ferment lactose. About 30 minutes after Hfr, $z2^-/str^s/pan^+$ bacteria are added to a culture medium containing F⁻, $z1^-/str^r/pan^-$ cells, the medium is diluted and plated on minimal medium containing streptomycin; 40 percent of the pan^+ colonies are able to ferment lactose, indicating they are z^+. Only 4 percent of pan^+ colonies from a reciprocal cross are z^+. What is the sequence of the mutant sites $z1$ and $z2$ relative to the *pan* locus? Show how you arrive at your answer.

27. Transduction crosses among seven proline-requiring mutants of *Salmonella typhimurium* give the results shown (after Miyake, T. and M. Demerec, Genetics 45: 755, 1960).

Recipient Strain	Donor strain						
	1	2	3	4	5	6	7
1	− 0	− 38	+ 521	+ 105	+ 1,920	+ 2,682	+ 12,342
2		− 0	+ 269	+ 194	+ 2,746	+ 2,898	+ 14,872
3			− 0	− 9	+ 1,369	+ 2,040	+ 8,236
4				− 0	+ 2,128	+ 2,440	+ 8,810
5					− 0	− 17	+ 5,388
6						− 0	+ 2,722
7							− 0

Key: Plus and minus signs indicate presence and absence of abortive transductants. Figures refer to the number of complete transductants.

a) How many functional units (genes) are represented by these seven mutants? Explain.

b) Is it possible to deduce the sequence of these mutants and the approximate distances between them? Explain.

28. *Highly resistant*, p^r, strains of *Diplococcus* can grow in the presence of 0.3 unit/ml of penicillin; *sensitives*, p^s, can grow only if the concentration is reduced to 0.01 unit/ml or less. Hotchkiss, R.D. (CSHSQB 16: 457, 1951) showed that *high-level resistance* can be acquired by a p^s strain in a series of three successive transformation steps. DNA from a p^r strain can transform p^s recipients only to *low resistance* (grow in 0.05 units/ml). These transformants could then be transformed with the p^r DNA to an *intermediate level* of resistance (tolerating 0.12 unit/ml). *High level* of resistance (tolerating 0.24 unit/ml) was achieved by treating the *intermediate-level* transformants with the p^r

DNA. Offer an explanation to account for these observations.

29. a) The six *histidine-requiring, his⁻*, mutants (41, 55, 134, 135, 150, and 712) in *Salmonella typhimurium* have been crossed in all possible pairwise combinations. The progeny, when screened for the *presence* or *absence* of *wild-types*, gave the tabulated results (based on data of Hartman, P.E. et al., J. Gen. Microbiol. 22: 323, 1960).

	41	55	134	135	150	712
41	-	-	+	-	-	-
55		-	+	+	-	-
134			-	+	-	-
135				-	+	-
150					-	-
712						-

Key: + = wild-types produced; - = no wild types produced

Draw a genetic map of these mutations.

b) Six *histidine-requiring* point mutations (1 to 6) were crossed in all possible pairwise combinations by transduction. Abortive transductants occurred in all crosses except 2 X 3 and 4 X 5. How many genes are involved? Why?

c) Only some of the crosses between the six point mutations and the six mutations studied in (a) produced *wild-type* recombinants, as shown in the following table:

	41	55	134	135	150	712
1	+	+	+	-	+	-
2	-	+	+	-	+	-
3	-	+	+	+	+	-
4	+	+	+	+	-	-
5	+	+	-	+	+	-
6	+	+	+	+	-	-

Key: + = wild-type recombinants; - = no wild type produced

Draw a map of the point mutations and draw lines below the map indicating the extent and end points of the six other mutations in (a).

d) Which of the following pairwise crosses by transduction would you expect to produce abortive transductants?
(1) 135 x 41   (2) 135 x 55   (3) 41 x 134   (4) 150 x 134

30. a) Bodmer, W.F. and A.T. Ganesan (Genetics 50: 717, 1964), and Lacks (J. Mol. Biol. 5: 119, 1962) found that only one strand of the donor transforming DNA segments replaces the corresponding segment of the recipient, by hybridizing (annealing) with the recipient's complementary strand. The product of integration, a physical "hybrid" duplex (one strand donor DNA and one strand host DNA for a certain segment of the genome) carries transforming activity. Two possibilities might be advanced regarding the nature of the hybrid duplex segment.

1. It remains genetically heterozygous, i.e., the two single strands remain unchanged, and either contains one particular donor strand or one or the other.
2. It becomes genetically homozygous; e.g., the recipient strand is lost in the following DNA synthesis or is subjected to a repair process rendering it complementary to the transforming strand. Such hybrids might occur as a consequence of integration of a unique strand or either strand of donor DNA fragments.

Guerrini, F. (reported by Fox, M.S., J. Gen. Physiol. 490: 183, 1966) treated d^+ (*sulfanilamide-sensitive, p-nitrobenzoic acid-utilizing*) diplococci with DNA from d^- (*sulfanilamide-utilizing, --nitrobenzoic acid-sensitive*) donors. Each transformed bacterium (d^-) produced a mixed clone containing d^+ and d^- bacteria. Which of the models of hybrid DNA structure are excluded by these observations and why?

b) Integration of a portion of a synapsed (donor) segment can theoretically occur in either of two ways:

1. By *breakage and reunion*, allowing exchange of genetic material between homologous segments of donor and recipient chromosomes.
2. By *copy choice*, where a daughter chromosome is formed by the alternate use of the host chromosome and the donor DNA as template.

Studies in the 1960's having a bearing on these alternatives in bacteria, using transforming DNA labelled with the isotope ^{32}P, have revealed that:

1. Incorporation is achieved in the absence of DNA synthesis (Fox, M.S., PNAS 48: 1043, 1962).
2. When the DNA is extracted from recipient bacteria immediately after incorporation, transforming ability shows a gradual decline. The curve of decline is of the same shape as that generally found for DNA inactivation by ^{32}P decay (Fox, 1962).

Which hypothesis do these results favour, and why?

31. Ravin, A. (J. Bacteriol. 77: 296, 1959) showed that *noncapsulated Diplococcus* which have arisen from a *capsulated* type by mutation can reacquire the *capsulated* state by transformation with DNA from the *capsulated* strain. Moreover the original *capsulated* as well as the *capsulated* transformants can in turn be transformed to *noncapsulated* types: all these types occur as true-breeding strains.

    a) Does the immigrant genetic material replace its homologous segment in the recipient, or is it added to the recipient genome? Explain.

    b) What do these results tell you about the ploidy of *Diplococcus*?

32. The pathway of tryptophan synthesis in *E. coli* is:

    CA ⟶ AA ⟶ PRA ⟶ CdRP ⟶ InGP ⟶ T

    PRA = N-(5'-phosphoribosyl) anthranilate
    CdRP = 1-(O-carboxyphenylamino)-1-deoxyribulose-5 phosphate
    InGP = indole-3-glycerol phosphate
    CA = chorismic acid
    AA = anthranilic acid
    T = tryptophan

    a) Anagnostopoulos, C. and I.P. Crawford (PNAS 47: 378, 1961) defined four types of *tryptophan-requiring* mutants in *Bacillus subtilis*. One mutant, *trp*⁻, responded only to tryptophan; two, *ind*⁻, to either indole or tryptophan; and one, *ant*⁻, to anthranilic acid, indole, or tryptophan. One of the *ind*⁻ mutants, *ind$_a$*⁻, accumulated CdRP and the other, *ind$_b$*⁻, anthranilic acid. Is the pathway of tryptophan synthesis the same as in *E. coli*? Explain.

    b) By means of transformation in *B. subtilis*, Anagnostopoulos and Crawford made a large number of crosses between these single-mutant strains and one unable to synthesize histidine, *his*⁻. Each strain was *mutant* at one locus and *wild type* at the other. The results of some of these two-factor crosses are shown in the table.

Donor	Recipient	%* of *wild-type* transformants
ind$_a$⁻	*trp*⁻	12.7
ind$_b$⁻	*trp*⁻	11.8
ant⁻	*trp*⁻	43.5
ant⁻	*ind$_a$*⁻	27.3
his⁻	*ind$_a$*⁻	44.9
trp⁻	*his*⁻	19.0
ind$_a$⁻	*his*⁻	26.3
ant⁻	*his*⁻	45.0

    *Percentage *wild-types* = $\dfrac{\text{number of prototrophs}}{\text{total number of transformants}} \times 100$

    Determine whether these five genes are linked and, if linked, whether the sequence corresponds to the order of the biochemical steps. Show how your results might relate to their operon hypothesis.

33. Heating double-stranded DNA to about 100°C causes the strands to separate. If denatured DNA is cooled rapidly, the strands remain separate, but if cooled slowly, the complementary strands reunite to form the renatured normal DNA duplex (Marmur, J.S. and D. Lane, PNAS 46: 453, 1960).

    a) In *Hemophilus influenzae* the genes specifying reaction to cathomycin and streptomycin are closely linked (capable of being carried by the same transforming fragment). Herriott, R.M. (PNAS 47: 146, 1961) heated the DNA from *cathomycin-resistant, streptomycin-sensitive* and *cathomycin-sensitive, streptomycin-resistant* to 100°C and then cooled them slowly. Some of the renatured double-stranded DNA molecules possessed the ability to transfer

both *cathomycin resistance* and *streptomycin resistance* into a bacterium sensitive to both antibiotics. The double-resistant transformed clones bred true. What is the inference regarding the constitution of the double-transforming DNA? Should such DNA be capable of yielding true-breeding double transformants? Explain.
b) Can you think of a way in which the experimental procedure developed by Marmur and Lane can be used in clarifying taxonomic relationships in the bacteria?

34. Nester, E.W. et al. (PNAS 49: 61, 1963) treated a *trp⁻his⁻tyr⁻* strain of *B. subtilis* unable to synthesize tryptophan, histidine, and tyrosine with DNA from the prototrophic strain *trp⁺his⁺tyr⁺* able to synthesize these amino acids. The number of colonies in each of the seven transformant classes is shown in the table.

Transformant class	No. of colonies
trp⁺his⁺tyr⁺	11,940
trp⁻his⁺tyr⁺	3,660
trp⁻his⁻tyr⁺	685
trp⁺his⁻tyr⁻	2,600
trp⁺his⁻tyr⁺	107
trp⁺his⁺tyr⁻	1,180
trp⁻his⁺tyr⁻	418

Illustrate the topographical information these data permit you to deduce regarding the region of the genome carrying the genes, *trp, his,* and *tyr*? Show your calculations.

Genetics of Viruses

35. When a nontransducing temperate phage infects a bacterium, it may take one of two distinct paths, depending on the conditions of infection. What are these paths, and what characteristics does each confer on the host cell? Illustrate. (See Jacob, F. and E.L. Wollman, Sci. Amer. 204: 92, 1961.)

36. What is meant when we say: The bacterial chromosome has temperate-phage memory, and temperate phage has bacterial-chromosome memory?

37. a) Explain what is meant by a phage "cross".
b) In what respects are the F factor and λ phage similar, and in what respects are they different?
c) Describe the lytic cycle of T-even phage.

38. In T2 and T4 phages the linkage maps are circular, but the chromosomes (DNA duplexes) are linear. Explain how this apparently paradoxical situation can exist.

39. Doermann, A.H. (CSHSQB 18: 3, 1953) infected *E. coli* cells with two strains of T4 phage, one a triple-mutant strain of genotype *m (minute), r (rapid lysis), tu (turbid)* and the other *wild-type* at each of the 3 loci. The number of plaques of each of the different kinds of lytic products were as shown.

Genotype	No. of plaques	Frequency
m r tu	3,467	0.335
m⁺r⁺tu⁺	3,729	0.361
m r⁺tu⁺	520	0.050
m⁺r tu	474	0.046
m r tu⁺	853	0.082
m⁺r⁺tu	965	0.093
m r⁺tu	162	0.016
m⁺r tu⁺	172	0.017

a) State with reasons why these genes are not inherited independently.
b) Determine: (1) The linkage distances between *m* and *r, r* and *tu*, and *m* and *tu*. (2) The sequence of the three genes.
c) Calculate the coefficient of coincidence and state its significance.

40. Certain phage, e.g., PL22, can transduce any gene in the donor genome. Others, e.g., λ, transduce only one or a few genes in a particular region of the genome. Account for these facts.

41. In T2 (and T4) phage the chromosomes (DNA molecules) are circularly permuted and terminally repetitious (Thomas, C.A., Jr. and L.A. MacHattie, PNAS 52: 1297, 1964). Explain what is meant by "circularly permuted and terminally repetitious".

42. a) Distinguish between virulent and temperate phage.
b) Discuss the life cycle of λ phage and compare and contrast it with that of phage T4.

43. By early 1964 the linkage map of T4 phage was shown to be:

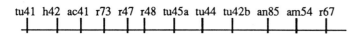

tu41 h42 ac41 r73 r47 r48 tu45a tu44 tu42b an85 am54 r67

This information is compatible with either of two alternatives:

(1) The linkage map is linear and rodlike (viz. two-ended). In this case *r67* is closer to *ac41* than to *h42*.
(2) The linkage map is circular, formed by joining the ends shown in the above map. In this case *r67* is closer to *h42* than to *ac41*.
Streisinger, G. et al. (Genetics 51: 775, 1964) crossed a *r67h42ac41+* phage with one of genotype *r67+h42+ac41*, and selected for *h42ac41* recombinants. About 65 percent of these were *r67* and 35 percent *r67+*. Which of the above hypotheses is correct and why?

44. In the bacteriophage T2, r^+ (*wild-type*) produce *fuzzy-edged* plaques, whereas *r* mutants produce *sharp-edged* plaques; h^+ and *h* determine the ability to infect certain *E. coli* strains (h^+ phages infect only strain B; *h* can infect both B and B/2). In 1949 Hershey, A.D. and R. Rotman (Genetics 34: 44) carried out the first detailed study of genetic recombination in "crosses" between $h\,r^+$ and various h^+r strains of T2. The results of crosses, involving the 3 mutants *r1*, *r7* and *r13* are presented in the table below.

Cross	Step	h^+r^+	hr^+	h^+r	hr
$hr^+ \times h^+r1$	Input	0	53	47	0
	Yield	12	43	34	12
$hr^+ \times h^+r7$	Input	0	49	51	0
	Yield	5.9	56	32	6.4
$hr^+ \times h^+r13$	Input	0	49	51	0
	Yield	0.74	59	39	0.94

a) Is T2 phage haploid or diploid?
b) How many genes specify plaque type?
c) Are these genes linked to the *h* locus?
d) If linked, illustrate the conceivable arrangements of *h*, *r1*, *r7* and *r13*.
e) What type of gene interaction is involved among the *r* mutants?

45. Hershey, A.D. and R. Rotman (Genetics 34: 44, 1949) concluded that T2 bacteriophage has three linkage groups, as shown below.

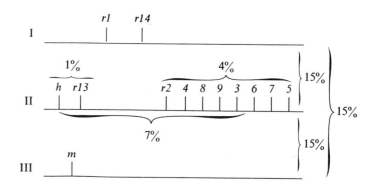

In 1960 Streisinger, G. and V. Bruce (Genetics 45: 1289) made 3-factor crosses involving two linked genes, e.g., *r2* and *r7*, and a third e.g., *r1*, apparently unlinked to the other two and therefore apparently belonging to a different linkage group. By using different concentrations of the two infecting (parental) phages in mixed, two-phage infections and forcing premature lysis of the bacterial cell to limit the frequency of crossing-over they obtained the tabulated results, which are typical of all other such crosses:

Minority parent	Majority parent	Multiplicities of infection	Recombinants selected	Frequency of r allele of minority parent among selected recombinants
$r2r7^+r1$ x $r2^+r7^+r1^+$		1:>10	$r2^+r7^+$	Most $r1$; few $r1^+$
$r2^+r7r1^+$ x $r2r7^+r1$		1:>10	$r2^+r7^+$	Few $r1^+$; most $r1$

Do the results of Streisinger and Bruce verify Hershey and Rotman's conclusions? Explain and give the sequence of the genes if possible.

46. Edgar, R.S. et al. (Genetics 47: 179, 1962) carried out a series of mapping experiments with r (rapid-lysis) mutants of bacteriophage T4D. Out of 194 such mutants, 64 failed to grow on λ-lysogenic strains of E. coli K12. Two-factor crosses among these mutants showed a very low percent (0.16, 0.40, etc.) of wild-type recombinants. In a few crosses no wild-types were obtained. When 74 other mutants were crossed to one or more "tester" strains, r48, rEDb50, r67, the results were as tabulated. Of the 64 mutants which failed to grow on K12, one, rEDdf41, never reverted to r^+ and gave no recombinants with any of the other 63 in the group or with the 7 mutants which produced few wild-types with rEDb50. All other mutants reverted to r^+ at a low frequency.

Cross			% wild-types
65 of these r mutants	x	r48	Extremely low
	x	rEDb50 and r67	Extremely high
7 r mutants	x	rEDb50	Extremely low
	x	r48 and r67	High
2 r mutants	x	r67	Extremely low
	x	r48 and rEDb50	High

a) At how many loci do these r mutants map? Explain.

b) What is the nature of (i) the rEDdf41 mutant and (ii) other r mutants? Explain.

47. a) In λ phage, the host-range mutant h reproduces in bacteria resistant to wild-type (h^+) phage; temperature-sensitive (ts) phage are able to grow and develop to lysis at 30°C but not at 42°C; the plaque-type mutant c makes clear plaques while c^+ (wild-type) makes turbid plaques. A three-factor cross is made between $tshc^+$ and ts^+h^+c by infecting bacteria growing at 30°C. After the cells have been lysed, the lysate (of progeny phages) is diluted and then plaque-assayed for ts^+h recombinants by incubation at 42°C on h^+-resistant bacteria. The proportion of turbid, c^+ plaques appearing under these plating conditions is found to be 0.95.
(i) What is the map order of the three markers?
(ii) Is there evidence of linkage?

b) The assay plates described above are incubated at 30°C instead of 42°C. Clear plaques are then spotted onto another assay plate seeded with λ-resistant indicator bacteria and the plates incubated at 42°C to determine the fraction of the h c recombinants that are temperature-sensitive, i.e., are ts and will not lyse the indicator cells at 42°C.
(i) If the fraction is again 0.95, show that the same order of the three markers is obtained.
(ii) What additional information can be obtained by this method which was not obtained by plating at 42°C?

c) Can you devise another, more direct way of analyzing the proportion of h c recombinants carrying the ts marker?

48. a) The single-stranded DNA phage φX174 contains approximately 5,500 nucleotides (Sinsheimer, R.L., Brookhaven Symp. Biol. 12: 27, 1959). If all this DNA codes for proteins and the average polypeptide contains 200 amino acids, how many different polypeptides could this DNA specify?
b) (i) The phage T4 chromosome is a single long continuous DNA duplex of molecular weight about 1.2×10^8. Since the average molecular weight of each nucleotide is 357 (assume 360) and base pairs are separated by 3.4 Å, what is the length of the T4 chromosome?

(ii) The total genetic map of T4 is about 2,500 recombination units long. If two h mutants with mutations in adjacent bases are crossed, with what frequency do you expect h^+ recombinants?

49. Four amber mutants of phage T4 (*am-11, am-12, am-13* and *am-14*) grow on strain CR63 of *E. coli* but not on strain B. Three temperature-sensitive mutants (*ts-7, ts-8* and *ts-9*) grow on both strains at 25°C, but on neither at 42°C. Strain B is infected at 42°C by the mutant strains in pairs, with the results shown (after Epstein, R.H. et al., CSHSQB 28: 375, 1963; Edgar, R.S. et al., Genetics 49: 635, 1964).

Strain			Burst size (ave. yield of phage per infected cell)
wild-type			300
ts-7	x	*ts-8*	250
am-11	x	*am-12*	1
am-11	x	*am-13*	250
am-13	x	*ts-8*	<1
ts-7	x	*ts-9*	30
am-11	x	*am-14*	1
am-11	x	*ts-9*	1
am-11	x	*ts-8*	250

a) In how many different genes did the seven mutations occur?
b) Can any of these genes undergo an *amber* as well as a *temperature-sensitive* type of mutation?
c) Is the protein determined by either gene more likely to be a monomer than a dimer, or is the reverse true?

50. In λ phage, c^+ (*wild-type*) forms *turbid* plaques and *c* forms *clear* plaques. A *suppressed-sensitive* (*sus* or *amber*) mutant strain of λ $csus_1sus_2^+$, and a second *sus* mutant, $c^+sus_1^+sus_2$, are crossed at 37°C by growing them together in a permissive host strain of bacteria (in which *sus* or *amber* mutants carrying the chain-terminating codon UAG in the middle of the gene can grow; in this host, codon UAG is translated as sense, in the non-permissive host, as nonsense). The proportion of *clear* to *turbid* plaques is found to be 20:1. What is the order of the three phage markers c, sus_1, and sus_2? Note: The three markers are closely linked.

51. A large number of *ts* (*temperature sensitive*) and *sus* (*amber* or *suppressor-sensitive*) mutants of bacteriophage S13 have been isolated and classified into seven complementation groups, six of which (A to F) are considered here. To establish the linear order of the mutants, 15 three-factor crosses were carried out by Tessman, E.S. (Virology 25: 303, 1965). The results of these crosses were as follows:

Cross			Selected progeny	Fraction of selected progeny carrying wild-type allele of outside (unselected) marker
sus8(F)*ts2*(A)	x	*ts4*(A)	*ts*$^+$	0.15
sus12(C)*ts4*(A)	x	*ts2*(A)	*ts*$^+$	0.25
sus13(C)*ts4*(A)	x	*sus12*(C)	*sus*$^+$	0.3
sus12(C)*ts4*(A)	x	*sus13*(C)	*sus*$^+$	0.8
sus12(C)*ts3*(D)	x	*sus13*(C)	*sus*$^+$	0.23
sus12(C)*ts6*(D)	x	*ts3*(D)	*ts*$^+$	0.17
sus12(C)*ts3*(D)	x	*ts6*(D)	*ts*$^+$	0.76
sus5(E)*ts3*(D)	x	*ts6*(D)	*ts*$^+$	0.07
sus10(E)*ts3*(D)	x	*ts6*(D)	*ts*$^+$	0.3
sus10(E)*ts3*(D)	x	*sus5*(E)	*sus*$^+$	0.3
sus5(E)*ts9*(B)	x	*sus10*(E)	*sus*$^+$	0.13
sus5(E)*ts9*(B)	x	*ts11*(B)	*ts*$^+$	0.20
sus10(E)*ts11*(B)	x	*ts9*(B)	*ts*$^+$	0.90
sus10(E)*ts9*(B)	x	*ts11*(B)	*ts*$^+$	0.19
sus8(F)*ts11*(B)	x	*ts9*(B)	*ts*$^+$	0.14

a) Analyze the data and establish the order of the markers in each cross.
b) Make a composite genetic map from all the data, and carefully note any peculiarities of this map.

52. *Wild-type* recombinants in the T4 phage cross $am85^+tu44am54$ x $am85tu44^+am54^+$ occur much less frequently than in either $am85tu44^+am54$ x $am85^+tu44am54^+$ or $am85tu44am54^+$ x $am85^+tu44^+am54$.

What is the sequence of the genes? Show how you arrive at your answer.

53. The table below shows results obtained by Kaiser, A.D. (Virology 1: 424, 1955) in two-factor λ-phage crosses involving the genes c, s, co_1 and mi:

Parents			Progeny*			
			ab^+	a^+b	a^+b^+	ab
smi^+	x	s^+mi	647	502	65	46
sc^+	x	s^+c	808	566	19	20
co_1mi^+	x	co_1mi	459	398	17	25
cmi^+	x	c^+mi	1,213	1,205	84	75
sco_1	x	$s^+co_1^+$	46	53	1,615	1,774

*ab^+ = mutant at first locus, wild-type at second locus; a^+b = wild-type at first locus, mutant at second locus; a^+b^+ = wild-type at both loci; ab = mutant at both loci.

a) Determine the percentage recombination in each cross and draw a linkage map showing the relative positions of the genes and the distances between them.

b) A three-factor cross involving c, co_1, and mi was also performed by Kaiser. The progeny phenotypes and genotypes are given below for the two least frequent recombinant types of plaques (classes) from a total of 6,600 progeny examined. What is the sequence of mi, c and co_1?

Parents			Recombinant phenotypes
$c\ co_1^+mi^+$	x	c^+co_1mi	8 $c^+co_1^+mi^+$; 1 $c^+co_1^+mi$

54. You are given the following three strains of *E. coli* K12:
Wild-type (prototrophic) grows on minimal medium.
Mutants I and II (both auxotrophic) need substances A and B for growth on minimal medium. The requirement for A in both strains is produced by mutation a, but the requirement for B is produced in mutant I by mutation b_1 and in mutant II by mutation b_2.

The genotypes of the strains may therefore be written:
Wild-type $a^+b_1^+b_2^+$; *Mutant I* $ab_1b_2^+$; *Mutant II* $ab_1^+b_2$
Explain how to produce and positively identify a double mutant of genotype $a^+b_1b_2$ using the generalized transducing phage P1. Assume that these markers are all cotransducible with high frequency and are in the order a-b_1-b_2.

55. You are given the following three phage strains:
(i) λ $tsA\ c^+h^+$, a *temperature-sensitive* mutant which grows at 30°C but not at 42°C.
(ii) λ$tsA^+c\ h^+$, a *clear* plaque mutant.
(iii) λ$ts\ A^+c^+h$, a *host-range* mutant
and bacterial indicator strains C600 λ^S and CR63 λ^R. Describe how you would proceed to isolate the triple mutant λ$ts\ A\ ch$.

56. When bacteria are mixedly infected with T2 r^+ and T2 r phage, which produce *fuzzy-edged* and *sharp-edged* plaques, respectively, and the lysate diluted (so that each phage produces its own separate plaque) and plaque-assayed, approximately 2 percent of the plaques are *mottled* (partly r^+ and partly r). The remaining 98 percent are either *sharp* or *fuzzy*. When these *mottled* plaques are picked, the phage suspended in broth, and then diluted and plaque-assayed, not only are *fuzzy* and *sharp* plaques obtained in about equal numbers but about 2 percent of the plaques are again *mottled*. Offer an explanation to account for the origin of the *mottled* plaques.

57. The c mutant strains of T2 and T4 phage require the amino acid tryptophan to adsorb to cells of *E. coli*; other strains of T2 and T4 are c^+ (*tryptophan-independent*). When host cells are infected with a mixture of c^+ and c phage, about half of the tryptophan-independent progeny are found upon subsequent testing to be c. Explain how this comes about.

58. Streisinger, G. (Virology 2: 388, 1956) found that some of the progeny viruses produced by mixed infection with T2 and T4 phage were capable of infecting both the B/2 and the B/4 strain of *E. coli*. The viruses produced in the next

generation were pure T2 or pure T4. Note: T2 can infect only B/4; T4 can infect only B/2.
a) Account for these events.
b) What proportions of the progeny of a mixed infection of *E. coli* with equal amounts of T2 and T4 would you expect to be capable of infecting both B/2 and B/4?

59. Genetic recombination in bacteriophage corresponds to the production of a DNA molecule derived partly from one and partly from another parent. This may occur by copy choice, breakage and reunion, or breakage and copy choice. In 1961 experiments were conducted by Meselson, M. and J.J. Weigle (PNAS 47: 857) and by Kellenberger, G. (PNAS 47: 869) to determine whether there are any DNA genomes that are entirely derived form one parent in a recombinant phage. Crosses were made between two strains of the temperate λ phage, carrying alternate alleles of two genes. One strain was labelled with the isotopes ^{13}C and ^{15}C, and the other was not. It was found that discrete amounts of the original parental DNA appeared in the recombinant phages.
a) Which is the most plausible mechanism for recombination on the basis of these results? Which is the least plausible? Are any of the mechanisms excluded? Explain.
b) Are the copy-choice and breakage explanations of recombination in phage mutually exclusive? Explain and illustrate.

60. λ phage has two linear linkage maps, the prophage and the vegetative maps:
Prophage map   *susj, susB, susA, susr, susQ, i*
Vegetative map   *susA, susB, susj, i, susQ, susr*
(Campbell, A.M., Virology 20: 344, 1963 and Virology 14: 22, 1961, respectively)
Explain how these two are interrelated.

61. In phage T4, *r+* (*wild-type*) strains produce *small* plaques on both *E. coli* strains B and K. The *r* mutants in the rII region, which consists of two contiguous genes A and B, produce *large* plaques on B but *no* plaques (no progeny) on K when they infect these hosts individually. When strain K is infected with both *r75* and *r101*, no progeny are produced, but when infected with both *r75* and *r89*, many *small* plaques (*r+* phenotype) are produced. Explain why plaques are formed in the latter infection but not in the former. What phenotype would you expect in mixed infections by *r101* and *r89* and why?

62. Edgar, R.S. and I. Lielausis (Genetics 49: 649, 1964) isolated a large number of *temperature-sensitive* mutants of bacteriophage T4D. They form plaques at 25°C but, unlike the *wild-type* strain, cannot form plaques at 42°C. A total of 382 mutations were studied.
a) To map the mutations complementation tests were first performed. Why?
b) What procedure would you follow after complementation tests to locate the mutations precisely on the linkage map?

63. When a new phage is obtained, it is usually assumed that a single mutational event (e.g., one base substitution) has given rise to the *mutant* phenotype. Occasionally, however, a double *mutant* is obtained involving two separate mutational events at different sites on the chromosome, each capable of producing the mutant phenotype. These double mutants can be recognized by certain properties.
a) How would the frequency of reversion to *wild-type* of **a** double-point mutant compare with that of a single-point mutant genotype?
b) Suppose that the *double mutant* in question is crossed with a set of *single-point mutants* (1 to 10) whose order on the genetic map is 1,2,3,...,10. Next suppose that *wild-type* recombinants are obtained at a frequency of 10 percent in biparental crosses with 1 and 10; 5 percent with 2 and 9; 2 percent with 3 and 8; and <0.1 percent with 4, 5, 6, and 7. Analyze these data and describe the most probable location of the two mutations in the *double mutant*. (Hint: If a mutant is double, the above are all three-factor crosses).
c) Using the above example, describe how to separate the two mutations in the *double mutant* by means of a cross and describe how to differentiate the various genotypes produced in the cross given the additional information that a two-factor cross between 4 and 7 produces *wild-type* recombinants at a frequency of 10 percent.

64. a) Certain T2 phage mutants fail to produce coat protein when grown in strain B of *E. coli* and thus form no plaques on this host strain. These mutants are able to produce coat protein in a permissive strain C and thus will grow and form plaques. An investigator isolates 10 such mutants,

designated 1 to 10 and carries out complementaion studies by testing the mutants in pairs (trans) using the spot test, with the results shown in the following table:

	1	2	3	4	5	6	7	8	9	10
1	no	many	few	few	few	few	few	few	few	many
2		no	few	many	many	many	many	many	few	few
3			no	many	many	many	many	many	few	no
4				no	few	few	few	few	few	many
5					no	few	few	few	few	many
6						no	no	no	few	many
7							no	no	few	many
8								no	few	many
9									no	few
10										no

*No, few and many refer to the number of plaques produced.
(i) Which indicator bacterial stain did the investigator use in the spot test? Explain.
(ii) How many genes appear to be involved in the production of coat protein in T2?
(iii) In how many genes did the mutations in the 10 mutant strains occur?
(iv) Do there appear to be any hot spots? If so, at what sites?
b) The investigator measures the reversion frequency of each mutant and finds the results shown below:

1) 10^{-6} 2) 2×10^{-6} 3) 10^{-5}
4) 10^{-7} 5) 5×10^{-7} 6) 10^{-6}
7) 2×10^{-6} 8) 3×10^{-6} 9) 0
10) 0

(i) How were these results on reversion frequencies obtained?
(ii) What do the reversion frequencies allow the investigator to conclude about each of the mutant genotypes?
(iii) Draw a linear map summarizing the data in (a) and (b), indicating the extent to which the sequence is arbitrary.
c) A single gene H, determines a certain h conditionally lethal mutant phenotype in this phage. A number of H-gene mutants are available, some of which are deletions. From the latter, the investigator selects a set of deletion standards which divide the H gene into three regions, a-b, b-c, and c-d as shown below:

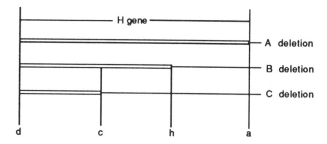

He now wishes to place nine mutants suspected to contain point mutations (numbered 1 to 9) in the H gene by spot-testing each of these mutants against the deletion standards with the results shown below:

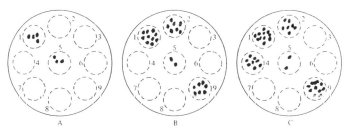

Each circle (broken lines) circumscribes the area to which a particular mutant is localized. A to C: deletion standard used.

(i) What is your interpretation of the results on plate A? Quantitate your answer where possible.
(ii) Localize the 9-point mutations within the regions a-b, b-c and c-d.
d) How do you suppose the experimenter decided his mutants 1 to 9 were point mutations?

65. In T4, gene 23 codes for the synthesis of head protein. A strain with a nonsense mutation in this gene produces an incomplete polypeptide chain. How would you detect and maintain such a mutant strain?

66. a) *Temperature-sensitive*, *ts*, alleles of at least 56 genes of T4 phage have been isolated. They constitute a class of conditional lethal mutants that allow the phage to propagate at 25°C but not at 37°C to 42°C (Edgar et al., Genetics 49:

635, 1964). Provide an explanation in molecular terms to account for the conditional lethal nature of their action.

b) Suppose you isolate three *temperature-sensitive* mutants in T4 phage. Outline the procedure you would follow to determine (1) whether the mutations are in the same or different genes and (2) the percent recombination between any two mutant sites. Indicate the results expected when the mutations are in the same gene and in different genes.

c) Do the same assuming that the three mutants are of the *amber* type.

Chapter 13
Genotype-Environment Interactions; Expressivity, Penetrance, Phenocopies and Pleiotropism

1. In the latter part of the nineteenth century Weismann performed experiments with a purebred strain of mice with *normal* tail length. He cut the tails off some and mated them. This procedure was repeated each generation for a number of generations. The progeny in all cases developed tails of *normal* length. What is the significance of this experiment?

2. a) Discuss the statement that mental characters are largely but not wholly determined by heredity.
b) A character is known to be inherited. Does this mean that the genotype alone is responsible for its expression? Cite an example to support your view.
c) Distinguish between congenital defects and hereditary defects.
d) Of what importance in medicine is a clear understanding of the concept of the norm of reaction?

3. Certain rats are *resistant* to rickets, i.e., they show no ill effects when fed a diet deficient in vitamin D. Other rats are *susceptible*; they suffer from the disease when fed the vitamin D deficient diet.
a) Suppose that 75 percent of the rats in a certain group fed a vitamin D-deficient diet develop rickets. Would the disease be attributed to heredity or to the environment? Would it be due to both? Explain.
b) Suppose that in a true-breeding *rickets-susceptible* line of rats 45 percent of the individuals are affected. Should the disease be attributed to heredity or environment?

4. An individual shows a condition caused by an external environmental agent, e.g., *diphtheria*, caused by the bacterium *Corynebacterium diphtheriae*. Does this mean that his genetic constitution plays no part in determining the expression of the character? Explain.

5. a) Distinguish between hereditary and nonhereditary diseases.
b) Many people regard heritable diseases as inborn and incurable traits against which there is no relief and nonheritable infectious diseases and even accidental mutilations as much less horrible because they can be cured. How would you explain to such people that their views are baseless, using *phenylketonuria* and *galactosemia* as examples of heritable diseases and *malaria* and *measles* as examples of nonheritable ones?

6. a) Distinguish between monozygotic and dizygotic twins.
b) How would you determine whether twins of like sex are monozygotic or dizygotic?
c) Why are monozygotic twins especially favourable for comparing the relative roles of heredity and environment?
d) Identical twins reared apart show a greater mean pair difference in IQ tests than in body weight when the values are compared with the population means. Suggest reasons for this.
e) What genetic information may be furnished by human-twin data that could not be obtained from studies of other types of data, such as pedigree and population studies?
f) In studies involving the comparisions of monozygotic and dizygotic twins, all the dizygotic twins investigated are of the same sex. Why are unlike-sexed twins not used in such comparisons?
g) Discuss the limitations of the twin method for genetic analyses.
h) Can twin studies by themselves tell us anything about genes? About genetic recombination?

7. Fraternal (nonidentical) pairs of twins may be of three types: both male, both female, or male and female. What are the expected frequencies of these types in a large population?
a) Distinguish between concordance and discordance.

b) When studying qualitative characters, one way of using twin data is to study the percentage of concordance between the members of twin pairs of like sex.
(i) What conclusions can be drawn when the percentage concordance is the same for dizygotic as for monozygotic twins?
(ii) What conclusions can be drawn if concordance is 100 percent for monozygotic twins and 45 percent for dizygotic twins?
(iii) Explain why concordance in identical twins may often be less than 100 percent.

8. In 1990 Bouchard T.J. et al. (Science 250: 223) reported on a thorough and extensive 10-year study of more than 100 sets of monozygotic (MZ) and dizygotic (DZ) twins that were separated in infancy and reared apart. Analysis of their data enabled them to make the following conclusions:
(1) "General intelligence or IQ is strongly affected by genetic factors". About 70% of the variation in IQ is associated with genetic variation.
(2) "The institutions and practices of modern Western society do not greatly constrain the development of individual differences in psychological traits".
(3) MZ twins separated early in life and reared apart "are so similar in psychological traits because their identical genotypes make it probable that their effective environments are similar".
They also showed that with respect to multiple measures of occupational and leisure-time interests, personality and temperament, as well as social attitudes, MZ twins reared apart are about as similar as are MZ twins reared together.
With reasons, indicate whether you agree with the conclusions drawn by these workers.

9. In humans, *galactosemics* are unable to metabolize galactose and consequently are mentally deficient; the condition is caused by a recessive allele, g at an autosomal locus.
a) When *galactosemics* are fed galactose-free diets, they apparently become normal in most respects. What important principle does this illustrate?
b) When two such *normals* mate, would you expect *galactosemic* children? Explain.

10. A very talented woman musician, wishing to transmit some of this talent to her child, spends much time practicing, playing, and studying musical masterpieces during her pregnancy. As her child matures he shows great musical talent. Does this prove that the mother influenced her child through her activities before birth? Explain.

11. Recent advances in various branches of medicine have enabled the effects of an increasing number of lethal and semilethal genes in humans to be circumvented by providing affected individuals the proper environment; e.g., *phenylketonurics* develop normally on a phenylalanine-free diet. Discuss the moral and genetic issues arising from such a practice from the point of view of the individual and society.

12. Some individuals affected with *porphyria variegata* may suffer mild *porphyria* only (skin abrasions and blistering) or the acute form, in which more variable features of the syndrome are expressed. Acute *porphyria* apparently appears only after the use of drugs such as barbiturates and sulfonamides. The basis of the disorder is a defect in porphyrin metabolism. Porphyrin titers are high in feces in both the mild (quiescent) and acute phases (Dean, G. Br. Med. Bull. 25: 48, 1969). The mode of inheritance of this disease is indicated in the following pedigree modified from the one presented by Dean, G. ("*The Porphyrias: A Story of Inheritance and Environment*", Pitman, London, 1963).

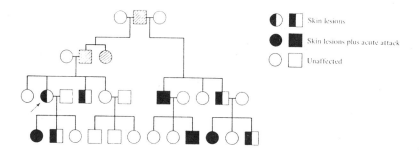

a) What is the mode of inheritance of this rare trait and the chance of the children of the propositus expressing it?

b) What counsel should the children of an *affected* parent be given? What additional counsel would you suggest to those found to be carriers?

13. In 1960 a sudden increase in the incidence of babies with anomalies of limb development (absence or markedly imperfect development of arms or legs but with hands and feet present) was noted in Germany and other European countries. These anomalies resemble the *phocomelia* phenotype in birds caused by a recessive allele at an autosomal locus. The inheritance of the condition is not definitely known in humans. Many of the mothers were found to have taken thalidomide, a drug used to ease the distress of pregnancy. The table shows the total number of births and number of *phocomelialike* births at 18 obstetric hospitals in Hamburg, Germany, from January 1960 to October 1962 (data from Lenz, W.D., Proc. 2d Int. Conf. Congenital Malformations, pp. 263-276, 1964).

	1960	1961	1962 Jan-July	1962 Aug-Oct	Total
Total births	19,052	19,917	13,326	5,542	57,837
Phocomelialike malformations	28	60	40	2	130
History:					
Thalidomide	13	46	33	2	94
No evidence of use	0	5	1	0	6
No history	15	9	6	0	30

What conclusions would you make from these data regarding the importance of the genotype in causing the effects? The importance of the environment?

14. a) A strain homozygous for a given allele at a specific locus shows variable expressivity for the trait. State whether you think its penetrance would be likely to be complete and why.
b) By what criteria can one decide whether an allele of a gene is fully penetrant or not?
c) Describe how a dominant trait may skip generations.

d) Suggest why there seem to be many more loci with alleles that show incomplete penetrance and variable expressivity in humans than in experimental organisms.

15. If a gene has one primary phenotypic effect of a biochemical and perhaps enzymatic nature, how can it affect the expression of two or more characters, i.e., be pleiotropic?

16. a) In what way does the concept of phenocopy production influence the theory and practice of clinical medicine?
b) What advice, if any, should medical practitioners give patients with inherited anomalies who are known to be phenocopies of "normal"?

17. a) Why are genes with alleles that exhibit complete penetrance and constant expressivity valuable in genetic studies?
b) How can phenocopies be used to study gene action? What value do they have for investigations of this kind?

18. *Manic depression* is a relatively rare but severe mental illness that has been demonstrated to have a genetic basis (Gill, M. et al., J. Med. Genet. 25: 634, 1988). Expression of the condition in members of a family varies from one individual to another with respect to its many symptoms: mood, restlessness, irritability, pressure of speech, increased energy and delusions of grandeur. The pedigree below from Gill, et al. is typical of the mode of inheritance for this behavioural disorder:

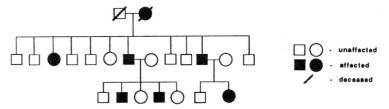

a) What is the most likely mode of inheritance of *manic depression*? Explain.
b) Is the data sufficient to determine whether the genotype for this illness is completely penetrant? Explain.
c) Suggest possible explanations for the variable expressivity of this disorder.

19. *Himalyan* rabbits, $c^h c^h$, are all *white* at birth; as they mature, the extremities become *black*. These differences reflect the effects of temperature upon the action of an enzyme produced by this genotype that transforms a colourless substance into black pigment. The enzyme is active only at temperatures below 34°C. The temperature of the body proper is about 34°C; at the extremities it is about 27°C (Danneel, Ergeb. Biol. 19: 55., 1941).
a) Does the c^h allele have more than one primary effect? Explain.
b) Do the effects of the c^h allele differ in different parts of the body?
c) Explain how the *Himalayan* phenotype is produced in the *Himalayan* breed.
d) May the allele be termed pleiotropic? Explain.

20. *Camptodactyly* (stiff little finger) vs. *normal* in humans is monogenically determined; the allele for *camptodactyly* is dominant (Moore, W.G. and P. Messina, J. Hered. 27: 27, 1936). However, *normal* couples sometimes have children who are *camptodactylous* in one hand. Explain these observations, giving the probable genotypes of the parents and the genotype of the *camptodactylous* children.

21. In the silkworm, *Bombyx mori*, the *Gr* locus is in the region of chromosome 2 where the cluster of chorion structural genes is located. Eggs laid by females homozygous for the mutant allele Gr^B have thin and wrinkled shell. Eggs laid by $Gr\_$ females have *normal* shells. Nadel, M.R. et al. (Cell 20: 649, 1980) found that in worms homozygous for Gr^B, there was a near or complete absence of a large number of chorion proteins from the eggshells. Offer two explanations to account for the pleiotropic consequences of the mutant allele.

22. In humans, *hemolytic jaundice* is caused by a dominant allele at an autosomal locus with about 10 percent penetrance (Gates, R.R., "Human Genetics", vol. 1, Macmillan, New York, 1946). A woman heterozygous for the allele marries a *normal* man.
a) If the first child is heterozygous, what is its chance of expressing the trait?
b) If the couple had 10 children, what is the expected number of *affected* children?

23. In a group of irradiated mice, Gruneberg, H. (Symp. Soc. Exp. Biol., Cambridge, 2: 155, 1948) isolated a line with a wide range of abnormalities of eyes, ears, and paws (clubfoot; too many or too few digits; or fused digits). He concluded from the breeding data shown that the traits were determined by a single pair of alleles, *H* vs. *h*.

Parents	Offspring
$Hh \times Hh$	589 *normal* : 109 *affected*
$Hh \times hh$	272 *normal* : 156 *affected*
The 156 *affected* intercrossed	132 *normal* : 595 *affected*
The 132 *normals* intercrossed	424 *normal* : 106 *affected*

The abnormality was found to be caused by the extrusion in the eleven-day-old embryo of an abnormally large amount of cerebrospinal fluid, which migrated posteriorly in the form of liquid vesicles, or blebs (Bonnevie, K., J. Exp. Zool. 67: 443, 1934).
Their temporary resting on specific protrusions of the embryo such as the eye or limb buds, caused the local abnormalities at birth. In some cases the blebs could miss these obstructions and be resorbed by the embryo.
a) Explain how the embryological observations might be related to the breeding data through reduced penetrance and variable expressivity.
b) Explain briefly how you would determine by breeding experiments whether reduced penetrance and variable expressivity were the result of the essentially random developmental effect described above or of modifying genes.
c) If positive and negative modifying alleles were the cause of the reduced penetrance and variable expressivity, outline a breeding program whereby you might increase penetrance to 100 percent or reduce it to zero.

24. In the domestic fowl, breeding experiments with homozygous *normal*, *TT* and homozygous *tremor*, *tt* strains gave the

tabulated results (modified after Hutt, F.B. and G.P. Child, J. Hered. 25: 341, 1934):

Parents	Offspring	
	Generation	Phenotype
normal x normal		All normal
normal x tremor	F_1	All normal
F_1 x F_1	F_2	423 normal : 45 tremor
F_1 normal x tremor (F_2)	Testcross	182 normal : 38 tremor
The 38 tremor intercrossed		190 normal : 110 tremor
The 190 normal intercrossed		370 normal : 210 tremor

Note: Some *tremor* fowls shake continuously and so violently that they have great difficulty in eating; in others the continuous *tremor* is barely perceptible.

a) State which allele shows incomplete penetrance and which shows variable expressivity. Give reasons for your answer.

b) If modifying genes are responsible for the variations in expression, is it possible with suitable breeding procedures to establish strains with no penetrance or with 100 percent penetrance? Explain.

25. A certain homozygous variety of the Chinese primrose (*Primula sinensis*) produces *white* flowers when the temperature during a critical period during early floral development is above 86°F and *red* ones if the temperature is below this level. Since different flowers develop at different times, it is possible, by changing the temperature during inflorescence development, to obtain both *red* and *white* flowers on a single plant.

a) Does the allele act differently in *red* and *white* flowers? Formulate an explanation for the flower colour differences described.

b) Is the allele pleiotropic? Fully penetrant? Constant in expressivity? Explain.

Another variety, homozygous for another allele at the same locus as the first, bears *red* flowers regardless of temperature during flower development.

c) Would you expect the dominance relationship to be independent of temperature? Outline the genotypic and phenotypic results you would expect among the F_1 and F_2 generations of a cross between the two varieties grown at temperatures below and above 86°F.

26. The pedigree below (from Bell, J., Treas. Hum. Inher. 2: 269, 1928) shows individuals affected with *blue sclera* (bluish, usually thin outer wall of the eye) and those with *brittle* (fragile) bones.

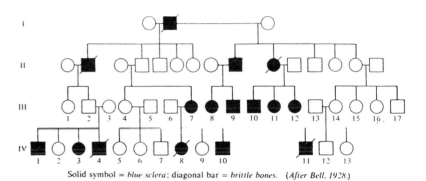

Solid symbol = *blue sclera*; diagonal bar = *brittle bones*. (After Bell, 1928.)

a) Is *blue sclera* determined by the same gene as *brittle bones*?

b) Is *blue sclera* determined by an autosomal or sex-linked gene? Is it dominant or recessive?

c) What is the percent penetrance of the allele? State the progenies from which you extracted your information and show your calculations.

d) How would you explain the fact that individuals with *blue sclera* frequently do not have *brittle bones*?

e) Give the probability of occurrence of a *blue-sclerotic* child from each of the following matings:
    (1) IV-3 x IV-7   (2) IV-4 x IV-5
    (3) IV-3 x III-9   (4) III-8 x IV-12

27. In mice, *normal* vs. *pituitary dwarf* is determined by a single pair of alleles *Dw dw* (Gruneberg, H., "The Genetics of the Mouse" 2d ed., M. Nyhoff, The Hague, 1952). Growth rate of *dwarfs* falls behind that of *normals* shortly after birth, and by the seventeenth day it stops. Thereafter they lose weight, and some die. The survivors continue to grow

slowly eventually reaching about a quarter of the normal adult weight. At this stage their tails, snouts and ears are unusually short and their thyroid, thymus, and pituitary glands are small (De Beer, G.R. and H. Gruneberg, J. Genet. 39: 297, 1940).

Normal young surgically deprived of the anterior pituitary grow and develop exactly like *dwarfs*; *dwarf* young surgically implanted with pituitaries from *normal* young approach *normals* in growth rate, adult size, glandular and morphological development, and fertility (Smith, P.E. and E.C. McDowell, Anat. Rec. 46: 249, 1930).

a) Using appropriate terms, explain the above observations.

b) How would you determine whether a *dwarf* arising in a laboratory stock was a phenocopy or homozygous for the allele, *dw*?

28. In a *normal* (homozygous) breed of fowl, a *rumpless* bird appears. The abnormality is characterized by the absence of caudal vertebrae and of tail structures such as muscles, feathers, and the preen gland (Landauer, W. J. Am. Nat. 89: 35, 1955).

a) Is one justified in stating from this information alone that *rumplessness* is a genetic trait? Support your answer.

b) Outline an experiment to distinguish between heritability and environmental inducement (phenocopy) of the trait.

Chapter 14
Euploidy: Haploidy and Polyploidy

1. a) Why are autotriploids highly sterile even though they frequently show greater vigour, yield and vitamin content than their progenitors?
   b) If autotriploids are highly or completely sterile, how is it possible to maintain the Keizerskroon tulip and the Baldwin apple which are known to be autotriploids?

2. a) Why have autotriploids been unsuccessful in establishing themselves in nature?
   b) Would you expect to find natural autotriploids in organisms that are: (1) Cross-fertilizing? (2) Self-fertilizing? (3) Asexually reproducing? Explain.

3. A teratogen is present at the time of conception of a human embryo. As a result the centromeres of all the chromosomes divide and the chromatids separate. However the nuclear membrane does not disappear and the zygote does not divide. The compound has no effect on subsequent cell divisions. What chromosome number and kinds would the embryo possess?

4. Explain which would be the easiest and which the most difficult to obtain in a given species of plant:
   a) Triploids if you have only diploids and tetraploids.
   b) Diploids if you have only haploids.
   c) Tetraploids if you have only diploids and tetraploids.

5. In humans, approximately 1% of all zygotes are autotriploids or autotetraploids. However, such polyploidy in our species is incompatible with life (Kajii, T. and N. Niikawa, Cytogenet. Cell Genet. 18: 109, 1977; Pettenati, M.J. et al., Am. J. Hum. Genet. 24: 23, 1986).
   a) Illustrate how such polyploids may arise.
   b) Offer a plausible explanation for their incompatibility.

6. Briefly discuss the significance and role of polyploidy in evolution explaining whether autopolyploidy or allopolyploidy has been the more important in speciation.

7. What percentage of the gametes of autotriploids that have 15 and 18 chromosomes, respectively, are likely to be n? What would be the expected percentages of the sum of n and 2n gametes for each triploid? Show how you derive your answers.

8. Autotriploids ($2n=3x=21$) of *Triticum monococcum* form V-shaped trivalents at diakinesis-metaphase I. At anaphase I, 2 of the homologues of each trivalent segregate to one pole and the third moves to the opposite pole. Kuspira, J. et al. (Can. J. Genet. Cytol. 28: 867, 1986) observed the following anaphase I distributions in 250 meiocytes of triploid *T. monococcum*:

Anaphase I distributions	Number of cells
10 - 11	143
9 - 12	76
8 - 13	29
7 - 14	2

Do the data indicate whether: (i) the orientation of the trivalents at metaphase I was random or not? (ii) the homologues of the different trivalents segregate randomly or not at anaphase I? Explain.

9. Rhoades, M.M. (J. Genet. 33: 355, 1936) found a triploid corn plant when he crossed the $glglWs_3Ws_3$ strain with the $GlGlws_3ws_3$ strain. In backcrosses with the former parent the triploid plant produced 89 $Glgl$ and 20 $glgl$ offspring. In backcrosses with the latter it produced 42 Ws_3ws_3 and 90 ws_3ws_3.
   a) What was the genotype of the triploid plant?
   b) Which parent provided the 2n gamete to form the triploid? Justify your answer.

10. In a self-fertilizing plant homozygous for the allele *A* of a given gene, one of the alleles mutates to the recessive form, *a*. Would the recessive trait be most likely to appear if the plant was a diploid, an autotetraploid or an allotetraploid? Explain.

11. The New World cotton species *Gossypium hirsutum* and *G. barbadense* have a 2n chromosome number of 52 (13 large and 13 small pairs). The American and Old World species *G. thurberi* (small chromosomes) and *G. herbaceum* (large chromosomes) each have a diploid chromosome number of 26. Cytological analysis of meiosis in various hybrids yields the data shown (Beasley, J.O., Genetics 27: 25, 1942).

Cross	Hybrid: metaphase I pairing
G. hirsutum x *G. thurberi*	13 small bivalents + 13 large univalents
G. hirsutum x *G. herbaceum*	13 large bivalents + 13 small univalents
G. thurberi x *G. herbaceum*	13 large univalents + 13 small univalents

a) Interpret these results with respect to the phylogenetic relationships of these species.
b) How would you determine whether your interpretation is correct? Illustrate the results expected.

12. Polyploids are often said to be conservative in the evolutionary sense, i.e., they do not allow the phenotypic expression of mutant recessive alleles. In allopolyploids, for example, this would be true for recessives of functionally identical genes, one in each of the different genomes. What are the reasons for this? Illustrate your answer with an example from an autotetraploid or an allotetraploid and compare the results with those expected in a diploid.

13. Three species in the genus *Brassica; campestris, nigra* and *oleracea* are known to be diploid and have the genome constitutions AA, BB and CC, respectively. The chromosome numbers of 3 other species and the number of bivalents and univalents in F_1 hybrids of various crosses among the 6 species are as follows (based on work of U, N., Jpn. J. Bot. 7: 389, 1935; Frandsen, K.J., Dansk. Bot. Arkiv, Bd. 12, Nr. 7, 1, 1947):

Species and/or F_1 hybrid	Chromosome number	Number of bivalents	Number of univalents
B. juncea	36	18	0
B. carinata	34	17	0
B. napus	38	19	0
F_1 from *juncea* x *nigra*	26	8	10
F_1 from *napus* x *campestris*	29	10	9
F_1 *carinata* x *oleracea*	26	9	8
F1 from *juncea* x *oleracea*	27	0	27
F1 from *carinata* x *campestris*	27	0	27
F1 from *napus* x *nigra*	27	0	27

a) What are the haploid chromosome numbers of the 3 diploid species? Explain.
b) Are *juncea, carinata,* and *napus* autopolyploids or allopolyploids? Give the genome constitutions of *juncea, carinata* and *napus*. Explain your answer.

14. The whiptail lizard, *Cnemidophorus uniparens*, is a parthenogenetically reproducing all-female autotriploid (3n=69) species (Cuellar, O., J. Morph. 133: 1, 1971). In 1977 Cuellar, O. (Evol. 31: 24) showed that skin grafts are permanently accepted if the donor and host belong to the same clone. Using electrophoresis, Dessauer, H.C. and C.J. Cole (J. Hered. 77: 8, 1986) revealed extensive genetically determined polymorphisms in each individual for many proteins, e.g., for adenosine deaminase, a monomeric enzyme, each individual revealed 2 forms of the protein and 6 forms of the dimeric enzyme peptidase-A were detected in the same individuals. Moreover, all females from all generations in a given clone possessed identical protein polymorphisms.
a) Are all females in this species (i) heterozygous? (ii) genetically identical? Explain.
b) Propose and illustrate 2 mechanisms, one pre-meiotic and the other meiotic, that will account for all the observed facts.

15. In the donkey (*Equus asinus*) the diploid chromosome number is 62, and in the horse (*E. caballus*) it is 64. The chromosome complements of the two species are morphologically different (Trujillo, J.M. et al., Chromosoma 13: 243, 1962). In crosses between male donkeys and mares, male mules are always sterile, but in rare instances females are fertile. When these fertile female mules are crossed with stallions, the offspring express only the traits of the horse (Anderson, W.S., J. Hered. 30: 549, 1939) and have 63 chromosomes.
a) Explain the complete sterility of the male and most female mules. (See Chandley, A.C. et al., Cytog. Cell Genet. 13: 330, 1974.)
b) Offer an explanation to account for the fertility of the rare female mules and the horselike nature of their offspring from backcrossing to stallions.

16. In the wheat genus *Triticum* a genome possesses 7 chromosomes. The types of chromosome associations and their numbers observed at diakinesis-metaphase I in interspecific hybrids derived from crosses involving the species *monococcum* (Einkorn wheat), *turgidum* (durum wheat) and *aestivum* (bread wheat) are as follows (Kihara, H., Mcm. Coll. Sci. Kyoto Imp. Univ. 1: 1, 1924; Kerby, K. and J. Kuspira, Genome 29: 722, 1987):

T. turgidum x *T. monococcum*	7 bivalents + 7 univalents
T. aestivum x *T. monococcum*	7 bivalents + 14 univalents
T. aestivum x *T. turgidum*	14 bivalents + 7 univalents

a) What is the somatic chromosome number of each of the 3 species? Explain.
b) Which of these species is (are) polyploids? Is it or are they allopolyploids or autopolyploids? Explain.
c) Discuss the phylogenetic relationships of the 3 species.

17. The A genome of the allohexaploid bread wheat, *Triticum aestivum* (2n=6x=42; AABBDD) is derived from an Einkorn wheat, probably *Triticum monococcum*. Its D genome originated from *Aegilops squarrosa* (Kerby, K. and J. Kuspira, Genome 29: 722, 1987). *Triticum urartu* along with 5 different *Aegilops* species (see table below) have each been implicated as the source of the B genome. The degree of taxonomic relatedness of 2 genomes can be ascertained by determining the differences in melting temperatures, ΔTm's, between homoduplexes and heteroduplexes derived from hybridization of polynucleotide strands of DNA from the 2 species. This parameter, ΔTm, gives a direct measure of the percentage of mispaired bases in the heteroduplexes. Nath, J. et al. (Biochem. Genet. 21: 745, 1983) hybridized tritium-labelled DNA from *T. aestivum* with unlabelled DNA from itself as well as the 6 putative B genome donors. They also hybridized labelled DNA from *Ae. squarrosa* and a synthetic AADD tetraploid with unlabelled DNA from *T. urartu*. Their results are shown below:

Species whose DNA was hybridized	$\Delta Tm(°C)$	% mispaired bases
T. aestivum and *T. aestivum*	0.44	0.66
T. aestivum and *T. urartu*	2.87	4.31
T. aestivum and *Ae. searsii*	3.20*	4.80
T. aestivum and *Ae. sharonensis*	3.94	5.91
T. aestivum and *Ae. speltoides*	4.30	6.45
T. aestivum and *Ae. bicornis*	4.48	6.72
T. aestivum and *Ae. longissimum*	5.23	7.85
Ae. squarrosa and *T. urartu*	4.63	6.95
Synthetic AADD tetraploid and *T. urartu*	1.41	2.12

*The ΔTm of 3.20°C for *Ae. searsii* was significantly different at the 5% confidence level from the next lowest, 3.94°C.

a) On the basis of these data, which of the 6 species is (are) eliminated as a source of the B genome in *T. aestivum*? Explain how you arrive at your conclusion. Is (are) the genome(s) of this (these) species closely related to the A or the D genome? How would you proceed to confirm this close relationship? What results would you expect in your experiment?
b) Which of the remaining species is the most likely source of the B genome in bread wheat? Explain.

18. *Turnera ulmifolia*, is a species of perennial weeds native to Central America and portions of South America. Its varieties *elegans* and *intermedia* are interfertile. Using horizontal starch gel electrophoresis, Shore, J.S. (Hered. 66: 305, 1991) studied the mode of inheritance of alcohol dehydrogenase (Adh) a dimeric enzyme in *intermedia* and that of aconitase (Aco), a monomeric enzyme, in *elegans*. The results for each enzyme were as follows:

(i) Alcohol dehydrogenase phenotypes in parents (P_1 and P_2) and their progeny

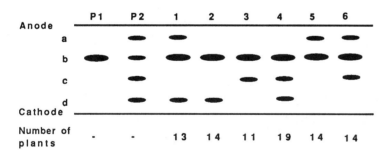

(ii) Aconitase phenotypes in parents (P_1 and P_2) and their offspring.

a) With reasons for your answers and using your own symbols indicate for each enzyme:
(i) Which parent is homozygous and which is heterozygous.
(ii) Whether the genotypes involved show disomic or tetrasomic inheritance.
b) On the basis of these data are the 2 varieties diploids or tetraploids? If the latter, are they likely to be allotetraploids or autotetraploids? Explain.

c) How would you proceed to confirm your answer to (b) and what would you expect with each procedure used?

19. An autotetraploid of genotype *AAaa* forms bivalents only. Will it show chromatid segregation with respect to these alleles? Illustrate your answer.

20. Blakeslee, A.F. et al. (Bot. Gaz. 76: 329, 1923) obtained the results shown for *purple* vs. *white* flowers, *Pp*, and *spiny* vs. *smooth* capsules, *Ss*, among the offspring of autotetraploid *Datura*.

Parents	Progeny	
	Dominant	Recessive
purple x *white*	160	1
purple x *purple*	1,280	0
purple x *white*	905	179
purple x *white*	696	682
purple x *purple*	7,547	2,619
spiny x *smooth*	257	6
spiny x *smooth*	518	137
spiny x *smooth*	3,383	118

a) Determine the genotypes of the parents of each cross.
b) Which of the 2 genes gives the better fit to chromosome segregation? (Refer to specific crosses as evidence to support your answer.)

21. The following findings are those of Becak, M.L. et al. (Chromosoma 19: 188, 1966; Chromosoma 23: 14, 1967; Chromosoma 31: 377, 1970). Two morphologically indistinguishable, fully fertile, cryptic South American frog species, *Odontophrynus cultripes* and *O. americanus*, possess 22 and 44 chromosomes respectively in their somatic cells. Meiocytes of the former species reveal 11 bivalents while those of the latter show 11 quadrivalents. Moreover, *O. americanus* has twice as much DNA in its somatic cells as does *O. cultripes*. The two species are reproductively isolated acoustically such that the females

in each species hear the sex call only of the males of the same species. Viable and partially fertile autotriploids also occur in *O. cultripes*.
a) Why is it likely that *O. americanus* is a derivative species of *O. cultripes*? On the basis of the given information outline and illustrate a plausible mechanism for its origin.
b) Suggest two approaches, one cytological and the other molecular, that would enable you to substantiate your explanation in (a). Illustrate the results expected with each.
c) The female:male sex ratio in both species is 1:1. Suggest a common plausible sex-determining mechanism for both species that will give rise to equal numbers of both sexes.

22. In the tomato (*Lycopersicon esculentum*), colour of the fruit flesh is determined by a gene close to the centromere with the allele *R* for red flesh dominant to *r* for *yellow* flesh. Both diploid and autotetraploid plants when self-fertilized or testcrossed generated the results shown below (modified from Lindstrom, E.W., J. Hered 23: 115, 1932):

| | Progeny | |
Parents	Red	White
red x red	362	11
red x red	98	34
red x yellow	158	32
red x yellow	134	141

a) For each cross were the parents both diploid or both tetraploid? Could one have been 4n and the other 2n? Give reasons for your answers.
b) Where it is impossible to decide on the basis of phenotypic ratios alone whether the parents were diploid or tetraploid, what other criteria would you employ to determine their ploidy level?

23. The mealybug *Planococcus citri* has chromosomes that are extremely small, thus making it impossible to distinguish the longitudinal axis of a chromosome from its broad axis (Chandra, H.S., Genetics 47: 1441, 1962). Outline an experiment that would enable you to determine whether the sequence of meiotic divisions is standard (first reductional, second equational) or inverse (first equational, second reductional).

24. Heterozygous Chinese primrose plants, when self-fertilized (SF) or testcrossed, produced the results shown (Somme, A.S., J. Genet. 23: 447, 1930). If the chromosomes carrying these 3 genes segregate at random, explain these results by giving the complete genotypes of the heterozygous parents, the gametic types they produce and their frequencies, and the approximate distance of each gene from its centromere.

Parents	Progeny
Palm vs. *fern leaf*	
palm (SF)	293 *palm* : 74 *extra lobes* : 10 *fern*
palm x fern	27 *palm* : 49 *extra lobe* : 18 *fern*
extra lobes (SF)	92 *palm* : 149 *extra lobe* : 76 *fern*
normal vs. *primrose queen eye*	
normal x primrose queen	88 *normal* : 17 *primrose queen*
normal (SF)	191 *normal* : 8 *primrose queen*
normal x primrose queen	48 *normal* : 40 *primrose queen*
Green vs. *red stigma*	
green x red	1,175 *green* : 223 *red*
green (SF)	1,659 *green* : 52 *red*
green x red	240 *green* : 212 *red*

25. Spinach (*Spinacia oleracea*) is a dioecious diploid; *staminate* and *pistillate* plants occur in a 1:1 ratio. Sex differences in spinach are probably due to a single pair of alleles with one sex being *Aa* and the other *aa*. Janick, J. (Am. Soc. Hortic. Sci. 66: 361, 1955) induced male and female tetraploids with colchicine. The progeny of the original tetraploid *pistillate* x tetraploid *staminate* cross consisted of 63 *staminate* and 10 *pistillate* plants. One 4n F_1 *staminate*

plant crossed with an F_1 4n *pistillate* plant produced 60 *staminate* and 20 *pistillate* progeny; crossed with a 2n *pistillate* plant, it produced 36 *staminate* and 9 *pistillate* plants. Four other 4n F_1 *staminate* plants were crossed with diploid *pistillate* offspring. One of these produced 23 *staminate* and 27 *pistillate* progeny. Similar results were obtained in the other three crosses. 4n *staminate* plants from the previous progenies were crossed with ten 4n *pistillate* sibs. Two of these crosses produced 79 *staminate* and 13 *pistillate* and 65 *staminate* and 12 *pistillate* plants respectively; 17 gave a 1:1 ratio of plants with the 2 phenotypes, e.g., in 1 cross 56 *staminate* and 52 *pistillate* plants were produced.
a) Explain these results by indicating:
(i) The sex genotypes of the diploid parents, colchicine-induced tetraploids, and *staminate* tetraploids in the first, second and third generations.
(ii) Which sex is heterogametic and the allelic relationships of the heterogametic and homogametic determining alleles.
(iii) Whether disjunction of the allele pair and therfore chromosomes carrying the sex-determining allele pair is random.
b) Are these results typical of chromosome segregation? If so, would the chromosomes involved necessarily pair to form quadrivalents? If so, what conclusions re: disjunction and gene-centromere distance would you make?

26. In 1928 Karpechenko, G.D. (Z. Abstamm.-Vererbungsl. 48: 1) crossed the diploid cabbage, *Brassica oleracea* (2n=18) with the diploid radish, *Raphanus sativus* (2n=18). The resulting highly sterile hybrid, with 18 univalents at meiosis, had the foliage characteristic of the radish and the tough root of the cabbage. It produced a few seeds from which vigorous, fertile plants, having 18 bivalents at metaphase I were obtained.
a) Explain: (i) why the first-generation plants were highly sterile, (ii) why the second-generation plants were fertile and (iii) how the plants with 18 bivalents were produced.
b) Can the second-generation plants be said to belong to either of the parental species *B. oleracea* or *R. sativus*? Give reasons for your answer.
c) Would you expect the second-generation plants to breed true for all traits? Explain.

27. By 1931 preliminary data had suggested that *Dahlia variabilis* was a hybrid octaploid, also called a double autotetraploid (2n=8x=64), derived from the doubling of the chromosome number of a sterile tetraploid hybrid (2n=4x=32) between two tetraploid species. *Ivory* vs. *white* flower colour in this species is determined by the allele pair *Ii*, and *yellow* vs. *nonyellow* flower colour is due to the allele pair *Yy*. *Y* is epistatic to both *I* and *i*. Lawrence, W.V.C. (J. Genet. 21: 125, 1931) obtained the tabulated results from a series of crosses.

	Progeny		
Parents	Yellow	Ivory	White
yellow x ivory	248	43	17
white x yellow	28	12	15
white x yellow	20	11	6
yellow x ivory	34	34	0
yellow x yellow	42	5	8
white x yellow	69	0	0
ivory x white	0	57	23

Show whether these results support the conclusion from the preliminary work.

28. *Helianthus tuberosus* is a sunflower species, that exhibits 70 percent fertility. True-breeding *tall, purple*-flowered and *dwarf, white*-flowered lines are intercrossed. The *tall, blue*-flowered F_1's when crossed with the *dwarf, white* line produced the following progeny:

Tall, blue	26	*Dwarf, blue*	5
Tall, bluish-white	100	*Dwarf, bluish-white*	19
Tall, white	24	*Dwarf, white*	6

Explain these results, indicating:
a) The number of allele pairs that determine each pair of traits.

b) Why the phenotypes were obtained in the proportions shown. Include in your explanation the basis for the F_1 gamete types and their proportions.
c) Whether you would classify the species as a diploid, an autopolyploid or an allopolyploid.
d) Whether the genes determining the two characters are linked or syntenic.
e) Why the species exhibits 30 percent sterility.

29. In alfalfa (*Medicago sativa*), certain varieties are *susceptible* to the root-knot nematode (*Meloidogyne haple*), a soil-infesting eelworm that causes severe damage in mild climates. Among a number of *resistant* plants from commercial fields in California, Goplen, E. found one plant (M-9) to be *immune* to all races of the nematode. This plant was cloned, and a number of the cloned plants were self-fertilized. The S_1 (self-fertilized) generation consisted of 319 *resistant* and 119 *susceptible* plants. When M-9 was crossed with a *susceptible* clone, 46 *resistant* and 55 *susceptible* plants were obtained; 36 S_1 plants were self-fertilized. Among the 36 S_2 progenies the segregation patterns shown were obtained (data from Allard, R.W., "*Principles of Plant Breeding*," Wiley, New York, 1960):

Type of S_2 Family	Number observed
Nonsegregating *resistant*	1
Segregating 21:1 to 35:1	3
Segregating approximately 3:1	20
Nonsegregating *susceptible*	12

Give a complete cytogenetic explanation of these results.

30. In sweet peas two different true-breeding diploid *white*-flowered strains when crossed produced *purple*-flowered F_1's. Two of these F_1's when self-fertilized produced a total of 98 *purple*- and 72 *white*-flowered plants. Other F_1's of this cross had their chromosome number doubled. State the gametic genotypes you expect to be produced by these tetraploids and their expected proportions if the gene loci are near the centromeres. What types and proportions of F_2 offspring would you expect if these tetraploids were self-fertilized?

31. The fertile allopolyploid *Nicotiana tabacum* probably originated from a highly sterile hybrid derived from a cross between *N. tomentosa* and *N. sylvestris*. If *N. tabacum* has 48 chromosomes in the nuclei of its somatic cells and *N. sylvestris* has a haploid number of 12 chromsomes:
a) What is the haploid chromosome number of *N. tomentosa*? Explain how you arrive at your answer.
b) Are bivalents expected at meiosis in *N. tabacum*? If so, how many and why?
c) Is *N. tabacum* an autopolyploid or an allopolyploid? Explain.

32. Chromosomes at metaphase I-anaphase I in a hybrid between two phenotypically different plants (one with a known somatic chromsome number of 4) behave as illustrated:

a) What is the somatic chromosome number of the second plant?
b) Are the parents of the hybrid likely to be members of the same species or not? Why?
c) What proportion of the meiotic products (e.g., microspores) of the hybrid would you expect to have the same chromosome number as its somatic cells?

33. Diploid species A and B are both true-breeding for *white* flowers. The F_1 hybrids derived from crosses between these two species are sterile and have *purple* flowers. Cytological examination shows that there is no chromosome pairing in these hybrids. Hybrids whose chromosome number has been doubled are fertile. State the types of offspring expected, with regard to colour, if these doubled hybrids are self-fertilized, giving the expected phenotypic ratio if possible. Assume that a dominant allele at each locus is required for *purple* flowers.

34. In white clover, either of the nonallelic dominant genes, W_1 and W_2 on different (nonhomologous) chromosomes determines *white* flower colour. Only the genotype $w_1w_1w_2w_2$ determines *red* flowers. A *red*-flowered tetraploid is crossed with a *white*-flowered tetraploid of genotype $W_1W_1W_1W_1W_2W_2W_2W_2$, and the F_1's are intercrossed. What are the expected proportions of *white* to *red* flowered plants in the F_2 if segregation of the alleles of both genes is two by two and chromosomal (not chromatid)?

35. *Frageria bracteata* and *F. helleri* are closely related diploid strawberry species each with 14 chromosomes. *F. bracteata* has *white* and *F. helleri pink* flowers. An F_1 plant with 14 bivalents was obtained by Yarnell, S.H. (Genetics 16: 455, 1931) from a cross between these two species. Upon self-fertilization it produced 7 F_2 plants, all with *pink* flowers. These 7 plants were backcrossed to *F. bracteata* and also self-fertilized. The results obtained are shown below:

Mating		Offspring	
F_2 plant	Fertilized by	*Pink*-flowered	*White*-flowered
1	*F. bracteata*	17	17
2	*F. bracteata*	23	29
3	*F. bracteata*	15	4
4	*F. bracteata*	9	8
5	*F. bracteata*	16	15
6	*F. bracteata*	12	3
7	*F. bracteata*	42	0
2	Self-fertilized	30	7
5	Self-fertilized	13	3
6	Self-fertilized	33	1
7	Self-fertilized	8	0

a) Analyze these data, to determine:
(i) the genotypes of the F_1 and the 7 F_2 plants.
(ii) The mode of pairing and segregation of the chromosome pairs which carried the *Pp* (*pink* vs. *white*) allele pair.

b) The F_1 plant showed 14 bivalents. Is this type of pairing expected from your genetic analysis? Explain what is expected and the segregation ratio that should follow.

36. In the bread wheat (*Triticum aestivum*) the F_2 segregation ratios of 63:1 and 15:1 occur frequently, but in barley (*Hordeum vulgare*) 63:1 ratios have not been recorded and 15:1 are rare. Account for these differences.

37. The Amazon molly fish (*Poecilia formosa*) is an all-female species which normally reproduces by gynogenesis after mating with males of related species such as *P. sphenops* (2n=46) and *P. vittata* (2n=46). Rasch, E.M. (J. Exp. Zool. 160: 155, 1965) estimated the average amount of DNA per nucleus by cytophotometry in Feulgen-stained tissue sections of *P. formosa, P. sphenops, P. vittata*, and hybrids between *P. formosa* and the latter two species as tabulated:

Organism	Nuclear class	DNA, Feulgen per nucleus	
		Mean	Standard error
P. formosa	2n	0.57	0.01
P. sphenops	2n	0.56	0.01
P. vittata	2n	0.69	0.01
P. formosa x P. *sphenops*	2n	0.83	0.01
P. formosa x P. *vittata*	2n	0.92	0.01

Account for the higher DNA content in cells of the hybrids and explain how you would proceed to verify your explanation.

38. Muntzing, A. (Hereditas 23: 371, 1937) crossed true-breeding *pubescent* lines of *Galeopsis tetrahit*, a species of hemp nettles which has 32 chrosomes, with lines true-breeding for *glabrous* stems. All F_1 plants had *pubescent* stems. Of the F_2's 1,404 were *pubescent* and 105 were *glabrous*.
a) On the basis of these results only, should the species be classified as a diploid, autopolyploid or an allopolyploid? Explain.

b) The species *G. speciosa* and *G. pubescens* each have 16 chromosomes. How would you proceed to determine whether the conclusion above is correct?

39. Autopolyploid plants are often larger, more vigorous and produce more seed and / or green matter than their diploid parents. If this is so, why do breeders not simply double the chromosome number of existing desirable diploid varieties?

40. The pentaploid species *Rosa canina* (2n=5x=35, AABCD) breeds true (Tackholm, G., Acta. Hort. Berg. 7: 97, 1922). The chromosomes of its 2 A genomes form bivalents; the remaining chromosomes occur as univalents in both the megaspore- and microspore-mother cells (meiocytes). The microspores from every microsporocyte form an unordered tetrad. Less than 2% of these spores form functional pollen grains (male gametophytes). The megaspores (nuclei) within each ovule form a linear tetrad.

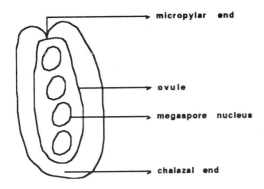

Whereas pollen fertility is very low, ovule fertility is almost always complete (100%).
On the basis of the above information propose a mechanism that will permit *Rosa canina* to breed true. What chromosome number would the endosperm possess?

Chapter 15
Aneuploidy

1. a) Deletion or addition of a single chromosome to the chromosome complement in a diploid organism usually produces a much more pronounced phenotypic effect than it does when added to the chromosome complement of an individual of a tetraploid species. Why?
   b) Trisomy for a chromosome may affect the expression of many different characters; e.g., in humans, individuals trisomic for chromosome 13 express among others, the following congenital anomalies: cerebral defect, cleft palate, harelip, apparent anophthalmia, simian creases, "trigger thumbs", polydactyly and heart defect (Patau, K. et al. Lancet 1: 790, 1960). Why should trisomics have multiple effects on the phenotype?

2. Explain briefly the arrangement of the following in the order of the greatest to the least genic imbalance that each is capable of producing:
   a) Monosomic   b) Tetrasomic   c) Nullisomic
   d) Primary trisomic   e) Secondary trisomic
   Assume the same chromosome is involved in each case.

3. a) In humans several inherited pathological syndromes such as multiple congenital anomalies and Down syndrome have been shown to be caused by trisomy. It has been suggested that a further search for such conditions should be made among traits that are dominant and rare. Explain why this is a good suggestion, referring to the frequency of occurrence and transmission characteristics of trisomics.
   b) Which type of chromosome mutation is responsible for most of the clinically detectable diseases, malformations, and syndromes? Discuss. (See Boue, A. et al., Adv. Hum. Genet. 14: 1, 1985; Hassold, T.J., TIG 2: 105, 1986.)
   c) The majority of individuals with trisomy 21 (Down syndrome) are due to meiotic nondisjunction in the mother. Suggest possible reasons for this occurrence.

4. Illustrate the following meiotic events in a primary trisomic (2n+1=5) plant:
   a) Pairing at zygotene and pachytene.
   b) Configurations at diakinesis and metaphase I.
   c) Segregation at anaphase I.
   d) The chromosome constitution of secondary meiocytes.
   e) Metaphase II and anaphase II.
   f) Chromosome constitutions of the meiotic products (megaspores and microspores). Show the proportions of the various types of spores.

5. a) How might you distinguish between a primary and a secondary trisomic cytologically?
   b) The illustration shows the types of chromosome configurations possible in trisomic plants at diakinesis (chiasmata not shown). One of these cannot occur in a primary trisomic. Which is it and why?

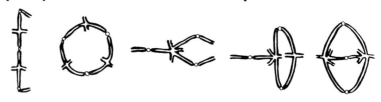

6. a) In monosomics of bread wheat (*Triticum aestivum*; 2n=6x=42) the transmission frequencies of n and n-1 gametes are, on the average, as follows: 25 percent and 75 percent respectively through the female and 96 percent and 4 percent through the male (Sears, E.R. Am. Nat. 87: 245, 1953).
   (i) Show the types and proportions of progeny expected from self-fertilization of a monosomic.
   (ii) What are possible reasons for the differences in transmission frequencies of n and n-1 gametes in the two sexes?
   b) The transmission of n+1 gametes in trisomic plants rarely reaches the expected 50 percent on the female side and is much lower (0 to 3 percent) on the male side. What are some of the possible reasons for the low transmission on the male side? For variations in transmission on the female side? Include in your answer events that occur at meiosis as well as post-meiotic factors.

7. a) *Nystagmus* in humans is an X-linked recessive trait, characterized by involuntary eye movement and impaired vision. Two males with Klinefelter's syndrome (XXY), both of *normal* parents, have *nystagmus*. One has it in both eyes; the other in only one eye. Trace the probable course of events in each case, showing how they differ.
b) *Normal* colour vision vs. *colour blindness* in humans is determined by the X-linked alleles *C* and *c*. A *colour-blind* woman and a *normal* man have a *colour-blind* daughter, contrary to the usual expectation. Explain how such a female can arise and show by chromosome constitutions of parents and daughter whether the abnormality occurred in the father or mother.

8. In a plant which has seven chromosomes in its somatic-cell nuclei, all meiocytes have three bivalents and one univalent at metaphase I. Is the plant monosomic or trisomic? Explain.

9. An individual with Turner's syndrome (AAXO) and afflicted with the X-linked recessive trait *nystagmus* (represented by solid symbols) is found in the third generation of a family. Is it the paternal or maternal X chromosome that is missing in the aneuploid female? Explain.

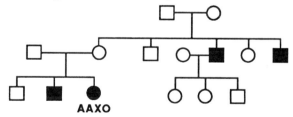

10. In the mouse, *scurfy, sf,* is a recessive X-linked allele that causes death of both sexes before reproductive age. Russell, W.L. et al. (PNAS 45: 554, 1959) were able to obtain progeny from *scurfy* females by transplanting their ovaries into females of a *normal* line. The recipient females when mated with *normal* males produced the offspring tabulated below. The pooled results of six matings are shown.

Scurfy sons - 19
Normal sons - 0
Normal daughters, transmit cause of *scurfy* - 16
Normal daughters, do not transmit cause of *scurfy* - 10

Possible explanations for the origin of *scurfy* females which gave rise to the above results are:
1. High rate of mutation of *Sf* to *sf*
2. Sex-reversal. *Scurfy* females are of genotype X^{sf}/Y, converted to a female phenotype through the action of other factors.
3. *Scurfy* females are homozygous *sfsf* caused by nondisjunction in *Sfsf* mothers and are $X^{sf}X^{sf}Y$.
4. *Scurfy* females are genetically hemizygous $X^{sf}X^o$ (where o denotes a spontaneous deletion of a portion of the X^{Sf} chromosome in heterozygous ($X^{Sf}X^{sf}$) females.
5. *Scurfy* females are monosomic X^{sf}/O due to nondisjunction in the male.

a) The data from ovarian-transplant offspring disprove three of these hypotheses. Which are they and why?
b) Which remaining explanation do you favour and why?
c) What genetic and other tests would you make to confirm or reject your explanation?

11. *Nicotiana tabacum* is an allotetraploid (genome formula SSTT; 2n=4x=48) which arose by doubling of the chromosome number of a hybrid derived from a cross between the diploid species *N. sylvestris* (SS; 2n=24) and *N. tomentosa* (TT; 2n=24). Clausen, R.E. and D.R. Cameron (Genetics 29: 447, 1944) crossed a monosomic *N. tabacum* plant with the *sylvestris* species and obtained some hybrids with 36 chromosomes and other hybrids with 35-chromosomes. The latter consistently showed 12 bivalents and 11 univalents at metaphase I.
a) Was this aneuploid monosomic for a chromosome in the S or the T genome? Why?
b) How many bivalents and univalents would you expect if the aneuploid was monosomic for a chromosome in the other genome? Illustrate.

12. For each of the aneuploids in the following pedigrees indicate (i) the parental source of the X or X's and (ii) explain whether nondisjunction occurred during the first or second division.
a) The *Xgª blood antigen* (solid symbol) is determined by *Xgª*, a dominant X-linked allele.

b) *Glucose-6-phosphate dehydrogenase deficiency* (solid symbol) is a recessive X-linked trait.
c) *Colour blindness* (solid symbol) is a recessive X-linked trait.
d) *Red-green colour blindness* (solid symbol) is an X-linked recessive trait.

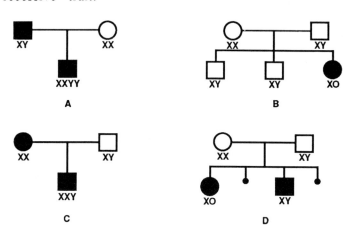

13. a) If both members of a pair of twins from *normal* parents suffer from *multiple congenital anomalies*, are they likely to be identical or fraternal? Explain.
b) In a pair of otherwise identical twins, one member is *normal* and the other has *Down* syndrome. Suggest how this difference between the twins could arise.

14. In general, the consequences of trisomy and tetrasomy in animals are much more serious than in plants. For example, in humans, only 4 of the possible 23 trisomics have been discovered, and of these only trisomics involving chromosome 21 (Lejeune, J. et al., C.R. Acad. Sci. 248: 1721; Bull. Acad. Natl. Med., 143: 256,1959) and the X and Y chromosomes (Jacobs, P.A. and J.A. Strong, Nature 183: 302, 1959) are compatible with life. The others have numerous congenital abnormalities which result in death at an early age (Patau, K. et al., Lancet 1: 790, 1960; Edwards, J.H. et al., Lancet 1: 787, 1960). In contrast, complete series of trisomics have been established in diploid plants like the tomato (Rick, C.M. et al., Can. J. Genet. Cytol. 6: 93, 1964) and Datura (Blakeslee, A.F., J. Hered. 25: 81, 1934), and in which tetrasomics are also viable. Why should plants and animals differ in the effects of extra chromosomes?

15. The types of sex mosaics shown below have been found in humans (autosomal number in all individuals and in all tissues examined is 44).

Chromosome number	Sex-chromosome mosaicism	Sex phenotype
45/46	XO/XX	Turner's syndrome
46/47	XX/XXY	Klinefelter's syndrome
45/47	XO/XYY	Amenorrhea
46/47/48	XX/XXY/XXYY	Hermaphrodite

a) Using any one of the mosaic types, illustrate a common mechanism by which these individuals may originate.
b) Explain how X-linked genes may be used to determine whether the event occurs in the male or the female parent.
c) Do you think that such individuals would show a sharp mosaicism for primary and secondary male and female characters, as is true in insect gynandromorphs? Explain your answer.

16. In the medaka fish (*Oryzias latipes*) sex reversal is accomplished in either direction by feeding appropriate sex hormones to the larvae (Yamamoto, T., Genetics 48: 293, 1963). An incompletely sex-linked pair of alleles *R* vs. *r* determines *orange-red* vs. *white* skin, *R* being dominant to *r*. A cross by Yamamoto (1963) between a heterozygous, sex-reversed female, X^rY^R, and a heterozygous X^rY^R male produced the following progeny:

```
72   orange-red males
 2   white males
23   white females
 3   orange-red females
```

Note: This fish has a primitive X-Y mechanism in which the X and Y chromsomes are apparently completely homologous. This chromosome pair carries the *Rr* allele pair and the *M m* (sex-determining) allele pair. Crossing-over may occur between the two allele pairs in both sexes.

a) Which of these four progeny classes are expected and which are exceptional? Explain.

b) Since the locus is incompletely sex-linked, the following explanations for exceptional individuals may be given:

(i) Crossing-over in the region between Rr and the sex-determining allele pair (thus changing X^r and Y^R to X^R and Y^r).

(ii) Mutation of R to r and r to R in either parent.

(iii) Spontaneous sex reversal of some of the progeny, e.g., X^rX^r *white* sons; X^rY^R *orange-red* daughters.

(iv) Nondisjunction.

Show diagrammatically how the exceptional individuals could arise by each of the four means listed above.

c) Progeny tests of exceptional offspring gave the results shown in the following table:

Parents		Offspring			
		White		Orange-red	
Female	Male	♀	♂	♀	♂
White (X^rX^r) x exceptional *white*		20	14	0	0
exceptional *orange-red* x red (X^rY^R)		6	0	7	11

From this information, determine which of the above explanations can be eliminated and show which represents the most satisfactory explanation.

17. In barley, *long* vs. *short* rachilla hairs are determined by a single pair of alleles, S vs. s. Tsuchiya, T. (Jpn. J. Bot. 17: 177, 1959) self-fertilized an F_1 plant trisomic for chromosome 7 and heterozygous at the S locus. The 2n+1 F_2 progeny consisted of 45 plants with *long* rachillas and 2 with *short* ones. The 2n F_2 plants when self-fertilized produced 100 *long* : 9 *short* F_3 progeny. Using your own symbols and diagrams, provide a satisfactory cytogenetic explanation for these results.

18. In *Datura stramonium* (2n=24), *purple* vs. *white* flower colour is determined by a single pair of alleles Pp located on the third smallest ("poinsettia") chromosome (Blakeslee, A.F. and M.E. Farnham, Am. Nat. 57: 481, 1923). The allele P for *purple* is dominant to p for *white*. In plants trisomic for this chromosome, n+1 pollen rarely ever functions, and only about 25 percent of the functional eggs are n+1.

a) What is the expected ratio of *purple* to *white* from the crosses shown below if P is close to the centromere?

Female		Male
PPp	x	PPp
Ppp	x	Ppp
PPp	x	pp
pp	x	PPp
pp	x	Ppp

b) Would the phenotypic ratio be the same if P is 50 or more map units from the centromere? Illustrate, using the cross PPp female x pp male.

c) Why do phenotypic ratios among progenies from trisomics differ from those expected in disomic inheritance?

19. a) What are the features by which B chromosomes can be distinguished from normal A chromosomes?

b) Rats (*Rattus rattus*) are diploid organisms whose somatic cells possess a normal complement of 38 chromosomes. Standard cytological analysis leads to the finding of a vigorous and fertile male with the normal complement plus two additional chromosomes.

(i) Offer a plausible explanation for the phenotype and fertility of this 2n+2 male.

(ii) Outline the experimental approaches you would use to confirm or refute your explanation, indicating the results expected if your hypothesis is (i) correct (ii) incorrect.

20. Morgan, L.V. (Biol. Bull. 42: 26, 1922) found that when a *yellow*-body *Drosophila* female was crossed with a *gray*-body male, male and female offspring were present in equal proportions with all the female offspring resembling the mother and all the sons (fertile) their father. The same results were obtained in subsequent generations. Propose a cytogenetic explanation to account for these results.

21. In spinach (*Spinacia oleracea;* 2n=12), sex is determined by a single pair of alleles. In 1959 Janick, J. et al. (J. Hered. 50: 47) established a complete series of primary trisomics and then crossed the *male (staminate)* plants in each of the trisomic lines with diploid *female (pistillate)* plants to determine which chromosome carried the sex-determining gene. Note: Rarely do n+1 gametes function on the male side. The results obtained were as shown:

Trisomic line used as male parent	Number of progeny	
	Male	Female
Savoy	387	384
Oxtongue	169	173
Star	191	217
Curled	218	226
Reflex	74	143
Wild	1,116	1,148

a) Which chromosome carries the sex-determining gene? Explain.
b) What was the chromosome constitution and genotype of the *male* parent trisomic for the sex-determining gene?
c) If the *staminate* trisomics of the type mentioned in (b) arose from a disomic x trisomic cross, state with reasons whether the *male* or the *female* parent was trisomic.

22. a) In organisms such as the mouse, XO individuals are healthy fertile females, whereas OY individuals are inviable. What types and proportions of offspring can be expected from XO female x XY male matings?
b) Cattanach, B.M. (Genet. Res. 3: 487, 1962) found that in such matings the XO offspring were only one-third as frequent as the XX. He further found that:
1. The litter sizes of XO and XX females were the same.
2. The sex ratios among the offspring of XO and XX females were the same.
3. There is no postnatal inviability of the XO class, at least up to three weeks.

What do Cattanach's observations suggest concerning the unexpected frequency of XO in such matings and its probable cause?

23. In tomatoes (*Lycopersicon esculentum*; 2n=24), Lesley, J.W. (Genetics 17: 545, 1932) determined the location of the following pairs of alleles by self-fertilizing three different plants, each trisomic for a different chromosome (A,B or I) and heterozygous for the following allele pairs:

Dd standard vs. dwarf growth habit   *Aa* purple vs. nonpurple stems
Ll green vs. virescent plant colour   *Yy* yellow vs. nonyellow fruit skin
Rr red vs. yellow fruit flesh

His results are tabulated below:

Parent	Allele pair studied	Progeny			
		Diploid		Trisomic	
		Dominant	Recessive	Dominant	Recessive
Trisomic A	*Dd*	71	44		
Trisomic B	*Dd*	19	6	27	9
	Ll	23	2	35	0
	Rr	9	3	13	4
	Aa	20	7	26	10
Trisomic I	*Yy*	68	24	6	1
	Rr	90	12	7	0

State with reasons:
a) Which of the allele pairs are on chromosome A, which are on B, and which are on I.
b) The genotypes of the three parents.
c) The percent transmission of n and n+1 gametes on the female side (no n+1 gametes function on the male side).

24. In mice, *male antigen* is produced by males, but not by females. It could therefore be determined either by a Y-linked or by an autosomal gene (hence present in both sexes but functionally suppressed by two X chromosomes of the female). Celada, F. and W.I. Welshons (Genetics 48: 131, 1963) found that XX and XXY animals differ antigenically whereas XX and XO do not. Do these observations permit you to make an unequivocal decision? Explain.

25. In variety C.I. 12633 of bread wheat (2n=6x=42), *resistance* to stem rust is governed by the dominant alleles of two linked duplicate genes. Nyquist, W.E. (Agron. J. 49: 222, 1957) crossed C.I. 12633 with each of the 21 monosomics of Chinese Spring, a variety *susceptible* to stem rust. The F_1 monosomics and one *normal* (euploid) F_1 were self-fertilized. The reactions to races 11, 17 and 56 in F_2 plants and F_3 families are tabulated below:

	F_2 progenies			F_3 families	
Chromosome	*Resistant*	*Susceptible*	X^2	*Resistant* and Segregating	*Susceptible*
1	68	15	0.232	16	4
2	16	3	0.083		
3	43	8	0.028	19	1
4	75	14	0.010		
5	57	6	1.314	36	4
6	73	15	0.057	17	3
7	109	28	2.133	19	1
8	78	17	0.239		
9	83	14	0.027	19	1
10	86	12	0.584		
11	86	7	3.970	16	4
12	69	12	0.001	19	1
13*	131	3	17.084	24	
13**	58	20	5.301		
14	31	6	0.013		
15	80	16	0.026		
16	43	8	0.028	17	3
17	74	15	0.037		
18	70	9	0.748		
19	94	20	0.209	19	1
20	65	10	0.138	18	2
21	54	5	1.744		

* Plant from monosomic F_1 plants ** Plants from *normal* F_1 plants

a) Which chromosome pair carries the duplicate pairs of alleles? Diagram the cross to show why you reach the conclusion you do.
b) What is the distance between the two linked genes? Indicate the results used in calculating map distance.

26. Reader, S.M. and T.E. Miller (Euphytica 53: 57, 1991) transferred an allele for *resistance* at the locus for reaction to powdery mildew from wild tetraploid wheat *Triticum dicoccoides* (2n=4x=28; AABB) to bread wheat, *Triticum aestivum* (2n=6x=42; AABBDD). Crosses between *resistant* plants in the cultivar Maris Nimrod and the *susceptible* cultivar Chinese Spring produced *resistant* F_1's and 44 *resistant* and 12 *susceptible* F_2 plants. Monosomic F_1's from crosses between the complete series (n=3x=21) of monosomics in Chinese Spring and *resistant* plants in Maris Nimrod were selected and self-fertilized. The F_2 results are shown below:

Monosomic line	*Resistant* plants	*Susceptible* plants	X^2
1A	18	2	2.400
1B	18	2	2.400
1D	16	4	0.267
2A	2	2	*
2B	13	7	1.067
2D	17	3	1.067
3A	31	8	0.429
3B	15	5	0.000
3D	16	4	0.267
4A	39	1	10.800++
4B	40	5	4.629+
4D	13	6	0.428
5A	16	4	0.267
5B	15	1	3.000
5D	17	3	1.067
6A	4	2	*
6B	24	14	2.842
6D	2	0	*
7A	17	3	1.067
7B	15	5	0.000
7D	15	5	0.000

* Calculation of X^2 not valid due to low population size. Level of significance of differences between observed and expected ratio of 3 *resistant* to 1 *susceptible* plants is denoted by: += p< 0.05; ++=<0.005.

Note: the single *susceptible* F₂ plant in the F₂ from monosomics for chromosome 4A possessed 20 bivalents and a telocentric univalent.

a) Outline the rationale for using monosomics in cytogenetic analysis.
b) Which chromosome carries the gene for reaction to powdery mildew? Explain taking into consideration all the given information.

27. Clausen, R.E. and D.R. Cameron (Genetics 35: 4, 1950) found that crosses between plants of the varieties *Purpurea* and *Chinchao* in *Nicotiana tabacum* (2n=4x=48) produced *green* F₁'s and F₂'s segregating 626 *green* and 44 *white* seedlings. A cross between F₁'s and a heterozygous Purpurea plant gave 87 *green* and 14 *white* offspring. Each of the 24 monosomics in Purpurea was crossed with Chinchao. The F₁ monosomics, all *green*, were self-fertilized. The number of seedlings in each phenotypic class in the F₂ for each of the 24 families as well as for one of the *normal* (euploid) crosses are shown:

Monosomic line	Green	White	Monosomic	Green	White
A	87	7	M	88	3
B	85	8	N	92	5
C	94	5	O	88	8
D	83	4	P	196	13
E	92	8	Q	94	5
F	77	7	R	91	7
G	86	0	S	88	3
H	87	8	T	66	24
I	79	6	U	72	7
J	79	6	V	94	3
K	94	4	W	86	7
L	-	-	Z	86	7
Normal x Normal	90	7			

Give a cytogenetic explanation of these results, including in your explanation the mode of inheritance of chlorophyll production and the location of the gene or genes concerned.

28. In humans, about 0.5 percent of the population are heteromorphic for chromosome 1, the homologue carrying the *uncoiler* allele Unl being significantly longer than that carrying unl because of uncoiling in the long arm. The pedigree shows the segregation of the abnormal and normal homologues of chromosome 1 and the co-dominant alleles Fy^a and Fy^b at the *Duffy* blood-group locus.

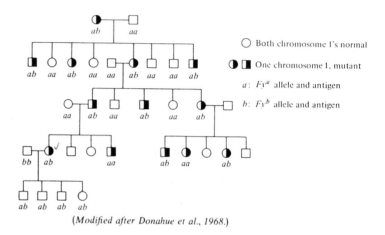

(*Modified after Donahue et al., 1968.*)

Is the *Duffy* locus likely to be on chromosome 1? Explain.

29. In bread wheat, *Triticum aestivum* (2n=6x=42; AABBDD), there are 3 variants, CL_1, CL_2 and CL_3 of the polypeptide present in lectins. These proteins are dimers and are present in 6 different active forms (CL_{11}, CL_{22}, CL_{33}, CL_{12}, CL_{13}, CL_{23}) in seed embryos. Using ion-exchange chromatography and agglutination assays, the different lectins can be separated, detected and quantitated. Stinissen, H.M. et al. (Theor. Appl. Genet. 67: 53, 1983) determined the elution patterns of these proteins in (i) the bread wheat cultivars Chinese Spring (CS) and Hope (H), (ii) nullisomic-tetrasomic (NT) lines in CS, and (iii) a chromosome substitution line of CS in which a chromosome pair 1B was replaced by a homologous pair from Hope. Their results are presented in figures 1 and 2. NT lines are those in which a specific chromosome pair, e.g., 1A, is missing and another chromosome, e.g., 1D, is present 4 times and not just 2 times (as a chromosome pair). Such lines are

follows: CS N1A T1D (nullisomic for chromosome 1A-tetrasomic for chromosome 1D).

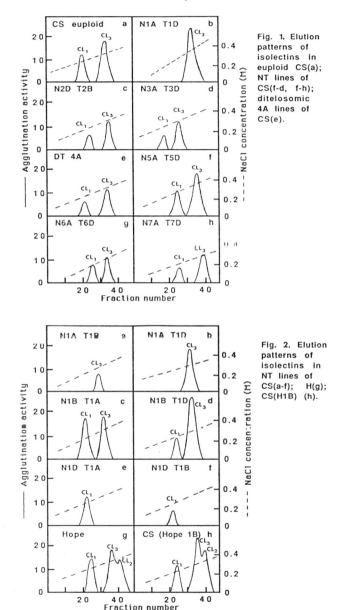

Fig. 1. Elution patterns of isolectins in euploid CS(a); NT lines of CS(f-d, f-h); ditelosomic 4A lines of CS(e).

Fig. 2. Elution patterns of isolectins in NT lines of CS(a-f); H(g); CS(H1B) (h).

a) Which of the lectins does euploid CS produce? Explain. Why does CS not produce the remaining type?

b) What is the minimum number of genes in *T. aestivum* that code for lectins? Explain.
c) Which chromosomes carry these genes? Explain.
d) Why was the CS (H 1B) chromosome substitution line used in addition to the NT ones and what information does it provide?

30. In bread wheat, *Triticum aestivum* (2n=6x=42, AABBDD), the gene $Lr32(=L)$ for *resistance* to leaf rust is on chromosome 3D. Kerber, E.R. (Crop Science 28: 178, 1988) generated monotelosomic F_1's which possessed a normal 3D chromosome from the *resistant* (LL) line and a telocentric chromosome for either the long (q) arm or the short (p) arm of chromosome 3D from appropriate aneuploids of the *susceptible* (ll) Chinese Spring line.
i) Self-fertilization of F_1 monotelodisomics with a chromosome telocentric for the q arm of 3D produced plants with the following kinds of chromosome constitutions and phenotypes:
42 chromosomes including a pair of normal 3Ds- *resistant*.
42 chromosomes including a normal 3D and a 3D q arm telocentric-*resistant*.
41 chromosomes the univalent being a 3D q arm telocentric-*susceptible*.
42 chromosomes including a pair of 3D q arm telocentrics - *susceptible*.
ii) The F_1 monotelodisomics that had a chromosome telocentric for the p arm of chromosome 3D, when testcrossed with Chinese Spring produced progeny with chromosome constitutions and phenotypes that were as follows:
42 chromosomes including a pair of normal 3Ds-29 *resistant* and 14 *susceptible*.
42 chromosomes including a normal 3D and a 3D p arm telocentric - 1 *resistant* and 12 *susceptible*.

With the aid of appropriate diagrams show which arm of chromosome 3D carries the L locus and the distance of this gene from the centromere.

31. In corn (2n=20) two aneuploids occur; one a tetrasomic (2n+2=22), the other a double trisomic (2n+1+1=22). How would you distinguish between the two cytologically?

Chapter 16
Chromosome Aberrations

1. The frequencies of radiation-induced single-break chromosomal aberrations seem to be in direct proportion to the dosage of radiation over a very wide range, but this is not true of two-break aberrations. Explain.

2. Why are chromosomal aberrations considered to have less significance than gene mutations for subsequent generations?

3. Many recessive mutations are minute deletions. Therefore the loci of the genes involved are lost rather than altered. What kinds of experiments and observations could be made to decide whether a particular mutation is the result of an alteration in the gene or the loss of a short segment of the chromosome carrying the gene?

4. a) Why are inversions referred to as crossover suppressors? What term would be more appropriate to describe the effect?
   b) Show how (i) paracentric and (ii) pericentric inversions can act as recombination suppressors.

5. A standard strain of a diploid plant, e.g., corn, is homozygous for the dominant alleles at 10 loci in the sequence A B C D E F G H I J. Another strain is homozygous for the corresponding recessive alleles a b c d g f e h i j.
   a) Diagram the configuration of synapsing chromosomes at pachytene in the hybrid between these strains.
   b) Diagram the configurations you would observe at anaphase I and anaphase II of meiosis and show the types of spores that the hybrid would form if the inversion were a paracentric one and:
   (i) There was one crossover within the inversion loop.
   (ii) There was one crossover between the inversion loop and the centromere.
   (iii) There were two crossovers within the inversion loop involving (a) all four chromatids (b) the same two non-sister chromatids.
   c) Answer the questions in (b) if the inversion was a pericentric one.

6. Explain why gametophyte (e.g., pollen) lethals are common in plants whereas gametic lethals are unknown in animals.

7. a) What are transposable elements (transposons)?
   b) Transposable elements are known to induce deletions, inversions, duplications and translocations (McClintock, B., Carnegie Inst. Wash. Yearbook 53: 254, 1954; Shapiro, J. (Ed.) *Mobile Genetic Elements*, Academic Press, New York; Fedoroff, N.V., Cell 56: 181, 1989).
   With the aid of diagrams explain how (i) Ac and Ds in *Zea mays* or (ii) Copia in *Drosophila melanogaster* can give rise to deletions and inversions.
   c) *Alu* repeat sequences in humans (i) are about 300 base-pairs long (ii) contain the sequence AGCT and (iii) are repeated about 300,000 times in the human genome. Ariga, T. et al. (Genomics 8: 607, 1990) demonstrated that these repeat sequences result in partial deletions within the C1 inhibitor gene.
   Suggest a plausible mechanism whereby these repetitive elements can generate such deletions.

8. What may be an important role of chromosomal duplications in evolution?

9. A geneticist has a strain of diploid plants; the gene orders for the homologues of two of the chromosome pairs in this strain are A B ● C D E F and K L ● M N O respectively. He subjects the seed to X-ray treatments and induces a reciprocal translocation between two non-homologues. The resulting translocation chromosomes are A B ● C N O and K L ● M D E F.
   a) Diagram the pachytene configuration in the translocation heterozygotes.
   b) Show the metaphase I orientations and anaphase I segregations when these are not directed and the types of spores resulting from each.

c) Show diagrammatically the chromosome constitutions and genotypes of offspring expected in a testcross of an F_1 heterozygous for the translocation and for the genes shown. Would you expect some of the progeny to be semisterile? Why?

10. Are viruses able to cause chromosome breakage? Explain.

11. Discuss the role of heterochromatin in karyotype evolution involving changes in chromosome number. Refer to specific examples if possible.

12. All chromosomes in a certain diploid plant except one are *normal*. This *mutant* chromosome has, as a result of chromosome breakage, genes a b g h i j in the order shown, whereas the gene content and gene order of its *normal* homologue of this chromosome is A B C D E F G H I J.
a) Show how these two chromosomes would pair at pachytene.
b) If such a plant was self-fertilized, show the types and proportions of zygotes expected.
c) Deletion heterozygotes do not have the same probability of occurrence in plants and animals. In which kingdom are they more likely to be found and why?

13. In *Drosophila melanogaster*, at the X-linked *yellow* locus the allele Y for *gray* body is dominant to y for *yellow* body. Deletion of the segment of the X chromosome carrying this locus occurs naturally and can be induced by various mutagens. In both YY and yy strains, when a female (XX) is homozygous for such a deletion (does not carry the *yellow* locus on either X), she is, like individuals with the y allele, phenotypically *yellow* (Ephrussi, B., PNAS 20: 420, 1934). What does this tell you about the allele y, knowing that it is an allele, and not a deficiency?

14. a) In mice, *normal* vs. *waltzing* gait is determined by a single gene; v, the allele for *waltzing*, is recessive to V for *normal*. In matings of true breeding *normals* with *waltzers*, Gates, W.H. (Genetics 12: 295, 1927) found a single *waltzing* female among several hundred *normal* F_1's. When the F_1 *waltzer* was mated with two different males of the *waltzer* stock, 11 progeny were obtained, all *waltzers*. When mated with *normal* males, she produced 13 *normal* progeny and no *waltzers*. Three females of this *normal* progeny mated with two of their brothers produced 60 progeny, all *normal*. In a mating of one of these females with a third brother, 4 *normals* and 2 *waltzers* appeared in a litter of 6. Possible explanations for the unexpected appearance of the original *waltzer* F_1 female are:
1. Mutation of a dominant to a recessive allele in the normal parent.
2. Mutation of a suppressor of v.
3. Nondisjunction of the chromosome carrying V in the *normal* parent.
4. A deletion, in the *normal* parent, of a portion of the chromosome carrying V.
Which two explanations are eliminated by the breeding data and why?
b) Painter, T.S. (Genetics 12: 379, 1927) cytologically examined the chromosomes of two *waltzers* from the progeny of this female. They had 40 chromosomes as did *normal* and *waltzer* mice, but one pair was heteromorphic: one chromosome of this pair being abnormally short.
(1) What was the cause of the original *waltzing* female? Explain and illustrate.
(2) What was the chromosome constitution of the *normal* male and female sibs of the cross F_1 *waltzer* x *normal* male that gave only *normal* progeny? Why?

15. In corn, $Bmbm$, $Btbt$, V_3v_3 and $Bvbv$ determine the expression of *green* vs *brown* midrib, *soft* vs. *brittle* endosperm, *green* vs. *virescent* seedling, and *normal* vs. *half-normal* height, respectively. In 1933 Stadler, L.J. (Univ. Mo. Res. Bull. 204, p. 3) crossed a $bmbmbtbtv_3v_3bvbv$ strain as female parent with a $BmBmBtBtV_3V_3BvBv$ strain. A few plants in the progeny expressed all four recessive traits, and half their pollen was defective. When crossed as males with the parental quadruple recessive strain, all progeny expressed the recessive traits but none had defective pollen. Progeny with defective pollen showed a short buckled region in the long arm of one chromosome 5 at pachytene. Discuss these observations, giving reasons for your answers, with respect to:
a) The condition causing the recessive F_1 phenotypes.
b) The transmission frequency of this condition through male and female gametophytes.

16. In corn, a plant heterozygous for a deletion of a portion of the short arm of chromosome 9 is also heterozygous at the *sh* locus located in the short arm of this chromosome. The deleted chromosome carries the recessive allele *sh* (*shrunken* endosperm), and its *normal* homologue carries the dominant allele *Sh* (*full* endosperm). When a *shsh* plant was fertilized with pollen from this heterozygote, 14 percent of the kernels were *shrunken*. Explain the appearance of these recessive types in view of the fact that the deleted chromosome is not transmitted through the pollen.

17. The following data are from Rees, H. and R.N. Jones (Nature 216: 825, 1967):
    (i) The chromosomal DNA content of the onion, *Allium cepa*, is about 27 percent greater than that of *A. fistulosum*.
    (ii) The chromosomes of *A. cepa* are correspondingly larger than those of *A. fistulosum*.
    (iii) In hybrids between the two species all bivalents at metaphase I are heteromorphic; the homologues of each chromosome pair are unequal in length. Each of the bivalents shows at least one loop which is formed by one of the homologues only.
    Both species have the same chromosome number. If *A. fistulosum* is ancestral to *A. cepa*, what is the probable basis for the difference in both chromosome size and DNA content between the two species? Elaborate.

18. Russell, L.B. and W.L. Russell (J. Cell Comp. Physiol. 56(Suppl. 1): 169, 1960) examined the F_1 progeny of a group of irradiated and unirradiated *wild-type* mice mated with a strain of mice homozygous for the recessive alleles of several genes, including *d* (*dilute*) and *se* (*short ear*), which are on the same chromosome, 0.16 map unit apart; 15 individuals among several hundred F_1's were *dilute, short ear*. Of the possible mechanisms listed below that might produce these mutants explain which should be accepted and which should not:

1. Simultaneous mutation of *D* to *d* and *Se* to *se*.
2. Inactivation of *D* and *Se* through position effect.
3. Nondisjunction in the *wild-type* parent producing the monosomic *d se*/O.
4. Deletion of chromosome segment carrying *D* and *Se*.
5. Nondisjunction in both parents leading to an individual homozygous for *d* and *se*.

19. In the mouse, the *Sp* (*splotch*) locus is located in band C4 of chromosome 1. The X-ray induced mutation Sp^r is a cytogenetically detectable deletion of band C4 which is lethal when homozygous. Epstein D.J. et al. (Genomics 10: 89, 1991) used molecular probes of different genes, known to be closely linked to the *Sp* locus, to characterize the chromosomal segment deleted in the Sp^r mutant. Southern blots of genomic DNA from a Sp^rSp^+ heterozygote and its Sp^+Sp^+ littermate were sequentially hybridized with the probes of 9 different genes. The results for 4 of these hybridizations involving probes for genes *Vil, Des, Acrg* and *Akp3*; were as follows:

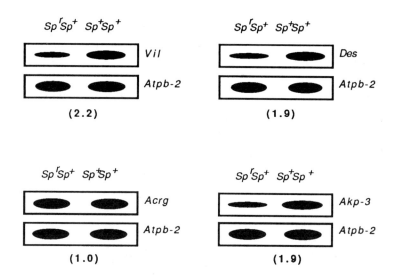

The gene probes and the internal hybridization standard (control) (*Atpb-2*) are indicated next to each pair of blots. The numbers in parantheses represent the ratio of band intensities in the 2 mice for each of the 4 probes.

a) Which of these 4 genes is (are) located in band C4 of chromosome 1? Which is (are) outside this region? Outline the rationale for your decision.

b) For the gene(s) next to the deletion how would you proceed to determine the distance between the deletion and the gene locus? Show the results expected.

20. In humans, the allele pair *Xgxg* (*presence* vs. *absence* of Xg^a blood antigen) is on the X chromosome. Lindsten, J. et al. (Nature 197: 648, 1963) presented the results of a study of two families in which *Xg^a* antigen (solid symbols) distribution was determined in the various members. The two propositae both displayed chromosome mosaicism involving two kinds of cell lines: one line had 45 chromosomes including a normal single X, and the other, had 46 chromosomes, including a normal X and an isochromosome for the long arm of the X.

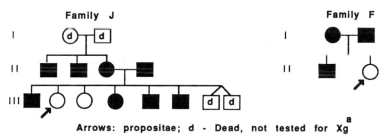

Arrows: propositae; d - Dead, not tested for Xg^a

a) Is the isochromosome of maternal or paternal origin?
b) Is the *Xg* locus on the long or the short arm of the X chromosome?

21. a) In *Drosophila melanogaster*, *bar* eye is caused by a tandem duplication of a segment of the X chromosome (*bar* females have the duplication in at least one X). In *nonbar* flies this segment is present only once in each X; in *double bar* it is present in triplicate in at least one X. In the offspring of homozygous *bar* females one out of 1,600 males is *nonbar*, and the offspring of *double bar* revert to *nonbar* at almost the same frequency (May, H.G., Biol. Bull. 33: 361, 1917; Zeleny, C. J., Genet. Physiol. 2: 69, 1919; J. Exp. Zool. 34: 203, 1921). Show diagrammatically how these reversions occur.
b) In 1923 Sturtevant, A.H. and T.H. Morgan (Science 57: 746) testcrossed females of the constitution *F B fu / f B Fu* and found three *nonbar*, two *wild-type* (*nonforked, nonfused*), one *forked, fused* among the male offspring. In 1925 A.H. Sturtevant (Genetics 10: 117) crossed *F B Fu / f B fu* females with *f B fu / Y* males, and classified the 18,999 progeny as shown:

	Bar				Nonbar			Double bar
	F Fu	f fu	F fu	f Fu	F Fu	F fu	f Fu	f Fu
Males	5,218	4,160	93	124	1	2	1	1
Females	5,413	3,749	94	140	0	0	2	1

F = *nonforked*, *f* = *forked* bristles; *Fu* = *nonfused*, *fu* = *fused* veins
(i) Show diagrammatically how this confirms your answer to (a) by using *forked* and *fused* as marker genes for the segments on either side of bar.
(ii) Illustrate how both *nonbar* and *double bar* may originate from homozygous bar females by a single meiotic event.

22. The pedigree, from Brogger, A. (Hereditas 62: 116, 1969), shows the phenotypes of the parents (I-1 and I-2) and of two of their sons, who are heterozygous for a deletion of a portion of the short arm of a chromosome 13. The alleles at each of the 4 loci are codominant.

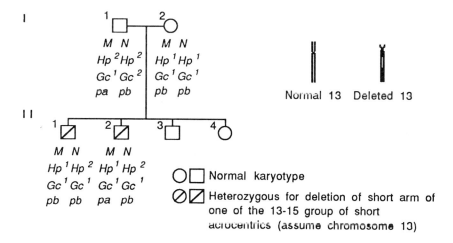

Is it possible to determine which, if any, of these genes is on chromosome 13? Explain.

23. In humans, the δ and β hemoglobin polypeptide chains, each 146 amino acids long, are coded for by genes δ and β respectively. The partial amino acid sequences of these polypepetides in *normal* individuals are as follows:

```
      1   9   12   22   50      86 87   116 117   125 126   146
δ-NH₂ Val...Thr....Asn....Ala.......Ser..Ser Gln...ArgAsn......Gln Met.....His-COOH
β-NH₂ Val......Ser....Thr....Glu......Thr....AlaThr.....HisHis ......ProVal.....His-COOH
```

In 1962 Baglioni C. (PNAS 48: 1880) discovered a mutant hemoglobin called Lepore. It was also 146 amino acids long and had the following amino acid sequence:

```
      1   9   12   22   50      86 87   116 117   125 126   146
-NH₂-Val.....Thr....Asn....Ala...Ser..... SerGln........HisHis .......Pro Val......His-COOH
```

a) What does this mutant hemoglobin amino acid sequence tell us about the location of the δ- and β- genes in the human genome?
b) What is the most likely mechanism by which such mutant genes and therefore hemoglobins arise? Illustrate the mechanism.

24. a) Using a hypothetical chromosome A B C D • E F G H, show how ring chromosomes can be produced which possess a centromere.
b) Show how a ring chromosome with a centromere may increase, decrease, or remain the same in size during mitotic divisions.
c) If the ring carries a dominant allele and is present in a plant along with normal chromosomes homozygous for the recessive allele, what phenotypic effects are expected?

25. In *Drosophila melanogaster*, the paracentric inversion *scute-4* involves most of the X from just to the right of the *scute* locus to a point between *carnation* and *bobbed*. The progeny of female inversion heterozygotes do not contain single recombinant types in three-point crosses for these genes with a standard chromosome line.
a) What are the two possible explanations for failure of the single recombinant to occur?
b) Single crossovers occur within the inverted segment, but no single recombinant is recovered and there are no zygotic abortions. Explain how this can happen.

26. Ferguson-Smith, M.A. et al. (Hum. Genet. 34: 35, 1976) measured the production of the red blood cell enzyme *adenylate kinase (AK)* in 7 patients heterozygous for deletions or duplications of specific segments of chromosome 9 (shown below). The following results were obtained:

Patient	Chromosome 9 Aberration		AK activity
	Type	Region	
1	Terminal deletion	pter-p21	*normal*
2 and 6	Terminal deletion	pter-q11	*normal*
3 and 4	Duplication	pter-q12	*normal*
7	Duplication	q31-qter	*Increased 43%*
5	Duplication	q11-q33	*normal*

a) Is the *AK* locus on chromosome 9? In what region? Explain how you arrive at your decision.
b) Indicate a current technological procedure that you might use to localize the *AK* locus more precisely.

27. In corn, the allele pairs *Cc* (*coloured* vs. *colourless* endosperm), *Shsh* (*full* vs. *shrunken* endosperm) and *Wxwx* (*starch* vs. *waxy* endosperm) are at loci 26, 29 and 54 repectively, in the short arm of chromosome 9. B. McClintock in 1941 (Genetics 26: 234) found a plant with a reversed duplication of the short arm of chromosome 9 in a strain homozygous for the dominant alleles *Wx, Sh, C*. Plants heterozygous for this chromosome and a normal 9 carrying the recessive alleles *c, sh,* and *wx* were crossed with plants homozygous for the recessive alleles and normal chromosome 9 as female parents. Many of the kernels had a

variegated aleurone (outer layer of endosperm) showing *purple* and *white* spots.

a) Illustrate the chromosomal and genotypic constitutions of the meiotic products of a meiocyte after crossing-over occurs between the reverted segment of duplicated chromosome 9 and its homologue as shown.

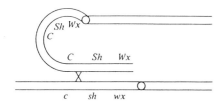

b) McClintock (1941) showed that the genotypic variegation in the endosperm of the testcross progeny was due to a breakage-fusion-bridge-breakage cycle. Explain with illustrative diagrams how this cycle, which occurs in the gametophytes and the endosperm tissue, gives rise to the genotypic and phenotypic variegation.

28. In *Drosophila melanogaster*, the allele pairs *Pp* (*purple* vs. *red* eyes) and *Mm* (*normal* vs. *miniature* wings) are located on chromosome 2 and 3 respectively. Males heterozygous for both pairs of alleles were individually crossed with *pp mm* females. The offspring of most testcrosses comprised 4 phenotypic classes; *purple, normal*; *purple, miniature*; *red, normal*; and *red, miniature*. In a few progenies only *purple, normal* and *red, miniature* individuals were present in approximately equal numbers.

a) Suggest an hypothesis to account for these results.
b) Indicate what cytological observations would confirm your hypothesis.
c) Would the same results be obtained in the reciprocal of this cross? Explain.

29. In rabbits the allele pairs *Bb* and *Ww* determine *black* vs. *white* and *wavy* vs. *straight* hair respectively. The map distance betweeen *B* and *W* in strains with the normal karyotype is 27 map units. In one particular strain (A) *BBWW* x *bbww* crosses revealed that the distance between these two loci is 9 map units. When strain A was crossed with a strain possessing the standard segmental arrangement in all chromosomes, approximately 24 percent of the progeny aborted (24% semisterility).

a) What is the cause of the reduction in map distance between the *B* and *W* loci in strain A? Explain.
b) How would you verify your explanation cytologically? Explain.

30. In *Drosophila melanogaster*, the allele pairs *Stst* (*red* vs. *scarlet* eye), *Srsr* (*normal* vs. *stripe* thorax), *Ee^s* (*gray* vs. *sooty* body), *Roro* (*smooth* vs. *rough* eye) and *Caca* (*red* vs. *claret* eye) are on chromosome 3 as follows:

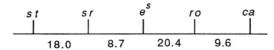

In 1926 Sturtevant, A.H. (Biol. Zentralbl. 46: 697) crossed *wild-type* females heterozygous at all these loci with males homozygous for the recessives alleles of all these genes. The progeny fell into eight phenotypic classes, as shown.

Phenotype	Number
Red, normal, gray, smooth, red	2,214
Scarlet, stripe, sooty, rough, claret	2,058
Red, stripe, sooty, rough, claret	219
Scarlet, normal, gray, smooth, red	238
Scarlet, stripe, gray, smooth, red	4
Red, normal, sooty, rough, claret	3
Scarlet, stripe, sooty, smooth, red	1
Red, normal, sooty, rough, red	1

Show how an inversion can account for these results and locate the appropriate position of the breakage point with respect to the gene loci.

31. Corn plants heterozygous for an inversion in chromosome 4 show no ovule abortion (Morgan, D.T., Genetics 35: 153, 1950). How can you account for this?

32. In 1974 Lamm, L.U. et al. (Hum. Hered. 24: 273) using Q-banding techniques showed that some individuals in a 3-

generation family shown below had a normal karytotype; others in this family had a normal chromosome 6 and a mutant chromosome 6. The difference between the 2 chromosomes is shown below:

These investigators also studied the inheritance of 6 alleles (Hl^1, HL^2, HL^3, HL^4, HL^5, HL^6) at the HL-A (=HL) locus in this family, each of which produces a specific serological haplotype or antigen. The chromosome 6 constitutions and genotypes of the individuals in this family are as shown below:

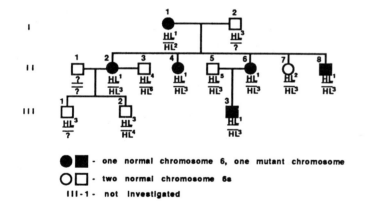

a) What is the nature of the mutant chromosome? Explain.
b) Is the HL-A locus on chromosome 6? Explain.
c) Would you expect abortions or miscarriages in matings involving individuals heterozygous for the chromosome mutation? Why?

33. a) It is found that an individual heterozygous for a long inversion produces more double- than single-recombinant progeny. Explain why this occurs.

b) A certain paracentric-inversion heterozygote in barley shows approximately 25 percent pollen abortion but no ovule abortion. Explain.
c) Why are *Drosophila* females heterozygous for a pericentric inversion less fertile than those heterozygous for a paracentric inversion?

34. A different sequence of the genes on chromosome 9 is found in each of four strains of corn, as shown below:

Strain 1 D^+ yg_2 bz sh_1 c bp wx
Strain 2 D^+ bp c yg_2 bz sh_1 wx
Strain 3 D^+ yg_2 c bp bz sh_1 wx
Strain 4 D^+ yg_2 c sh_1 bz bp wx

Given that the first strain is the ancestral one, account for the sequence of events involved in the production of the other 3 strains.

35. In 1936 Sturtevant, A.H. and T. Dobzhansky (PNAS 22: 448) pointed out that overlapping inversions permit inferences regarding the phylogeny of different races with respect to the gene arrangements in a given chromosome. In *Drosophila pseudoobscura* the *Pike's Peak* and *Arrowhead* strains differ from the *standard* by one inversion each:

Standard   1 2 3 4 5 6 7 8 9 10 11 12 13
Arrowhead   1 2 3 4 9 8 7 6 5 10 11 12 13
Pike's Peak 1 2 7 6 5 4 3 8 9 10 11 12 13

What are the possible phylogenetic sequences? Elaborate.

36. In barley, approximately half the progeny of a self-fertilized translocation heterozygote are semisterile, whereas all the progeny of a self-fertilized inversion heterozygote are fully fertile. Explain why the progeny of the former but not the latter are semisterile.

37. In *Datura*, the allele pairs *Pp* (*purple* vs. *white* flowers) and *Ss* (*spiny* vs. *smooth* capsules) are usually independently inherited. When the F_1's (all fully fertile) from a certain cross between *PPSS* and *ppss* were testcrossed, however, two

of the F_1 plants gave the following progeny that were all *fully fertile*:

PS 80   *ps* 90   *Ps* 16   *pS* 14

Suggest an hypothesis to account for these results and explain why the F_1 parents of the above progeny and all the progeny are fully fertile.

38. In *Drosophila*, the *Aa* and *Bb* pairs of alleles determine the expression of *normal* vs. *abrupt* wing and *gray* vs. *black* body, respectively. In each of two different populations, homozygous strains *normal, gray* and *abrupt, black* are found. The F_1's of crosses between the two homozygous strains within each population were testcrossed and produced the results shown below: Population 1 is the ancestral population.

Phenotype	Number in each class	
	Population 1	Population 2
Normal, gray	2,205	2,230
Normal, black	20	3
Abrupt, gray	24	2
Abrupt, black	2,245	2,265

Calculate the percent recombination between the 2 pairs of alleles in each strain and provide a cytogenetic explanation for the difference. Explain how you would verify your explanation cytologically.

39. In *Drosophila melanogaster*, the dominant allele, *Adh*, at the *alcohol dehydrogenase* (ADH) locus on chromosome 2, enables flies to convert ethanol to acetaldehyde. Flies homozygous or hemizygous for the null (recessive) allele, *adh*, fail to convert ethanol to acetaldehyde. When these flies are reared on a medium containing ethanol they die. In 1981 Robinson, A.S. and C. van Heemert (Theor. Appl. Genet. 59: 23) crossed heterozygous males with *adh adh* females; the larvae produced by this cross were placed on a 4.0% ethanol medium. They found that only the male larvae developed into adults. The female larvae died.

a) Offer a plausible explanation by contrasting the results obtained in this cross with those expected.
b) Suggest a practical application of the system you proposed in (a) if it could be generated in the mosquito or any other pest.

40. Anderson, E.G. (Am. Nat. 68: 345, 1934) crossed a *semisterile* plant heterozygous for a reciprocal translocation and the alleles *Pl* (*purple* plant colour) and *pl* (*green* plant colour) with a *fully fertile* plant with the standard segmental arrangement and homozygous for *pl*. The offspring consisted of the following phenotypes:

Semisterile, green   55   *Semisterile, purple*   141
Fully fertile, purple 69   *Fully fertile, green* 137

How far from the *pl* locus is the interchange (translocation) point?

41. In standard lines of corn the locus *An* for ear type is on chromosome 1, and locus *Ra* for tassel type is on chromosome 7. Working with a *semisterile* line of corn, Burnham C.R. (Genetics 33: 5, 1948) found that in crosses with *fully fertile* lines it produced progeny in a ratio of 1 *fully fertile*: 1 *semisterile* and in the phenotypic distributions as indicated below:

Trait pair	Semisterile		Fully fertile	
normal vs. *anther ear*	6	*normal*	196	*normal*
	188	*anther*	10	*anther*
normal vs. *ramosa tassel*	190	*normal*	2	*normal*
	4	*ramosa*	204	*ramosa*

a) Formulate a plausible hypothesis to explain why these genes, which belong to two different linkage groups, fail to recombine at random with *semisterility* vs. *full fertility*.
b) Give the genotypes and chromosome constitutions of the (i) *semisterile* and (ii) *fully fertile* lines.

c) What cytological configuration would you expect to see if chromosomes of *semisterile* plants were examined at pachytene? At metaphase I?
d) Show one possible location of each gene.
e) Illustrate how the aborted pollen originates.

42. In corn, the allele pairs *Cr cr* (*normal* vs. *crinkly* leaves) and *Dd* (*normal* vs. *dwarf* plant height) are on chromosome 3; *Ww* (*waxy* vs. *starchy* endosperm) and *Cc* (*purple* vs. *white* aleurone) are on chromosome 9. Their positions relative to each other and the centromere, together with distances in map units, are as follows:

```
Chromosome 3  ●————————————Cr——D
                      20        1.5

Chromosome 9  ●————————W————————————C
                      20              20
```

Line 1 is homozygous for all four dominant alleles and line 2 is homozygous for all four recessive alleles. Pollen of line 1 was irradiated with X-rays and used to pollinate line 2. The following results were obtained:

1. One F_1 had *crinkly leaves*. When self-fertilized, 1/4 of the offspring showed this trait.
2. A second F_1 plant had *crinkly* leaves and was a *dwarf*. When testcrossed with line 2, all the progeny were *dwarf* and had *crinkled* leaves.
3. A third F_1 plant, dominant for all characters, was *semisterile*. When testcrossed to line 2, it yielded offspring with the expected percent recombination between *Cr* and *D* (1.5%) but no recombinants between *W* and *C*.
4. A fourth F_1 plant dominant for all characters was also *semisterile*; when testcrossed to line 2 and the progeny classified for *Cr* vs. *cr* and *W* vs. *w*, it was found that these allele pairs were linked.

Describe the most probable nature of the change (if any) in the hereditary material of the pollen grains that lead to each of the 4 F_1 plants described above. Indicate what you would expect to observe if you studied meiosis in each F_1.

43. In a series of crosses between various homozygous translocation lines, each of which differs from the standard line (2n=22) by a reciprocal translocation, the following configurations are found at metaphase I in the F_1 hybrids:

Translocation 1 x Translocation 2 ⟶ F_1 ring of 6
Translocation 3 x Translocation 4 ⟶ F_1 2 rings of 4
Translocation 5 x Translocation 6 ⟶ F_1 bivalents only
Translocation 7 x Translocation 8 ⟶ F_1 ring of 4

For each cross determine the number of chromosome pairs that the two translocation lines have in common.

44. The non-familial type of *Down* syndrome is due to trisomy of chromosome 21. Individuals with the familial type, where more than one child may be *affected* and where the mother maybe young, have 46 chromosomes (Polani, et al., Lancet 1: 721, 1960; Carter, C.O. et al., J. Genet. 54: 462, 1960). The mother or, in rare instances, the father of such *affected* children is found to have 45 chromosomes, one of which represents a translocation between chromosome 21 and another autosome (in most cases 13, 14 or 15). Such mothers (2n-1=45) are referred to as "carriers". The origin of such a reciprocal translocation is illustrated.

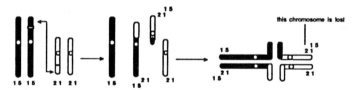

Show the types of gametes a "carrier" mother can form and explain why she can produce a *normal* (2n=46), a *Down* syndrome (2n=46), or a child with 45 chromosomes. Why would some of her offspring die early in pregnancy or shortly after birth?

45. In animals, e.g., the mouse, a large proportion of the zygotes and embryos formed by individuals heterozygous for a reciprocal translocation abort. In higher plants, e.g., barley, however, very few, if any, zygotes or embryos formed by such individuals abort. What is the reason for this difference?

46. In *wild-type Haplopappus gracilis* (2n=4), a reciprocal translocation occurs between the nonhomologous standard

chromosomes 1.2 and 3.4 in two plants. It is found that the 152 progeny of a cross between these two plants show a ring of four chromosomes at metaphase I and 48 show two bivalents. What were the chromosome constitutions of these two plants? Illustrate.

47. A standard plant has the chromosomes 1.2, 3.4, 5.6, 7.8, 9.10, 11.12, 13.14, 15.16. It is X-rayed, and a reciprocal translocation occurs between chromosomes 7.8 and 13.14 so that the new chromosomes are 7.14 and 13.8. The homozygous translocation line 7.14, 13.8 is X-rayed and further reciprocal translocations are produced.

(i) One produces a ring of four with the 7.14, 13.8 translocation line and only bivalents with the standard line.
(ii) Another produces a ring of four with the 7.14, 13.8 lines and two rings of four with the standard line.
(iii) A third produces a ring of four with the standard line and a ring of six with the 7.14, 13.8 line.

What reciprocal translocations are present in these new lines?

48. In humans, some individuals are heterozygous for reciprocal translocations and phenotypically *normal*. Matings between such individuals and those with a normal chromosome complement can produce families in which the offspring may (i) be phenotypically and chromosomally normal (2n=46), (ii) suffer from diverse congenital anomalies but still have 2n=46 chromosomes (iii) abort spontaneously (iv) abort spontaneously, be stillborn or alive with multiple congenital defects and possess 45 or 47 chromosomes (Lindenbaum, R.H. and M. Bobrow, J. Med. Genet. 12: 29, 1975; Pihko, H. et al., Hum. Genet. 58: 129, 1981; Therman, E., *Human Chromosomes*, 2nd ed., Springer-Verlag, New York).
Assuming the reciprocal translocation involves chromosomes 4 and 11, with the aid of diagrams, account for the different types of offspring that can be produced by such matings.

49. a) In the genus *Carex*, chromosomes possess diffuse centromeres. The various species constitute an aneuploid series with the haploid chromosome numbers 30, 31, 32, 33, 34, 35, 36, etc. (Davies, E.W., Hereditas 42: 349, 1956).
(i) Outline two mechanisms that could account for the origin of this series.
(ii) Suggest how information from extensive linkage-group studies might be used to decide which mechanism is the prevalent one with regard to the differences between any two species.
(iii) How might studies of DNA content help to determine the mechanism involved?

b) In the African pigmy mouse (*Mus minutoides*), in which all chromosomes possess localized centromeres, the basic karyotype has 36 acrocentrics. Various subspecies show a variety of karyotypes down to 2n=18, with all chromosomes metacentric (Matthey, R., Rev. Suisse Zool. 73: 585, 1966). There appears to be only one mechanism that can account for these results. Which is it and why?

50. The Chinese Muntjac (*Muntiacus reevesi*) has a diploid chromosome number of 46. The Indian muntjac (*Muntiacus muntjak vaginalis*) with 20% less repetitive DNA than its Chinese relative has the lowest diploid chromosome number of all mammals - 6 in females and 7 in males. Surprisingly the interspecific hybrids derived from crosses between the 2 species are viable (Shi, L.M. et al., Cytogenet. Cell Genet. 26: 22, 1980).
a) It has been suggested that the chromosomes of the Indian muntjac could have evolved from those of its Chinese counterpart. Suggest and illustrate a plausible mechanism(s) that could account for this large disparity in the chromosome number of the two species.
b) Show how you would expect the chromosomes of the hybrid to behave during meiosis. Indicate whether the gametes would possess a balanced or unbalanced genetic constitution.
c) G-banding and *in situ* hybridization with clones possessing highly repetitive sequences have been used to obtain data bearing on the mode of karyotypic evolution of the Indian muntjac chromosomes (Lin, C.C. et al., Chromosoma 101: 19, 1991). What results would you expect with each of these techniques if applied to the 2 muntjac species and the F_1 hybrids?

d) Could genetic studies of various characters in the 2 species also provide data bearing on the evolutionary relationship of the 2 species? Elaborate.

51. The chromosome constitutions of 4 primate species, and that of humans for comparison as reported by Hamerton, J.L. and H.P. Klinger (New Sci. 18(341): 483, 1963) are tabulated below:

Species	Metacentric chromosomes		Acrocentric chromosomes		
			Autosomes		
	Autosomes	Sex	Large	Small	Sex
Homo sapiens	34	X	6	4	Y
Pan troglodytes troglodytes (northern chimpanzee)	34	X	8	4	Y
P. t. paniscus (pigmy chimpanzee)	36	X	8	2	Y
Gorilla gorilla gorilla	30	X,Y	12	4	
Pongo pygmaeus (orangutan)	26	X,Y	16	4	

a) On the basis of numbers and kinds of chromosomes involved in the evolutionary changes, explain how you would relate the 4 primate species in paired comparisons with humans (as distant, close, very close).
b) Suggest how the human karyotype might have arisen from a primitive man-ape population having basically 48 chromosomes.

52. A cross between two plants in *Crepis fuliginosa* (2n=6), each with a ring of six chromosomes at metaphase I, produces 98 progeny with a ring of six, 52 with three bivalents and 50 with a ring of four and a bivalent. If the chromosome arms in the three nonhomologous chromosomes in the standard line are designated 1.2, 3.4, and 5.6, give the arm constitution of each of the six chromosomes in the parents of the cross.

53. The chromosome complements of six *Drosophila* species reproduced from Lewis, K.R. and B. John, "*Chromosome Marker*", J. and A. Churchill, Ltd., London, 1963, are shown below. *D. trispina* is the putative ancestor.

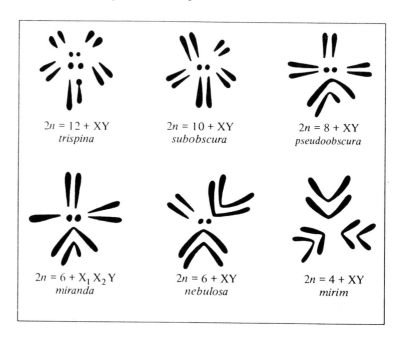

$2n = 12 + XY$ trispina
$2n = 10 + XY$ subobscura
$2n = 8 + XY$ pseudoobscura
$2n = 6 + X_1 X_2 Y$ miranda
$2n = 6 + XY$ nebulosa
$2n = 4 + XY$ mirim

a) Explain how each of these types of changes could come about and the conditions (e.g., chromosome and centromere type, importance of heterochromatin) under which the changes in chromosome number could occur.
b) Indicate genetic experiments to test your explanation.

54. The house mouse (*Mus musculus*; 2n=40) has 20 acrocentric pairs of chromosomes, and the tobacco mouse (*M. poschiavinus*; 2n=26) which has 7 metacentric and 6 acrocentric pairs of chromosomes are easily hybridized and produce healthy but semisterile offspring (Gropp, A. et al., Cytogenetics 9: 9, 1970; Tettenborn and Gropp, Cytogenetics 9: 272, 1970). These hybrids form 5 bivalents, 7 chain trivalents and possess an homomorphic XX bivalent in females and a heteromorphic XY pair in males.
a) *M. poschiavinus* appears to be a species of more recent origin than *M. musculus* and therefore may have arisen from the latter. What appears to be the most plausible mechanism of origin of the *M. poschiavinus* karyotype from that of *M. musculus*? What is the basis for your

explanation? Discuss fully. Note that the tobacco mouse has less DNA than the house mouse.

b) How would you proceed to genetically and cytologically verify your hypothesis? Indicate the results expected.

55. *Truncate Drosophila* (lacking wing tips) when mated together always produce some normal offspring. *Plum*-eyed flies mated together always produce some *nonplum* offspring. However, when *truncate, plum* flies are intermated, the offspring are always *truncate, plum* only. Explain all three results.

56. The segmental arrangements of five chromosome complexes in *Oenothera* are as follows:

hookeri	1.2	3.4	5.6	7.8	9.10	11.12	13.14
flavens	1.4	3.2	5.6	7.8	9.10	11.12	13.14
velans	1.2	3.4	5.8	7.6	9.10	11.12	13.14
Johansen	1.2	3.4	5.6	7.10	9.8	11.12	13.14
acuens	1.4	3.2	5.6	7.10	9.8	11.12	13.14

a) For each of the following combinations of complexes give the ring configurations and the chromosomes involved and the number of chromosome pairs forming bivalents at metaphase I of meiosis:

(1) *hookeri* x *velans* (2) *flavens* x *velans*
(3) *acuens* x *Johansen* (4) *hookeri* x *acuens*
(5) *flavens* x *Johansen* (6) *velans* x *acuens*

b) *Rigens*, the egg complex of *muricata*, gives the following ring configurations and chromosome pairs:

With *hookeri*	Ring of 6 and four pairs
With *flavens*	Ring of 4, ring of 6 and two pairs
With *velans*	Ring of 8 and three pairs
With *acuens*	Ring of 4, ring of 8 and one pair

Show with illustrations:
(i) The chromosome configurations in a *rigens-Johansen* hybrid.
(ii) The segmental arrangements in the *rigens* complex.

57. In *Drosophila melanogaster*, *lg* is a recessive allele on the second chromosome that causes larvae to develop to a *large* size and then die. In a stock of flies true-breeding for *curly* wings the adult progeny size is one-half that produced from *wild-type* matings. Of the embryos that do not develop into adults, half die as giant larvae. Why does the stock breed true for *curly* wings?

58. Reciprocal crosses between *Oenothera chicaginensis* and *O. cockerelli* produce a single, unique class of progeny in each direction; the progeny of reciprocal crosses are different. Reciprocal crosses between *O. lamarckiana* and *O. grandiflora* produce several hybrid phenotypes called twin or multiple hybrids in each direction; the results of reciprocal crosses are identical in this respect. Progeny of all crosses mentioned above breed true when self-fertilized (Cleland, R.E., Bot. Rev. 2: 316, 1936).

a) For each of the two sets of reciprocal crosses, show whether the lethals are gametic, zygotic or both, and give reasons for your statements.

b) In which of the two sets of reciprocal crosses would you expect seed sterility? How much? Why?

59. Darlington, C.D. and A.E. Gairdner (Genetics 35: 97, 1937) found in the bellflower, *Campanula persicifolia* (n=8), that translocation heterozygotes discovered in *wild* stocks when self-fertilized bred true. This was not true for translocation lines produced in cultivated varieties, which upon self-fertilization produce some progeny homozygous for the interchange and some homozygous for the standard segmental arrangement. Why do the *wild* and not the *cultivated* translocation stocks breed true for the ring configuration?

Chapter 17
Chemistry, Structure and Replication of Genetic Material and Chromosomes

Note: Questions on the molecular structure and function of centromeres and telomeres are included in Chapter 25.

1. a) To carry genetic information any molecule must be structurally unique. What genetic properties must be accounted for in such a molecule?
   b) What kinds of proof are there that DNA is the carrier of genetic information in bacteria and most viruses?
   c) What lines of evidence indicate that DNA is the carrier of genetic information in eukaryotes?

2. Mirsky, A.E. and H. Ris (J. Gen. Physiol. 30: 117, 1951) have shown that the main constituents of chromosomes of eukaryotes are DNA, RNA and proteins (mostly basic) in the approximate percentages (by weight) of 14:14:72. Boivin, A. et al. (C.R. Acad. Sci. 226: 1061, 1948) and Mirsky, A.E. and H. Ris (Nature 163: 666, 1949) used the Feulgen test for measuring the quantity of DNA per nucleus in different cells of several animal species and obtained the data shown below:

	DNA (picograms)		
Animal	Nuclei (2n) of erythrocytes	Nuclei (2n) of hepatic cells	Nuclei (n) of sperm
Domestic fowl	2.34	2.39	1.26
Carp	3.49	3.33	1.64
Brown trout	5.79		2.67
Toad	7.33		3.70

The quantities of the other chromosome components varied with cell type, depending on the cell's function and its metabolic state. Discuss these findings with regard to the chemical nature of genes.

3. a) What are the major protein components of eukaryotic chromosomes? What are their functions?
   b) What is a nucleosome? A linker? What is the chemical compostion of each?
   c) The nucleofilament is about 100 Å in diameter, the solenoid is about 250-300 Å in diameter. Is H1 histone necessary for the structure of the nucleofilament? The solenoid structure? Discuss.

4. Although the nucleic acid in some DNA-containing viruses is single-stranded, the DNA of eukaryotes must be double-stranded. Why?

5. The schematic illustration along both template strands of a replicating DNA duplex at a replicating fork is incorrect in some respects. Identify the errors and redraw the illustration indicating the errors you have corrected.

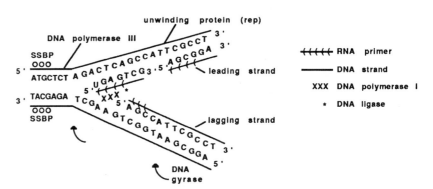

6. a) A single nucleic acid-pentanucleotide chain contains the base sequence G-A-U-C-G. State whether it is from DNA or RNA and why?
   b) The DNA base ratios from cattle and rat are the same: 28A : 22G : 22C : 28T. Would you expect hybridization between single strands of the two species? Explain.

7. DNA replicates bidirectionally and in basically the same way at the point-of-origin of replication (O) in bacterial chromosomes and each such point in eukaryotic chromosomes. The events on both sides of this point are identical. Consider those on the left side (half) of O in the segment of a DNA duplex shown below. The first event in replication, unwinding of the duplex at O, is shown.

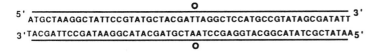

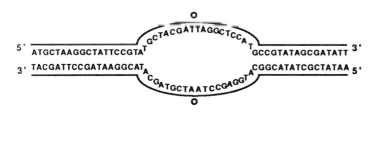

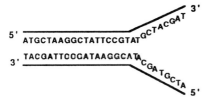

a) Which protein is responsible for the first event?
b) Explain with the aid of illustrations how the duplex replicates. In your explanation indicate the functions that each of the enzymes and proteins; DNA gyrase, SSBP (single-stranded binding protein), DNA polymerase I, DNA ligase, DNA polymerase III and primase, performs and how each carries out its function. Label such features as the template, leading and lagging strands, primers, direction of synthesis, Okazaki fragments, replication fork and parental DNA.
c) Why is DNA replication virtually error-free?

8. Electron microscopy of eukaryotic chromosomes (chromatin) at interphase shows that the primary structure is a nucleofilament (Finch, J.T. and A. Klug, PNAS 73: 1897, 1976) which has the appearance of beads on a string (Olins, A.L. and D.E. Olins, Science 183: 330, 1974)

The beads are about 100 Å in diameter and about 50 Å deep; the strand which runs the full length of the chromosome and interconnects the beads is about 20-25 Å in diameter.
a) What are (i) the beads and (ii) the regions between them called?
b) What is the chemical compositon and structure of (i) the beads and (ii) the regions between them? Illustrate.
c) The fully extended DNA molecule in each chromosome is shortened (condensed) by approximately 7 times in the nucleofilament. How does this condensation occur?
d) Studies by many workers (see Kornberg, R.D., Ann. Rev. Biochem. 46: 931, 1977; Felsenfeld, G. and J.D. McGhee, Cell 44: 375, 1986) suggest that at least 3 further levels of coiling and/or folding are involved to form the metaphase chromosomes, in which the initially extended DNA molecule is condensed by a factor of 8,000 to 10,000. The first of these coilings produces a solenoid structure, the second a supersolenoid (unit) structure and the third level gives rise to the quaternary structure observed at metaphase. (i) What factors are involved in producing these structures? (ii) How do they produce each level of coiling? (iii) How much is the DNA duplex condensed at each level?
e) Which of these 4 levels of coiling would you expect to observe in (i) euchromatin at G1 and S, (ii) euchromatin at G2, (iii) heterochromatin, (iv) euchromatin that is being transcribed? Explain.

9. *Escherichia coli* possesses one circular chromosome consisting of one continuous DNA double helix. Cells are grown for many generations on a medium in which the only nitrogen source is ^{15}N. Thereafter the cells are transferred to a medium containing only ^{14}N and allowed to go through three cell divisions, i.e., DNA replications. The DNA

density is measured by density-gradient centrifugation just before transfer and after each cell division on ^{14}N medium. Disregarding the circularity of the chromosome and using a solid line for labelled chains and a broken line for unlabelled ones, if replication were (1) semiconservative and (2) conservative, show the kinds of chromosomes expected after:
a) Continued replication in ^{15}N.
b) One generation of replication in ^{14}N.
c) Two generations of replication in ^{14}N.
d) Three generations of replication in ^{14}N.

10. A student was asked to illustrate a DNA dinucleotide containing cytosine and guanine. His incorrect drawing is shown below:

a) How many errors did the student make? Identify them in the student's illustration.
b) Number all the carbon and nitrogen atoms in the corrected illustration.
c) Contrast the corrected illustration with that of a pair of nucleotides possessing the bases guanine and cytosine.

11. a) What is heterochromatin? What are its chemical and cytological properties?
b) Discuss the various kinds of supporting evidence that indicate that heterochromatin, regardless of whether it is a permanent feature of a genome or a transient one (as in B chromosomes), is genetically inert or almost so.
c) Does heterochromatin replicate at the same time as euchromatin? Cite evidence to support your answer.

d) Why is constitutive heterochromatin not likely to code for proteins?

12. What are the chemical units of DNA that correspond to the genetic units of replication, recombination, mutation and function? Outline one form of evidence for each statement. Are the chemical units the same in both eukaryotes and prokaryotes? Discuss data that support your answer.

13. Mirsky, A.E. and H. Ris (J. Gen. Physiol. 30: 117, 1951) found that the amphibian *Amphiuma* contains about 20 times as much DNA as does a certain toad (also an amphibian) and 70 times as much DNA as is found in the cells of ducks, geese and domestic fowl. Do you think that *Amphiuma* contains 20 and 70 times as many genes respectively as do the toads and birds mentioned? If not, what is a more plausible explanation?

14. Sinsheimer, R.L. et al. (J. Mol. Biol. 1: 43, 1959; Sci. Am. 207: 109, 1962) showed that the DNA of mature phage $\phi X174$ is single-stranded and that immediately after entry into a susceptible host it displays the following features:
1. It undergoes several replications.
2. Limited replication can occur in the presence of the protein-synthesis inhibitor chloramphenicol.
3. It is of a lower density than the DNA of mature $\phi X174$ and about the same as that of its host, *Escherichia coli*.
4. Upon heating, it is denatured in a manner similar to double-stranded DNA.
5. It is more resistant to inactivation by ultraviolet irradiation than the DNA of mature viruses.

In chloramphenicol no mature progeny viruses are formed, whereas in normal infection these are present as early as 8 minutes after infection. These, like the parents, are single-stranded and possess the same base composition. These results suggest a sequence of events after infection which leads to the formation of single-stranded progeny phage. Illustrate this sequence.

15. a) Show whether it is possible to obtain an inversion in a single-stranded DNA molecule.
b) The two strands of a DNA (or RNA) double helix can be separated and subsequently reannealed (Schildkraut, C.L. et

al., J. Mol. Biol. 3: 595, 1961). Of what value is this technique in testing affinities between species?

c) If the base sequence of a polynucleotide chain is G-C-A-G-T-C-T-A-A, what is the sequence of the bases on the complementary chain? Why?

d) A sample of DNA contains 20% thymine. A second sample contains 20% guanine. Is it possible for the 2 samples to have originated from the same species? Explain.

e) Prior to the early 1940's, the prevalent theory was that proteins were genetic material. Discuss some of the reasons for this.

f) Why is primer synthesis more error-prone than synthesis of DNA chains?

g) Why are the genomes of RNA viruses small? Is there a relationship between the fact that RNA genomes are small and are never duplex circles?

h) Why is ATPase necessary for the enzyme helicase to perform its function in DNA replication?

16. a) Diagram and label a eukaryotic chromosome at mitotic metaphase as it would be seen under a light microscope after standard staining with acetocarmine or Feulgen. Include all the features that a chromosome could possibly possess.

b) During the period 1968-1973 numerous new staining procedures, namely Q-, G-, C, and R-banding were developed to study mitotic and meiotic chromosomes in eukaryotes (Caspersson, T. et al., Exp. Cell Res. 49: 219, 1968; Sumner, A.T., Cancer Genet. Cytogenet. 6: 39, 1982; Burkholder, G.D. in *Chromosome Structure and Function*, 18th Stadler Genetics Symposium, Plenum Press, pp. 1-52, 1988). Why was there the need and thus the concerted effort to develop these new banding procedures?

c) Diagram the chromosome shown in (a) as it would appear under the microscope after (i) Q-banding (ii) G-banding (iii) C-banding and (iv) R-banding.

d) For each of the new banding protocols indicate whether it stains euchromatin and/or heterochromatin.

e) Describe the mechanism by which each technique mediates its specific banding pattern.

f) What is the advantage of the R- and G-banding procedures over that of Q-banding?

g) Renaturation studies of denatured DNA fragments from nuclear chromosomes in different eukaryotic species have identified 4 classes of DNA sequences (Britten, R.J. and Kohne, Science 161: 529, 1968; Davidson, E. and R.J. Britten, Quart. Rev. Biol 48: 565, 1973).

(i) What are these 4 classes of DNA called?

(ii) Diagram the chromosome shown in (a) once again, showing the possible locations of these 4 classes of DNA sequences relative to each other, centromeres and other chromosome features. Indicate in which of these classes of DNA sequences structural and non-structural genes are most likely located? Which of these classes of DNA sequences are absent from prokaryotes?

h) Yunis, J.J. and O. Prakash (Science 215: 1525, 1982) upon comparing high resolution G-banding patterns of human chromosomes with those of chimpanzee, gorilla and orangutan chromosomes found that 18 of the 23 pairs of human chromosomes have virtually identical G-banding patterns to those of the common ancestor. The patterns of the remaining 5 human chromosome pairs were very similar to those in the great apes. What is the implication of these findings?

17. Ogur, M.S. et al. (Arch. Biochem. 40: 175, 1952) determined the amount of DNA present in the nuclei of cells from four different strains of a certain species of yeast (*Saccharomyces cerevisiae*) to be as shown below:

Strain	DNA per cell, (pgms)
1	2.26 ± 0.23
2	4.57 ± 0.60
3	6.18 ± 0.54
4	9.42 ± 1.77

Offer an explanation to account for the differences in DNA content among the four strains and suggest a way of testing your explanation.

18. a) The (A+T)/(G+C) ratio in the DNA of *E. coli* is 1.00. Is this information sufficient to decide whether the DNA is likely to be double- or single-stranded? If not, state the kind of information that is required to determine the strandedness of DNA.

b) T2 DNA is about 52μm long (Cairns, J. J., Mol. Biol. 3: 756, 1961). The distance between base pairs is 3.4 Å; 660 is the average molecular weight of a nucleotide pair. If the molecule is a double helix throughout its length, what is its molecular weight?

19. Alfert, M. and H. Swift (Exp. Cell Res. 5: 455, 1953) determined the following relative amounts of DNA in nuclei of various cells of the annelid worm *Sabellaria*:

First polar body	127 ± 3
Sperm	61 ± 1
Male pronucleus	133 ± 7
Prophase of first cleavage	263 ± 10
Telophase of first cleavage	124 ± 3

a) Show how these values relate to expected levels of ploidy and account for any major anomalies (viz. double or half the expected value).
b) For each stage, what is the C-value (i.e., C, 2C, 3C etc.)? How many chromosomes do you expect to see at each stage (assume 2n=10)?

20. The base ratios of nucleic acids from eight different species in both the eukaryotes and prokaryotes are presented in the table.

Species	T	C	U	A	G	A+T or U / G+C	A+G / C+T or U
1	31	19		31	19		
2		19	31	31	19		
3	19	31		19	31		
4						1.00	1.26
5	32	18		25	25		
6		25	32	23	20		
7						1.26	1.00
8						1.00	1.00

For each species state with reasons whether its nucleic acid is: a) DNA or RNA b) Single or double stranded.

21. Tessman, I. (Virology 9: 375, 1959) found that *in vitro* treatment of the *wild-type* strains of two different DNA phages (A and B) with nitrous acid produces mutants. The mutation tested in the A phage was that for *rapid* lysis (r^+ to r) and that tested in the B phage was for *host range* (h^+ to h). Viable phage of each type were seeded very thinly on each of many plates to ensure that *mutant* plaques would be completely isolated from *wild-type* plaques. The majority of plaques were *wild-type*. The *mutants* of phage A always occurred in *mottled* plaques containing a mixture of *mutant* and *wild-type* phage, whereas all *mutants* of phage B arose in plaques that contained *mutants* only. Offer an explanation to account for the difference in response of the two phages to the mutagen.

22. Bollum, F.J. (J. Biol. Chem. 234: 2733, 1959) found that DNA polymerase from calf thymus glands would not catalyze the *in vitro* synthesis of DNA unless the primer DNA (of high molecular weight) was first heated to 99°C. What is the significance of this finding?

23. At any base pair site in a DNA duplex any one of four possible base-pair combinations A-T, T-A, G-C, C-G can occur. Calculate the number of possible sequences for a DNA duplex 10^3 base pairs long if no restriction is placed on the relative frequencies of the four kinds of base pairs.

24. The vaccinia and M13 viruses have the base compositions shown:

	A	T	G	C
Vaccinia	29.5	29.9	20.6	20.0
M13	23.3	32.8	21.1	19.8

a) Offer an explanation to account for these differences.
b) How would you test your explanation?

25. In an RNA duplex 500 of 5,000 bases are uracil. Calculate the approximate molecular weight of the duplex and determine the proportions of the bases present.

26. The proportions of bases in DNA of yeast, which is double-stranded, are 31A : 18G : 18C : 31T. Would you expect the (A+G)/(C+T) and (A+T)/(G+C) ratios to be the same? Why?

27. The following results were obtained in transformation experiments in *Pnemococcus pneumoniae* involving encapsulated (*smooth, S*) and unencapsulated (*rough, R*) strains:
    a) Griffith, F. (J. Hyg. 27: 113, 1928) heat-killed *SIII* bacteria and injected them with living *RII* types into living mice. Many of the mice died of pneumonia; their blood contained not only *RII* but also *SIII* bacteria, each of which bred true-to-type. *R* or *S* of any type, when heat-killed and injected alone, did not multiply. In 1931 Dawson, M.H. and R.H.P. Sia (J. Exp. Med. 54: 681) obtained the same results *in vitro*.
    (i) Discuss mutation as a probable cause of the genetic change (*RII* to *SIII*).
    (ii) What alternative explanation can be given to account for the heritable change?

    b) Alloway, J.L. (J. Exp. Med. 55: 91, 1932) produced *in vitro* hereditary changes of the kind described in (a) by exposing *R* bacteria to extracts of *S* cells from which all traces of structural components (mitochondria, ribosomes, etc.) had been removed. In 1944 Avery, O.T. et al. (J. Exp. Med. 79: 137) separated the extract from disrupted *S* cells into various chemical fractions and determined the ability of each fraction to cause specific hereditary changes in the *R* cells. It was found that the fractions obtained after treating the different fractions with proteases, RNase and enzymes that destroy polysaccharides continued to show this capability. When the filtrate was treated with DNase, it lost its capacity to change *R* into *S* cells. What is the chemical component responsible for the hereditary changes and why? Illustrate the probable mechanism by which the living SIII bacteria are produced.

28. Nucleic acids extracted from the phage MS2 can be used to infect bacterial protoplasts (bacteria with capsule removed) and to obtain mature phage particles from this infection. This infectivity is destroyed by ribonuclease but not by deoxyribonuclease (Davis, J.E., Science 134: 1427, 1961). When the infecting nucleic acid or phage is labelled, no label is ever found in the progeny phage, which are identical in all respects to parental particles (Davis, J.E. and R.L. Sinsheimer, J. Mol. Biol. 6: 203, 1963). What inferences are possible from this information with respect to:
    (i) Kind of nucleic acid in MS2 chromosomes?
    (ii) How the phage chromosome replicates?

29. In 1957 Fraenkel-Conrat, H. and B. Singer (Biochem. Biophys. Acta 24: 540) separated the protein and RNA of the normal and the *Holmes-Ribgrass* (*HR*) strains of tobacco mosaic virus. Artificial reciprocal hybrids were then made by combining *HR* protein with *normal* RNA and *normal* protein with *HR* RNA. These and the separated fractions were applied to leaves of *Nicotiana tabacum* and *N. sylvestris* plants each susceptible to both types of viruses. The results of these experiments are shown below:

Infection		
Protein	RNA	Infection capacity
HR	*Normal*	+
Normal	*HR*	+
HR	None	-
Normal	None	-
None	*HR*	+
None	*Normal*	+

    The protein constitution in progeny viruses was always identical to the strain from which the nucleic acid was obtained.
    Can you conclude from this experiment that:
    (i) Protein may be a carrier of genetic information? Explain.
    (ii) RNA may be a carrier of genetic information? Explain.

30. Three mechanisms - conservative, dispersive and semiconservative - were suggested as a means for replication

of DNA (Delbruck, M. and G.S. Stent "*The Chemical Basis of Heredity*" pp. 699-736, Johns Hopkins, Baltimore. 1957). Meselson, M. and F.W. Stahl (PNAS 44: 671, 1958) grew an *E. coli* culture for several generations on a medium containing labelled nitrogen (^{15}N). These bacteria were then transferred to a medium containing only unlabelled nitrogen (^{14}N) and allowed to grow for a number of generations (one DNA replication = one division). Just before transfer and after each of the first three divisions, samples of the culture were removed, DNA was extracted, and its density determined by density-gradient centrifugation. A schematic presentation of their results is illustrated below:

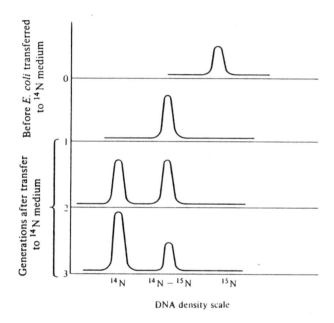

a) Which mechanism of DNA replication do these data favour? Which do they definitely rule out? Explain fully.
b) DNA extracted from the ^{14}N-^{15}N (hybrid) band was heated to 100°C, centrifuged and its density determined. Two bands, one heavy and one light, were observed. What mechanism of replication is supported by these results?
c) What information is necessary to prove that the most likely method is the correct one?

31. a) Why were ^{35}S and ^{32}P chosen by Hershey and Chase in their experiment with T2 phage to determine the most likely candidate substance for genetic material?
b) *E. coli* cells are infected with T2 phage labelled with ^{35}S. About 10 minutes after infection the portions of the phage that fail to enter host cells are separated from the latter by shaking the culture in a Waring blender. The cells and supernatant fluid are then tested for radioactivity. Where would you expect most of the label to be found? Explain.

32. Taylor, J.H. et al. (PNAS 43: 122, 1957) grew broad bean (*Vicia faba*; 2n=12) seedlings for 8 hours in a medium containing radioactive (^{3}H) thymidine to allow root-tip cells in interphase to go through one complete cycle of DNA synthesis and terminate in prophase of division. The seedlings were then washed and placed in a nonradioactive solution of thymine, containing colchicine, which delays division of the centromere and inhibits spindle formation. Because of the latter effect, daughter chromatids, after separating, are incorporated in the same anaphase nucleus, with the result that the chromosome number is doubled upon completion of each mitosis. Some of the roots were kept in colchicine for 10 hours before fixing, staining, squashing, and autoradiographing. Others were kept in colchicine for 34 hours before treatment. Root-tip cells at c-metaphase after 10 hours in colchicine had the normal diploid number of 12 chromosomes; all were labelled, equally and uniformly, in both chromatids. Many cells at c-metaphase after 34 hours in colchicine contained 24 chromosomes. All chromosomes were labelled, but only one of the two sister chromatids of each was radioactive.
a) Interpret these results in terms of the mechanism of DNA replication if the chromosome is uninemic (one DNA double helix per chromosome in interphase before duplication).
b) Would the interpretation of the method of DNA replication necessarily be the same if the chromosome were polynemic? Illustrate your explanation with a binemic chromosome (two double helices per chromosome in interphase before duplication)?
c) Some root-tip cells after 34 hours in colchicine had 48 chromosomes. What would be the labelling pattern of the chromosomes and chromatids in these cells?

d) A few of the metaphases from root-tips after 34 hours in colchicine had 48 chromosomes. How many chromosomes in the complement would be completely unlabelled if the chromosome contains a single DNA duplex? How many of the other chromosomes would have a labelled and an unlabelled chromatid? Illustrate.

33. a) When heavy DNA (containing ^{15}N) and light DNA (containing ^{14}N) from different cultures of *Diplococcus pneumoniae* are density-gradient-centrifuged, two bands of DNA are formed in the tube. However, if the same mixture is heated to 100°C and cooled slowly before centrifuging, the results are somewhat different. The two bands are formed as before and in the same position in the tube as before, but in addition, an intermediate band, representing the mean of the densities of the other two bands, is formed. Account for (i) the two bands in the mixtures and (ii) the band of intermediate density.
b) When labelled DNA from *D. pneumoniae* and unlabelled DNA from *E. coli*, another bacterial species, are mixed, heated to 100°C, slowly cooled, and then centrifuged, only two bands are formed in the cesium chloride density gradient. Account for the absence of an intermediate band.
c) What conclusions can be drawn from (a) and (b) regarding the nature and specificity of DNA base pairing and base sequence?

34. Meselson, M. and J.J. Weigle (PNAS 47: 857, 1961) used λ phage, whose DNA is double-stranded, to show that crossing-over occurs by breakage and reunion of DNA molecules. In subsequent experiments, *E. coli* growing in a light medium (^{12}C, ^{14}N) were infected with heavy (^{13}C, ^{15}N) λ of genotype *AB* and heavy λ *ab*. Most λ DNA molecules replicated in the normal semiconservative manner. A few, however, did not do so but became enclosed in new protein shells. The DNA of these rare progeny was completely heavy. Some of the rare completely heavy DNA progeny possessed the genotype *Ab*, others the genotype *aB*. Are crossing-over and DNA replication independent phenomena? Explain.

35. Okazaki, R. et al. (PNAS 59: 598, 1968) studied DNA replication in *E. coli* using the folowing pulse-labelling experiments with ^{3}H-thymidine: If ^{3}H-thymidine was present for only a few seconds before DNA was extracted from the cells and then denatured, a significant quantity of the newly synthesized DNA was present in the form of single-stranded fragments, about 1,000 nucleotides long. If the pulse of ^{3}H-thymidine was terminated after transferring cells to a nonlabelled medium and permitting growth of cells (and DNA replication) for a few minutes before DNA extraction, they found a lower frequency of these short fragments. Most of the radioactivity was found in long single-stranded fragments. How might these observations be related to the fact that all known DNA polymerases synthesize new chains only in a 5' to 3' direction?

36. a) Unlike other proteins which are synthesized throughout interphase, why are histones produced only during the S phase?
b) Five types of histones (H1, H2A, H2B, H3, H4) are synthesized in most eukaryotic species. Would you expect the genes for these proteins to be present in single or multiple copies per genome?
c) For each type of histone the amino acid sequence is highly conserved among different eukaryotic species, e.g., histone 4A of peas and cattle differ by only 2 of the 102 amino acids present in these proteins. What conclusion can you draw from this fact?
d) In 1979, Riley, R. and H. Weintraub (PNAS 76: 328) reported on the distribution of old and newly sythesized histones during replication of DNA in the presence of cycloheximide (prevents protein, including histone, synthesis). Electronmicrographs of replicating DNA showed the following:

The dark spots (beads) are histones. Are the old and new histones distributed conservatively or semiconservatively? Explain. Which of the daughter strands of the replicating duplex is the lagging one? Explain.

37. Following the original *in vitro* synthesis of DNA, Lehman, I.R. et al. (PNAS 44: 1191, 1958) set out to answer the following questions:
a) What is the role of primer DNA? Is it simply extended in length during synthesis by the terminal addition of nucleotides, or does it serve as a template for the synthesis and accumulation of molecules identical to itself?
b) Does synthetic DNA show the equivalence of adenine to thymine and guanine to cytosine that characterizes natural DNA?
c) Does the base composition of primer DNA influence the composition of the synthetic product?
Using DNA from different organisms as primer, they determined and compared the base composition of the newly synthesized DNA with that of the DNA primer to be as follows:

$\frac{A+T}{G+C}$		$\frac{A+G}{T+C}$		Source of primer DNA
Primer DNA	Synthetic DNA	Primer DNA	Synthetic DNA	
0.49	0.48	1.01	0.99	*M. phlei*
0.97	1.01	0.98	1.01	*E. coli*
1.25	1.29	1.05	1.02	Calf thymus
1.92	1.90	0.98	1.02	Phage T2
-	>40	1.00	1.03	A-T copolymer*

* A synthetic DNA containing A and T in alternating sequence.

Varying the rate of synthesis or the deoxynucleotide concentration of the reaction mixture did not change these values. The data are sufficient to answer all the questions posed. What are the answers? Explain briefly.

38. a) What is satellite DNA? What is a characteristic property of such DNA and how was this property determined?
b) How would you determine its presence in somatic cells of a given eukaryotic species?

c) Kadouri, A. et al. (PNAS 72: 2260, 1975) isolated whole cell DNA from the cucumber, and centrifuged its DNA fragments in a CsCl gradient. Three bands of DNA were formed, the buoyant densities (ρ) of which are shown in Figure A:

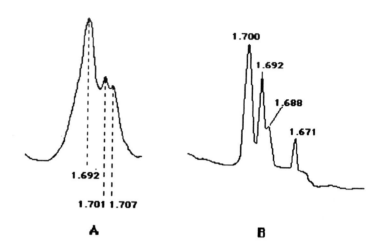

(i) Which of these buoyant densities represents the main band? Satellite bands? Explain. What is the most probable source of DNA in each buoyant density? Explain.
(ii) Using the formula $\rho=1.660+0.098$ (GC) where GC is the mole fraction of guanine plus cytosine, determine the % GC in each of the 3 buoyant densities.
d) What technique would you use to localize the satellite DNA in a genome? Outline the essential features of this technique and indicate the most likely locations of these sequences.
e) Figure B is from Gall, J.B. (CSHSQB 38: 417, 1973). It represents chromosomal DNA of *Drosophila melanogaster* that was fragmented and centrifuged. How many satellite bands are there? What is the significance of this information?
f) Would you expect satellite bands if DNA of *E. coli* was fragmented and centrifuged? Explain.

39. Swift, H. (Physiol. Zool. 23: 169, 1950) measured photometrically the amount of DNA during mitosis in Feulgen-stained pronephros cells and erythrocyte nuclei of a freshly hatched *Ambystoma opacum* larva. Interphase

nuclei of dividing tissues showed a large range of DNA values, intermediate between two classes I and II. All early prophase nuclei measured fell into class II, and all telophases fell into class I.
a) When is DNA synthesized in these cells?
b) Does its quantity and distribution in the various stages conform to expectations on the theory that genes are linear segments of DNA?

40. In 1969 De Lucia, P. and J. Cairns (Nature 224: 1164) isolated a mutant strain (*pol-A1*) of *E. coli* that had less than 1 percent of normal level of DNA polymerase I. The mutant survived and replicated DNA normally, but unlike *wild-type E. coli* it was sensitive to ultraviolet light (possessed a defective repair mechanism). Subsequently, Gross, J.D. and Y. Gross (Nature 224: 1166, 1969) showed that the recessive mutant gene *pol-A1* which affects DNA polymerase activity is of the amber nonsense type and located between *met E* and *rha*, at approximately 75 minutes on the *E. coli* chromosome. De Lucia and Cairns then isolated five additional *pol* mutants, all of which are allelic to *pol-A1*. Kelley, W.S. and H.J. Whitfield (Nature 230: 33, 1971) studied one of these, *pol-6*. This temperature-sensitive mutant contains approximately 20 percent of *wild-type* DNA polymerase I activity, which is partially destroyed by preincubation at temperatures which have little effect on the *wild-type* form of the enzyme. Whereas purified DNA polymerase I from *wild-type* strains has the same activity at 37° and 52°C and maximum activity at 55°C, the pure form of the enzyme from *pol-6* cells is less active at 52°C than at 37°C, and its optimum activity occurs at 45°C. The *wild-type* and *pol-6* forms of this enzyme also differ in several other ways. All *pol* mutants excise dimers and undergo genetic recombination normally. Kornberg, T. and M.L. Gefter (PNAS 68: 761, 1971) isolated a second DNA polymerase (II) from *pol-A1* cells which is the same as that which Okazaki, R.T. et al. (Nature 228: 223, 1970) found in a cell-free fraction from cells from the same mutant strain which can synthesize DNA rapidly.
a) Are the *pol* alleles in the structural gene for DNA polymerase? Explain.
b) If so, is the enzyme involved in DNA replication? Explain.
c) Is it involved in DNA repair? If so, at what stage?

41. a) The lecanoid system of chromosome behaviour is peculiar to certain related groups of insects. Both sexes develop from fertilized eggs and are diploid (in a typical species, the mealy bug *Pseudococcus nipae*, 2n=10 for both sexes). Except that centromeres are diffuse, mitotic chromosomal behaviour is normal; however, unusual phenomena are observed in these insects:
1. In females, meiosis is inverse. It produces haploid gametes each with a euchromatic set of chromosomes.
2. In males, one entire haploid set becomes heterochromatic before the blastula stage and remains so in all subsequent cells. Meiosis produces four gamete nuclei, two contain heterochromatic sets, and two contain the euchromatic sets. Only the euchromatic derivatives form sperm; those containing the heterochromatic set disintegrate.
(i) How would you prove genetically that the heterochromatic chromosome set in the lecanoid system comes from the father and is largely or completely inert? Assume that in the mealy bug two true-breeding lines are available, one homozygous for the dominant alleles at the loci under study, the other homozygous for the recessive alleles at the loci.
(ii) How might such a system transform into a true haplodiploid one?

b) In 1961 Brown, S.W. and W.A. Nelson-Rees (Genetics 46: 983) carried out radiation studies with *Pseudococcus citri*; typical results are shown below:

Experiment 1: Progeny of X-irradiated ♀ * x untreated ♂		
X-ray dose, R	Average number of survivors per mother after maternal treatment	
	♂	♀
Control	172.4	295.1
1,000	59.9	45.7
2,000	26.4	40.6
4,000	1.4	4.1
8,000	0.2	0.0

Experiment 2: Progeny of X-irradiated ♂** x untreated ♀		
X-ray dose, R	Average number of survivors per mother after paternal treatment	
	♀	♂
Control	163.8	212.8
2,000	129.0	114.0
8,000	188.0	41.0
16,000	275.0	3.0

*In male progeny all chromosome breaks occurred in the euchromatic set.
**In male progeny all chromosome breaks occurred in the heterochromatic set.

(1) Discuss the bearing of these data on the hypothesis that:
(i) The heterochromatic set is of paternal origin.
(ii) The heterochromatic set is genetically inert?
(2) Is heterochromatin a permanent characteristic of a chromosome or chromosome segment in the lecanoid system? Explain.

42. The appearance of the polytene chromosomes in a diploid (2n=12) species of *Drosophila* is shown below:

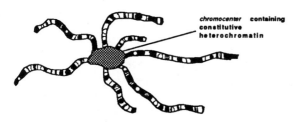

Illustrate the 12 chromosomes in a normal somatic (mitotic) cell of the species indicating (i) the position of the centromere in each chromosome, (ii) whether each chromosome consists of euchromatin and/or heterochromatin and (iii) if both types of chromatin are present, their distributions in the chromosomes.

43. Laird, C.D. (Chromosoma 32: 378, 1971) measured the rates of renaturation of denatured DNA in *Mus musculus* (mouse), *Ciona intestinalis* (sea squirt) and *Drosophila melanogaster* (fruit fly) and found that in each of the species most of the nucleotide sequences are present only once per sperm. What does this imply regarding the number of DNA duplexes per chromosome before S phase and the number of such molecules per chromatid after S until anaphase of mitosis and anaphase II of meiosis? Elaborate.

44. Bivalents at diplotene of meiosis in the oocytes of amphibians and birds are called lampbrush chromosomes.
a) Callan, H.G. and H.C. Macgregor (Nature 181: 1479, 1958) treated isolated unfixed lampbrush chromosomes from the newt *Triturus cristatus* with trypsin, pepsin, and ribonuclease, first individually and then in various combinations. Although matrix material surrounding the axis of the chromosome was dissolved, the chromosome thread in the loop remained continuous. After treatment with deoxyribonuclease these submicroscopic fibrils broke, and the lateral loops fragmented into small pieces. Which of the three macromolecules appears to be stable and indispensable to the maintenance of structural integrity and therefore to form the backbone of the chromosome? Explain.
b) Gall, J. (Nature 198: 36, 1963) showed that DNase causes breakage in the interchromomeric regions as well. Breakage kinetic studies in *T. viridescens* to determine the number of subunits per chromosome shows that there are two subunits in the loops and four in the interchromomeric regions. Studies by Miller, O.L. (Natl. Cancer Inst. Monogr. 18: p. 79, 1965) revealed a fine fibril approximately 40 Å in diameter. What can you infer from these data regarding the number of DNA duplexes (continuous or discontinuous) per chromatid (or chromosome) at anaphase, telophase and interphase before replication?

45. The viscoelastic technique permits extraction of complete (unbroken) DNA molecules from eukaryotic chromosomes and the measurement of the largest (longest) DNA molecules in solution. Kavenoff, R. and B. H. Zimm (Chromosoma 41: 1, 1973) isolated DNA from chromosomes of *wild-type D. melanogaster* as well as from (i) a strain homozygous for a pericentric inversion for chromosome 3 which changes the arm ratio from about 1:1 to about 7:1 and (ii) a strain

heterozygous for a translocation of chromosome 3 which carries about 60% of the X chromosome attached to virtually the tip of its long arm. The translocated chromosome 3 is therefore increased in length by approximately 37%. Their results were as follows:

Strain	Chromosomal Constitution	Molecular weight of largest DNA molecule	DNA content of each chromatid of largest chromosome*
Wild-type	2 ⟩⟩ ‖ ⟨⟨ 3 (4, X)	$41 \pm 3 \times 10^9$	43×10^9
Inversion homozygote	⟩⟩ ‖ ⟨⟨	$42 \pm 4 \times 10^9$	43×10^9
Translocation heterozygote	⟩⟩ ‖ ⟨⟨	$50 \pm 6 \times 10^9$	60×10^9

*These values are from Rudkin, G.T. In vitro 1: 12, 1965 and are based on photometric measurements of individual Feulgen-stained metaphase chromosomes.

a) Does the size of a DNA molecule depend on the morphology of the largest chromosome? Explain.
b) Does (do) each arm of the largest chromosome possess its own DNA molecule(s) or is/are the molecule(s) continuous through the centromere?
c) These data permit an almost unequivocal decision as to whether normal eukaryotic chromosomes are uninemic or polynemic. What is this decision and why?

46. a) The temperature at which a DNA sample denatures, its melting temperature, Tm, can be used to determine the percent of its nucleotide pairs that are GC and AT.
(i) What is the basis for this determination?
(ii) What would a low melting temperature for a given DNA sample indicate?

b) Suppose you have 4 strains of T7 virus: a *wild-type* and 3 *mutant* strains, one with a deletion, another with an inversion and still another with a duplication. What would you expect to see with an electron microscope if the DNA of the *wild-type* strain was denatured and allowed to reanneal with denatured DNA from each of the 3 *mutant* strains?

47. a) Is there a relationship (correlation) in eukaryotes between genome size (total amount of DNA per chromosome set=C value) and biochemical, physiological and morphological complexity (number of genes)? Explain. What term is applied to this relationship?
b) In many plant genera, closely related species with the same chromosome number, which are hybridizable and produce at least partially fertile hybrids, possess significantly different amounts of DNA. For example in the sweet pea genus *Lathyrus*, each of the three diploid species whose chromosome complements are shown below have a somatic chromosome number of 14 chromosomes and the F_1 hybrids between any 2 of the 4 species form 7 bivalents at diakinesis and metaphase I. There is a 3-fold difference in total chromosome size between *L. hirsutus*, with the largest chromosomes, and *L. angulatus*, with the smallest chromosomes. The variation in size is very closely correlated with DNA content (Narayan, R.K.I. and H. Rees, Chromosoma 54: 141 1976; Rees, H. and M.H. Hazarika In *Chromosomes Today* 2: 157, 1969).

The chromosome complements of three *Lathyrus* species.
A) *L. hirsutus*; B) *L. ringitanus*; C) *L. angulatus*.

i) Is it likely that *L. hirsutus* has three times as many different kinds of genes as *L. angulatus*? If not, offer a plausible cytogenetic explanation to account for the difference in DNA content between these two and other closely related species with differences in DNA content.
ii) In which fraction of total chromosomal DNA (unique, middle repetitive, highly repetitive or palindromic

sequences) would you expect to find all or most of these DNA differences? Explain.

iii) Would you expect these changes in DNA quantities to affect one, some or all the chromosomes in the genome? Explain. Would you expect the bivalents in the F_1 hybrids to possess the same morphological appearance as those formed during meiosis within each of the species? Explain.

iv) Briefly discuss and illustrate one intrachromosomal mechanism that can cause such large differences in DNA content.

48. A collection of T2 DNA molecules, each depicted by two parallel lines that denote the complementary chains of a double helix, is illustrated.

```
 1   2   3   4   5   6   7   8   9   0   1   2
 1¹  2¹  3¹  4¹  5¹  6¹  7¹  8¹  9¹  0¹  1¹  2¹
         3   4   5   6   7   8   9   0   1   2   3   4
         3¹  4¹  5¹  6¹  7¹  8¹  9¹  0¹  1¹  2¹  3¹  4¹
                 5   6   7   8   9   0   1   2   3   4
                 5¹  6¹  7¹  8¹  9¹  0¹  1¹  2¹  3¹  4¹
```

(After Thomas, C.A. Jr., J. Cell Physiol. 70(Suppl. 1): 13, 1967.)

Show how circular DNA molecules can be generated after treatment of these DNA molecules with exonuclease III.

49. Denaturation of DNA is accompanied by an increase in its optical density (O.D.) measured at 260 nm. This "hyperchromatic" shift is usually characterized by a melting temperature, Tm, at which denaturation is 50% complete. In the figure shown below are absorbance-temperature curves of various samples of cattle DNA. Sample C is total "native" cattle DNA. Sample B comprises unique sequences only which have been denatured and re-annealed before being subjected to the melting experiment shown. Sample A comprises the moderately repetitive fraction of cattle DNA which was also denatured and re-annealed prior to the melting analysis shown.

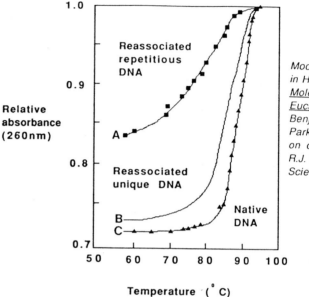

Modified from Fig.2.24 in Hood, L.E. et al. *Molecular Biology of Eucaryotic Cells*, W.A. Benjamin, Inc., Menlo Park, Calif.,1975 based on data from Britten, R.J. and Kohne, D.E., Science 161: 529, 1968.

a) What is the Tm for each DNA sample?

b) Describe how you would isolate (i) the unique DNA component and (ii) the moderately repetitive fraction.

c) Would you expect the melting profiles of a native, moderately repetitive sample and a native, unique sample to differ? Explain.

d) How would you explain the large differences in the melting profiles of the reassociated, repetitive and unique DNA samples.

The reassociation curves, for cattle-thymus DNA (+, ■ and ●) and an internal standard, *E. coli* DNA (O) are shown in the Figure below:

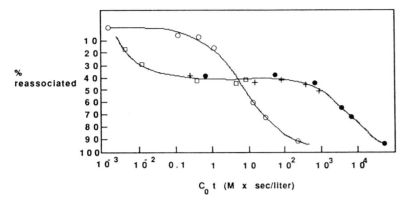

This figure is modified from Fig. 3 in Britten, R.J. and Kohne, D.E., Science 161: 529, 1968.

e) How many frequency classes of DNA sequences are present in cattle and what fraction of the DNA does each class represent?

f) How many times are the most rapidly reassociating sequences repeated with respect to the most slowly reassociating sequences?

g) Given that the *E. coli* genome size is 4.5×10^6 nucleotide pairs, what is the complexity of the most prevalent class of cattle DNA?

h) If you assume that the most prevalent class is unique DNA, what estimate do these data provide for the size of he cattle genome?

Chapter 18
Mutation and Repair

1. a) In microorganisms the proportion of induced mutations that can revert is higher with ultraviolet as a mutagen than with ionizing radiation or radiomimetic compounds. What conclusions can be drawn from this observation?
   b) Induced mutations occasionally appear in *Neurospora* and bacterial cultures several cell generations after treatment with a mutagen. Offer three explanations for this observation.

2. a) In a true-breeding *wild-type* (*red*-eyed) strain of *Drosophila melanogaster*, only males are occasionally found that have one *red* and one *white* eye. If *white* eye colour is due to a recessive mutant allele, why is only one eye *white*, and why did the mutant allele express itself in spite of being recessive?
   b) A man has one *brown* and one *blue* eye. What are the possible causes of this mosaicism?

3. An *rII mutant* in phage T4 reverts to produce a *pseudo-wild-type* strain. What tests would you employ to determine whether the *pseudo-wild-type* phenotype is due to a mutation: (i) in another gene, (ii) at a second site in the same gene, or (iii) in the same nucleotide-pair site as the first mutation?

4. When 1,000 plates of chloramphenicol agar are each plated with 1,000 cells of chloramphenicol-sensitive strain of bacteria, one plate is found with a single colony on it which is *resistant* to the drug.
   a) What is the frequency of mutation at the chloramphenicol locus?
   b) Explain how you would determine whether this mutation was postadaptive or preadaptive.

5. Explain:
   a) Why geneticists find most mutations to be deleterious.
   b) Why, nevertheless, the mutation process is considered to be the basis of evolutionary progress.
   c) Why most mutations are recessive.
   d) Why one mutation in a gene may cause a drastic phenotypic effect and another mutation in the same gene may have only a slight effect.

6. In an experiment using the *ClB* method, a recessive lethal allele is induced in a gene on the X chromosome of a sperm that fertilizes an egg carrying a *ClB* chromosome. Why does this F_1 zygote not die? What should be the results in the progeny of crossing such a F_1 female with a *wild-type* male?

7. The homopteran insects, the plant genus *Luzula* and some other organisms which have chromosomes with diffuse centromeres (Brown, S.W. and W.A. Nelson-Rees, Genetics 46: 983, 1961; Hughes-Schrader, S., Adv. Genet. 2: 127, 1948) are all less sensitive to radiation and chemical mutagens than organisms with localized centromeres. Explain why this should be so.

8. A mutant allele occurs at a certain locus. How would you determine whether the locus in question encodes a protein without assaying for the protein?

9. Kunkel, T.A. et al. (Nature 291: 349, 1981) showed that subsequent to depurination of ϕX174 DNA, a DNA polymerase can copy past the apurinic sites and this enhances the rate of base-pair substitutions.
   What mechanism would you expect to be involved in depurination which subsequently results in increased mutations?

10. Answer each of the following as briefly as possible:
    a) Which type of mutation, one induced by a base analogue or that induced by proflavine, would you expect to be more deleterious to an organism and why?
    b) What evidence is there that ionization caused by X-rays need not occur in the gene itself to cause mutation?
    c) What are the possible mechanisms by which a gene may change to many different allelic forms?
    d) Is there a known finite dosage level of ionizing radiation below which gene mutations are not induced? Explain.
    e) Why are sex-linked lethal mutations easier to detect than autosomal lethals?

f) Why is the X chromosome more useful than the autosomes in studying the induction of recessive mutations?

11. Certain reports state that oxygen decreases radiation damage. Others state the opposite. Suggest a reason for these conflicting reports.

12. The *mutD* locus in *E. coli* codes for the epsilon (ε) subunit of the polymerase III holoenzyme (Scheuermann, R.H. and H. Echols, PNAS, USA 81: 7747, 1984). In *mutD* mutants, the frequency of transitions, transversions and frameshift mutations is enhanced about 10^5 fold above the spontaneous rate (Cox, E.C., Ann. Rev. Genet. 10: 135, 1976).
a) What is the most likely function of the epsilon subunit?
b) Discuss the mechanism by which the mutant alleles at the *mutD* locus cause mutations at such a high rate.

13. The human immunodeficiency RNA virus (HIV) responsible for AIDS exhibits extreme genetic variation and a high rate of spontaneous mutations. These features of the virus present a serious problem in designing an appropriate treatment for AIDS. The enzyme reverse transcriptase (RNA dependent DNA polymerase) under *in vitro* conditions introduces base-substitution errors (using ϕX174 DNA as a template) at a frequency of 1 per 2,000 bases, which is much higher than that for other DNA polymerases.
In light of this information, suggest a plausible mechanism for the hypermutability of the AIDS virus.

14. There are two ways in which base-analogue incorporation during DNA synthesis can lead to mutation: (i) the analogue may initially be incorporated opposite the correct base but an error occurs in the choice of a partner during one of the subsequent replications (error during replication), or (ii) the analogue may pair with the wrong base during its incorporation (error during incorporation) (Freese, E., J. Mol. Biol. 1: 87, 1959).
a) Illustrate these two processes to show how transitions can be induced in either direction; e.g., A-T to G-C or G-C to A-T.
b) Explain how BU and other base analogues are able to cause two-way changes.

15. According to the slipped mispairing model (Streisinger, G. et al., CSHSQB 31: 77, 1986) base-pair additions and deletions (frameshift mutations) are a result of mispairing in regions of DNA where the stability of the duplex is reduced, e.g., near single-strand breaks during replication and crossing-over.
a) Using the following short segment of duplex DNA for discussion purposes, illustrate how frameshift mutations can be generated according to this model:

```
T G C A A A G C T
A C G T T T C G A
```

Intercalating agents such as the acridine dye proflavine are mutagenic.
b) In the context of your answer to (a), propose a plausible mechanism whereby these agents may effect frameshift mutations. Compare your mechanism with that of Streisinger, G. and J. Owen (Genetics 109: 633, 1985).
c) Spontaneous frameshift mutations are also known to be associated with the formation of transient hairpin structures which are formed during replication. How may these structures result in the formation of base-pair additions and/or deletions?

16. Three repair mechanisms are known in *E. coli* for the repair of DNA damage (pyrimidine dimer formation) after exposure to ultraviolet light: (i) photoreactivation (Wulff, D.L. and C.S. Rupert, Biochem. Biophys. Res. Commun. 7237, 1962); (ii) excision (dark) repair (Howard-Flanders, P. and R.P. Boyce, Radiat. Res. Suppl. 6: 156, 1966); and (iii) postreplication repair (Rupp, W.D. and P. Howard-Flanders, J. Mol. Biol. 31: 291, 1968). Compare and contrast these mechanisms, indicating how each achieves repair and how the events occurring in each may lead to gene mutations.

17. McClintock, B. (PNAS USA 36: 344, 1950) was the first to show that transposable elements cause gene mutations and reversions. In *Zea mays*, plants of genotype *CC* produce *purple* pigment in kernel cells. McClintock demonstrated that *C* can mutate to *c*, thus preventing pigment synthesis

and resulting in kernels that are completely *colourless* (*white*) in most *cc* plants. In some *cc* plants, some cells in some kernels produce *purple* pigment; these kernels possess a *variegated* phenotype with *purple* spots on a *white* background.

The transposable elements *Ds* and *Ac* are involved in effecting these forward and reverse mutations. How do they do so? Illustrate. See Fedoroff, N.V. Cell 56: 181, 1989 for an excellent discussion of how these transposons function.

18. A *normal*-winged *ClB Drosophila* female crossed with an irradiated *normal*-winged male produces a large progeny amongst which one *short*-winged F_1 *ClB* female is found. She is crossed with an F_1 *normal*-winged male and produces the F_2 progeny shown:

♀				♂	
ClB		*Non-ClB*			
Normal	Short	Normal	Short	Normal	Short
66	58	64	60	64	58

a) Discuss the suggestion that the *short*-winged F_2's may have been phenocopies.
b) If the change is genetic, is it autosomal or sex-linked? Recessive or dominant?
c) What kinds of changes in the genetic material could produce this inherited change? How would you distinguish between these alternatives?
d) A true-breeding strain of *short*-winged flies is established and treated with X-rays. Among 50,000 progeny, 1 *normal* revertant appears and gives rise to a true-breeding revertant strain which in crosses with the original *normal* strain gave the following results:
F_1 *normal*
F_2 134 *normal* : 31 *short*
Did the reverse mutation occur at the original or at a suppressor locus? —Explain.

19. A study of vital statistics in a certain country reveals that 50 children among 735,000 born of *normal* parents were *abnormal*: 11 were *brachydactylic* (a dominant trait, involving short fingers and toes), 36 were *albino* (a recessive trait, involving complete lack of epithelial pigment), and 3 had *aniridia* (a dominant trait, absence of iris). Assume that none of these traits affects the reproductive fitness (ability to marry and have children) of individuals.
a) For each of these traits, state whether or not a mutation rate can be determined and why.
b) Estimate the mutation rate for each trait by the direct method where this is possible.
c) Suggest two possible sources of error in your estimates.

20. a) A *normal* X-ray technician in an industrial plant whose wife is also *normal* has a son with *Duchenne muscular dystrophy* (X-linked). There are no cases of this disease in his ancestry or in that of his wife for the past four generations. He sues the industrial plant for failing to provide proper protection from radiation, claiming that his son's abnormality is the result of an induced mutation. Show whether his claim is justified.
b) A man and a woman both work in an atomic-energy plant, where they are exposed to small amounts of radiation daily. They marry and produce a child who is affected with *Tay-Sachs* disease, an autosomal disorder. They find no cases of this disease in their ancestries and believe that this abnormality in their child resulted from the effects of radiation on their gametes. Evaluate this belief.

21. In the housefly, a diploid species, the difference in eye colour between two true-breeding strains is due to a single pair of autosomal alleles; *B* for *black* is dominant to *b* for *white*. Of 250,000 F_1 progeny from a cross between the two strains, 4 have *white* eyes and the remainder all have *black* eyes.
(i) One F_1 *white* when crossed with a fly from the true-breeding *white* strain produced 68 *black* and 71 *white* eyed offspring.
(ii) Another *white* F_1 when crossed with an individual from the true-breeding *white* strain produced 33 *black* and 98 *white* eyed flies.

(iii) The remaining two F_1 *whites* in crosses with true-breeding *white*-strain individuals produced *white* offspring only, all of which produced true-breeding strains, in each of which reversion to *black* occurred.

a) Explain, with reasons, the cause of each of the *white* F_1 flies.
b) Calculate the mutation rate (1) per gamete and (2) from *B* to *b*.

22. In an appropriate genetic background multiple alleles at the *A* locus of maize show serial dominance in the determination of aleurone (kernel) colour as follows: *A* and A^b give *deep* colour, a^p gives *pale*, and *a* gives *colourless*. These aleurone phenotypes are sometimes associated with a pattern effect, *dotted* (*deep* spots on a *pale* or *colourless* background).
a) Rhoades, M.M. (J. Genet. 33: 347, 1936) found an exceptional self-fertilized ear of Black Mexican sweet corn, a true-breeding *deep* variety, which revealed kernels in a phenotypic ratio of 12 *deep* : 3 *dotted* : 1 *colourless* (the *dotted* had *colourless* backgrounds). Formulate an hypothesis that will account for this observation and for *dotted* being the result of somatic mutation.
b) The plants from the *colourless* kernels when self-fertilized or crossed with plants of any other *colourless*-variety produced plants with *colourless* kernels only. When crossed with true-breeding *dotted*-bearing plants, all progeny were *dotted*-bearing. When $a^p a^p$ plants (*pale*) were crossed with true-breeding *dotted* ones from the progeny shown in (a), the F_1's were *dotted* on a *pale* background and the progeny from F_1 x $a^p a^p$ parent segregated 3 *pale* : 1 *dotted* on a *pale* background. Two-thirds of the *pale* plants bred true. On self-fertilization the other third segregated 3 *pale* : 1 *colourless*. The plants with *dotted* kernels on a *pale* backgound segregated 6 *pale* : 6 *dotted* on *pale* background : 3 *dotted* on *colourless* background : 1 *colourless*. The same results were obtained with *A* and A^b as with a^p. What can you conclude from these data regarding the effect of alleles at the *dotted* locus on the mutability of *a*, a^p, A^b and *A* alleles?
c) The *dotted* kernels, on ears of *aa* plants have small spots of aleurone colour which are fairly uniform in size and distributed at random over the aleurone layer. What does this suggest regarding the timing and physiological conditions of mutation induction? Suggest a mechanism by which coloured cells can be produced in kernels of an *a a* plant.

23. Several *methionineless Neurospora crassa* strains have been shown to be determined by mutant alleles of the same gene. In each, an occasional colony is found that grows in the absence of methionine and breeds true for this reversion. Possible causes of this phenotypic reversion to *wild-type* are:
(i) Mutation at a suppressor locus.
(ii) True reverse mutation of the mutant allele to *wild-type*.
(iii) Contamination of the culture with *wild-type Neurospora*.
(iv) Physiological "adaptation".
Outline an experimental procedure to determine which of these explanations is the correct one.

24. Stadler, L.J. (PNAS 15: 876, 1929) measured the occurrence of mutant types in self-fertilized progeny of diploid (2n=14), tetraploid (2n=28), and hexaploid (2n=42) species of wheat and oats exposed to varying doses of X-rays. Typical results (for a single dosage level) are shown:

Species	Chromosome number (n)	Relative frequency of occurrence of induced visible *mutants*
Avena brevis	7	4.1 ± 1.2
A. strigosa	14	2.6 ± 0.6
A. sativa	21	0
Triticum monococcum	7	10.4 ± 3.4
T. turgidum	14	1.9 ± 0.5
T. aestivum	21	0

Show how the frequency of mutants is related to chromosome number. Elaborate.

25. In 1946 Mitchell, H.K. and M.B. Houlahan (Am. J. Bot. 33: 31) obtained an *adenineless* (*ade*⁻) mutant in *N. crassa* by ultraviolet induction. Kolmark, G. and M. Westergaard

(Hereditas 35: 490, 1949) found that the rate of reversion to *wild-type* could be increased by X-rays, ultraviolet light and nitrogen mustard. A sample of the results obtained in crosses between different revertants and the *wild-type* strain are shown below:

Culture number	Treatment	Asci with 8 *wild-type* ascospores	Asci with 4 *wild-type* and 4 *ade⁻* spores
1	ultraviolet	4	0
2	light	0	3
8		6	2
13	X-rays	5	0
18		0	6
23		1	4
26	nitrogen	3	1
27	mustard	3	0
34	none	5	0
35		2	1

a) Show why these results indicate that gene mutation is the cause of the reversions.
b) The haploid mycelium and the macroconidia (asexual spores) by which this organism reproduces are multinucleate. Explain the results for cultures 1, 13, 27 and 34.
c) Describe two ways in which gene mutation may lead to reversion.
d) What bearing do these results have on Stadler's (PNAS 30: 123, 1944) conclusion that X-ray-induced mutation consists of the loss of the gene?

26. Neel, J.V. and H.F. Falls (Science 114: 410, 1951) reported that in a Michigan population, 49 children born to *normal* parents out of a total of 1,054,984 births were affected with *retinoblastoma*, a dominant trait.
a) What is the mutation rate per gamete for this affliction? Explain.
b) What is the mutation rate per gamete: if the dominant allele of only one gene is involved? If dominant alleles at either of two loci can cause the abnormality?

27. The percentage of X-linked lethals in *D. melanogaster* increases in direct proportion to the amount of radiation at dosage levels up to approximately 4,000 R. At higher dosage levels the percentage of detectable lethal mutations falls below linear expectations. Account for these results.

Note: Biochemical abbreviations used in some of the following questions in this chapter are:
AP- 2-aminopurine   BU- 5-bromouracil   EES- Ethyl ethanesulfonate
HA- Hydroxylamine   HX- Hypoxanthine   EMS- Ethyl methanesulfonate
I- Inosine   NA- Nitrous acid   P- Proflavine
UV- Ultraviolet light

28. Nitrous acid (NA) is highly mutagenic for phage, bacteria, and tobacco mosaic virus (Tessman, I. Virology 9: 375, 1959). It acts directly on the bases of DNA changing adenine to hypoxanthine, cytosine to uracil, and guanine to xanthine.
a) Illustrate the chemical reaction for each of these changes and suggest how they are related to NA mutagenicity.
b) Why does nitrous acid not have any affect on thymine?
c) Would you expect xanthine to be mutagenic? Explain.
d) NA acts directly on the bases and does not cause pairing mistakes during incorporation or subsequent replications, as the base analogues do. Nevertheless, it has been shown that mutations induced by base analogues such as BU and AP can be reverted by NA and vice versa. Thus NA appears capable of inducing transitions in both directions (A-T ↔ G-C). Show how it may do so.

29. Suggest an explanation for the observations of Miller, J.H. (J. Mol. Biol. 182: 45, 1985) that (i) ultraviolet light can result in a mutation if the base-pair sequence is
   5' - C - T -3'   but not if it is   5' - T - C* - 3'
   3' - G - A -5'   3' - A - G - 5'
and (ii) the base designated by an asterisk leads to a transition type of mutation.

30. Irradiation of cells with ultraviolet light damages DNA (causes pyrimidine dimers and other lesions). These lesions elicit the SOS repair system which leads to an increase in the frequency of mutations (Little, J.W. and D.W. Mount, Cell 29: 11, 1982; Miller, J.H. and K.B. Low, Cell 37: 675, 1984).
a) Why should induction of the SOS system result in an increase in mutation frequency?

b) Most of the mutations resulting from this error-prone system are base-pair substitutions of the transition type. Explain in terms of the *umuCD/pol II* replication bypass system.

31. 5-Bromouracil (BU) an analogue of thymine that may occasionally pair with guanine, is mutagenic (Freese, E., J. Mol. Biol. 1: 87, 1959; "*Molecular Genetics*", pt 1, pp. 207-269, Academic, New York, 1963). Dunn, D.B. and J.D. Smith (Nature 174: 304, 1954) grew phage on bacteria in a medium containing BU, and found that the analogue replaced thymine in the phage DNA. Phage with BU incorporated in their DNA were allowed to infect bacteria growing on medium containing thymine. Among the progeny a few (many more than would occur spontaneously) were *mutant* in phenotype. These and other studies imply that the incorporation of BU itself does not constitute a mutation. On the basis of these findings and taking all the above facts (as well as the formulae of thymine and BU) into consideration:
a) Illustrate why replacement of thymine by BU does not constitute a mutation.
b) Suggest and illustrate a possible mechanism by which BU causes mutations.

32. a) A single strand of a DNA duplex, which contains the bases A and T in an alternating sequence is treated with EES, AP, BU, NA and P. Which of the mutagens can cause mutations in this duplex? Why?
b) A specific mutation in the *tryptophan synthetase* A-gene of *E. coli*:
(i) Can be induced by NA.
(ii) Can be reverted to *wild-type* by AP but not by NA.
(iii) Cannot be induced to revert by P.
A single base-pair is involved. What kind is it and why?

33. Hydroxylamine (HA) is a chemical mutagen that reacts almost exclusively with C (or 5-hydroxymethylcytosine), causing the HA-reacted base to pair with A. Thus HA predominantly induces the base-pair transition G-C to A-T or G-HMC to A-T (Freese, E. et al., PNAS 47: 845, 1961).
a) In 1963 Drake, J.A. (J. Mol. Biol. 6: 268.) found that 53 of 99 ultraviolet-induced rII mutations of phage T4 were reverted with base analogues. Of these, 47 were reverted by AP and BU but not by HA. The remaining 6 reverted with all 3 mutagens. Which of the following changes does ultraviolet light primarily induce:

(1) A-T to G-C? (3) A-T to C-G?
(2) G-C to A-T? (4) C-G to A-T?

Explain.
b) Most of the other UV-induced mutations were reverted by proflavine but did not respond to base analogues. What was the probable mutagenic effect of ultraviolet light in these cases? Illustrate.

34. A study of forward and reverse host-range mutations induced *in vitro* by HA, EMS and NA, in the single-stranded DNA phage S13 (Tessman, I. et al., J. Mol. Biol. 9: 352, 1964) gave the data shown:

Mutational change	Relative Success in induction by					
	HA		EMS		NA	
	F	R	F	R	F	R
h^+ to $h_i 1$	0	++	+	++	+	++
h^+ to $h_i 2$	0	0	0	++	++	++
h^+ to $h_i 1$	0	0	++	0	++	++
h^+ to $h_i 2$	0	0	++	+	++	++
h^+ to $h_i 65$	0	0	++	0	++	++
$h_l UR48$ to $h_l UR48s$	++	-	++	-	++	-

Key: ++=high; +=low, 0=zero induction frequency; F=forward; R=Reverse

None of the h^+ revertants shows recombination with any of the others. Assume (i) that HA acts primarily on cytosine, and EMS on guanine, to produce 7-alkylguanine and (ii) that only transitions are involved in these mutations. Show:
a) The base change for forward mutation for each mutant.
b) Which transitions are induced by NA.
c) Which bases (other than guanine) are acted on by EMS.

35. In *E. coli*, *lexA*$^+$ codes for a SOS repressor (Little, J.W. et al., PNAS USA 78: 4199, 1981), *uvrB*$^+$ specifies the synthesis of one of the proteins involved in excision repair (Sancar, A. and G.B. Sancar, Ann. Rev. Biochem. 57: 29, 1988) and *hisB*$^+$ encodes a protein involved in histidine biosynthesis (Ames, B.N. et al., In *Regulation of Nucleic Acid and Protein Biosynthesis*, Elseveir, Amsterdam, 1967).
What would be the effect of each of the *lexA*$^-$, *uvrB*$^-$ and *hisB*$^-$ mutant alleles on the induction of mutations by an activated form of the mutagen benzopyrene?

36. The revertability of phage T4 rII mutations induced by various chemical mutagens is shown below:

Mutations induced by	Relative Frequency of reversions induced by				
	NA	HA	BU	AP	EES
NA	+++	-	++	++	-
HA	-	0	++	+++	-
BU	+++	0	+	+++	0
AP	+++	++	+++	+	++
EES	-	-	++	++	+

Key: +++ = complete reversion; ++ = high reversion rate; + = low reversion rate; 0 = no or virtually no reversion.
BU induced mutations in both directions; HA acts on cytosine, altering its pairing properties; EES removes guanine preferentially; all reversions were true back mutations to *wild-type*.

For each of the mutagens, state with reasons:
a) Whether transitions, transversions or both are induced.
b) Whether the transitions or transversions occur in both directions or only one.
c) If in both directions, whether they occur predominantly in one direction.

37. Exposure of DNA to low pH causes depurination, i.e., removal of A and G (Tamm, C.M. et al., J. Biol. Chem. 195: 49, 1952), and occasionally backbone breakage (Tamm, C.M. et al., J. Biol. Chem. 203: 673, 1953).

a) *Wild-type* phage T4 lysozyme has the amino acid sequence Lys-Ser-Pro-Ser-Leu-Asn-Ala in a portion of its chain. Mutants with mutations induced by low pH show the following amino acid sequences:

Mutant 1 Lys-Gly-Pro-Ser-Leu-Asn-Ala
Mutant 2 Lys-Ser-Pro-Ser-Leu-Tyr-Ala
Mutant 3 Lys-Ser-Pro-Ser-Leu-His-Ala
Mutant 4 Lys-Val-His-His-Leu-Met

Using the code word assignments in the Appendix show the possible kinds of changes caused by depurination.

b) Treatment of these mutants with certain mutagens produced revertants with the amino acid sequences shown below:

Mutant	Treated with mutagen	Amino Acid sequence in revertant
1	HA	Lys-Ser-Pro-Ser-Leu-Asn-Ala
1	P	Lys-Ser-Pro-Ser-Leu-Lys-Cys
2	BU	Lys-Ser-Pro-Ser-Leu-Asn-Ala
2	P	Lys-Ser-Pro-Ser-Leu-Met-Leu
3	HA	Lys-Ser-Ser-Ser-Leu-Asn-Ala
4	P	Lys-Val-His-His-Ile-Asn-Ala

Do these results corroborate or refute your explanation in (a)?

38. Zamenhof, S. and S.B. Greer (Nature 182: 611, 1958) showed that heating *E. coli* strain W6 to 60°C can be mutagenic. Assuming that transitions, transversions and deletions can occur by heating, what is the most likely effect of this treatment?

39. *E. coli* possesses an allele at a certain locus which confers resistance to T1 phage irradiated with ultraviolet light. When a thymine analogue (bromodeoxyuridine) is substituted for thymine in the DNA of T1 phage, this

protection is removed. What is the product and mechanism of action of the gene?

40. Krieg, D.R. (Genetics 48: 561, 1963) used EMS to generate reversions of mutations induced in the rII region of the T4 phage. The table shows a selection from his results:

Mutagen used to produce original mutant	Designation of original mutant	Frequency of revertants per 10^6 progeny phage	
		With EMS	No EMS
AP	114	290.0s	2.8s
		30.0t	0.9t
	275	50.0	1.0
	70	0.5	0.0
BU	90	23.3	0.8
		5.6	0.4
	7	0.6	0.01
	19	0.18	0.07
EMS	126	147.0	0.35
	34	22.0	0.4
		36.0	15.0
	30	0.3	0.0
P	85	60.8	1.7
		76.0	27.0
	28	0.7	0.3
	83	0.07	0.10

Key: s = standard plaques; t = tiny plaques

Outline a possible mechanism of mutagenic action for this chemical, showing which types of base-pair alterations are favoured or possible. Note: AP and BU cause A-T ↔ G-C transitions; proflavine causes base-pair deletions and additions.

41. A *mutator* gene in *E. coli* increases the mutation rate of other genes (Treffers, H.P. et al., PNAS 40: 1069, 1954) including the *tryptophan synthetase A* gene. At some amino acid sites in the *A* protein different mutations may involve the substitution of any of several different amino acids for the original, making it possible to determine whether the mutator gene favours certain amino acid substitutions and hence to designate the type of base-pair change it favours. Yanofsky, C. et al. (PNAS 55: 274, 1966) studied five tryptophan *mutants*, each showing a single amino acid difference from *wild-type A* protein. Each *mutant A* gene was placed in a mutator background, and reversion to *wild-type* was studied. Revertants were either full (F) or partial (P). Reversion of all *mutants* was at a much greater rate in the mutator background. The results of the study, together with the data on amino acid substitutions and the corresponding RNA codons, are shown in the table below:

Mutant	Characteristics of revertants	Peptide	From	Code word	To	Code Word
A223	F	TP4	Ile	AUU	Ser	AGU
				AUC		AGC
A78	F	TP6	Cys	UGU	Gly	GGU
				UGC		GGC
A58	P	TP6	Asp	GAU	Ala	GCU
				GAC		GCC
A23	F	CP2	Arg*	AGA	Ser	AGC
						AGU
A46	F	CP2	Glu	GAA	Ala	GCA

*No reversions of Arg (AGA) to Ile (AUA) were observed

a) Using the RNA codons assigned to the amino acids, determine whether:
(i) The effect of the mutator gene is specific.
(ii) It is unidirectional in its action.
(iii) It induces transversions or transitions.
b) Suggest two mechanisms by which such mutator genes could produce their effects.

42. *Wild-type* (r⁺) T4 phage can grow on *E. coli* strains B and K; rII mutants are able to form plaques on B only. Levisohn, R. (Genetics 55: 345, 1967) found that of 16 *rII* mutants known to have G-C at the mutated site, 11 showed a high reversion rate when plated directly on strain K; 5 reverted only after exposure to hydroxylamine which reacts with 5-hydroxymethylcytosine of T4. In which of these two classes

of mutants did *C* occur at the mutated site of the transcribed strand?

43. In T4 phage the enzyme lysozyme, a product of e^+, can occur in a variety of different *mutant* forms. The amino acid sequence in a specific peptide of the *wild-type* enzyme and the corresponding sequences in 4 different naturally occurring mutants and their revertants are shown below:

	Amino Acid sequence of	
	Wild-type and *mutants*	*Revertants*
Wild-type	Lys-Ser-Pro-Ser-Leu-Asn-Ala	
Mutant A	Lys-Val-His-His-Leu-Met-Arg	Lys-Val-His-His-Leu-Asn-Ala
Mutant B	Lys-Ser-Pro-Ile-Lys-Cys-Ala	
Mutant C	Lys-Ser-Pro-Cys-Leu-Asn-Ala	1. Lys-Ser-Pro-Ser-Leu-Asn-Ala
		2. Lys-Ser-Pro-Cys-Leu-Asn-Thr
Mutant D	Lys-Gln-Ser-Ile-Thr...	Lys-Gln-Ser-Ile-Thr-Asn-Ala

a) Classify the forward mutations as nonsense or missense.
b) What kind of mutation is responsible for each forward mutation? Explain.
c) Two true-breeding mutant strains, *e* and *e'*, arise naturally. In *e*, the protein formed is incomplete; in *e'* it is complete but has one arginine residue of the *wild-type* substituted by lysine. Both mutants can revert to produce *wild-type* strains, which fail to produce recombinants when crossed with the original *wild-type*. Explain.

44. Thymine dimers block DNA synthesis *in vitro* and *in vivo* (Setlow, R.B. et al., Science 142: 1464, 1963). Certain strains of *E. coli* are ultraviolet *radiation-resistant* and even in the dark can recover to resume DNA synthesis. *Sensitive* cells cannot recover. Setlow et al. showed that thymine dimers are conserved in whole cells after ultraviolet irradiation. Boyce, R.P. and P. Howard-Flanders (PNAS 51: 293, 1964) studied the fate of thymine dimers in DNA during incubation after ultraviolet light irradiation in a *resistant* strain (AB1157) and an ultraviolet-*sensitive* strain (AB1886) of *E. coli* K12. The two strains differ with respect to alleles at one locus. They separated the DNA into an acid-insoluble fraction (corresponding to macromolecular polynucleotides) and an acid-soluble fraction (corresponding to thymine and thymine dimers).
(i) Thymine dimers were identified in the acid-insoluble fractions of both strains before dark incubation.
(ii) After dark incubation, the acid-insoluble fraction (hydrolyzed) of the irradiated ultraviolet-*resistant* strain showed less thymine dimers than before incubation. Moreover, as the thymine dimers disappeared from the acid-insoluble fraction, they appeared in the acid-soluble fractions.
(iii) In the irradiated ultraviolet *sensitive* strain, which cannot synthesize DNA in the dark, the dimers remain in the acid-insoluble fractions; none are detectable in the acid-soluble fraction.
Photoreactivation involves the splitting of dimers. A different mechanism occurs here, which permits *resistant* strains to recover and resume DNA synthesis. What is this gene-determined mechanism that occurs and what events follow in the *resistant* strains but not the *sensitive* ones to produce the original kind of DNA? Illustrate.

45. Ultraviolet irradiation completely blocks colony formation in a *sensitive* strain (B1) of *E. coli* but not in a *resistant* strain (B/r^s). It also permanently inhibits DNA synthesis in B1 but only temporarily in strain B/r^s. The following additional data are from Setlow, R.B. et al. (Science 142: 1464, 1963):
(i) Inhibition of DNA synthesis was reduced by photoreactivation with blue light in the *sensitive* but not in the *resistant* strain.
(ii) Inhibition immediately after irradiation is complete in both strains, but the *resistant* strain rapidly recovers its ability to synthesize DNA.
(iii) Immediately after ultraviolet irradiation considerable amounts of thymine dimers are formed in the two strains.

a) Show how these observations are consistent with the concept that ultraviolet light is an excitatory rather than an ionizing radiation.
b) Assuming that DNA synthesis is under the control of an enzyme produced by a single gene, suggest how the gene-determined differences in recovery and the effects of photoreactivation may be related.

In the *resistant* strain, thymine dimers are not split in appreciable numbers during the time between inhibition and recovery of DNA synthesis. Photoreactivation of the *sensitive* strain, on the other hand, results in a considerable decrease in the number of thymine dimers.

c) Explain how this information modifies (or changes) your answer to (b) and suggest a satisfactory mechanism of gene-determined recovery.

46. Ultraviolet light (UV) induces DNA lesions that are primarily thymine dimers. In cells of *normal* humans, systems remove or repair such damage without detrimental consequences. Exposure to UV light of individuals afflicted with the autosomal recessive disorder *xeroderma pigmentosum* (*XP*) results in all or many of these lesions not being repaired (excised) and thus leading to skin cancer. Studies by many investigators (Cleaver, J.E. and D. Bootsma, Ann. Rev. Genet. 9: 19, 1975; Cleaver, J.E., Carcinogenesis 11: 875, 1990) have revealed the following: Cultured fibroblasts from any 2 individuals can be fused to form heterokaryons in which the 2 nuclei share a common cytoplasm. Heterokaryons derived from fibroblasts of *normal* and *XP* patients express normal DNA repair. The results shown below are from studies of heterokaryons formed from cultured fibroblasts of 7 different *XP* patients:

XP patients	1	2	3	4	5	6	7
1	0						
2	+	0					
3	+	+	0				
4	+	+	+	0			
5	+	+	+	+	0		
6	+	+	+	+	+	0	
7	+	+	+	+	+	+	0

+=repair; 0=lack of repair as determined by lack of unscheduled DNA synthesis.

a) What is the implication of repair or lack thereof in the different heterokaryons?
b) What is the minimum number of genes involved in the repair of UV-induced DNA lesions? Explain.

c) In *E. coli* excision repair is a 3-step process involving excision of the lesion, repair synthesis and ligation (Haseltine, W.A., Cell 33: 13, 1983) determined by at least 5 genes. The excision repair system in humans also appears to be a 3-step process. What role(s) can you ascribe to the genes identified in (b) in the excision pathway in humans?
d) Assuming that the genes identified in (b) are the only ones involved and that the enzyme catalyzing excision of the lesion is the only multimeric enzyme in the system, how would you modify your answer to (c)?
e) Sutherland, B.M. (Bioscience 31: 439, 1981) demonstrated that somatic cell cultures of some XP patients contained no or little photoreactivation enzyme activity. What bearing does this information have on your answers to (c) and (d)?

47. Westmoreland, B.C. et al. (Science 163: 1343, 1969) separated the strands of the DNA duplexes of *wild-type* λ (λ$^+$) and the *double mutant* λ b2b5 by heating to 100°C and then hybridized the complementary strands of the two strains. An interpretive drawing of the electron micrograph of a heteroduplex formed between strand 1 of λ$^+$ and a complementary strand of λ b2b5 is shown.

a) What is the nature of the 2 mutations in λ b2b5? Explain.
b) Explain how they can be used in genetic studies.

48. Ultraviolet irradiation at 2600 to 2800 Å of DNA causes the formation of thymine dimers (Beukers, R. and W. Berends, Biochem. Biophys. Acta 49: 181, 1961; Beukers, R. et al., Rev. Trav. Chim. 79: 101, 1960). Short-wavelength (approximately 2400 Å) ultraviolet irradiation of thymine dimers causes reconversion to thymine. In 1949 Kelner, A. (J. Bacteriol. 58: 511) showed that the killing and mutagen action of ultraviolet light may be reversed by exposure of irradiated organisms to visible light. In 1960 Rupert, C.S. (J. Gen. Physiol. 43: 573) discovered an enzyme in baker's yeast which when added to ultraviolet-treated *E. coli* or other bacteria in the presence of visible light restored about

10 percent of the activity of DNA. Wulff, D.L. and C.S. Rupert (Biochem. Biophys. Res. Commun. 7: 237, 1962) showed that while samples of irradiated DNA incubated with photoreactivating enzyme in the dark and samples incubated with heat-inactivated enzyme in the light both show the same amount of thymine dimers as present in the untreated irradiated DNA, incubation of irradiated DNA with this enzyme in the presence of visible light destroys over 90 percent of the dimers present and restores the normal DNA structure. The DNA damage (dimers) is repaired in one step. What appears to be the effect of the enzyme?

Chapter 19
The Gene: Its Genetics and Interallelic Complementation

1. From about 1915 to the early 1950's it was commonly held that allelic genes were incapable of undergoing recombination and complementing each other (in heterozygotes or heterokaryons). Now we know that in some cases they can do both. What basis remains for calling genes allelic?

2. a) What tests should be performed to determine whether two independently isolated *mutants* with identical phenotypes are due to mutations in the same or different genes?
   b) Discuss the evidence indicating that functional and recombinational units are not materially equivalent.
   c) Is complementation of two recessive mutant phenotypes evidence that the two mutations are nonallelic? Is absence of complementation proof of allelism?

3. *Hemophilia A* (classic hemophilia) and *hemophilia B* (Christmas disease) are both X-linked conditions in which the blood exhibits clotting failure. Would you expect a mixture of blood from A and B *hemophiliacs* to show normal clotting capacity? Explain.

4. a) In which of the following situations would you expect intragenic (interallelic) complementation between mutant alleles:
   (1) A gene determining synthesis of a dimeric enzyme.
   (2) A gene determining formation of a monomeric enzyme.
   (3) A gene specifying a tRNA molecule.
   (4) A gene determining phage head-protein synthesis.
   b) Would it matter whether the mutant alleles were due to missense or nonsense mutations? Explain.

5. Explain which of the following you would not equate with the term (i) gene and (ii) allele:
   a) Recon? b) Cistron? c) Muton?

6. Two mutant alleles occurring in a gene coding for arginine in *Neurospora crassa*, which specifies a dimeric protein, do not complement, whereas two others do. Explain these observations, showing what you would expect in the F_1 of a cross between one of the first pair of mutants and one of the second pair.

7. a) Why is it that abortive transduction is useful in complementation studies and complete transduction is not?
   b) Why is transduction useful in fine-structure analysis?

8. Some eukaryotic genes possess coding information that is discontinuous; that is, they possess both exons and introns. Does this render the one-gene one-polypeptide hypothesis untenable? Discuss.

9. Five different strains of *Salmonella typhimurium*, some of which are *cysteine-requiring* auxotrophs, when plated on minimal medium and separately transduced with phage from each of the other strains give the results shown:

Recipient Strain	Donor Strain				
	A	B	C	D	E
A	L	L	L	L	L
B	L,S	0	L,S	L,S	L,S
C	L,S	L,S	0	L	0
D	L,S	L,S	L	0	L
E	L,S	L,S	0	L	0

Key: L=*large* colonies only; S=some *large* colonies and some *minute* ones; 0=no colonies

a) Classify each of the strains as auxotrophic or prototrophic for *cysteine* requirement.

b) Explain whether or not the mutations in the auxotrophic strains are functionally allelic.

c) What further tests would you conduct to test your classification of the mutations as allelic or not?

10. Brenner, S. (PNAS 41: 862, 1955) showed that 10 independent, *tryptophan-requiring* auxotrophic mutants of *Salmonella typhimurium* fell into 4 groups according to the steps blocked in the pathway of tryptophan synthesis. In 1956 Demerec, M. and Z. Hartman (Annu. Rev. Microbiol. 13: 377, 195) made transduction crosses between some of these mutants. Their results are given in the following table:

Recipient	Donor	Number of *wild-type* recombinants
trp-10	*wild-type*	1,822
	trp-1	4
	trp-3	270
	trp-4	602
	trp-7	7
	trp-8	208
	trp-9	12
	trp-10	0
	trp-11	0
trp-11	*trp-1*	22
	trp-2	240
	trp-3	280

For each, explain: (i) whether the 2 mutants involved are determined by allelic genes and (ii) if allelic whether the mutant alleles were due to mutations at identical or different sites.

11. a) Distinguish between continuous and discontinuous genes.
b) Contrast an exon with an intron.
c) Speculate as to the mode of origin of discontinuous genes.
d) Can there be any evolutionary advantage to organisms and species having genes with exons and introns?
e) You have a large number of individuals that are homozygous for different alleles of a given gene, e.g., different alleles for *dwarfism* in mice. Would analysis of genetic data derived from crosses involving these mutant strains permit you to determine whether the gene was continuous or not? Explain.

12. Phage from *wild-type*, (*adenine-thymine requiring, ath$^+$*) and (*adenine-requiring, Ad*) mutants of *Salmonella typhimurium* were used to infect two *ath* mutants, seeded on adenine pantothenate-enriched minimal medium at concentrations of 1×10^7 for *ath* A-4 and 5×10^7 for *ath* C-5 bacteria per plate. Progeny phenotypes were determined after 4 days incubation. Figures represent average numbers of colonies per plate (Ozeki, H., Carnegie Inst. Wash., D.C., Publ. 612, p. 97 1956).

Recipient	Colony size	Donor						
		Wild type	Ath A 2	Ath A 8	Ath A 10	ath B-6	ath C-5	ath D-12	ad D-10
ath A-5	large	116	63	15	0	119	160	130	116
	minute	313	0	0	0	622	341	268	125
ath C-5	large	72	74	70	92	84	0	67	56
	minute	955	731	878	658	1,371	0	109	285

Note: all *mutants* can revert to *wild-type*.

a) Explain how the *large* colonies originate and why there are so many in some transductions and very few or none in others.
b) Why do *minute* colonies arise, and what is the significance of their occurrence?
c) Which of these mutant genes are functionally allelic? Why?
d) Which of the *mutants* appear to be due to mutations at identical sites (homoalleles)?
e) Which condition, (*wild-type* or *mutant*) is dominant?
f) What is the minimum number of genes involved in adenine-thymine formation? Explain.

13. The *mutants* in the *r*II region of phage T4 form sharp plaques on *E. coli* strain B and no plaques on *E. coli* strain K. *Wild-type* phage form plaques with rough edges on both strains. Benzer, S. (PNAS 41: 344, 1955) infected strain K with various *r*II mutants, two at a time, and plaque-assayed the lysates obtained on *E. coli* B. Typical results are shown.

These data are representative of all *mutants* in the *rII* region.

	47	51	101	102	104	106
47	-	+	-	+	-	-
51		-	+	-	+	+
101			-	+	-	-
102				-	+	+
104					-	-
106						-

Key: +=plaques formed; -=few, if any, plaques formed

a) Does the *rII* region consist of one, two or more genes? Explain.
b) What results would you expect if *wild-type* (r^+) and any one of the *mutant* phage infected *E. coli* K?

14. a) Three *temperature-sensitive* mutants are induced in *wild-type* phage T4 by treatment with 5-bromouracil. These reproduce at 25°C, producing much larger plaques than *wild-type* T4, but do not produce at 42°C.
(i) Indicate how you would determine whether any two mutant genes are functionally allelic; show expectations.
(ii) Indicate how you would determine the frequency of *wild-type* recombinants.
(iii) Show how you would calculate the percent recombination.

b) Two mutants are discovered in phage T2 which, although they can infect *E. coli* B, cannot reproduce because they produce an incomplete head protein. They do, however, grow on *E. coli* CR63 and produce some head protein. Outline the hosts you would use and the methods you would employ to make the determinations outlined in (a).

15. In snapdragon (*Antirrhinum majus*), *white*-flowered mutant strains 1, 2, 3 and 4 were obtained by treating the pollen of a standard *magenta*-flowered variety with ultraviolet light. The phenotypes of the hybrids from crosses beween these mutants are shown. The F_2's in all crosses except 3 x 4 segregated in a ratio of 9 *magenta* : 7 *white*.

Cross	Hybrid phenotype
1 x 2	*magenta*
(1 or 2) x (3 or 4)	*magenta*
3 x 4	*white*

a) What is the minimum number of genes that determines flower colour in *Antirrhinum*? Explain.
b) Are the genes linked or not? Explain.
c) Are they acting in the same or different biochemical pathways?

16. In *Drosophila melanogaster*, the eye colours *carmine*, *coral* and *white* are determined by sex-linked mutant genes recessive to *wild-type*. The results of two crosses beteween true-breeding strains with these phenotypes are shown below:

Cross		Offspring	
♀	♂	♀	♂
white	x carmine	wild type	white
white	x coral	light coral	white

a) Which of the mutant phenotypes are due to allelic genes? Why?
b) What results would you expect in the cross *coral* female x *carmine* male? *Carmine* female x *coral* male?

17. The alleles *w* and w^{co} at the sex-linked *white* locus (position 1.5) in *D. melanogaster*, cause *white* and *coral* eyes respectively. Females heterozygous for these alleles and *sc* (*scute*, position 0.0), *ec* (*echinus*, position 5.5) and *cv* (*crossveinless*, position 13.7) in coupling phase, viz., *sc w ec cv / Sc w^{co} Ec Cv*, were crossed with *sc w ec cv / Y* males. The phenotypes of all but 3 of 21,067 progeny

examined conformed to expectations. The 3 unexpected female progeny had *wild-type* eyes and were of genotype *sc W Ec Cv / sc w ec cv*.
a) Why can mutation be ruled out as a cause of the *wild-type* phenotype of these 3 females?
b) What process is probably responsible for the formation of the *W* (*wild-type*) alleles in these three females? Explain.
c) Where would you place *w* and w^{co} on the linkage map relative to *sc* and *ec*. Why?

18. Roper, J.A. (Nature 166: 956, 1950; Adv. Genet. 5: 208, 1953) in studies of three *biotin-requiring* mutants (bi_1, bi_2, bi_3) of *Aspergillus nidulans*, found that:
(1) The *mutant* strains did not respond to the immediate precursor of biotin, pimelic acid.
(2) Prototrophic recombinants occurred in all possible crosses; their frequencies, for the $bi_1 \times bi_3$ and $bi_1 \times bi_2$ crosses, were approximately 1 per 5,000 progeny and 1 per 2,000 progeny, respectively.
(3) *Prototrophs* always showed recombination for markers on either side of the biotin locus.
(4) The sites at which the mutations responsible for bi_1, bi_2, bi_3 occurred were all within a chromosomal segment approximately 0.2 map unit long.
(5) Mycelia heterokaryotic for bi_1-bi_2, bi_1-bi_3 or bi_2-bi_3 were *biotin-requiring* as were the resulting diploid mycelia.

a) Diagram the chromosomal basis of these findings, showing the sequence of the bi_1, bi_2 and bi_3 mutant sites.
b) According to Pontecorvo, G. (Adv. Enzymol. 13: 121, 1952) the unit of function (= gene) has many sites arranged in a linear array. These sites can mutate independently, and crossing-over can occur between them. Discuss the relationship of Roper's findings to Pontecorvo's hypothesis and define the terms gene and allele.

19. The F_1 of a cross between two different true-breeding *golden* strains of swordfish is backcrossed to each of the parental strains. About 1 in every 5,000 progeny in each backcross is *wild-type* (*olive-green*). Was the F_1 *wild-type* in phenotype or not? Explain.

20. a) In *E. coli* the *ara* mutants (unable to utilize arabinose) are located between the *thr* and *leu* loci, close to *leu* and far from *thr* (Lennox, E.S., Virology 1: 190, 1955). To determine the sequence of the nonidentical mutant sites *ara*-1, *ara*-2 and *ara*-3 with respect to each other and *thr* and *leu*, Gross, J. and E. Englesberg, (Virology 9: 314, 1959) made four-factor reciprocal crosses by transduction. The recipient parent in each case was thr^- leu^- and the donor was thr^+ leu^+. Each *ara mutant* was used as donor in one cross and as a recipient in the reciprocal. Ara^+ transductants were selected on medium lacking arabinose, and containing threonine and leucine. These transductants were then scored for both thr^+ and leu^+. The results are tabulated below:

recipient	Donor	$\dfrac{ara^+ leu^+}{\text{Total } ara^+} \times 100$	$\dfrac{ara^+ thr}{\text{Total } ara^+} \times 100$
$ara\text{-}1^-$	$ara\text{-}2^-$	64.4	1.2
$ara\text{-}2^-$	$ara\text{-}1^-$	17.4	7.4
$ara\text{-}1^-$	$ara\text{-}3^-$	26.1	6.4
$ara\text{-}3^-$	$ara\text{-}1^-$	52.4	2.4
$ara\text{-}2^-$	$ara\text{-}3^-$	14.3	9.5
$ara\text{-}3^-$	$ara\text{-}2^-$	65.8	2.8

What is the order of the *ara* mutant sites relative to each other and to the *thr* and *leu* loci? Explain with the aid of diagrams.

b) When *ara*-3 is infected with phage grown on *ara*-1 and *ara*-2, in addition to *large* colonies many *minute* colonies are formed. *Minute* colonies are also formed when *ara*-1 is infected with phage grown on *ara*-2 and *vice versa*. What additional information do these data provide? Explain. Illustrate the arabinose region schematically as completely as possible.

21. *Wild-type*, mot^+, *Salmonella typhimurium* possess flagella and are motile. *Paralyzed* mutants, mot^-, also possess

flagella but are nonmotile. Complementation of two mutants is indicated by a linear trail of isolated colonies leading from a region of circumscribed growth on agar medium. Lack of complementation is indicated by failure of trails to appear when one mutant is transduced by another.

1. Results of complementation studies of eight paralyzed mutants using abortive transduction are shown below (Enomoto, M., Genetics 54: 715, 1966):

	210	222	225	244	246	261	279	300
210	-	+	+	+	-	+	+	-
222		-	-	+	+	-	+	+
225			-	+	+	-	+	+
244				-	+	+	-	+
246					-	+	+	-
261						-	+	+
279							-	+
300								-

Key: += complementation; -= no complementation

2. Seven nonreverting *mot* mutants crossed with each other in all pairwise combinations by transduction give the results shown below. All these *mutants* produced *wild-type* recombinants when crossed with mutants 244 and 279.

	238	292	282	277	290	253	293
238	-	+	+	+	+	-	+
292		-	+	+	-	+	+
282			-	+	+	+	-
277				-	+	-	-
290					-	+	+
253						-	+
293							-

Key: + = *wild-type*, (*mot*$^+$) recombinants produced; - = *no wild-type* recombinants.

3. The following results were obtained when mutants 238, 253, 290 and 292 were crossed by transduction with the reverting mutants 210, 222, 225, 246, 261 and 300.

	210	246	300	222	225	261
238	-	+	+	+	+	+
292	+	-	-	-	-	-
290	+	+	+	-	-	+
253	+	+	+	+	+	+

Key: + = *wild-type* (*mot*$^+$) recombinants produced; - = no *wild-type* recombinants.

Note: (1) *mot* 292 failed to produce motile *wild-type* recombinants with all 27 mutants that failed to complement it. (2) Reverting mutants produced *wild-type* recombinants in all pairwise crosses with each other. (3) Percent recombination between mutants 246 and 261 is 2.75 and 3.82 between mutants 300 and 261.

Referring to the data that support your contention:
a) How many complementation groups (genes) do these mutants represent?
b) Which genes, if any, are contiguous?
c) Which mutants are the result of multisite (deletions) and which are due to single (point) mutations?
d) Map the *Salmonella* genome (n=1) as completely as possible.

22. In *E. coli*, locus z is closely linked with locus i. In the cross $z_2^- i^+ \times z_1^- i^-$ most of the *wild-type*, z^+, recombinants are also i^+. What is the order of the z_1 and z_2 mutations relative to i?

23. The ovalbumin gene in fowl occupies a region of the genome consisting of approximately 8,490 base pairs but it codes for the ovalbumin protein (egg white) which is only 386 amino acids long (Dugaiczyk, A. et al., Nature 274: 328, 1978; Chambon, P., Sci. Amer. (May) 244: 60, 1981).
Account for the differences in gene and protein sizes.

24. In *Bacillus subtilis*, Carlton, B.C. (J. Bacteriol. 91: 1795, 1966) used transformation to map mutant sites in the *trp* B locus, closely linked to the *anth* locus. Double- and single-mutant strains for the *trp* B mutants B4, B6, B7 and B14 were crossed reciprocally and the percentage of *anth*$^+$ and *trp*$^+$ recombinants determined, with the results shown:

Recipient Parent	Donor Parent	Percent *anth*$^+$ among *trp*$^+$ recombinants
anth$^-$B4$^-$B6$^+$	*anth*$^+$B4$^+$B6$^-$	12.0
anth$^+$B4$^+$B6$^-$	*anth*$^-$B4$^-$B6$^+$	90.4
anth$^-$B4$^-$B7$^+$	*anth*$^+$B4$^+$B7$^-$	17.4
anth$^+$B4$^+$B7$^-$	*anth*$^-$B4$^-$B7$^+$	53.0
anth$^-$B6$^-$B7$^+$	*anth*$^+$B6$^+$B7$^-$	50.9
anth$^+$B6$^+$B7$^-$	*anth*$^-$B6$^-$B7$^+$	71.9
anth$^-$B6$^-$B14$^+$	*anth*$^+$B6$^+$B14$^-$	57.8
anth$^+$B6$^+$B14$^-$	*anth*$^-$B6$^-$B14$^+$	98.4
anth$^-$B7$^-$B14$^+$	*anth*$^+$B7$^+$B14$^-$	29.0
anth$^+$B7$^+$B14$^-$	*anth*$^-$B7$^-$B14$^+$	73.6

a) Order the four *trp* B mutants relative to each other and *anth*.
b) How would you determine whether a mutant is point (single) or multisite in nature?

25. Partially denatured DNA duplexes incubated in the presence of complementary RNA result in hybridization of the single-stranded RNA with the complementary strand of the denatured DNA duplex. Binding of the RNA with the complementary DNA strand results in a displacement of the other DNA strand of the duplex. When observed under the electron microscope the displaced DNA strand forms a "R loop". R loops can be used to study gene structure at the molecular level - a procedure known as R-loop mapping. Using a lambda (λ) clone carrying the mouse β gene Tilghman, S.M. et al. (PNAS 75: 1309, 1978) carried out R-loop mapping involving the denatured DNA of the β gene with both its primary transcript and its messenger RNA. Electron micrographs revealed the following:

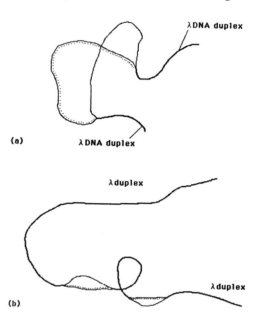

Electron micrographs of R-loops formed between denatured β gene and its primary transcript (A) and mRNA (B); (a) and (b) are drawings of the electron micrographs A and B respectively. (From Tilghman, S.M. et al., PNAS 75: 1309, 1978).

a) Explain why it can be concluded that the β gene in the mouse is discontinuous?
b) How many exons and introns does the β gene possess?
c) Are the introns excised before or after primary transcript formation? Explain.
d) In the drawings label the R-loop(s), introns, exons, DNA-RNA hybrids and the β-gene anti-sense strand.

26. Ishikawa, T. (Genetics 47: 1147, 1962) studied 308 independently obtained *adenine-requiring* mutants in *Neurospora crassa*. All were unable to convert inosine monophosphate (IMP) to adenosine monophosphate succinate (AMPS) and either lacked the enzyme AMPS synthetase or had a defective form of it. Pairwise crosses in all possible combinations, with rare exceptions, yielded only

auxotrophic progeny. Among the progeny of crosses between *wild-type* and any *mutant*, one-half were *wild-type*, the other half *mutant*.

a) Is AMPS synthetase composed of one or more kinds of polypeptide chains?

b) Two of the mutants, number 32 (induced by ultraviolet light) and number 161 (induced by nitrous acid), never produced *wild-type* recombinants when crossed. Explain in molecular terms. Would you expect the polypeptide chains produced by these mutants to contain the same amino acid substitution if the mutation was missense?

27. a) What are overlapping genes?

b) Could you determine their existence from analysis of data derived from crosses between strains with *mutant* alleles of the same gene that possess mutations at different sites? Explain.

c) Are overlapping genes expected from the evolutionary point of view? Explain. Why are overlapping regions of genes tolerated?

28. The phage φX174 genome is a single-stranded, circular DNA molecule 5,375 nucleotides long (Sanger, F. et al., Nature 265: 687, 1977). The chromosome contains 9 genes that code for 9 different proteins which *in toto* contain approximately 2,050 amino acids (Barrell, B.G. et al., Nature 264: 34, 1976). Account for this incongruity in DNA length and number of amino acids.

29. On the basis of recombination data, DNA and protein sequencing, Barrell, B.G. et al. (Nature 264: 34, 1976) obtained the following results in the phage φX174:

(i) Genes D and E are completely linked.

(ii) Genes D and E code for proteins 152 and 90 amino acids long respectively.

(iii) *Amber* (nonsense) mutations in gene D make full-length E proteins and vice versa.

(iv) *Amber* mutations *am3* and *am6* in gene E occur in the *HaeIII* restriction fragment Z7 which is within gene D.

(v) The first three and last two amino acids in the protein synthesized by gene E are methionine, valine, arginine, lysine and glutamic acid respectively.

(vi) The base sequence and nucleotide number of segments of gene D and the sequence and number of amino acids specified by the codons in these regions necessary to answer the following questions, are shown below:

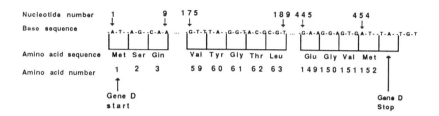

a) Why are genes D and E overlapping?

b) Scrutinize the sequence of gene D and with the aid of the coding dictionary in the appendix determine where gene E starts and where it ends in gene D.

c) Why do nonsense mutations in either of these genes not affect the synthesis of full-length proteins specified by the other?

30. Estimates by Chovnick, A. et al. (Am. Nat. 96: 281, 1962) place the molecular weight of *Drosophila melanogaster* xanthine dehydrogenase near 400,000, which is close to estimates for this enzyme in the chicken (liver) and other vertebrates. Assuming that the average molecular weight of amino acid residues is 100 and that the enzyme consists of one polypeptide chain specified by one gene, Ry, derive the length of this gene in terms of nucleotides.

31. The *ad-4* gene in *Neurospora crassa* specifies the synthesis of adenylo-succinase, which catalyzes the conversion of AMPS to AMP and fumarate. Woodward, D.O. et al. (PNAS 44: 1237, 1958) showed that each of the 123 allelic mutant genes of independent origin produce defective forms of this enzyme with activities less than 1 percent that of the *wild-type* enzyme. Certain pairs of mutants complement each other, with a partial restoration of enzyme acitivity, when combined in heterokaryons. These also show complementation *in vitro* (D.O. Woodward, PNAS 45: 846, 1959). The enzyme acivity in heterokaryons is always low, never exceeding 25 percent of *wild-type* activity. Assuming that adenylo-succinase is a dimer and that the allelic

mutants F4 and F39 specify defective polypeptide chains that are altered as illustrated below, show:
a) The types of enzymes and their proportions in heterokaryons between F4 and F39.
b) Which of these would express activity.

Allele	Location of protein defect (arrow)
F4	4 ↓
F39	39 ↑

32. Catcheside, D.G. and A. Overton (CSHSQB 23: 137, 1958) classified 40 *arginineless* (*arg-1*) mutants of *Neurospora crassa* into six groups, whose complementation relationships, determined by growth rate of heterokaryons on minimal medium, are shown.

	A	B	C	D	E	F
A	−	−	−	−	−	−
B		−	−	+	+	+
C			−	−	+	−
D				−	−	+
E					−	+
F						−

Key: + = complementing; − = noncomplementing

a) Draw a complementation map of the region represented by these *mutants*.
b) *Mutant A* does not complement any of the other *mutants*. Does this indicate that it is a large deletion? Explain.
c) Show whether you expect the complementation and genetic maps of this region to be colinear or not and why?

33. In *Escherichia coli* the enzyme alkaline phosphate whose structure is determined by a single structural gene *p* (Garen, A. and S. Garen, J. Mol. Biol. 7: 13, 1963) is composed of two identical subunits (polypeptides), each with a molecular weight of 40,000. These workers found that p^- (phosphate-negative) point *mutants* mapped at different sites in this gene. Each produced a functionally defective form of the enzyme with little or no activity. The *mutants* complemented each other in certain pairwise heterozygous combinations *in vivo*, but the resulting enzyme activity was much lower than in *wild-type* strains (see U9 and S33).

F' complementing strain		Enzyme acitvity
Episome	Chromosome	
$p^+(CRM^+)$*	$p^+(CRM^+)$	240
U9(CRM$^+$)	S33(CRM$^+$)	50

*CRM = cross-reacting material

By acidification, Schlesinger, M.J. and C. Levinthal (J. Mol. Biol. 7: 1, 1963) prepared monomers from the defective enzymes of the p^- mutants U9 and S33. The enzyme activity of the native mutant enzymes, a combination of these enzymes, enzyme monomers alone, and enzyme monomers together is shown in the table below:

Protein	Concentration, μg/ml	Enzyme activity, units/ml in 50 min.
U9 monomer	320	0.03
S33 monomer	320	0.30
U9 monomer + S33 monomer	320	5.00*
U9 native	200	0.20
S33 native	200	0.02
U9 native + S33 native	400	0.20

*Enzymatically active protein

The active enzyme of U9-S33 reaction mixture, separated by electrophoresis, forms bands in a position intermediate between those of the two parental CRMs. S33 CRM, which is

able to carry out the complementation reaction with U9, sediments at the same rate as the *wild-type* monomer (mol wt = 40,000). Discuss the bearing of these data on:
a) The hybrid-protein hypothesis of interallelic complementation.
b) Whether active enzyme is a hybrid.
c) Whether the material participating in complementation is a monomer.

34. In phage, rII deletion mutants can be mapped by a spot test. Mutants are added two at a time, to molten top agar seeded with a mixture of many K (λ) and a few B cells of *E. coli*. Benzer, S. (PNAS 45: 1607, 1959) tested hundreds of rII deletion mutants in phage T4 in this manner for their ability to produce r^+ recombinants. The results for eight such mutants are given in the table below:

	H88	B37	184	C51	782	C33	347	B138
H88	-	-	-	-	-	-	-	-
B37		-	+	+	+	+	+	+
184			-	-	+	-	+	+
C51				-	+	+	+	+
782					-	-	-	+
C33						-	+	+
347							-	+
B138								-

Key: + = recombinants formed; - = no recombinants.

a) Draw a topological representation of these *mutants*.
b) How would you test to determine whether the *mutants* are in the A or B gene (cistron) or both?

35. The rII region of phage T4 consists of two contiguous genes, A and B (Benzer, S. et al., *The Chemical Basis of Heredity*, pp. 70-93, Johns Hopkins, Baltimore; Benzer, S. Sci. Am. 206 (Jan.): 70, 1962). The r mutants in this region are individually incapable of reproducing within (and hence of lysing) strain K of *E. coli*, but mixed infection with certain of these *mutants* can infect and lyse *E. coli* cells. The five naturally occurring mutations in mutants 47, 51, 101, 102 and 104 were recombined in pairs to form *double-mutant* (cis) strains referred to as r^1r^2. The mutations are present in the same DNA strand; r^1 and r^2 represent any two mutations. Mixed infections of *E. coli* K with the five *single-mutant* strains, the various r^1r^2 strains, and *wild-type*, r^+, two at a time, in all possible combinations were performed and studied for burst size, which indicates the number of phage particles developed per host cell.

Strains used in mixed infection		Burst size
r^+	r^1r^2	253
104	102	251
102	101	247
51	102	0*
51	101	256
47	104	0
47	101	0
51	47	253
47	102	249
51	104	264
101	104	0

*0 refers to failure of phage development or, if some develop, defective phage that are incapable of lysing the host.

a) If the mutation in mutant 51 occurs in the B gene, in which of the genes does each of the other mutations lie and why?
b) All *mutants* are capable of reverting to *wild-type*, r^+, except 104. What is the probable nature of the mutation in this mutant?

36. You have 1,000 r mutants of T4 which map in the same region of the genetic map. How many pairwise crosses would be required to map all the *mutants*?

37. In *Neurospora crassa*, *pdx* and *pdxp* are *mutant* forms of the *wild-type* gene pdx^+ involved in determining pyridoxine synthesis. The gene locus is linked closely on either side by the markers *pyr* (*pyrimidine requirement*) and *co* (*colony type variant*). In crosses between $pyr^+pdxp\,co^-$ and $pyr^-pdx\,co^+$ (Mitchell, M.B., PNAS 41: 935, 1955) the vast majority of tetrads showed the expected 2:2 segregation

ratio of pyr^+ from pyr^-, pdx^- from $pdxp^-$ and co^+ from co^-, but four unexpected asci contained pdx^+ spores. Three of these are shown:

Spore Pair	Ascus 1	Ascus 2	Ascus 3
1	$pyr^+pdxp^-co^-$	$pyr^+pdxp^-co^-$	$pyr^+pdx^+co^+$
2	$pyr^+pdx^+co^-$	$pyr^+pdx^+co^-$	$pyr^+pdxp^-co^-$
3	$pyr^-pdxp^-co^+$	$pyr^-pdx^-co^+$	$pyr^-pdx^-co^+$
4	$pyr^-pdx^-co^+$	$pyr^-pdx^-co^+$	$pyr^-pdx^+co^-$

In addition Mitchell showed that:
(i) pdx^+ spores were never found in the crosses $pdx^- \times pdx^-$ or $pdxp^- \times pdxp^-$.
(ii) Analysis of randomly collected pdx^+ spores gave the following distribution of genotypes for the outside markers:

$pyr^- co^+$ 13 $pyr^+ co^+$ 7
$pyr^- co^-$ 7 $pyr^+ co^-$ 5

This distortion of the 2:2 segregation (referred to as gene conversion) may be explained theoretically in more than one way. Show why these results cannot be explained (a) by mutation or (b) reciprocal crossing-over. Offer an explanation that accounts for all the observations.

38. a) The data in the table below for pairwise crosses between rII mutants of T4 are from Benzer, S. (PNAS 47: 403, 1961)

	1605	164	1589	196
1605	-	-	-	-
164		-	-	+
1589			-	-
196				-

+ = recombinants; - = no recombinants.

When K (λ) *E. coli* bacteria, on which the *mutants* do not grow, are mixedly infected with pairs of these deletion mutants and with two other rII mutants 250 and 187, the results shown below are obtained:

Mutants used in infections		Consequences
1605 x 164	1605 x 196	Few, if any
1589 x 164	1589 x 196	phage produced
164 x 196	164 x 187	Normal number
250 x 196	250 x 187	of phage produced
250 x 1605	187 x 1605	Few, if any
250 x 1589	187 x 1589	phage produced

Analyse these data for the sequence and extent of the deletions and the number of genes in this region.

b) The data in the table below are the results of crosses between rII mutants in T4 phage:

	1	2	3	4	5
A	-	+	+	+	+
B	-	-	+	+	+
C	-	-	-	+	+
D	-	-	-	-	+
E	-	-	-	-	-

Key: + = recombinants; - = no recombinants

(i) Is it possible to determine from these data whether the mutants in the horizontal or vertical column are deletion mutants? Explain.
(ii) Draw a topological representation of the mutations in the vertical column.
(iii) What circumstantial evidence would indicate that these mutants involve deletions?

39. Amber mutants of T4 phage can infect and multiply on strain CR63 of *E. coli* K12, a permissive host. On the nonpermissive strain B the infection is abortive; i.e.,

reproduction is blocked at some stage in development. *E. coli* B is infected with four amber mutants two at a time. The burst size (phage yield / number of cells infected) is then calculated for both strains CR63 and B. The normal burst size is 200 per cell. The results are as follows:

	Burst size on CR63				Burst size on B			
	am^1	am^2	am^3	am^4	am^1	am^2	am^3	am^4
am^1	0	2	200	200	0	1	4	3
am^2		0	200	200		0	5	4
am^3			0	2			0	1
am^4				0				0

a) Construct a genetic map, enclosing markers in the same complementation group with parentheses.
b) How many of the phage from crosses am^1 x am^3, am^1 x am^4 and am^3 x am^4 are recombinants? Wild-type in genotype? Explain.

40. In the fungus *Sordaria fimicola* crosses between *gray*-spored strains and *white*-spored *wild-type* strains, made by Kitani, Y. and L.S. Olive, (Genetics 57: 767, 1967), yielded asci with the types of linear spore arrangements as illustrated.

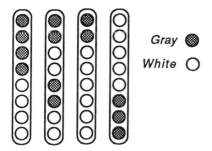

Gray ●
White ○

Outside markers in these asci always showed regular 4:4 segregations.
a) State the kind of recombination event that can account for each segregation pattern.
b) The *gray* and *white* spore colours are determined by alleles g^+ and g which possess the base-pairs A-T and G-C respectively at position 99. Their other base-pair sequences are identical.
(i) Diagram the pair of homologues carrying these alleles at pachytene. Specifically show the base-pair at position 99 in each chromatid of each homologue.
(ii) The phenomenon responsible for the formation of the latter two asci involves any 2 non-sister chromatids and heteroduplex formation. Using one chromatid of each homologue show how heteroduplexes form and the molecular mechanism(s) by which they can be resolved to produce the 5:3 and 6:2 distributions of alleles and therefore spore pattern arrangements.

41. Wild-type λ phage, c^+, form *turbid* plaques; c^- mutants form *clear* plaques. Two *suppressor-sensitive* (*amber*) mutants of λ, $c^-Su1^-Su2^+$ and $c^+Su1^+Su2^-$ are crossed at 37°C by being grown together in a permissive host strain of bacteria. In the permissive host UAG is translated as sense; in the nonpermissive strain as nonsense. The cells are incubated to lysis and the lysate diluted and plaque-assayed on the permissive and nonpermissive hosts.

	Permissive host (plaque type)	Nonpermissive host (plaque type and number)
Parental:		
$c^-Su1^-Su2^+$	clear	none
$c^-Su1^+Su2^-$	turbid	none
Recombinant:		
$c^+Su1^-Su2^+$	clear	none
$c^-Su1^+Su2^+$	clear	clear 10
$c^-Su1^-Su2^-$	clear	none
$c^+Su1^-Su2^+$	turbid	none
$c^+Su1^+Su2^-$	turbid	turbid 90
$c^+Su1^-Su2^-$	turbid	none

a) What is the order of the three markers c, $Su1$ and $Su2$?
b) Is there evidence of linkage? Explain.

Chapter 20
Biochemical Genetics

1. Discuss the evidence for the "one gene-one enzyme" hypothesis and explain why it has now been modified to the form "one gene-one polypeptide".

2. Several instances of single-gene mutations each causing a double growth-factor requirement are known. In *Neurospora crassa*, for example, a single mutation results in a nutritional requirement for the amino acids methionine and threonine. Show how such double growth factor requiring mutants would be distinguished from mutants with mutations in each of two genes and discuss the possible mechanisms underlying the double requirement.

3. The majority of *mutant* proteins differ from *wild-type* (normal) proteins by having single amino acid substitutions at specific residue sites. Some of these substitutions have no effect on enzyme function, others affect enzyme activity slightly, still others affect it severely and others cause its complete cessation. Offer an explanation to account for this variability of effect of amino acid substitution.

4. a) In what respect can genetically determined diseases such as *diabetes mellitus* and *phenylketonuria* be "cured" (alleviated)?
   b) Do such cured individuals have any moral obligation to society? Discuss.
   c) Discuss the advantages and disadvantages of curing or alleviating inborn errors of metabolism.

5. In one strain of *Neurospora crassa*, an enzyme catalyzing one of the steps in the chain of reactions leading to the formation of thiazole is lacking. This strain is also unable to synthesize thiamine. Explain how this is possible.

6. In some cases a particular gene mutation results in the loss of a specific enzyme activity. Does this mean that the enzyme is not produced? Explain.

7. a) A tetrapeptide contains the amino acids histidine, tyrosine, leucine and phenylalanine. Enzymatic degradation produces three kinds of dipeptides: Tyr-Leu, Tyr-His and Phe-His. What is the amino acid sequence in this tetrapeptide?
   b) The *A* polypeptide (267 residues) of *E. coli tryptophan synthetase*, when treated with cyanogen bromide, breaks into five fragments. The N-terminal one, called F-1 fragment, with residues 1 through 83, can be further fragmented by trypsin (which splits bonds at the carboxyl side of arginine and lysine) and chymotrypsin (which hydrolyzes bonds on the carboxyl side of phenylalanine, tyrosine, tryptophan and leucine). The peptides containing the first 30 residues of this chain possess the amino acid sequences shown below. In each peptide the N-terminal amino acid is presented to the left. Methionine is the N-terminal amino acid in the *A* chain (Guest, G.R., J. Biol. Biochem. 242: 5442, 1967).

Peptide number	Amino acid sequence
	Tryptic peptides
9	*Met-Gln-Arg
14	Tyr-Glu-Ser-Leu-Phe-Ala-Gln-Leu-Lys
19	Glu-Arg
27	Met-Gln-Arg-Tyr-Glu-Ser-Leu-Phe
21	Lys
12	Glu-Gly-Ala-Phe-Val-Pro-Phe-Val-Thr-Leu-Gly-Asp-Pro-Gly-Ile
10	Lys-Glu-Gly-Ala-Phe-Val-Pro-Phe-Val-Thr-Leu-Gly-Asp-Pro-Gly-Ile
	Chymotryptic peptides
5	Met-Gln-Arg-Tyr
27	Met-Gln-Arg-Tyr-Glu-Ser-Leu-Phe
36	Lys-Glu-Arg-Lys-Glu-Gly-Ala-Phe
37	Lys-Glu-Arg-Lys-Glu-Gly-Ala-Phe-Val-Pro-Phe
8	Ala-Gln-Leu-Lys-Glu-Arg-Lys-Glu-Gly-Ala-Phe-Val-Pro-Phe

* N-terminal amino acid sequence

Determine the sequence of the 30 amino acids.

8. In *Drosophila melanogaster* the genotypes *Bw_St_*, *bwbw St_*, *Bw_stst* and *bwbwstst* result in *red* (*wild-type*), *brown*,

scarlet and *white* eye colours respectively. Outline a hypothetical biochemical pathway that would explain this type of gene interaction and demonstrate why, according to your explanation, each genotype shows its specific phenotype.

9. The *red-lilac-blue* flower colours in higher plants are due to anthocyanins, molecules consisting of an anthocyanidin to which one or two sugar residues are attached through an oxygen. If only one sugar is attached, it is always joined to the anthocyanidin at the 3 carbon position. If two sugars are attached they are always present at the 3 and 5 carbon positions.

Anthocyanidin

1. Wit, F. (Genetica 19: 1, 1937) working with the china aster (*Callistemma chinensis*) obtained the results shown below:

Cross	F_1	F_2
lilac x *deep-pink*	almost-*lilac*	294 *lilac* : 91 *deep-pink*
blue x *lilac*	*blue*	282 *blue* : 105 *lilac*
blue x *deep-pink*	*blue*	195 *blue* : 64 *deep-pink*

The anthocyanins in *blue*, *lilac* and *deep-pink* varieties had *delphinidins*, *cyanidins* and *pelargonidins* respectively. *Delphinidins* have hydroxyl groups at the 3', 4' and 5' positions on the lateral benzene ring, and *cyanidins* have these groups at the 3' and 4' positions on this ring. *Pelargonidins* have such a group at the 4' position only. In all these molecules the sugars are located at the 3 and 5 carbon positions.

2. Further crosses by Wit between true-breeding lines gave the results shown below:

Cross	F_1	F_2
blue x *salmon-pink*	*blue*	135 *blue* : 16 *salmon-pink** 31 *slaty-blue** : 43 *deep pink*
lilac x *salmon-pink*	*lilac*	51 *lilac* : 13 *deep-pink* 17 *slaty-lilac* : 4 *salmon pink*

*Salmon-pink, *slaty-blue* and *slaty-lilac* are lighter in colour than *pink*, *blue* and *lilac* respectively.

The anthocyanins in the *blue*, *slaty-blue*, *pink* and *salmon-pink* F_2's were as follows: Blue: *delphinidin*; Pink: *pelargonidin*;

Slaty-blue:

Salmon-pink:

R= sugar molecule

a) What is the genetic basis for:
(i) The differences among *blue*, *lilac* and *deep-pink* strains?
(ii) The phenotypes in each of the last two crosses?
b) Account biochemically for the phenotypic differences that occur in each of the crosses.

c) What would be the formula of the anthocyanin molecule for *slaty-lilac*? Why?

10. In corn, a cross-fertilizing species that can be self-fertilized, *wild-type* plants have *purple*-coloured stems. Six true breeding mutant lines are established, two with *brown* stems and four with *red* stems. The results of crosses among these mutants are shown below:

Cross	F_1	F_2
brown 1 or 2 × red 1, 2, 3 or 4	purple	9 purple : 3 red : 3 brown : 1 white
brown 1 × brown 2	brown	All brown
red 1 or 2 × red 3 or 4	purple	9 purple : 7 red
red 1 × red 2	reddish-blue	1 reddish-blue : 1 red
red 3 × red 4	blue	5 purple : 4 blue : 7 red

a) How many different *wild-type* genes mutated to give the 6 mutant strains? Explain by providing the genotypes for the P_1, F_1 and F_2 phenotypes of all crosses.
b) Are the genes linked or not?

Enzymes specified by each of the genes revealed the following results:
1. In cross 2 (*brown* 1 × *brown* 2), none of the F_1's or F_2's showed any enzyme B activity.
2. In cross 4 (*red* 1 × *red* 2), the F_1's showed some enzyme R activity; one-half the F_2's showed the same amount of enzyme R activity as the F_1's; others did not show any.
3. In cross 5 (*red* 3 × *red* 4), four different kinds of enzymes (A, B, R, P) were detected, one identical to that in *purple* plants. In the F_2 one-fourth of the individuals produced the same kinds of enzymes as the F_1's.
c) How many kinds of polypeptide chains do these enzymes possess, and how many of each kind of chain (one or more) are present?
d) Show whether the genes act in the same or different biochemical pathways.

11. Ribonuclease contains one polypeptide chain only (124 amino acids); insulin contains two: A, with 21 amino acids and B, with 30 amino acids. Alkaline phosphatase has two chains of the same kind, whereas human adult hemoglobin contains four polypeptide chains of two kinds, α and β. Explain whether these proteins will have a:
a) Primary structure?   c) Tertiary structure?
b) Secondary structure? d) Quaternary structure?

12. The alleles C, c^k, c^d, c^r, c^a in the *albino* series of the *pink-eyed sepia* breed of guinea pig have the effects shown on the relative amounts of melanin pigmentation in the hairs (Wright, S. Genetics 44: 1001, 1959). Classify each allele as dominant, hypomorphic or amorphic and give a biochemical explanation for the different effects of these alleles.

Genotype	% melanin pigment	Phenotype	Genotype	% melanin pigment	Phenotype
$C\_$	100	Full	$c^d c^r$	19	Intermediate
$c^k c^k$	88	Intermediate	$c^d c^a$	14	Intermediate
$c^k c$	65	Intermediate	$c^r c^r$	12	Intermediate
$c^k c^r$	54	Intermediate	$c^r c^a$	3	Intermediate
$c^k c^a$	36	Intermediate	$c^a c^a$	0	White
$c^d c^d$	31	Intermediate			

13. Bateson, W., et al (Rep. Evol. Comm. R. Soc. 2: 1-55, 80-99, 1905) crossed a true-breeding *white*-flowered variety of the sage *Salvia horminum* with a true-breeding *pink*-flowered variety. The F_1's had *purple* flowers and when self-fertilized produced 255 *purple*, 92 *pink* and 114 *white*-flowered F_2's. The same investigators obtained F_1's with *purple* flowers from a cross between true-breeding *white*-flowered varieties of sweet pea (*Lathyrus odoratus*). The F_1's when self-fertilized produced 382 plants with *purple* and 269 with *white* flowers. After stating the type(s) of gene interaction involved, outline a genetically determined biochemical system to account for each set of results.

14. Srb, A.M. and N.H. Horowitz (J. Biol. Chem. 154: 129, 1944) obtained by X-ray or ultraviolet irradiation, 15 mutants of *N. crassa* which lacked the ability to synthesize arginine. Crosses between *wild-type* and each *mutant* gave a 1:1 ratio of *wild-type* : *arginineless*. The mutant alleles in eight of these genes were found to be allelic with one of the seven mutants (A to G) to be considered here.
1. Crosses between either of the mutants A, B, C or D and either of the mutants E, F or G produced *wild-type* and

arginineless progeny in a 1:3 ratio. The same results were obtained in crosses between either mutants E or F and mutant G.

2. Strains A to D grew on a minimal medium if ornithine, citrulline or arginine was added. Strains E and F could not utilize ornithine but grew if the other two compounds were supplied. Strain G grew only if arginine was added to the minimal medium.

With reasons for your answers, state: (i) the minimum number of genes involved in arginine synthesis, (ii) whether they act in the same or different pathways and (iii) if in the same pathway, the sequence of biochemical reactions.

15. A *scarlet*-eyed mutant appears in a *wild-type (red-eyed)* strain of *Drosophila melanogster*. Crossed with *wild-type* flies, it produces *wild-type* F_1's and F_2's in the ratio 3 *wild-type* : 1 *scarlet*. Two separate revertants to *wild-type* subsequently appear in a true-breeding *scarlet* strain, and by subsequent breeding within this strain, a true-breeding progeny strain of each reversion is obtained. The first, when crossed with *wild-type* flies, produces F_1's that breed true for *wild-type*. The second, similarly crossed, produces *wild-type* F_1's and F_2's that segregate 15 *wild-type* : 1 *scarlet*.

a) Why does the second and not the first revertant segregate for eye colour in the F_2?

b) Outline one possible mode of action of the gene or genes involved in the reversion to *wild-type* observed in the second cross.

16. In *Neurospora crassa*, methionine is an amino acid essential for growth. It is synthesized via a chain of chemical reactions each mediated by a single enzyme under the control of a single gene, as shown below (Fischer, G.A., Biochem. Biophys. Acta 25: 50, 1957):

$$\xrightarrow{\text{gene A}} \text{cysteine} \xrightarrow{\text{gene B}} \text{cystathionine} \xrightarrow{\text{gene C}} \text{homocysteine} \xrightarrow{\text{gene D}} \text{methionine}$$

Strains, incapable of synthesizing methionine, result from mutations at any one of the four loci. Three *mutant* strains of independent origin showed the following characteristics:

1. The first *mutant* will grow if supplied with cystathionine or homocysteine or methionine but not cysteine.
2. The second *mutant* grows when either homocysteine or methionine is provided but not if supplied only cysteine or cystathionine or both.
3. The third *mutant* grows only if methionine is supplied.

a) At which step does a metabolic block appear to occur in each of the *mutant* strains? Explain.

b) How would you determine experimentally whether a block caused by a mutant gene occurs before or after a given reaction in the metabolic pathway?

17. Four mutant strains of *N. crassa* were tested for their ability to grow on minimal medium (MM) when various substances were added. The following results were obtained:

Mutant	MM + Pyrimidine	MM + Thiamine	MM + Cmpd B	MM + Thiazole
a	-	+	+	-
b	+	+	+	-
c	-	+	-	-
d	-	+	+	+

+ = growth; - = no growth

a) Why do these data indicate that a branched biochemical pathway is involved?

b) What is the sequence of biochemical reactions in this pathway? Explain.

c) Which step does each mutant gene block? Explain.

18. In individuals affected with type I *osteogenesis imperfecta* (*OI*), collagen-rich tissues and blood vellels degenerate, bones weaken and fracture easily, sclera become blue, deafness sets in and a heart attack or stroke usually causes premature death. Onset of the disease is as late as age 35 to 45 years.

Type I collagen is a heterodimer composed of two $\alpha 1$ polypeptides and one $\alpha 2$ polypeptide. In *normal*

individuals these three polypeptides form a triple-helical structure which is necessary for the protein to function normally. Cohn, D.H. et al. (PNAS 83: 6045, 1986) studied a heterozygous individual with *OI*. They found that the difference between the *normal* and *mutant* alleles proα1$^+$ and proα1$^-$ at the locus coding for the α1 polypeptide involved a GC to AT substitution in codon 988. This resulted in the substitution of cysteine for glycine at the corresponding position in the α1 polypeptide.
a) What is the most likely effect of the mutant allele on collagen structure?
b) Why should the mutant allele have pleiotropic effects?
c) Why is this disease due to a dominant allele? Explain.
d) What percent of all type I collagen molecules are expected to be abnormal in heterozygotes? Explain.

19. *Vermilion*, *v*, and *cinnabar*, *cn*, are recessive alleles at different loci in *D. melanogaster* which produce similar phenotypic effects (*bright red* eyes); both phenotypes lack the brown pigment xanthommatin. In 1936 Beadle, G.W. and B. Ephrussi (Genetics 21: 225) transplanted imaginal disks (larval tissues giving rise to the adult eye) from *Drosophila* donor larvae of one genotype into host larvae of a different genotype. Explain the results shown in the table, stating which eye primordia develop autonomously (by gene action in the eye transplant itself) and which develop non-autonomously (by gene action in the hemolymph).

Donor larvae		Host larvae		Colour of transplant
Genotype	Phenotype	Genotype	Phenotype	
CnCnVVWW	wild-type	CnCnVVww	white	wild-type
CnCnVVww	white	CnCnVVWW	wild-type	white
CnCnVVWW	wild-type	CnCnvvWW	vermilion	wild-type
CnCnVVWW	wild-type	cncnVVWW	cinnabar	wild-type
CnCnvvWW	vermilion	CnCnVVWW	wild-type	wild-type
cncnVVWW	cinnabar	CnCnVVWW	wild-type	wild-type
cncnVVWW	cinnabar	cncnVVWW	cinnabar	wild-type

20. An ultraviolet-treated culture of *Neurospora crassa* yields an *inositolless* mutant. After lengthy culture the mutant regains the ability to grow on inositolless medium. Possible explanations for this reversion to the *wild-type* trait are:
1. Reverse mutation of the *mutant* allele to the *wild-type* allele.
2. Contamination of the *mutant* cultures with *wild-type* Neurospora.
3. A mutation at another locus to an allele, which suppresses the effect of the *mutant* allele at the first locus.
4. Mutation at the same locus but at a site different from the original mutation.
5. Physiological adaptation involving no genetic change.

Outline a series of experiments to determine which of these explanations is the correct one.

21. a) In 1944 E.L. Tatum et al. (Arch. Biochem. 3: 477) showed that two independently arising *tryptophanless Neurospora crassa* mutants yielded *wild-type* recombinants when crossed. Strain A grew on minimal medium if indole or anthranilic acid was added; strain B grew only if indole was supplied and accumulated large amounts of anthranilic acid in the medium. The uptake of indole was greatly increased if serine was also present. Under these circumstances tryptophan appeared in the medium. In 1948 H.K. Mitchell and J. Lein (J. Biol. Chem. 175: 481) discovered a third *mutant* strain C, which required tryptophan for growth and could not utilize indole. Cell-free extracts of this mutant, unlike similar extracts prepared from *wild-type* strains, lacked the enzyme tryptophan synthetase and were unable to synthesize tryptophan even in the presence of indole and serine (Yanofsky, C. Bacteriol. Rev. 24: 221, 1960). Strain C produced *wild-type* recombinants with each of the first two *mutants*. Discuss the implications of these findings with regard to the metabolic pathway of tryptophan synthesis in *Neurospora*, indicating how many gene loci there are and how they are involved in the process.
b) Yanofsky, C. and D.M. Bonner (Genetics 40: 761, 1955) studied 25 *tryptophanless mutants* whose requirements could not be satisfied by indole or anthranilic acid. All responded well to tryptophan. Each was unable to synthesize tryptophan synthetase or produced it in defective form. Crosses between any two of these *mutants* produced, with extremely rare exceptions, *tryptophanless*

progeny only. Discuss the genetic and biochemical implications of these findings.

c) Reversion to *tryptophan independence* occurred in many of the *mutant* strains. Several revertants in crosses with original *wild-type* produced only *wild-type* (tryptophan independent) progeny. Others produced asci segregating 6 *pseudo-wild-type* : 2 *mutant*. The *pseudo-wild-types* grew slower than *wild-types* on minimal medium lacking tryptophan and responded markedly to the addition of this amino acid to the medium. Account genetically for the two types of revertants and offer an explanation to account for the *pseudo-wild-type* phenotype.

22. a) In *Drosophila melanogaster*, flies homozygous for either *v* (*vermilion*) or *cn* (*cinnabar*) are unable to produce brown pigment and therefore have *bright-red* eyes. When imaginal disks of these flies are transplanted into *wild-type* hosts, they develop the *dull-red* (*wild-type*) eye colour. It has been shown that the *wild-type* host contributes a diffusable substance or substances required by the transplants to develop *wild-type* colour. Three explanations are possible for the failure of *cinnabar* and *vermilion* strains to develop *wild-type* colour:

1. The *wild-type* alleles at the *v* and *cn* loci may be structurally and functionally identical, and the alleles *v* and *cn* may therefore fail to produce the same precursor of brown pigment.

2. They may function in different ways so that *v* fails to bring about formation of *V* substance and *cn* the formation of *Cn* substance, both substances being precursors of brown pigment.

i) These two genes may be acting dependently, i.e., in different series of chemical reactions, thus:

```
Precursor A  --V-->  V substance  \
                                    >  brown pigment
Precursor F  --Cn--> Cn substance /
```

ii) The genes may be concerned with different steps in the same pathway so that the product of one acts as a precursor for the formation of the other; thus:

Precursor → *Cn* substance → *V* substance → brown pigment

Outline the results expected in reciprocal transplants between *v v* and *cn cn* flies under each of the three conditions, 1, 2(i) and 2(ii), giving reasons for these results.

b) When reciprocal transplants are made, it is found that *cinnabar* disks in *vermilion* hosts develop *cinnabar* eye colour but *vermilion* disks develop *wild-type* colour in *cinnabar* hosts. Which hypothesis in (a) do these results support? Using these facts, explain the results of thereciprocal transplantation.

23. Giles, N.H. et al. (PNAS 43: 305, 1957) studied 21 *adenine-requiring mutants* of independent origin in *Neurospora crassa*. All the *mutants* were blocked in the terminal step in adenine biosynthesis, involving the splitting of adenosine monophosphate succinate (AMPS) to adenosine monophosphate (AMP), and found to lack or have impaired activity for the AMPS-splitting enzyme, adenylosuccinase. Crosses of *mutants* in all possible combinations, with rare exceptions, yielded only auxotrophic progeny. *Wild-type* x any *mutant* produced *wild-types* and *mutants* in a 1:1 ratio.
a) What is the apparent genetic basis for enzyme control?
b) How many kinds of polypeptide chains does the enzyme possess? Explain.

24. Baglioni, C. and V.M. Ingram (Nature 189: 465, 1961) showed that adult individuals who can synthesize all four of the chains α^A, α^G, β^A, and β^C produce only four types of hemoglobin molecules in approximately equal numbers:

Hemoglobin A = $\alpha_2^A \beta_2^A$ Hemoglobin C = $\alpha_2^A \beta_2^C$
Hemoglobin G = $\alpha_2^G \beta_2^A$ Hemoglobin X = $\alpha_2^G \beta_2^C$

Note: No hybrid molecules, e.g., $\alpha_2^A \beta^A \beta^C$, were formed.
What conclusions can be drawn from these data regarding the relationship, chain synthesis, dimerization and assembly of dimers into hemoglobin molecules?

25. A sample of the results obtained by Jonasson, L.M.V. et al. (Theor. Appl. Genet. 66: 349, 1983) in *Petunia hybrida* is as follows:

(i) Plants of genotype $m1m1m2m2f1f1f2f2$ do not modify the *delphinidin* and *cyanidin* molecules whose composition and structure is as follows:

Delphinidin

R = sugar molecule

Cyanidin

(ii) Plants of genotype $M1\_m2m2f1f1f2f2$, $m1m1M2\_f1f1f2f2$, and $M1\_M2\_f1f1f2f2$ have a methyl (CH_3) group in place of the hydroxyl (OH) at the 3' position.

(iii) Plants of genotype $m1m1m2m2F1\_f2f2$, $m1m1m2m2f1f1F2\_$ and $m1m1m2m2F1\_F2\_$ have a methyl group at the 3' and 5' positions.

a) Why is the *delphinidin* molecule unaffected by the genotype in (i)?

b) Are the four loci regulatory or structural? Explain.

c) What are the roles of the functional alleles at these loci? Explain.

d) How would you determine whether the first two loci produce identical or different polypeptides? Are the latter two loci likely to produce the same protein as the first two? Explain. Are the latter two loci likely to produce identical or different polypeptides?

e) The difference between *delphinidin* and *cyanidin* is due to alleles of one gene: $D\_$ produces *delphinidin* and dd *cyanidin*. $ddm1m1m2m2f1f1f2f2$ plants have *magenta* flowers; all other genotypes at these five loci produce *purple* flowers. What phenotypic ratio would you expect among the progeny obtained by self-fertilization of the following plants (assume all plants are $m2m2f1f1$):

(i) $DdM1m1F2f2$? (ii) $ddM1m1f2f2$?
(iii) $DdM1m1f2f2$? (iv) $ddm1m1F2f2$?

f) Some petunia flowers are completely *white*. What is the minimum number of genes involved in determining flower colour in these species? Explain.

26. Whether one individual accepts or rejects tissues from another depends on the genotypes of the donor and the host at each of several *histocompatibility* (H), loci. Each histocompatibility antigen is determined by a single, autosomal, codominant allele at an H locus. A recipient will accept a graft only if he carries all H genes present in the donor.

a) For the cases in the table indicate whether the recipient will accept or reject skin from the donor.

Donor	Recipient
Inbred strain A	Inbred strain B
Inbred strain A	F_1 hybrid from inbred A x inbred B
F_1 hybrid	Either inbred strain A or B
F_2 hybrid	F_1 hybrid
Either inbred A or B or F_1 hybrid	F_2 hybrids

b) Show why acceptance of grafts from either inbred parent strain by F_1's rules out the possiblility that H antigens are determined by recessive genes.

c) It is possible to establish a series of inbred lines each differing from the common inbred progenitor by a different single H allele at one H locus only. Otherwise these congenic lines are genetically identical. Results with four such lines are as follows:

line 1 x line 2	F_1 rejects graft from common inbred parent
line 1 x line 3	F_1 accepts graft from common inbred parent
line 4 x line 3	F_1 accepts graft from common inbred parent
line 1 or 2 x line 4	F_1 rejects graft from common inbred parent

How many H loci and how many alleles at each are involved in graft acceptance and rejection in these lines? Explain.

27. The enzyme tryptophan synthetase (TSase) catalyzes the following three reactions (Yanofsky, C. and M. Rachmeler, Biochim. Biophys. Acta 28: 640, 1958):

   Indole + L-serine → L-tryptophan   (1)
   Indoleglycerol phosphate → indole + triose phosphate   (2)
   Indoleglycerol phosphate + L-serine → L-tryptophan + triose phosphate.   (3)

   Reaction (3) is the physiologically important reaction (viz. it carries out the major tryptophan synthesis in the organism). TSase in *E. coli* consists of two different polypeptide chains, A and B. The normal B subunit can catalyze reaction (1) even in the absence of subunit A but does so very inefficiently. The normal A protein alone can catalyze reaction (2). Both subunits are required for the catalysis of any of the reactions at maximal rates. Reaction (3) occurs only in the presence of the AB complex (Crawford, I.P. and C. Yanofsky, PNAS 44: 1161, 1958).
   In a study of numerous mutations affecting the formation of TSase in *E. coli*, Crawford, I.P. and C. Yanofsky (PNAS 45: 1280, 1959) observed the following:
   1. In certain *mutants* which do not revert and do not recombine with two or more other tryptophan synthetase *mutants* none of the three reactions occurs.
   2. Some mutations affect the A protein only. The mutant A protein in these strains (in the presence of normal B protein) is effective in reaction (1) only (at the *wild-type* rate).
   3. Other mutations affect the B protein only. These mutant B proteins (in the presence of normal A protein) catalyze reaction (2) only (at a *wild-type* rate).
   4. The mutants producing defective A protein accumulate indoleglycerol phosphate. The mutants producing defective A proteins and defective B proteins revert naturally to tryptophan independence.

   Interpret these results with respect to genetic control of TSase synthesis and activity.

28. In *E. coli* the two very closely linked nonsense *mutant* genes *5972* and *9778* affecting tryptophan synthesis have the following effects (Ito, J. and C. Yanofsky, J. Biol. Chem. 241: 4112, 1966):

   1. Mutant 5972 cannot convert chorismic acid to anthranilic acid due to lack of anthranilate synthetase activity but can carry out all other reactions.
   2. Mutant 9778 lacks not only anthranilate synthetase activity but also phosphoribosyl anthranilate transferase activity required for conversion of anthranilic acid to N-5'-phosphoribosyl anthranilate (PRA).
   3. When extracts from the two mutants are mixed, anthranilate synthetase activity is restored.

   Did the two nonsense mutations occur in the same or different genes? What is the nature (number of kinds of polypeptides) of each of the enzymes? Explain.

29. Hemoglobins S (*Hb-S*) and C (*Hb-C*) in humans are specified by alleles of one gene (Hunt, J.A. and V.M. Ingram, Nature 181: 1062, 1958). Each differs from *normal*, *Hb-A* by one amino acid substitution in the β chain at position 6. Glutamic acid in the *Hb-A* β chain is replaced by valine and lysine in *Hb-S* and *Hb-C*, respectively. Hemoglobin Hopkins-2 (*HO-2*) differs from *normal* hemoglobin by an amino acid substitution in the α chain. Lysine replaces leucine at position 65. Certain individuals with both *HO-2* and *Hb-S* chains are known who have one parent like themselves and the other *normal*. The sibs of these *HO-2*, *Hb-S* individuals may be of four types for the hemoglobins they produce (Smith, E.W. and J.V. Torbert, Bull. Johns Hopkins Hosp. 102: 38, 1958):
   No Hb-S, no HO-2; No Hb-S, HO-2; Hb-S, no HO-2; Hb-S, HO-2
   Explain, what bearing these results have on the one-gene-one-enzyme hypothesis.

30. Hsia, D.Y. et al. (Nature 178: 1239, 1956) demonstrated that the levels of phenylalanine in the blood plasma at 1-, 2- and 4-hour intervals after feeding a standard dose of phenylalanine were about twice as high in *normal* parents of *phenylketonurics* as in persons with no history of *phenylketonuria* in their pedigrees. What do you conclude from these findings?

31. It is possible to determine whether a mutation which has resulted in loss of enzyme acitvity has also resulted in the loss of ability to synthesize the enzyme. This can be done

by injecting animals (e.g., rabbits) with the enzyme from the *wild-type* strain to obtain antibodies against the enzyme. Extracts from the *mutant* strains are then tested for reaction with these antibodies. Reaction indicates the presence of a protein immunologically identical to the *wild-type* enzyme. Such proteins are called CRM (cross-reacting material). Strains synthesizing CRM are termed CRM$^+$, and those not able to do so are called CRM$^-$ (Suskind, S.R. and E. Jordan Science 129: 1614, 1959). Mutations at structural gene loci can result in the formation of CRM$^+$ enzymes which are less effective in catalyzing the reaction controlled by the *wild-type* enzyme. Moreover, quantitative differences occur among the enzymes specified by these *mutant* alleles with respect to this function. Other mutant alleles fail to produce CRM.

a) Does failure to produce CRM indicate that no enzyme is formed? Explain.

b) Would missense and/or nonsense mutations be expected to produce each of these kinds of enzymes or no enzymes? Explain and illustrate.

32. The genes, polypeptides and enzymes involved in the five sequential enzymatic steps in the *tryptophan* pathway in *E. coli* are as follows (after Ito, J. and C. Yanofsky, J. Biol. Chem. 241: 4112, 1966; Goldberg, M.E. et al., J. Mol. Biol. 21: 71, 1966):

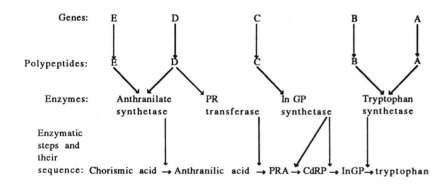

Nonsense mutants a, b, c, d and e are combined in the following merozygotic combinations:

c and d   d and d   a and b   e and d

c and b   d and a   b and b   e and c

Explain which of the merozygotes would show complementation (growth in minimal medium).

b) In *Aspergillus nidulans* the sequential metabolic pathway for tryptophan is the same as in *E. coli*. The manner of gene determination of the enzymes is, however, somewhat different, as revealed by the mutant studies summarized in the table (data from Roberts, C.F., Genetics 55: 233, 1967). Using a hypothetical set of polypeptides for these genes, show how the polypeptides participate in enzyme assembly and function in *A. nidulans*.

Complementation group	Growth response on minimal medium		Accumulation	Other facts
	Anthranilic acid	Indole		
A	+	+	None	Lack anth. synthetase; no interallelic complementation
B	-	- -	Indole glycerol; some anthranilic acid	Deficient in tryp. synthetase; no interallelic comp.
C Few	+	+	None	Defective only in anth. synthetase
Most	-	-	Anthranilic acid	Do not complement anthranilic acid-utilizing mutants; lack anth. synthetase, PR isomerase, and InGP synthetase activity
D	-	+	Anthranilic acid	Lack PR activity; no interallelic complementation
E	+	+	None	Do not lack any of the enzymatic activities tested

33. Schwartz, D. (PNAS 46: 1210, 1960) found a different form of a certain esterase in each of three true-breeding lines of *Zea mays*. Line *a* was characterized by a *slow* (S) migration rate to the negative pole at pH 8.6 during electrophoresis; lines *b* and *c* possessed *normal* (N) and *fast* (F) migration rates, respectively. Mixtures of enzymes from different lines showed two components each (d, e, f in zymogram

illustration below). F_1 hybrids from crosses between N and F lines and N and S lines form three components each. One of these in each F_1 is intermediate between the parental components in mobility. Zymograms (starch gels stained to reveal location of the various isozymes) of the various esterase types are illustrated. The results of similar tests on F_2's are also tabulated.

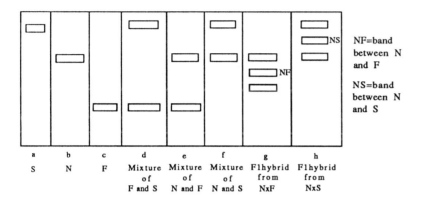

		F_2					
Cross	In F_1	Type	No.	Type	No.	Type	No.
$N \times F$	N, NF, F	F only	141	F, NF, N	292	N only	146
$N \times S$	N, NS, S	N only	21	N, NS, S	41	S only	20
$F \times S$	F, FS^*, S	F only	139	F, NS	314	S only	141

*FS band is in normal position

Describe the mode of genetic specification of this enzyme. Explain whether the enzyme is monomeric, dimeric or trimeric.

34. a) Lactose dehydrogenase (LDH) contains two types of polypeptide (or subunit) chains, A and B. It exists in five active forms, or isozymes, LDH-1 to LDH-5. LDH-1 and LDH-5 are pure isozymes, each with one kind of chain only A and B respectively. When their chains are mixed in equal proportions and are allowed to associate and reassemble spontaneously, the isozymes 1 to 5 appear in the respective proportions 1:4:6:4:1 (Markert, C.L. Science 140: 1329, 1963). How many subunits does each isozyme contain and why? Using symbols A and B, indicate the chain composition of each of the isozymes.

b) The zone-electrophoresis LDH-isozymic patterns (=phenotypes) on cylindrical starch gel from brain extracts of *Peromyscus maniculatus* individuals that are *homozygous normal*, *homozygous mutant* and *heterozygous* are illustrated.

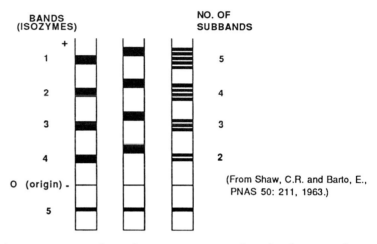

(From Shaw, C.R. and Barto, E., PNAS 50: 211, 1963.)

Matings among the three types produced the results tabulated.

	Offspring phenotype					
	Homozygous normal		Heterozygous		Homozygous normal	
Parental Phenotype	♂	♀	♂	♀	♂	♀
homozygous normal x homozygous normal	13	11	0	0	0	0
homozygous normal x heterozygote	12	15	11	16	0	0
heterozygote x heterozygote	2	1	0	5	2	2

(i) How many allele pairs are responsible for the phenotypic differences?
(ii) The A and B chains are specified by genes A and B, respectively. Of polypeptides A and B, do you expect one or both to exist in alternative forms and why?
(iii) Bands 1 to 4 share common polypeptide B not present in band 5, which contains polypeptide A only. If only one

polypeptide exists in alternative forms, which is it and why?

(iv) Give the genotypes of the three types of individuals and describe (or give) the tetrameric-polypeptide constitution of each band and subband in *homozygous normal*, *homozygous mutant* and *heterozygous* individuals. Give the relative proportions of the subunits in each isozyme in the heterozygotes.

35. The molecular weights of the A and B chains of *E. coli* tryptophan synthetase are 29,500 and 49,500 respectively. That of the entire enzyme is 159,000.

    a) If the average molecular weight of each amino acid is 110, approximately how many amino acids does each chain contain?

    b) How many chains does the enzyme contain? Explain.

Chapter 21.
Protein Synthesis: Transcription and Translation

1. Discuss the function of each of the following components in protein synthesis:

   Aminoacyl-RNA synthetase
   ATP
   DNA
   F_1, F_2, F_3 initiating factors
   GTP
   mRNA
   Peptide transferase
   Polysome

   R_1, R_2, S
   Ribosomal proteins
   rRNA
   30S ribosomal subunit
   50S ribosomal subunit
   tRNA
   Transfer factors I and II

2. a) What genetic attributes do RNA and DNA of eukaryotes share?
   b) RNA is of 3 types in both eukaryotes and bacteria. What are they? Where are they located in the cell? Where are they produced? What are their characteristics and functions?
   c) Viruses form only 2 types of RNA. Which type do the viruses not synthesize and why?
   d) Explain how an amino acid is activated and then attached to its specific tRNA.
   e) What is the evidence that a mRNA may be long enough to code for several proteins?
   f) Explain why polypeptides specified by a polygenic message are not coupled.

3. a) Compare and contrast:
   i) Pribnow box and Hogness box
   ii) Upstream and downstream sequences
   iii) Conserved and consensus sequences
   iv) Holoenzyme and σ subunit
   v) Leader and trailer sequences

   b) How would you proceed to demonstrate that the 2 complementary strands of a DNA duplex for a given gene carry different information?
   c) i) In eukaryotes, the primary transcripts of most genes, e.g., β- hemoglobin, undergo 3 major changes. What are these changes and what is the function of each?
   ii) The primary transcripts of other eukaryotic genes, e.g., histone genes, are modified in only 2 ways. Which of the 3 changes specified in (i) does not affect the primary transcripts of these genes? Explain why.

   d) Eukaryotes and prokaryotes differ with respect to 3 features: the transcription products of their genes, whether these products are modified and how they are modified (Darnell, J.E., Sci. Amer. 249 (Oct.) 90, 1983; Shatkin, A.J., Cell 9:645. 1976). What are the major differences between prokaryotes and eukaryotes in these respects and why do these differences exist?

4. According to the colinearity hypothesis, the specific base-pair content and sequence in a gene specifies the amino acid content and sequence in the corresponding polypeptide. Explain and illustrate how the sequence of base pairs (or bases, if the DNA or RNA is single-stranded) is *transcribed* into mRNA and how the mRNA is *translated* into the sequence of amino acids in the polypeptide chain. For discussion purposes use a hypothetical gene that codes for a hypothetical protein three amino acids long. State the functions of all of the components involved.

5. A tRNA with its attached amino acid is isolated and subsequently the amino acid is changed into another. Such a tRNA with its altered amino acid is introduced into a cell-free, protein-synthesizing system.
   a) Would you expect the altered amino acid to be incorporated into polypeptides?
   b) Would it be at the residue site occupied by the amino acid from which it was derived?
   c) What conclusion could you draw from this experiment?

6. The schematic representation on the next page of the relationships between aminoacyl-tRNAs and mRNA at the ribosome is incorrect in some respects. Redraw the illustration and indicate the errors you have corrected.

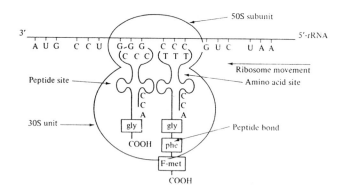

7. The first step in protein synthesis is the formation of an initiation complex (Nomura, M. et al., PNAS 58:1487, 1967). What is this complex, and what events follow to begin polypeptide synthesis?

8. a) What is the role of codons UAA, UGA, UAG in protein synthesis?
b) What effect does each of the following have on polypeptide synthesis: (1) Nonsense mutation? (2) Missense mutation? (3) Frame-shift mutation?

9. AUG and GUG are polypeptide-initiating codons. Why do many synthetic messenger RNAs without these codons (such as poly-U) nevertheless still manage to direct *in vitro* polypeptide synthesis?

10. One cell makes one kind of protein only. Others make a great variety of protein molecules. In which of these two kinds of cells would you expect a broad distribution of polysome sizes and why? Outline how you would proceed to support your expectation.

11. The tRNA in all species have the same overall base composition [(A + U)/(G + C) = 0.6], molecular weight (25,000), and sedimentation constants (4S). They also function nonspecifically; viz. tRNA plus amino acid-activating enzymes from *E. coli* mixed with ribosomes plus mRNA from rabbit (*Oryctolagus*) reticulocytes synthesize rabbit hemoglobin. Nevertheless, they hybridize specifically; e.g., *E. coli* tRNA hybridizes only with *E. coli* DNA (Goodman, H.M. and A. Rich, PNAS 48:2101, 1962). Account as completely as possible for these paradoxical properties of tRNA.

12. At any given time each functioning ribosome is attached to only one growing polypeptide chain. Suggest why.

13. If cell-free extract active in protein synthesis is centrifuged, the growing protein chains sediment attached to 70S ribosomes which contain mRNA. When such an extract is treated with minute amounts of ribonuclease, the growing protein continues to sediment with the 70S particles. Is the nascent protein attached to mRNA or directly bound to ribosomes? Explain.

14. Outline an experiment that would indicate whether tRNA molecules are chromosomal (DNA) in origin.

15. The electron micrograph shows a thin fiber, from *E. coli*, with attached strings of granules (each about 200 Å in diameter). DNase treatment destroys the fiber but not the strings of granules, whereas RNase removes the granular strings from the fiber. Free strings with granules are never observed (Miller, O.L. and B.R. Beatty, Science 164:955, 1969; Miller, O.L. et al., Cold Spring Harbor Symp. Quant. Biol., 35:505, 1970).

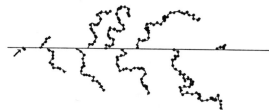

a) Which of these structures represents: (i) A portion of the *E. coli* chromosome?, (ii) Ribosomes?, (iii) mRNA?
b) Is the translation coupled with transcription? Explain.
c) May some of the granules be RNA polymerase molecules? If so, which ones and why?

16. Explain whether the following are true or false:
a) Each of the 20 amino acids has its own specific activating enzyme.
b) The ribosome is the site of protein synthesis.
c) Polypeptide chains grow at their carboxyl end.
d) The 23S RNA is not a dimer of 16S RNA.

e) Any ribosome may serve as a "workbench" for the synthesis of more than one kind of polypeptide chain.
f) That tRNA and associated amino acid-activating enzymes from one species when mixed with the ribosomes and attached mRNA from another species can bring about the synthesis of proteins of the kind directed by the mRNA proves that the code is universal.
g) If the code is degenerate, only 20 different tRNA molecules should exist.
h) The tRNA and rRNA, like mRNA, are formed on a DNA template.
i) Nascent polypeptide chain is bound to the 30S subunit of the 70S ribosome.
j) Peptide-bond formation on the 70S ribosome involves at least two binding sites.
k) All tRNAs carry amino acids.
l) One messenger RNA is formed for an entire operon.
m) There is only one kind of tRNA molecule capable of accepting any one amino acid.
n) Methylation does not change the coding properties of tRNA.

17. A DNA duplex without introns possesses the following base-pair sequence and codes for a polypeptide 6 amino acids long:

```
3' GCG TAA TAG TAC CCA GAC TGA TAC CAT AAA AAA GAT GTA TCG 3'
5' CGC ATT ATC ATG GGT CTC ACT ATG GTA TTA TTT CTA CAT AGC 3'
```

a) Explain which strand of the DNA duplex is the sense strand. In which direction is it being transcribed?
b) Are the ends of the 2 strands correctly labeled?
c) An inversion occurs in this DNA duplex. The same strand is transcribed as before and produces a polypeptide 11 amino acids long. Indicate the position of the breaks that led to the formation of the inversion. Draw the inverted duplex and label the initiation and termination codons. Show its mRNA and the amino acid sequence of its polypeptide.

18. a) Compare and contrast DNA replication with DNA transcription.
b) Distinguish between transcription and translation.
c) Which kinds of genes are transcribed but not translated?

d) A segment of a DNA duplex has the following base-pair sequence:

```
Strand A    3'    T A C G A T T G    5'
Strand B    5'    A T G C T A A C    3'
```

If Strand A serves as the template for transcription:
i) Give the sequence of bases of the mRNA.
ii) Which three bases of the template will be transcribed first? Explain.
iii) Which 3 bases of the mRNA will be translated first? Explain.

19. The following results obtained by Hurwitz, et al. (Biochem. Biophys. Res. Commun., 3:15, 1960) are from *in vitro* experiments from 2 different species with double-stranded DNA and its RNA products:

Species	DNA base ratio A + T/G + C	RNA base ratios A + U/G + C	A + G/U + C
E. coli	1.00	.96	.90
B. subtilis	1.36	1.30	1.08

a) Is it possible to determine from these data whether the RNAs of these species are copied from single- or double-stranded DNA?
b) Is the RNA itself single- or double-stranded? Explain.

20. Why was it known as early as the late 1940's that transcription and translation in eukaryotes are not coupled?

21. Why are mutations in rRNA genes, like those in structural genes that cause amino acid substitutions in polypeptide chains, unlikely to be deleterious?

22. Prokaryotic promoters are characterized by 2 conserved sequences at -10 and at -35. What are these sequences called and what is their function?

23. Messenger RNA molecules in prokaryotes are quickly degraded whereas those in eukaryotes are more stable and exist for a longer time. Suggest a reason for this difference.

24. The antibiotic actinomycin-D is known to inhibit DNA dependent RNA synthesis; when it is added to an *E. coli* culture protein synthesis declines at a steady rate until it terminates completely, approximately 20 minutes after the addition of the antibiotic.
a) Why does protein synthesis (translation) continue to occur after the antibiotic is added to the culture?
b) What is the average lifetime of an *E. coli* mRNA? Explain.

25. Each transfer RNA molecule has 4 specific recognition sites to carry out its role in translation of the genetic code in the mRNA. What are these sites and what roles do they play?

26. A gene contains 5 introns. Its DNA is denatured and hybridized with the gene's mRNA. How many R-loops would you expect to observe with an electron microscope? Why?

27. Offer a plausible explanation for the fact that genes coding for histones do not possess introns, i.e., they are not discontinuous.

28. Structural genes Z and Y are very closely linked. Each of these genes codes for a monomeric protein 200 amino acids long. A mutant is isolated which does not express the phenotypic activity of either of these loci. Although this mutant does not synthesize the proteins Z and Y it produces a protein that is 290 amino acids long. The sequence of the 110 amino acids beginning at the amino end of this protein is identical to that at the amino end of the Y protein. The remaining 180 amino acids have an identical sequence to that of the latter 180 residues in the Z protein.
a) Are the 2 genes usually transcribed separately? Explain.
b) Is the species a prokaryote or an eukaryote? Explain.
c) What is the most likely nature of the mutation producing this protein? Explain and be specific.

29. a) What is a ribozyme? What is self-splicing?
b) Both group I and group II introns are self-splicing. With the aid of diagrams, compare and contrast the mechanisms of intron excision and exon ligation in the 2 groups.

30. Transfer RNA genes possess a single intron about 15 nucleotides long. Knapp, G. et al. (Cell 14:221, 1978) and Greer, C.L. et al. (Cell 32:537, 1983), demonstrated enzymatic removal of the introns *in vitro* which did not involve ribozymes and spliceosomes. Propose and illustrate a plausible mechanism for intron excision and exon ligation in tRNA molecules based on the activity of the 2 enzymes you chose to splice these molecules.

31. In eukaryotes, messenger RNA precursors are processed in such a way that introns are removed and the exons ligated with the aid of snRNPs ("snurps") and spliceosomes. (Grabowski, P.J. et al., Cell 37:415, 1984; Padgett, R.A. et al., Science 205:898, 1984; Kanarska, M.M. and Sharp, P.A., Cell 49:763, 1987). Outline and illustrate this process, indicating the roles of snRNPs and spliceosomes. Label the initial transcript and the intermediate and final products of the process.

32. What roles do the TATA (Goldberg and Hogness) box, CCAAT sequence and enhancers perform in initiation of transcription in eukaryotes?

33. a) von Ehrenstein, G. et al. (PNAS 49:669, 1963) synthesized rabbit hemoglobin molecules in a cell-free system containing ribosomes from rabbit reticulocytes, together with the amino acids cysteine and alanine, and tRNAs for these amino acids from *E. coli*. They studied the effects of modifying this system on peptide 13 of the α chain of hemoglobin, which normally contains cysteine but not alanine, with the following results:
i) Cysteine attached to its tRNA was incorporated into peptide 13.
ii) When cysteine, while attached to its tRNA, was changed to alanine by Raney nickel only alanine was found present in peptide 13.
iii) Alanine bonded to its own tRNA was not incorporated into this peptide.

Do these observations answer the question: Is the mRNA code recognized by the tRNA or the amino acid?

34. When growing bacteria are infected with virulent phage, the synthesis of bacterial DNA stops immediately. Moreover net

synthesis of RNA also stops, unlike that in uninfected growing cells, while synthesis of proteins continues at its preinfection rate. Using isotopic labeling, Volkin, E. and L. Astrachan (Virology 2:149, 1956, In W.E. McElroy (ed.), "The Chemical Basis of Heredity," pp. 686-695, John Hopkins, Baltimore, 1957) showed that immediately after *E. coli* cells are infected by T2 phage, a very small amount of RNA is rapidly synthesized and just as rapidly destroyed (has a high turnover rate). The base compositions of this RNA, the total RNA of *E. coli*, and the DNA of the host and the T2 phage are tabulated below:

Base	RNA		DNA	
	Total *E. coli*	High turnover	*E. coli*	T2
Adenine	23	31	25	32
Cytosine*	23	18	25	18
Guanine	31	22	25	18
Thymine (DNA) or uracil (RNA)	23	30	25	32

*5-Hydroxymethylcytosine replaces cytosine in viral DNA.

a) What appears to be the function of this new RNA and why?

b) How does the T2 DNA appear to be related to this function? (See Hall, B.D. and S. Spiegelman, PNAS 47:137, 1961, for confirmation or refutation of your answer).

35. a) Somatic cells of the toad *Xenopus laevis* ($2n = 36$) have 2 nucleoli produced at corresponding nucleolus-organizer regions (segments of DNA) of a pair of homologous chromosomes. Elsdale, T.R. et al. (Exp. Cell Res., 14:642, 1958) discovered a mutant female ($2n = 36$) with somatic nuclei containing only one nucleolus. This characteristic was transmitted to its progeny. When toads with only one nucleolus per somatic nucleus were interbred, they produced *binucleolar*, *uninucleolar*, and *anucleolar* progeny in a 1:2:1 ratio. The *uninucleolar* toads lacked a secondary constriction on one of the two homologues. *Anucleolar* toads, which showed no secondary constrictions, died as tadpoles. Brown, D.D. and J.B. Gurdon (PNAS 51:139, 1964) showed that *anucleolar* embryos were unable to synthesize either the 18S or the 28S molecules of the ribosomes, which are known to differ in RNA base composition but to be synthesized in a coordinated manner (same time, same place in the cell). The survival of the *anucleolar* toads to the tadpole stage is made possible by the large pool of maternal ribosomes from the egg.

i) Offer a hypothesis that will explain the inheritance, the lethality, and the biochemical data.

ii) Offer two plausible genetic explanations to account for the phenotypic distributions in the cross of two *uninucleolar* individuals. Discuss the apparent significance of these results.

b) Wallace, H. and M.L. Birnstiel (Biochim. Biophys. Acta 114:296, 1966) found that the DNA of *anucleolar* tadpoles failed to hybridize with *wild-type X. laevis* rRNA whereas that of *uninucleolar* tadpoles hybridized with a quantity of rRNA intermediate to that annealed by the DNA of *binucleolar* and *anucleolar* tadpoles. The 18S and 28S rRNAs hybridize with 0.04 and 0.07 percent of homologous *wild-type X. laevis* DNA respectively. The 18S and 28S components differ in base sequences, annealing to different stretches of DNA, which together occupy approximately 1 percent of the genome. The potential number of nucleoli, the number of secondary constrictions, and the portion of the genome complementary in base sequence to rRNA all show a linear reduction in proportion to the dosage of the mutation. Do these results support your previous conclusion and how? If not, what is the nature of the mutant? Where are ribosomal genes for 18S and 28S located, and why?

36. a) In *Drosophila melanogaster* the approximate molecular weight of DNA per genome is 1.2×10^{11} (8×10^{11} nucleotide pairs). Moreover, 0.27 percent of the *wild-type* genome, specifically in the nucleolus-organizer region, hybridizes with rRNA.

i) Why must this molecular weight be divided by 2 to estimate the number of genes coding for rRNA synthesis?
ii) Approximately how many rRNA genes are there in a diploid cell?

b) DNA-RNA hybridization studies in the yeast *Saccharomyces cerevisiae* by Schweizer, E. et al. (J. Mol. Biol. 40:261, 1969) have indicated that 0.064 to 0.08, 0.8

and 1.6 percent of the nuclear genome (DNA) hybridizes with 4S tRNA, 18S and 26S rRNA, respectively. The amount of DNA per genome is 1.25×10^{10} daltons.

Type of RNA	Molecular weight	DNA hybridized	
		Percent of total DNA	Weight, daltons
4S	2.5×10^4	0.064 – 0.08	2×10^7 (max. value)
18S	0.7×10^6	0.8	2.0×10^8
26S	1.4×10^6	1.6	4.0×10^8

i) How many genes are there per genome for: (a) 18S rRNA? (b) 26S rRNA? (c) All tRNAs?

ii) Assuming there are approximately 60 different species of tRNA, on the average, how many genes are there for each tRNA species? *Note:* The genes for different RNAs differ in size.

37. According to the adaptor hypothesis (Crick, F.H.C., Symp. Soc. Exp. Biol., 12:138, 1958), each of the 20 amino acids is carried by an adaptor molecule to the template RNA; each adaptor recognizes the appropriate codon in the mRNA, thus correctly positioning the amino acid in the polypeptide specified by the gene. The following experiments concern the role of tRNA in genetic coding:

i) Amino acids bonded to tRNA molecules were obtained from cell-free extracts. Each amino acid was joined to a specific tRNA species (Hoagland, M.B. et al., J. Biol. Chem., 231:241, 1957).

ii) Chapeville, F. et al. (PNAS 48:1086, 1962), using a synthetic mRNA containing only uracil and guanine in an *in vitro* protein-synthesizing system from *E. coli*, found that this mRNA (poly-UG) templated the incorporation of phenylalanine and cysteine into polypeptides when these amino acids were attached to their own tRNA species. When cysteine was reduced to alanine by Raney nickel after it had become attached to its tRNA, poly-UG stimulated the synthesis of a polypeptide containing phenylalanine and alanine. When alanine was attached to its own tRNA, it was not incorporated into polypeptides synthesized on poly-UG.

a) Show how these data support Crick's adaptor hypothesis.

b) Show what results would have been expected if the code word in the mRNA was recognized by the amino acid itself (the tRNAs being merely carriers, bringing the amino acids to the ribosomes).

c) If $tRNA_{Cys}$ remains attached to an amino acid that has been catalytically converted from cysteine to alanine, what suggestions can be made regarding the specific manner in which this tRNA becomes attached in the first place?

38. In *Drosophila melanogaster,* the approximately 130 or more identical genes for each of the rRNA species 18S and 28S are located in the DNA (rDNA) of the nucleolus-organizer (NO) regions, which are in corresponding positions of the X and Y chromosomes (Ritossa, F.M. and S. Spiegelman, PNAS 53:737, 1965). This NO DNA, which is about 0.27 percent of the total *wild-type* genotype, is at the *bobbed, bb,* locus at which many hypomorphic alleles exist, each due to a partial deletion of the *bobbed* gene. Each of these alleles causes small bristles and poor development. The severity of the mutant phenotype depends on the extent of the *bb* deletion. After denaturing DNA and using the technique of DNA-RNA hybridization along with RNase to degrade any RNA that did not hybridize, Quagliarotti, G. and F.M. Ritossa (J. Mol. Biol., 36:57, 1968) studied flies homozygous for different hypomorphic alleles and *wild-types* to determine the sequence and ratio of genes for 18S and 28S rRNA. The ratio of DNA complementary to 18S and 28S in *wild-type* and different *mutant* flies is shown below:

	rRNA	28S rRNA	18S rRNA
	DNA % in hybrid (1)	DNA % in hybrid (2)	DNA % in hybrid (3)
Wild-type	0.373	0.247	0.122
ywbbS1	0.165	0.111	0.053
ywbbS2	0.205	0.138	0.066
carbbS1	0.254	0.170	0.082
carbbS2	0.168	0.109	0.053
UC03bbS1	0.185	0.123	0.060

a) Suggest the ratio between the number of genes for 18S rRNA and that for 28S rRNA and explain it.

b) What is the ratio between the amount of DNA complementary to 28S rRNA and that complementary to 18S rRNA?

c) The genes for the 2 rRNAs may be arranged as follows:

i) Genes for 28S rRNA are clustered in one block, and those for 18S rRNA are clustered in another block, the two blocks being adjacent to each other.

ii) The 2 kinds of genes are interspersed in the *bb* locus, e.g., in alternating sequence 18S-28S-18S-28S.... Which arrangement do these data support and why?

39. Rabbit reticulocytes synthesize α and β polypeptide chains of hemoglobin almost exclusively. Although both complete chains and incomplete ones at different stages of formation are present at any 1 time, consider the results that Dintzis, H.M. (PNAS 47:247, 1961) obtained with complete α chains. The results obtained when a culture of reticulocytes was exposed to radioactively labeled amino acids for 30 seconds and then for 3 minutes are shown below. After 30 seconds exposure to labeled amino acids, only chains already partly synthesized have time to be completed. Complete new chains can be produced in 3 minutes in the presence of labeled amino acids.

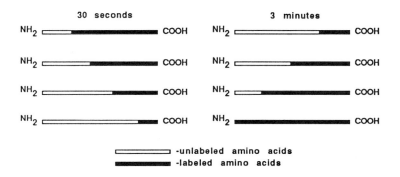

Theoretically polypeptides can be synthesized in 5 ways:

i) A number of separate peptides are assembled first and are then combined to form a long chain.

ii) Each chain is formed by the sequential addition of amino acids, starting at either the amino or the carboxyl end.

iii) Chain formation may begin internally and proceed to both ends.

iv) Chains form sequentially from the amino to the carboxyl end.

v) Chains form sequentially from the carboxyl to the amino end.

Which of the above methods of chain formation do the data support? Explain.

40. a) It has been shown Hurwitz, J. and J.J. Furth (Sci. Am., 206 (February):41, 1962) that the RNA polymerases of *E. coli* and mammals require all 4 ribonucleoside 5'-triphosphates (UTP, ATP, GTP, and CTP) simultaneously for *in vitro* synthesis of RNA. Deoxyribonuclease inhibits but DNA addition stimulates RNA formation. The base ratios of DNA and RNA in T2 phage, *E. coli* and calf thymus were found by Hurwitz, J. et al. (Cold Spring Harbor Symp. Quant. Biol., 26:91, 1961) to be as given below. Two synthetic DNA polymers were added to mixtures containing all 4 bases and RNA polymerase. One contained thymine (poly-T), the other contained adenine and thymine in alternating sequence (poly-AT). The synthetic RNA that was formed when poly-T was template contained adenine, whereas that produced when poly-AT was template contained uracil and adenine in alternating sequence. What appear to be the functions of DNA and of RNA polymerase in RNA synthesis (viz. what kind of substrate does the enzyme use, and how is templating performed)?

		Nucleotide incorporation in RNA, nanomoles				
	A+T					A+U
DNA	G+C	AMP	UMP	GMP	CMP	C+G
T2	1.86	0.54	0.59	0.31	0.30	1.85
Thymus	1.35	3.10	3.30	2.00	2.20	1.52
E. coli	1.00	2.70	2.74	2.90	2.94	0.93

b) It has been shown (Nirenberg, M. and J.H. Matthaei, PNAS 47:1588, 1961) that a synthetic mRNA containing uracil yields polyphenylalanine only as the product of *in vitro* protein synthesis, whereas a mixture of polyuracil and polyadenine chains yields very short polypeptides. Does this information help to elucidate the role of RNA polymerase more clearly? If so, how?

c) i) The base compositions of single-stranded φX174 DNA and the enzymatically synthesized RNA using this DNA as template are given in the table below.

		A	T(U)	G	C
φX174	DNA	0.25	0.33	0.24	0.18
	RNA	0.32	0.25	0.20	0.23

ii) With rare exceptions, in all RNA molecules in all species, the amounts of A and U differ, and the same is true for G and C.

iii) When 3'-deoxyadenosine is added to cells synthesizing RNA chains, this inhibitor is first phosphorylated to 3'-deoxyadenosine-P ~ P ~ P and then joined to the 3' end of RNA molecules. The latter event terminated RNA synthesis.

Is RNA single- or double-stranded? Does synthesis of RNA chains occur in the 3' to 5' or the reverse direction? Explain.

d) i) RNA polymerase contains 5 different polypeptide chains, α, β, β^1, ω and σ; two σ chains are present, along with one of each of the other chains (Travers, A.A. and R.R. Burgess, Nature 222:537, 1969). The complete enzyme (holoenzyme) is easily dissociated into two subunits, a *core* polymerase ($\alpha_2\beta\beta^1\omega$) and σ (Berg, D. et al., Fed. Proc., 28:659, 1969).

ii) The core enzyme can synthesize RNA chains starting anywhere along either strand of a gene (Sugiura, M. et al., Nature 225:598, 1970).

iii) By itself σ has no catalytic function, but the holoenzyme (σ + $\alpha_2\beta\beta^1\omega$) synthesizes RNA chains complementary to only one strand of each gene and begins synthesis at specific points in the DNA complex (Travers, A.A. and R.R. Burgess, 1969).

iv) Synthesis of all RNA chains appears to start with either A or G (Bremer, H. and R. Burner, Mol. Gen. Genet., 107:6, 1968).

v) Stop signals (specific base sequences) exist to terminate RNA synthesis at specific points along the DNA. When the *p* factor (a specific protein) is absent from cells, chain elongation is not terminated at stop signals (Roberts, J.W. Nature 224:1168, 1969).

vi) Once a gene has been transcribed, hydrogen bonds re-form between the complementary chains.

(1) What is the function of (a) the core enzyme, (b) σ, and (c) *p* factor?

(2) Illustrate the transcription of an RNA chain using a hypothetical DNA representing three consecutive genes.

41. In *Diplococcus pneumoniae* either of the 2 DNA strands, as well as the duplex. can bring about transformation. Moreover, 1 of the 2 chains of DNA contains sufficiently more of the heavier purine guanine and the heavier pyrimidine thymine than the other, so that the 2 chains, after denaturation and density-gradient centrifugation, form separate bands. Using DNA from a *novobiocin-resistant* strain, Guild, W.R. and M. Robinson (PNAS 50:106, 1963) obtained two fractions by this means. They found that the fraction containing the heavier chain required about 45 minutes at 25°C to begin transforming *novobiocin-sensitive* cells to resistant ones. The lighter fraction modified the phenotype of transformed cells almost immediately. The cell-generation time (time for one DNA replication) at 25°C is approximately 40 minutes. Analyze this information to show that mRNA is copied form only 1 of the strands of DNA although either can transform the cell's genotype.

42. tRNA molecules are approximately 80 nucleotides long with an average molecular weight of 2.5×10^4. In base-composition and DNA-RNA hybridization experiments in *E. coli*, in 1962, Giacomoni, D. and S. Spiegelman (Science 138:1328) obtained the results shown below:

RNA hybridized	Base composition in moles, %				Percent G+C	Purine/ pyrimidine
	C	A	U	G		
Total tRNA	27.2	20.6	18.2	34.0	61.2	1.23
RNase-resistant tRNA	27.2	18.6	19.0	35.2	62.4	1.20
rRNA	24.3	25.0	19.7	31.0	55.3	1.27
mRNA	24.7	24.1	23.5	27.7	52.4	1.07

i) *E. coli* DNA was saturated with tRNA at 0.025 percent of total DNA.

ii) In one DNA-RNA hybridization experiment tRNA from *E. coli* (^{32}P) and *Bacillus megaterium* (^{3}H) were incubated with DNA from *E. coli*. In another, the tRNA from *B. megaterium* was incubated with DNA from the same species and *E. coli*. The hybridization results are shown in the illustration.

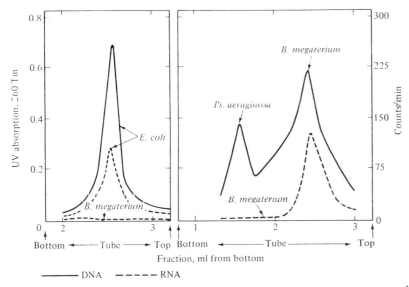

(Redrawn from S. Spiegelman, Hybrid Nucleic Acids, Sci. Am., May, 1964)

a) Indicate why these results show that tRNA molecules are specified by segments of DNA.

b) The *E. coli* genome has a molecular weight of 4×10^9. With a non-degenerate code what portion of total DNA should hybridize with tRNA? Is degeneracy indicated? Explain.

43. In the mature bacterial virus ϕX174, DNA is single-stranded and circular. After the virus infects a bacterial cell, this strand serves as a template for the synthesis of the complementary strand, resulting in the double-stranded replicating form (RF), also in a circle. Hayashi, M. et al. (PNAS 50:664, 1963) conducted hybridization tests between ^{3}H-labeled mRNA, transcribed from RF, and the DNA of the mature virus and RF (both labeled with ^{32}P). Although the single-stranded DNA of the mature virus, heated or unheated, did not hybridize with mRNA, the heat-denatured (chains-separated) RF DNA did. The base composition of mRNA, the DNA of ϕX174, and strand of RF complementary to the original strand were found to be given in the table.

Nucleic acid	Base ratio			
	C	A	U(T)	G
mRNA	18.5	23.8	33.1	23
ϕX174, DNA	19	25	33	23
ϕX174 DNA, complementary strand	23	33	25	19

Are transcribed messenger RNAs *in vivo* complementary to:
1) Both chains of DNA?
2) One chain only?
3) Both chains but only one at a time?

44. An individual homozygous for a mutant allele (β^-) possessed a AT → GC base-pair substitution at position -29 of the sequence for the β- globin gene (Antonarakis, S.E. et al., PNAS 81:1154, 1984). This individual produced 75% less β- globin mRNA than normal ($\beta^+\beta^+$) individuals and expressed mild β^- thalassemia as did heterozygous individuals. In a few other heterozygotes a different mutant allele (β^0) had a AT → GC base-pair substitution in the second base-pair position of the second intervening sequence (IVS-2). These heterozygotes suffer from severe β^0-thalassemia.

i) Which of the 2 mutant alleles β^- or β^0 affects transcription? RNA processing? Explain.

ii) In which of the 2 types of heterozygotes would you expect normal hemoglobin molecules to be synthesized? Discuss.

iii) Why are $\beta^+\beta^o$ individuals more severely affected than $\beta^+\beta^-$ individuals?

45. The 2 strands of the DNA of the virulent phage SP8 have different densities and therefore can be separated by centrifugation. The heavier (H) strand is richer in pyrimidines and the lighter (L) strand is richer in purines. Only the H strand hybridizes with mRNA isolated from SP8-infected bacteria, but both strands anneal with RNA synthesized *in vitro* using SP8 DNA as primer (Marmur, J. and C.M. Greenspan, Science 142:387, 1963).
a) Why does mRNA formed *in vivo* hybridize with the H strand only?
b) How does the RNA formed *in vitro* differ from that synthesized *in vivo*?

46. Two species of tRNA for each of the amino acids isoleucine and phenylalanine are detectable in light-grown but not in dark-grown *wild-type Euglena gracilis* cells and in the light- or dark-grown bleached mutant W$_3$BUL cells, which contain neither chloroplast DNA nor chloroplast structure. In addition, light-grown *wild-type Euglena* cells contain two aminoacyl-tRNA synthetases for each of these amino acids. Isoleucyl-tRNA synthetase II is present only in light-grown *wild-type* cells. The phenylalanyl-tRNA synthetases are present in dark-grown *wild-type* and W$_3$BUL cells as well as in light-grown ones. Only one of the two synthetases for each amino acid (light-inducible isoleucyl-tRNA synthetase II and phenylalanyl-tRNA synthetase I) is found in isolated chloroplasts. For each amino acid only the light-induced phenylalanine and isoleucine tRNAs can be acetylated by the chloroplast synthetases (Reger et al., PNAS 67:1207, 1970).
a) Which of these tRNAs and their aminoacyl synthetases are coded for by nuclear genes? Cytoplasmic genes?
b) If it is impossible to answer the question for a particular tRNA or its amino acylsynthetase on the basis of the above data, design an experiment to obtain this information. Show the results expected.
c) Which chloroplast synthetase is definitely synthesized in the cytoplasm? Explain.
d) Do chloroplasts possess a translational apparatus? Explain.

47. When a bacterium is transformed, only a segment of one strand of donor DNA appears to be integrated into the recipient DNA by displacing the corresponding segment of the recipient duplex. Working with *Diplococcus*, Guild, W.R. and M. Robinson (PNAS 50:106, 1963) found that one of the DNA strands was heavier than the other and that single strands, as well as duplexes, can transform recipient cells. Some *Diplococus* of a *novobiocin-resistant* strain, show the *resistant* phenotype almost immediately. These *resistant* segregants breed true. In contrast, transformation with the heavy strand does not alter phenotypic expression until after one generation of replication, at which time some of the bacteria show *resistance*. These cells also breed true.
a) State whether either strand or only a specific one can be incorporated into the recipient.
b) Explain why the expression of donor phenotype is immediate with the light strand but requires one replication with the heavy strand.

48. Brawerman, G. (Biochim. Biophys. Acta 61:313, 1962) found a unique species of RNA in *Euglena* chloroplasts, which was associated with the ribosomes. It was significantly higher in A + U and lower in G + C than cytoplasmic RNA (Brawerman, G. and H. Eisenstadt, Biochim. Biophys. Acta 91:477, 1964). Kirk, J.T.O. (Biochem. Biophys. Res. Commun. 14:393, 1964) found that all four nucleotide triphosphates are required for chloroplast RNA synthesis. Chloroplast treatment with actinomycin D or with deoxyribonuclease inhibited RNA synthesis. The DNA of the satellite and of chloroplasts is also characterized by a higher ratio of A + T to G + C.
a) Do these data suggest that chloroplast DNA acts as template for RNA synthesis?
b) How might you proceed to support your conclusion?
c) What results would you expect?

49. What is the significance of the following finding: Incorporation of radioactive amino acids by isolated chloroplasts of *Euglena* is inhibited by treating the chloroplasts with ribonuclease or with actinomycin D (Eisenstadt, J. and G. Brawerman, Biochim. Biophys. Acta 76:319, 1963)?

50. The human hemoglobin mutants Constant Spring (Hb-CS) and Wayne 1 (HB-W1) have α chains that are lengthened at the carboxyl end of the polypeptide. Consequently they have more than the normal number of 141 amino acids. The Hb-CS α chain is 172 amino acids long and has a residue sequence identical to that of normal α chains up to the amino acid in position 142 (Clegg, J.B. et al., Nature 234:337, 1971). The Hb-W1 α chain is 146 amino acids long and it differs from the normal α polypeptide beginning with amino acid position 139 (Seid-Akhaven, M. et al., PNAS 73:882, 1976). Single nucleotide pair substitutions are responsible for both of the variant α chains. Propose plausible explanations in terms of DNA nucleotide pairs and codon sequences to account for the 2 mutant α chains.

51. The illustration shows a model of polysome function in protein synthesis.

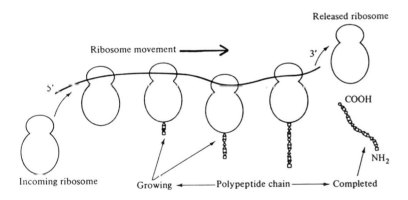

The model predicts that under protein-synthesizing conditions:
i) It should be possible to attach single ribosomes (70S) to polysomes.
ii) Single ribosomes should be released from polysomes.
iii) Polypeptide chains should be released from polysomes.

Discuss the bearing of each of the following experiments on these predictions:
a) Extracts of living cells in the process of protein synthesis incubated by Rich, A. et al. (Cold Spring Harbor Symp. Quant. Biol. 28:269, 1963) for varying periods and subjected to sucrose-gradient centrifugation showed at the beginning a large polysome peak and a smaller single-ribosome peak. As incubation proceeded, the number and size of polysomes decreased and the number of single ribosomes increased. At the end of 90 minutes of incubation most of the polysomes disappeared; only single ribosomes were present.
b) Goodman, H.M. and A. Rich (Nature 199:318, 1963) incubated a suspension of living HeLa cells for 1 1/2 minutes with ^{14}C amino acids. The cells were then chilled to stop protein synthesis. This process loaded the polysomes with ^{14}C amino acids that were joined into growing polypeptide chains. Next the cells were broken and the ribosomes and polysomes were isolated, using appropriate procedures, and resuspended in a fresh cell extract identical with that removed except that it contained normal (nonradioactive) amino acids. The suspension was incubated and radioactivity measured in the polysome fraction and soluble protein fraction (free of polysomes). As incubation proceeded, radioactivity decreased in the former fraction and increased in the latter.
c) A mixture of ^3H-labeled single ribosomes from HeLa cells and an unlabeled extract of polyribosomes plus single ribosomes from the same source was incubated briefly and subjected to sucrose-gradient centrifugation. Some of the tritium-labeled ribosomes had become attached to polysomes, as indicated by a test for radioactivity. Twice as many single labeled ribosomes were attached to polysomes composed of 5 ribosomes as compared to polysomes composed of 10 ribosomes when the total number of ribosomes in each fraction was equal (Goodman and Rich, 1963).

52. a) Terminator sequences in prokaryotes contain GC-rich inverted repeats upstream of the termination site of transcription. What is the mechanism by which these repeats lead to termination of transcription?
b) Rosenberg, M. et al. (Nature 272:414, 1978) studied λ phage mutants containing single base-pair substitutions in the GC-rich inverted repeat region. In mutants cnc^8 (TA → CG) and cnc^1 (AT → GC) the terminator was less efficient

than in normal λ, while in mutant cin^1 (GC → AT) the efficiency of the terminator was increased. Postulate plausible mechanisms for the activities of the *cnc* and *cin mutant* genes.

53. Isolated 50S ribosomal subunits can be induced to catalyze peptide-bond formation between peptidyl-tRNA and aminoacyl-tRNA in the absence of template and 30S subunits (Monro, R.E., Nature 223:903, 1969). Is peptidyl transferase activity associated with the 30S, 50S, or both subunits?

54. Rother, C. et al. (Curr. Genet. 11:171, 1986) cloned and sequenced the spinach (*Spinacia oleracea*) gene which codes for the protein plastocyanin, a member of photosynthetic electron transport chains. The gene, inclusive of the chain terminating codon was found to be 507 base-pairs long. Without carrying out R-loop mapping between the denatured DNA of the gene and its primary transcript and messenger RNA they concluded that the gene did not contain any introns.

What additional information must they have had, which is not provided here, that permitted them to make this conclusion? Discuss.

55. In *Tetrahymena thermophila*, the primary transcript (35S rRNA) for 3 of the 4 ribosomal RNAs is ~ 6,400 bases long. *In vitro* experiments carried out by Cech, T.R. (Cell 27:487, 1981) and Zaug, A.J. and Cech, T.R. (Nucl. Acids Res. 10:2823, 1982) in which the mixtures contained the primary transcript, a guanine nucleotide (GMP, DMP or GTP) and magnesium ions only, the 414 bases long intron was excised from the rRNA precursors. Mutations in certain regions of this sequence region prevent intron excision.
a) Why are proteins not involved in catalyzing intron excision from these transcripts?
b) Which molecular entity is or appears to be involved in intron removal? Explain.

56. a) What is R-loop mapping?
b) Dugaiczyk, A. et al. (PNAS 76:2253, 1979), Chambon, P. (Sci. Amer. (May) 244:60, 1981) conducted R-loop mapping of the ovalbumin gene in fowl (*Gallus domesticus*). An electron micrograph and a line drawing of the hybrid DNA-RNA molecule formed between the ovalbumin gene and its messenger RNA (mRNA) are shown below:

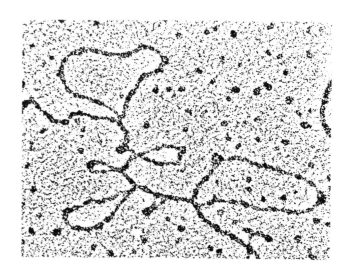

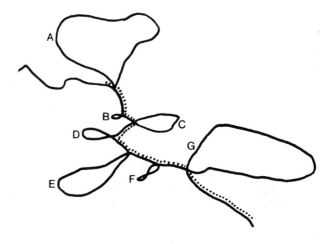

i) Why does this finding indicate that the gene is discontinuous?
ii) How many introns and exons does the gene possess? Explain. Show the linear sequence of these segments.
iii) In the line drawing label the R-loops, exons, and the ovalbumin gene's antisense strand.

c) When the denatured DNA of this gene is hybridized with the primary transcript only one R-loop is formed. Why does this indicate that the introns are excised after the primary transcript is formed?

57. a) The β chain of human hemoglobin is 146 amino acids long. What would be the length of the mRNA molecule required to direct the synthesis of this chain?
b) An enzyme of molecular weight of 300,000 consists of one polypeptide chain only. Assuming the average weight of an amino aid is 100, how many nucleotides are there in the gene coding for its synthesis?
c) The single-stranded DNA of bacteriophage φX174 has 4.5×10^3 nucleotides. How many proteins of molecular weight 30,000 could be coded for by this genome?
d) The DNA of *Aspergillus* has 4×10^7 nucleotide pairs. If an average gene contains 1,500 nucleotide pairs, how many genes does *Aspergillus* possess?

58. When 5'-AAA ... AC[(A7)nC]-3' was used as a synthetic messenger in the rabbit reticulocyte cell-free system, the polypeptide product was NH2-Lys-Lys ... Lys-Asn-COOH (Lamfrom, H. et al., J. Mol. Biol. 22:355, 1966). Is this mRNA read from the 3' or 5' end? Explain. Does the NH2 end or the COOH end of the polypeptide correspond to the 3' end of the mRNA?

59. Thach, R.E. et al. (Cold Spring Harbor Symp. Quant. Biol. 34:277, 1965) showed that the peptide synthesized under the direction of the messenger 5'-AAAUUU-3' is NH2-Lys-Phe-COOH and not NH2-Phe-Lys-COOH. In what direction does translation proceed along the messenger polynucleotide chain?

60. Kossel, H. et al. (J. Mol. Biol. 26:449, 1967), using a synthetic messenger RNA with the repeating sequence 5'(UAUC)n ... 3' in a cell-free amino acid-incorporating system from *E. coli* B, found that the messenger directed the synthesis of a polypeptide with the repeating tetrapeptide sequence Tyr-Leu-Ser-Ile. Chymotrypsin digestion (breaks peptide bonds on carboxyl side of tyrosine) produced tetrapeptides, NH2-Tyr-Leu-Ser-Ile-COOH. Show how those results established that the mRNA is read 5' to 3'.

61. Methionine tRNA of *E. coli* exists in 2 forms only, one of which can have its methionine formylated. In a cell-free system from *E. coli* only 2 of a large number of synthetic messengers (random poly-UG and random poly-UAG) were capable of incorporating methionine into polypeptide chains. When only the formylated Met-tRNA was present, both random poly-UAG and random poly-UG directed synthesis of polypeptides which contained methionine, but only in the start position (NH2 end of chain). The same results were obtained even when the formyl group was not attached to formylatable Met-tRNA. The nonformylated Met-tRNA responded only to random poly-UAG. The polypeptide so produced has methionine only in internal positions (Clark, B.F.C. and K.A. Marcker, J. Mol. Biol. 17:394, 1966; Sci. Am., 218 (January):36, 1968). Binding of Met-tRNA$_M$ and F-Met-tRNA$_F$ to ribosomes in the presence of various trinucleoside diphosphates gave the results tabulated below. Synthetic messengers poly-(GUG)$_n$ and poly-(AUG)$_n$, in which the same sequences were repeated over and over, lead to formation of polypeptides with formylmethionine in the start position. The rest of poly-(GUG)$_n$-directed chains contained valine, whereas the rest of the poly-(AUG)$_n$ polypeptides contained methionine.

Trinucleotide	Met-tRNA$_M$	F-Met-tRNA$_F$ or Met-tRNA$_F$
UAG	−	−
AUU	−	−
AAG	−	−
AUG	+	+
UUG	−	−
GUG	−	+
UGU	−	−
Key: + = binding; − = no binding		

a) What are the apparent functions in protein synthesis of each of the two species of tRNA that carry methionine?
b) What may be the role of the formyl group in protein synthesis?
c) Which are the chain initiating codons? Do they perform different functions within the gene? Explain.

62. The codons for polypeptide-chain initiation are AUG and GUG. Moreover, mRNA from the RNA phage f2 in a cell-free system can direct the synthesis of phage coat protein which has the *in vitro* N-terminal amino acid sequence F-Met-Ala-Ser-Asp-Phe-Thr. *In vitro* experiments by Nomura, M. and C.V. Lowry (PNAS 58:946, 1967) revealed the following:

i) Addition of f2 RNA to 30S ribosome particles stimulates binding of F-Met-tRNA$_F$ to these particles. Such stimulation does not occur with 70S ribosomal particles, 30S-50S complexes, or 50S subunits alone.
ii) f2 mRNA does not direct the binding of any other tRNA to 30S particles.
iii) f2 mRNA directs binding of all tRNAs (other than tRNA$_F$) to 70S or 30S-50S complexes (formed by addition of 50S to 30S subunits).
iv) When the synthetic mRNA containing A:U:G (1:1:1) in random sequence is used, (it contains the initiator codon AUG and also valine codons GUA, GUU, and GUG in addition to others) the 30S particles bind F-Met-tRNA$_F$ only. Addition of 50S particles results in the 30S-50S complex binding Val-tRNA and Met-tRNA$_M$ but not F-Met-tRNA$_F$.

a) An accepted hypothesis of the mechanism of polypeptide-chain initiation until 1968 was that 70S ribosomes bind to the initiator codon in mRNA and that the resultant mRNA-ribosome complex binds F-Met-tRNA$_F$. Do these results validate this hypothesis? If so, how?
b) Since ribosomes of polysomes are known to be 70S, what events appear to follow the initiation steps to permit protein synthesis to continue?

63. Esterification of amino acids to tRNAs is catalyzed by aminoacyl-tRNA synthetases, each kind being specific for one amino acid. The Ser-tRNA synthetase attaches serine to two different Ser-tRNA species. One responds to codons AGU and AGC; the other responds to codons UCA and UCG (Sundharadas, G. et al., PNAS 61:693, 1968). Does the anticodon serve as the recognition site for the synthetase? Explain.

64. The data shown below are based on results of Capecchi, M.R. (PNAS 58:1144, 1967), and Milman, G. et al. (PNAS 63:183, 1969). In an *in vitro* system containing the messenger AUGAAAUUUUAA, ribosomes, tRNAs, amino acids, and initiating and elongation factors, a polypeptide containing F-Met-Lys-Phe is synthesized but not released from the ribosome. When the protein factors R_1, R_2, and S are added, the results shown in the table are obtained:

Factor	Polypeptide chain
R_1 with or without S	Released from ribosome
R_2 with or without S	Released from ribosome
S alone	Not released

a) What appears to be the function of each of these factors?
b) Do the data imply there is a tRNA species for chain-terminating codons in nonsuppressor-containing, Su^-, hosts?

65. a) Malkin, L.I. and A. Rich (J. Mol. Biol. 26:329, 1967) showed that proteolytic enzymes acting on the growing polypeptide chain on reticulocyte ribosomes are unable to digest the 30 to 35 amino acids proximal to the peptidyl tRNA. What do these results suggest regarding the location of the tRNA bearing the nascent polypeptide chain?
b) Takanami, M. et al. (J. Mol. Biol. 12:761, 1965) showed that the fragment of mRNA bound to ribosomes is resistant to RNase digestion. Suggest why bound mRNA might be resistant to RNAse.
c) Kuechler, E. and A. Rich (Nature 225:920, 1970) found that if an initiation complex is formed involving *E. coli* ribosomes, R17 *wild-type* RNA, and F-Met-tRNA$_F$, and if the unprotected mRNA is digested with RNase, the remaining complex is still able to synthesize the N-terminal pentapeptide (F-Met-Ala-Ser-Asn-Phe) of the coat protein attached to ribosome-bound tRNA. How many codons is the initiator site from the point of entry of mRNA on the ribosome? Explain.

d) When an *amber* mutant of R17 virus carrying an *amber* codon in the seventh position of the coat-protein gene is used in an amino acid incorporating system, a free hexapeptide (F-Met-Ala-Ser-Asn-Phe-Thr) is produced. In a system deficient in asparagine a tripeptide (F-Met-Ala-Ser) is formed which is attached to tRNA bound to the ribosome (in this complex the *amber* codon is four codons from the peptidyl site occupied by tRNA containing the tripeptide). After RNase digestion of unprotected mRNA the ribosomes were added to an *in vitro* system with all amino acids. A N-terminal hexapeptide is formed and released. Is the peptidyl site the same distance from the point of entry of mRNA on the ribosome as the initiator site?

e) The portion of the R17 coat-protein gene mRNA protected by the ribosome from RNase digestion was isolated and sequenced by Steitz, J. (Nature 224:957, 1969) with the following results:

5'-AGAGCCUAACCGGGGUUUGAAGGAUGGCUUCUAACUUU-3'

Are the results in agreement with those of Kuechler and Rich? Illustrate by assigning amino acids to codons in the sequenced messenger.

66. The chain-initiating amino acid is formylmethionine. Livingstone, D.M. and P. Leder (Biochemistry 8:435, 1969) found an enzyme that removes the formyl group from the amino acid. Do you think the enzyme acts on N-F-Met-tRNA$_F$ before or after methionine forms the initial peptide bond with the second amino acid?

67. The base-pair sequence below represents the entire *wild-type* leader sequence of the *trp* operon for *E. coli* plus the first few codons of the *trpE* structural gene. The regions that encode potential secondary structures for a transcript are underlined and numbered 1 to 4. The coding sequence of the leader peptide is indicated by a dotted line, and the start of the coding sequence for the *trpE* is indicated by a double line. Single or double base-pair substitutions, deletions, or insertions of this sequence have been isolated and are indicated i - viii below the *wild-type* sequence. Describe how each of these mutations might be expected to influence the ability of *E. coli* cells to synthesize trytophan.

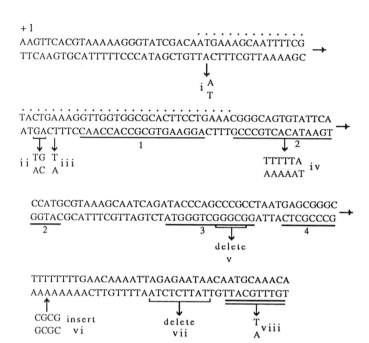

68. a) When streptomycin is added to a cell-free system containing ribosomes from a *streptomycin-sensitive*, *str-s*, bacterium, protein synthesis is inhibited and amino acids other than those dictated by the genetic code in the messenger are incorporated. In a similar system containing ribosomes from a *streptomycin-resistant*, *str-r*, strain, protein synthesis proceeds normally, and normal proteins are produced. The ribosome consists of 2 subunits; the larger one sediments on centrifugation as 50S, and the smaller one as 30S. Normal ribosomes can be reconstituted by mixing the subunits under appropriate conditions, and thus "hybrid" ribosomes, containing 30S particles from one strain of bacteria and 50S from another, can be produced. The effects of streptomycin on hybrid ribosomes from *streptomycin-resistant* and *-sensitive* bacteria are shown below. What roles do the 30S particles appear to play in translation?

	50S subunit	
30S subunit	From *str-s*	From *str-r*
From *str-s*	Inhibition; abnormal proteins	Inhibition; abnormal proteins
From *str-r*	No inhibition; normal proteins	No inhibition; normal proteins

b) Ribosomal proteins and rRNAs can also be separated from either subunit and can be similarly reconstituted to re-form the original particles (30S subunits contain about 20 proteins and 20 rRNAs). The data for the effect of streptomycin on reconstituted ribosomes prepared in this way are shown in the table below. Is it the rRNA or the ribosomal protein that is involved in the inhibition and translational error caused by streptomycin?

	From *resistant* strain	From *sensitive* strain
Protein of 30S	No inhibition; normal proteins	inhibition; abnormal proteins
rRNA of 30S	No inhibition; normal proteins	No inhibition; normal proteins

c) The P-10 protein of 30S subunits, derived from a *sensitive* strain, causes inhibition and translation error with streptomycin; that from *resistant* strains does not. Moreover, systems using synthetic messengers are not affected. What do you think the P-10 protein does in translation?

Chapter 22
Coding, Colinearity, and Suppressors

1. Gamow, G. (Nature 173:318, 1954) pointed out that since the genetic language contains only 4 letters, A, U (=T), C and G if all words (= codons) are of the same size, they must be at least 3 bases long.
   a) Show why codons cannot consist of 1 or of 2 bases.
   b) Since codons are 3 bases long, 64 different triplets can exist. Illustrate these using the branching method.
   c) How many of the 64 triplets will contain (1) no adenine, (2) at least one adenine?
   d) Why are many mutational sites expected within a gene?
   e) Explain how single base-pair substitutions in DNA are reflected in phenotypic changes.
   f) Which are the nonsense triplets and why are they so termed?
   g) Are nonsense mutations identical with stop codons?
   h) Explain why polypeptides specified by a polygenic message are not coupled.
   i) What is characteristic of polypeptide chains specified by genes carrying nonsense mutations? Do missense mutations have the same effect?

2. a) Explain what is meant by a degenerate code and illustrate your answer to show degeneracy in translation.
   b) Does most of the degeneracy in the code involve the first, second, or third base of a codon?
   c) Some amino acids like tyrosine and histidine are coded by 2 triplets only. Can the tRNA have inosine at the 5' end of the anticodon?
   d) A highly purified species of *E. coli* alanine-tRNA can recognize 3 alanine codons, GCU, GCC and GCA. Explain this on the basis of the wobble hypothesis.

3. a) Of the 2 molecules 5-bromouracil and proflavine which is more likely to produce leaky mutant alleles? Why?
   b) Do you expect *amber* mutant alleles in the same gene to complement each other? Why?
   c) Why is UAA more likely than UAG to be the regular chain-terminating triplet?
   d) If the mRNA codon for valine is GUA, what are the corresponding DNA and tRNA sequences?

4. A polysome contains 10 ribosomes 150 Å apart, held together by a monogenic mRNA. Approximately how many amino acids will this messenger specify in the corresponding polypeptide chain? Explain.

5. How would you determine in *Salmonella typhimurium* whether a revertant is the result of a suppressor mutation at a different locus, a suppressor mutation at a second site in the same gene, or a true reversion at the original mutation site?

6. Show how a mutation which changes the base sequence of a tRNA molecule might act as a suppressor mutation.

7. Missense mutations can revert as a result of mutation outside the codon in which the original mutation occurred, whereas reversions of nonsense mutations result from mutations within the nonsense codon. Why should this difference exist?

8. Suppressed strains produce 2 types of proteins: one that resembles that of the unsuppressed mutant strain and another with physical (although not always) and enzymatic properties characteristic of the *wild-type* protein. For example, in the missense *E. coli A*-gene mutant *A36*, A-type polypeptides have arginine in place of glycine at a specific position in peptide CP2. When *A36* carries a suppressor, $Su36^+$, some A polypeptides have arginine and others have glycine (as in *wild-type* A polypeptides) at this site (Brody, S. and C. Yanofsky, PNAS 50:9, 1963). Explain how this is possible.

9. a) Why can a given suppressor gene suppress mutations in a number of different genes?
   b) Suppressor alleles of certain genes misread nonsense codons, and each inserts a specific amino acid into the polypeptide chain at the position of the nonsense triplet. In *E. coli*, the suppressors $Su1^+$, $Su2^+$ and $Su3^+$ suppress only

the *amber* (UAG) codon, inserting serine, glutamine, and tyrosine, respectively, at the nonsense position. $Su4^+$ and $Su5^+$ suppress both *amber* and *ochre* (UAA) codons by inserting tyrosine and a basic amino acid (probably lysine) respectively. Account for the difference in behavior of the 2 classes of suppressors in the light of the wobble hypothesis.

c) UGA was the last nonsense codon to be discovered, and it was found that none of the 5 suppressors $Su1^+$ - $Su5^+$ could suppress this codon.

i) Explain how a nonsense codon interferes with the normal functioning of a gene. Indicate whether gene function is interrupted at the level of transcription or translation.

ii) Indicate why neither the suppressor of *amber* (UAG) nor that of the *ochre* (UAA) codons could suppress the UGA nonsense codon.

iii) Suggest the amino acids that one might expect the suppressor of the UGA codon to have inserted and indicate your reasoning.

10. One of Benzer's rII deletion mutants in phage T4, r1589, extends from within the A gene into the left end of the B gene as illustrated:

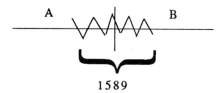

1589

a) This mutant complements with mutants of gene B but not those in gene A. What does this suggest vis-a-vis; (1) the activities of the gene products produced by translation of the portions of the A and B genes present in r1589 and (2) the left end region of the B gene?

b) Nonsense mutations mapping to the left of the r1589 deletion, when crossed with r1589 produce double mutants. Describe how these double mutants can be obtained and isolated. (Note that all rII mutants grow in *E. coli* B but not in *E. coli* K.)

c) The double mutants do not complement with B mutants, whereas r1589 by itself does. Explain this in terms of the direction of translation of the mRNA code and the effect of the nonsense mutation on the size (complete vs. incomplete) of the polypeptide product and punctuation between the A and B genes.

d) A deletion mutant missing 4 base-pairs and mapping in the A gene to the left of the r1589 segment is crossed with r1589 to produce a double mutant.

i) Would the double mutant have B activity (complement with B mutants)? *Note:* The deletion mutant by itself has B activity.

ii) What name has been given to mutations consisting of the insertion and deletion of a number of bases not divisible by 3?

iii) Suppose that the double mutant has B activity; how would you explain these results in terms of generation of nonsense codons by the deletion of 4 bases?

e) Mutant, r1589 shows no A activity but almost *wild-type* B activity in *E. coli* B. The addition or deletion of a single base pair in the A gene suppresses B activity. What hypothesis do these results support?

11. Caron, F. and E. Meyer (Nature 314:185, 1985) showed that the non-coding strand of DNA of the *wild-type* gene that specifies the normal G surface antigen of *Paramecium primaurelia* possesses numerous TAA and TAG codons scattered throughout its base sequence. Why does this information indicate that the genetic code is not universal?

12. Valine occupies position 181 in the *wild-type* A polypeptide of *E. coli* tryptophan synthetase. A 2-aminopurine-induced *mutant* has isoleucine in position 181; treatment of this *mutant* with 2-aminopurine gives a *mutant* with methionine at this position. Finally treatment of the second *mutant* with the same mutagen produces a *revertant*, with valine replacing methionine at this site. What information do these results provide about the fundamental properties of the genetic code? Explain.

13. Tsugita, A. and H. Fraenkel-Conrat (J. Mol. Biol. 4:73, 1962) studied alterations by nitrous acid (NA) in the 158-amino acid coat protein of tobacco mosaic virus. This chemical is known to effect single base-pair substitutions. Usually only one amino acid at a time is changed. Only in rare cases

were 2 amino acids altered in a mutant, and these were never at adjacent sites. Some of the amino acid substitutions in the polypeptides of 3 NA induced mutants are shown in the table below:

Strain	Amino acid position				
	1	11	20	81	156
Wild-type	NH_2-Ser	Val	Pro	Thr	Gly
Mutant 1	NH_2-Ser	Met	Pro	Thr	Gly
Mutant 2	NH_2-Ser	Val	Thr	Thr	Gly
Mutant 3	NH_2-Ser	Val	Thr	Thr	Leu

Explain which aspects of coding these results have a bearing on:
a) Codons consist of 3 consecutive bases.
b) The code is degenerate.
c) The code is nonoverlapping.
d) The codons are read from a fixed starting point.

14. At the Su locus in *E. coli* the suppressor allele $Su6^+$ causes the *amber* triplet UAG to code for leucine. Gopinathan, K.P. and A. Garen (J. Mol. Biol. 47:393, 1970) in experiments with fractionated tRNA from $Su6^-$ and $Su6^+$ strains showed that in $Su6^-$ strains there were 2 species of leucyl-tRNA which bind to ribosomes in the presence of leucine codon UUG. In $Su6^+$ strains only 1 of these tRNA species was present; the other was replaced by a homologous leucyl-tRNA which binds to ribosomes only in the presence of UAG.
a) What appears to have been the consequence of the suppressor mutation $Su6^-$ to $Su6^+$?
b) What is the most plausible mechanism for this transformation?
c) Why is a suppressor mutation that alters the codon recognition of a tRNA potentially lethal? Why is $Su6^+$ not lethal?
d) The normal amino acid sequence of a certain region of the *wild-type* T4 phage lysozyme molecule is:

... Lys-Ser-Pro-Ser-Leu-Asn-Ala ...

A frame-shift mutation was induced in the viral gene for this lysozyme by treating T4 phage with an acridine dye. The mutation involved the deletion of a base in the DNA codon for the serine residue nearest the NH_2-end of the fragment shown. A reversion of the *mutant* to the *wild-type* phenotype was obtained by treating the mutant with acridines and causing addition of a base in the alanine codon. The amino acid sequence in the lysozyme molecule of the *revertant* was found to be:

... Lys-Val-His-His-Leu-Met-Ala ...

i) Because of degeneracy, there are a large number of theoretically possible sequences for the *wild-type* mRNA which would give rise to the amino acid sequence in the *wild-type* polypeptide fragment shown. Calculate how many.
ii) Only one of these sequences was present in the T4 lysozyme mRNA molecule. This unique sequence can be determined, since upon deletion of a base in the serine codon and addition of a base in the alanine codon it must have given rise to a mRNA sequence in the *revertant*, which coded for the altered lysozyme sequence shown. Determine the *wild-type* mRNA sequence and indicate which base must have been deleted from the serine codon and which must have been added to the alanine codon to produce the revertant mRNA.

15. Both poly-UC (3:1) and poly-UG (3:1) messengers in which the bases are incorporated randomly, stimulate the incorporation of leucine into polypeptides but to only one-third the extent of phenylalanine, which is incorporated to the greatest extent. Weisblum, B. et al. (PNAS 48:1449, 1962) separated *E. coli* tRNA into two samples. Each sample was charged with ^{14}C leucine and the response of each to poly-UC and poly-UG was determined. Leucine attached to tRNA in sample 1 was incorporated into polypeptides by poly-UC, whereas leucine attached to tRNA from sample 2 was incorporated into polypeptides only when poly-UG was the messenger.
a) These results clearly indicate the role of tRNA in protein synthesis. What is it and why?
b) Is coding degeneracy involved? If so, provide an explanation of the mechanism involved.

16. a) Nirenberg, M.W. and J.H. Matthaei (PNAS 47:1588, 1961) included a different ^{14}C-labeled amino acid in each of 20 mixtures containing all amino acids. These were added to cell-free translation systems containing ribosomes, tRNAs, enzymes, synthetic messenger containing uracil only, and an energy source. The only mixture in which polypeptides were labeled was the one containing labeled phenylalanine. When poly-U was omitted from the system or paired with an adenine chain, no protein containing this amino acid was synthesized. Discuss the significance of this discovery from the coding point of view.

    b) Synthetic mRNA containing only adenine templates the *in vitro* synthesis of a polypeptide containing lysine only (Gardner, R.S. et al., PNAS 48:2087, 1962), and that containing only uracil leads to a polypeptide composed of phenylalanine only (Nirenberg and Matthaei, 1961). Since adenine and uracil (thymine) are complementary, what do these results imply regarding the transcription process?

17. Using a synthetic mRNA containing adenine and cytosine in a ratio of 3:7 incorporated at random, Jones, O.W. and M.W. Nirenberg (PNAS 48:2115, 1962) found that the proteins contained the following percentages of amino acids: aspartic acid 3.2%, glutamic acid 3.5%, histidine 11.7%, lysine 3.0%, proline 68.0% and threonine 10.5%. Do these percentages of amino acids approximate the values expected on the basis of random incorporation of bases into the messenger? Explain.

18. Where the synthetic messenger consists of more than one kind of nucleotide, the initial concentration of nucleotides, incorporated at random, determines the base ratio in the polymer. An experiment by Wahba, A.J. et al. (PNAS 49:880, 1963), incorporated the bases adenine (A) and cytosine (C) into 2 synthetic messengers in 5:1 and 1:5 ratios respectively as well as C and uracil (U) into a messenger in a 1:5 ratio. These three messengers were used in *in vitro* protein-synthesizing systems. The percent amino acid incorporations into the polypeptides in each of 3 experiments are shown on the next page.

	Percent amino acid incorporation		
	A - C		U - C
Amino acid incorporated	5:1	1:5	5:1
Asparagine	24.2	5.2	
Glutamine	23.7	5.3	
Histidine	6.5	23.4	
Leucine			22.2
Lysine	100.0*	1.0	
Phenylalanine			100.0*
Proline	7.2	100.0*	5.1
Serine			23.6
Threonine	26.5	20.8	

* The incorporation of an amino acid is given as percent incorporation of that amino acid whose incorporation is promoted to the greatest extent by the polymer. For example, incorporation of lysine is promoted to the greatest extent; it is therefore given a value of 100 percent. Other incorporations are relative to this.

a) After showing the different codons that are possible with A and C as well as with U and C and their probabilities in a synthetic messenger, assign the different codons to the amino acids that were incorporated. Explain your assignment.

b) Do these data indicate that the code is degenerate? If so, for what amino acids?

19. Position 210 in the *wild-type* A protein of *E. coli* tryptophan synthetase is occupied by glycine. In mutants *A46* and *A47* this glycine is replaced by glutamic acid and valine respectively. Both replacements are associated with loss of enzyme activity. *Wild-type* recombinants (glycine at 210) are not produced in crosses between these mutants. Crosses between mutants *A46* and *A23* (arginine replaces glycine at 210) produce 0.002 percent *wild-type* recombinants. Treatment of *A23* with mutagens produces new mutant strains. Those with serine in the same position produce a fully active enzyme, whereas those with threonine in the same position produce a partially active enzyme

(Helinski, D.R. and C. Yanofsky, PNAS 48:173, 1962; Guest, J.R. and C. Yanofsky, Nature 210:799, 1966).
a) What is the minimum number of (1) alternative forms that a mutable site can exist in, (2) adjacent mutable sites that specify a single amino acid?
b) Is a unique amino acid sequence required for enzyme activity? Explain.

20. Using randomly ordered synthetic polyribonucleotides as messenger, it has been shown that the codons for valine, leucine, and cysteine contain two Us and one G. Using the ribosome-binding technique (see Prob. 23), Leder, P. and M.W. Nirenberg (PNAS 52:1521, 1964) obtained the results shown.

Cell free system plus	Radioactive aminoacyl-tRNA bound to ribosomes, picomoles	
	^{35}S-Cys-tRNA	^{14}C-Leu-tRNA
No messenger	0.29	0.76
UGU	1.46	0.78
UUG	0.32	1.74
GUU	0.34	0.92
UUU	0.32	
UG	0.21	0.92
GU	0.34	0.86
UU	0.26	0.88

What is the base sequence of:
a) The codon for leucine? Explain.
b) Leu-tRNA anticodon when read from 5' to 3'?

21. The base sequence in the DNA strand complementary to mRNA is 5'-TACTAACTTAGCCTCGCATAC ... 3'
a) What amino acids are coded by this sequence?
b) An adenine is inserted in this strand after the first guanine from the left. The resulting polypeptide is 4 amino acids long. Where is the newly produced nonsense codon located? What is the amino acid sequence in the fragment?

22. Barrell, B.G. et al. (Nature 282:189, 1979) compared the base-pair sequence of the human mitochondrial DNA gene for cytochrome oxidase subunit II (COII) with the sequence of its 227 amino acids long protein. The sequences of some of the segments of the non-transcribed DNA strand of this gene and the corresponding amino acid sequences of the COII protein are shown below:

	1	2	16	17	65	66	85	86	163	164	165	221	227	228
Base sequence (Codon)	ATG	GCA	ATC	ATA	TCA	ACT	TAC	ATA	TGA	GCT	GTC	ATA	CTA	TAG
Amino acid	Met	Ala	Ile	Ile	Trp	Thr	Tyr	Ile	Trp	Ala	Val	Ile	Leu	

Is the genetic code in human mitochondrial DNA genes the same as that in nuclear structural genes? Explain.

23. The 6 codons for arginine fall into 2 groups: (1) AGA, AGG and (2) CGA, CGC, CGU, CGG (Morgan, A.R. et al., PNAS 56:1899, 1966). Söll, D. et al. (J. Mol. Biol. 19:556, 1966) separated 2 arginine tRNA species from yeast and charged them with radioactive (^{14}C) amino acid and tested them for binding to ribosomes in the presence of trinucleotides with the above codons. Arg-tRNA I binds to E. coli ribosomes in the presence of CGU, CGA and CGC, whereas Arg-tRNA II binds to ribosomes in the presence of AGA and AGG. Weisblum, B. et al. (J. Mol. Biol. 28:275, 1967) studied the transfer of arginine into the wild-type α chain of rabbit hemoglobin which has 3 arginine residues at positions 31, 92 and 141. The other arginine residues are in the β chain. tRNA I transfers its arginine to position 141; tRNA II transfers its arginine to position 31; neither species of tRNA transfers its arginine to position 92.

a) Is there a discrete tRNA species for the recognition of each codon, or can one tRNA species recognize more than one codon? Explain.
b) Which of the 6 codons occur at positions 31, 92 and 141 in the α chain of rabbit hemoglobin? Explain. In some mutant α chains arginine at 31 is replaced by lysine, arginine at 141 by histidine, and arginine at 92 by leucine or glutamine. Does this information confirm your codon assignments? Explain.
c) Which base occurs at the third position in the anticodon of tRNA that reads the three codons CGU, CGA, CGC?

d) Do these results provide support for the wobble hypothesis?

24. Nirenberg, M.W. and P. Leder (Science 145:1399, 1964) showed that if a trinucleotide messenger is added to a cell-free system containing ribosomes and the corresponding ^{14}C-labeled aminoacyl-tRNA, the charged-tRNA binds to the ribosome (pairs with the mRNA) and the resulting complex is retained on nitrocellulose membranes. Free aminoacyl-tRNAs do not absorb. The above procedure was repeated for each trinucleotide in 20 different media, all carrying the full complement of 20 amino acids but with a different ^{14}C-labeled amino acid in each. After each experiment the amount of ^{14}C radioactivity associated with the membrane was determined. The results of Leder, P. and M.W. Nirenberg (PNAS 52:420, 1964) using trinucleotides and other types of templates are given in the table.

Experiment	Type of RNA used as template	Activity of templates for ^{14}C-aminoacyl-tRNA picomoles of ^{14}C-aminoacyl-tRNA bound to ribosomes		
		^{14}C-Val-tRNA	^{14}C-Phe-tRNA	^{14}C-Leu-tRNA
1	None*	0.38	0.22	0.37
	Poly-U	0.23	4.73	0.27
	Poly-UG	2.65	1.93	0.24
2	None	0.40	0.22	0.62
	GUU†	1.11	0.27	0.56
	UGU	0.40	0.25	0.53
	UUG	0.37	0.25	0.44
3	None	0.18	0.22	0.69
	GU++	0.20	0.22	0.69
	UG	0.19	0.21	0.71
	UU	0.18	0.20	0.67

* Values represent background binding of ^{14}C-aminoacyl-tRNA to ribosomes in the absence of template RNA.
† Triplets of trinucleoside diphosphate type
++ Doublets of dinucleoside monophosphate type.

a) Can doublets as well as triplets code for amino acids? Explain.
b) Do the tested triplets and poly messengers code for either of the 3 amino acids? If so, which codons specify each amino acid?

25. Khorana and his colleagues have synthesized long RNA molecules with various repeating sequences of bases. Some of these, together with the amino acid sequence in the polypeptide chain or chains synthesized *in vitro*, are shown (Khorana, H.G. et al., Cold Spring Harbor Symp. Quant. Biol., 31:39, 1966; Kossel, H. et al., J. Mol. Biol. 26:449, 1967).

mRNA base sequence	Amino acid sequence in polypeptide or polypeptides
$(UC)_n$	Ser-Leu
$(UG)_n$	Val-Cys
$(AC)_n$	Thr-His
$(AG)_n$	Arg-Glu
$(UUC)_n$	(Phe-Phe), (Ser-Ser), (Leu-Leu)
$(UUG)_n$	(Leu-Leu), (Cys-Cys), (Val-Val)
$(AAG)_n$	(Lys-Lys), (Arg-Arg), (Glu-Glu)
$(CAA)_n$	(Gln-Gln), (Asn-Asn), (Thr-Thr)
$(UAC)_n$	(Tyr-Tyr), (Thr-Thr), (Leu-Leu)
$(AUC)_n$	(Ile-Ile), (Ser-Ser), (His-His)
$(GUA)_n$	(Val-Val), (Ser-Ser)
$(GAU)_n$	(Asp-Asp), (Met-Met)
$(UAUC)_n$	(Try-Leu-Ser-Ile)
$(UUAC)_n$	(Leu-Leu-Thr-Tyr)
$(GAUA)_n$	None
$(GUAA)_n$	None

a) Determine from the above results what the *in vitro* codons for each amino acid are and compare your results with the codons assigned by Khorana et al.
b) Assign (with reasons) each of the following sequences to its correct amino acid: CUU, UCU, UUC, CUC, ACA, CAC.
c) Why do $(GUA)_n$ and $(GAU)_n$ code for only 2 rather than 3 homopolypeptides?

d) Why do (GAUA)$_n$ and (GUAA)$_n$ fail to stimulate the synthesis of polypeptides? (Short peptides, 2 to 3 amino acids long, are formed to a small extent.)

26. Kossel, H. et al. (J. Mol. Biol. 26:449, 1967) showed that the synthetic messenger (UAUC)$_n$ with U at the 5' end and C at the 3' end of the chain directed the synthesis of the repeating tetrapeptide sequence NH$_2$-Tyr-Leu-Ser-Ile-COOH. The messenger (UUAC)$_n$ with U and C at the 5' and 3' ends, respectively, directed the incorporation of the repeating amino acid sequence NH$_2$-Leu-Leu-Thr-Tyr-COOH. Explain whether these results verify that:
(i) The direction of translation of mRNA is 5' to 3'?
(ii) The code is triplet and nonoverlapping?

27. UGA is a chain-terminating codon in mRNAs of structural genes and codes for tryptophan in the mitochondrion and the bacterium *Mycoplasma capricolum* (Yamao, F. et al., PNAS 82:2306, 1985). In the mRNAs of structural genes tryptophan is specified by UGG. AUA specifies isoleucine in the mRNAs derived from nuclear genes and methionine in the mitochondrion (Barrell, B.G., Nature 282:189, 1979). In mRNAs of structural genes, methionine is specified by AUG.
a) Is the genetic code completely universal? Explain.
b) Does the change in coding capacity involve a shift in recognition of the first, second or third positions of the codon in the mitochondrion and *M. capricolum*? Explain.
c) Suggest a possible reason for such changes in codon recognition.

28. Terzaghi, E. et al. (PNAS 56:500, 1966) compared lysozyme from a *wild-type* (e^+) phage T4 with that from a proflavine induced *pseudo-wild-type* revertant strain, *J42 J44*. All the peptides in the lysozymes from these strains were identical except for the following sequences of peptide 10:

Wild-type lysozyme   NH$_2$-Thr-Lys-Ser-Pro-Ser-Leu-Asn-Ala-COOH
Pseudo wild-type   NH$_2$-Thr-Lys-Val-His-His-Leu-Met-Ala-COOH

The mutations involved in producing the original *mutant eJ42* and the *pseudo-wild-type* strain were changes involving single base pairs.

a) How many mutations occurred to produce the original mutant *lysozymeless* strain? To produce the revertant strain from the original mutant?
b) What kinds of changes were they and why?

29. Mutations in the *e* gene of phage T4 affect the structure of the enzyme lysozyme which breaks down the bacterial cell wall, permitting lysis and plaque formation. Terzaghi, E. et al. (PNAS 56:500, 1966) and Okada, Y. et al. (PNAS 56:1692, 1966) isolated three proflavine-induced *e* mutants, *eJ17*, *eJ42* and *eJ44*. Crosses involving these mutants and their recombinants produced the results shown in the following table:

Cross	Recombinants in the progeny
eJ17 x eJ44	wild-type (plaques with large holes)
	pseudo wild-type 1 (plaques with small holes)
pseudo wild-type 1 x wild-type	eJ17; eJ44
eJ42 x eJ44	wild-type
	pseudo wild-type 2
pseudo wild-type 2 x wild-type	eJ42; eJ44

The amino acid sequences in the *wild-type* and the two *pseudo-wild-type* strains were identical except for the corresponding peptides (called A), which showed the sequences given in the table below:

Strain	Amino acid sequence in peptide A
e^+	NH$_2$-Lys-Ser-Pro-Ser-Leu-Asn-Ala-COOH
pseudo wild-type 1	NH$_2$-Lys-Val-His-His-Leu-Met-Ala-COOH
pseudo wild-type 2	NH$_2$-Lys-Val-His-His-Leu-Met-Ala-COOH

a) Are the *pseudo wild-types* single or double mutants? Explain.
b) What kinds of changes were induced in the DNA of the e^+ gene by proflavine to produce each of the mutants *eJ17*, *eJ42*

and *eJ44*? Show why the changes can only be of this type and not any other.

30. Brammer, W. et al. (PNAS 58:1499, 1967) determined the amino acid sequence of a portion of the *wild-type* A polypeptide of *E. coli* tryptophan synthetase and of the corresponding region in the polypeptide of a double-frame-shift mutant A9813PR8.

	Amino acid residue						
Polypeptide	173	174	175	176	177	178	179
Wild-type	NH_2-Thr-Tyr-Leu-Leu-Ser-Arg-Ala-COOH						
Double mutant	NH_2-Thr-Phe-Cys-Cys-His-Cly-Ala-COOH						

a) Determine the base sequence of mRNA for both the *wild-type* and the *double-mutant* polypeptides.
b) Which base was deleted and where? Which was added and where?

31. The *E. coli* gene A mutants *A23* and *A46* have glycine at position 210 replaced with arginine and glutamic acid, respectively, in the A polypeptide of tryptophan synthetase. In three-factor transduction crosses involving the *anth* locus at the operator end of the tryptophan-independent recombinants in the cross *anth⁻ A46* (donor) x *anth⁺* A23 (recipient), 14 were *anth⁻* and 3 *anth⁺*.
a) Did the mutations producing the mutant strains *A23* and *A46* occur in the same or different base pairs and why?
b) After assigning codons to glycine (in the *wild-type* and recombinant strains), arginine, and glutamic acid, orient the bases in the codons relative to the *anth* locus and peptide chain.
c) What is the molecular nature of the mutations in mutants *A23* and *A46*?

32. In phage T4 a series of mutant strains induced by hydroxylamine have amino acid substitutions at position 30 in lysozyme. Mutant A has serine; mutant B, derived from A, has lysine; and mutant C, derived from B, has phenylalanine. Only the first strain can revert naturally to *wild-type*. None of the crosses A x B, A x C, or B x C produce *wild-type* recombinants.
a) What amino acid is present at position 30 in the *wild-type* lysozyme? Explain.
b) Give the base sequences of the sense and antisense strands.

33. The antisense strand of the DNA of gene *A* has the codon GGA (glycine) at position 48 and UAU (tyrosine) at position 174. a) Since the genetic map, DNA, mRNA, and protein are colinear, indicate in the table below which end (1) of the DNA sense strand, (2) of the DNA antisense strand, and (3) of the mRNA, is the 5' and which the 3'. b) Which end of the protein would possess the amino end and which would contain the carboxyl group.

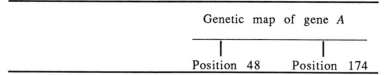

	Genetic map of gene *A*	
	Position 48	Position 174
DNA: Sense strand		
Antisense strand		
mRNA		
A protein		

34. Sarabhai, A. et al. (Nature 201:13, 1964) found that 10 base-analogue-induced mutants of phage T4 affecting the protein coat of the head of the virus grew on *E. coli* strain CR63 but not on strain B. No progeny were produced when *E. coli* B was infected with all possible combinations of these mutants, 2 at a time. Two- and 3-factor crosses were performed by infecting strain CR63 with a mixture of the parental phages in a 1:1 ratio. In each cross the proportion of *wild-type* recombinants to total progeny was determined and the percent recombination calculated. (*Note:* Designations *B17, H11,* and so on, refer to point mutations. Therefore, *B17* x *H11* is a 2-factor (point) cross and *B17* x *H11* + *B278* is a 3-factor cross.) The results of these crosses are shown on the next page.

Cross	Recombination, %	Cross	Recombination, %
B17 x H11	0.80	C137 x B278	0.33
B17 x H11 + B278	0.14	C137 x B272 + B278	0.28
		C137 x B278 + A489	0.06
B278 x A489	1.30	H36 x B278	0.46
B278 x B17 + A489	0.34	H36 x B272 + B278	
		H36 x B278 + A489	0.06
B272 X B278	2.20		
B272 x H11 + B278	0.32	H36 x C137	0.22
		H36 x C137 + A489	0.05
B272 x B17	1.30	C137 x H36 + A489	0.20
B272 x B17 + A489	0.24		
		C208 x A489	0.34
C140 x B17	0.27	C208 x B17	4.0
C140 x H11 + B17	0.04	C208 x B17 + A489	0.33
C140 x B17 + B272	0.20	C208 x B278 + A489	0.29
H32 x B272	0.5		
H32 x B17 + B272	0.36		
H32 x B272 + B278	0.05		

The defective proteins produced by each of the 6 mutants *B17, B272, B278, A489, C137* and *H32* and the protein produced by *wild-type* T4, upon infection of *E. coli* B, were as shown below:

Strain	Proteins produced
B17	NH$_2$
B272	NH$_2$-Cys*
H32	NH$_2$-Cys-HisT7C
B278	NH$_2$-Cys-HisT7C-TyrC12b
C137	NH$_2$-Cys-HisT7C-TyrC12b-TrpT6
A489	NH$_2$-Cys-HisT7C-TyrC12b-TrpT6-ProT2a-TrpT2
Wild-type T4	NH$_2$-Cys-HisT7C-TyrC12b-TrpT6-ProT2a-TrpT2-HisC6

*Cys, HisT7C, etc., designate the peptides in the protein.

a) Did the base analogues induce missense or nonsense mutations? Explain.

b) Why did the mutants grow on *E. coli* CR63 but not on *E. coli* B?
c) In how many genes did these mutations occur? Explain.
d) Map the 6 mutants, whose head proteins have been sequenced with respect to peptides, and discuss the bearing of these results on the *sequence hypothesis*, which states that the amino acid sequence of a protein is specified by the nucleotide sequence of the gene determining that protein.

35. a) Tsugita, A. (J. Mol. Biol. 5:284, 1962; J. Mol. Biol. 5:293, 1962) and Fraenkel-Conrat, H. (Sci. Am. 211 (October):47, 1964) found that treatment of the single-stranded RNA tobacco mosaic virus (TMV) with nitrous acid brought about the substitution of proline by leucine and serine by phenylalanine at several positions in the coat protein but never the reverse. Explain.
b) The *wild-type* coat protein of TMV contains proline at position 20. Treatment with nitrous acid produced mutant variants with amino acid substitutions at this position according to the scheme shown below:

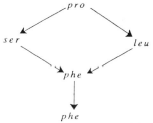

Treatment of the final *phe* mutant with nitrous acid cannot induce further amino acid substitutions at this position.
i) What are the codons for each of the amino acids shown? (Give correct base sequences.)
ii) Do these results indicate degeneracy? If so, for which of the codons shown?

36. Some rare (mutant) forms of human adult hemoglobin show single amino acid substitutions in the b chain, caused by different alleles of a single gene. Four of these types are shown.

Type of hemoglobin	Amino acid position	Change involved in amino acid substitution, *wild-type* → *mutant*
S	6	Glu → Val
C	6	Glu → Lys
E	26	Glu → Lys
G$_{San\ Jose}$	7	Glu → Gly

a) If the codons for lysine, valine, and glycine are AAG, GUG and GGG respectively, and a mutation involving a single base-pair change causes each of these rare forms, what is the codon for glutamic acid?

b) Which of these changes represent transitions and which represent transversions?

c) In the a chain, hemoglobins *I* and *Norfolk*, respectively, contain aspartic acid at positions 16 and 57 in the place of lysine and glycine at those positions.

i) What is the codon for aspartic acid?
ii) Did both changes occur via the same base-pair substitution mechanism?

37. The tRNA of *E. coli* contains two different methionine-accepting tRNAs, one (Met-tRNA$_F$) of which can have a hydrogen atom in the amino group of its methionine replaced by a formyl, CHO, group. The methionine in the other tRNA species (Met-tRNA$_M$) is, like the other amino acids on tRNAs, nonformylatable (Clark, B.F. and K.A. Marcker, J. Mol. Biol. 17:394, 1966). Of the trinucleotides tested for ability to bind amino acids to ribosomes they found the following.

1. Only AUG bound the methionine carried by Met-tRNA$_M$, whereas both AUG and GUG bound that carried by Met-tRNA$_F$, GUG also bound valine. Of the synthetic messengers tested for their ability to bring about synthesis of polypeptides containing methionine (in a cell-free *E. coli* system) only poly-UAG and poly-UG gave positive results (Clark, B.F. and K.A. Marcker, J. Mol. Biol. 17:394, 1966: Nature 211:378, 1966: Sci. Am. 218 (January):36, 1968).

2. Using Met-tRNA$_M$, methionine was incorporated only in the presence of poly-UAG, and only internally. Using Met-tRNA$_F$, methionine was incorporated into polypeptides with both synthetic messengers but only in the starting position; this was true regardless whether the methionine was formylated or not. Folmylation increased the rate of protein synthesis.

Note: Subsequent studies by Ghosh, H.P. et al. (J. Mol. Biol. 25:275, 1967) revealed that if poly-(UG)$_n$ with a repeating sequence or uracil and guanine was used as messenger, polypeptide chains were synthesized with the amino acid sequence F-Met-(Cys-Val)$_n$. Formylmethionine was in the N-terminal position only. The synthesis was dependent on the addition of Met-tRNA$_F$ (whether formylated or not). Poly-(AUG)$_n$ directed the synthesis of polymethionine, which required both Met-tRNA$_F$ and Met-tRNA$_M$.

a) With reasons for your answers state:
(1) Which is the initial codon or which are the initiating codons.
(2) The role of tRNA$_F$ and tRNA$_M$.
(3) The apparent role of the formyl group in protein synthesis.

b) Describe the system whereby polypeptide translation is initiated, including the role of codons and tRNAs, and suggest a role for formylation in polypeptide synthesis.

38. In 1966 Crick, F.H.C. (J. Mol. Biol. 19:548) suggested that during translation, while standard base pairing (A with U and G with C) occurs in the first 2 positions of the triplet, there may be some play (or wobble) in the pairing at the third base site so that a particular species of tRNA may pair with more than one codon; thus the base I at the 5' end of the anticodon can pair with the bases A, U, or C at the 3' end of a codon, G at the wobble position can pair with the U or C, while U can pair with A or G. Base C at the 5' end of the anticodon would pair with G and A only with U.

a) Show what codons can pair with the anticodons tRNA$_{Phe}$(GAA) and tRNA$_{Ala}$(IGC).

b) There are 2 kinds of serine tRNAs, tRNA$_{SerI}$(IGC) and tRNA$_{SerI}$(UGA). List the codons recognized by each of these tRNAs.

c) The anticodons of 6 tRNAs are tabulated. Predict the codons these tRNAs would pair with according to the wobble hypothesis.

Species and amino acid	Anticodon in tRNA 3' → 5'
Alanine	CGI
Valine	GAI
Phenylalanine	AAG
Tyrosine	AUG, AUC
Formylmethionine	UAC

d) Explain how one tRNA can recognize several codons provided they differ in the last base in the codon.

e) If an amino acid is coded for by all four bases in the third position, i.e., proline (CCU, CCC, CCA, CC), according to the wobble theory at least how many tRNAs must there be for this amino acid? *Note:* Codon-anticodon pairing occurs in an antiparallel fashion.

39. Garen, A. and O. Siddiqi (PNAS 48:1121, 1962) noted that *amber* mutants contain nonsense triplets within the gene causing premature polypeptide-chain termination at the residue specified by the nonsense codon. The same has been found to hold true for *ochre* mutants. Brenner, S. et al. (Nature 206:994, 1965) studied the production and reversion of rII *amber* and *ochre* mutants in phage T4 and also investigated head-protein *amber* mutants to define the amino acids connected to the *amber* triplet. Their experiments showed that:

1) No *amber* or *ochre* mutants reverted to *wild-type* when treated with hydroxylamine, NH_2OH, which reacts with cytosine and changes G-C base pairs to A-T pairs.
2) *Ochre* mutants can be converted into *amber* mutants by treating with 2-amino-purine, which causes (A-T ↔ G-C changes). *Ochre* mutants cannot be induced to mutate to *amber* with NH_2OH.
3) Both *amber* and *ochre* mutants can be induced by NH_2OH from *wild-type*.

These data indicated that the *amber* and *ochre* triplets (codons) in DNA have at least one A-T pair in common and that *amber* contains a G-C pair which corresponds to an A-T pair in *ochre*.

a) Analyze the data to show that this is so.
b) Is it possible to establish the orientation of these bases with respect to the sense and antisense strands of DNA? Why?

From previous studies it was known that rII genes express their functions before the onset of DNA synthesis. Therefore only changes in the sense strand of DNA will register a phenotypic effect in the first cycle of growth. Changes (mutations) in the antisense strand would not produce their mutant effects until the second cycle of growth. Brenner et al. treated *wild-type*, (r^+), T4 phage with NH_2OH and studied *amber* and *ochre* mutants in the rII region. Some of these mutants produced their effects before the first round of DNA replication. Others did not do so until after the first round of DNA replication.

c) These data permit a definite decision regarding the number of A-T pairs in the *amber* and *ochre* triplets and the orientation of these base pairs with respect to the sense and antisense strands of DNA.
(1) Analyze the data and give this information.
(2) Show the possible *amber* and *ochre* triplets in the mRNA.
Note: A(C → U) change in the sense strand will produce a defective mRNA before replication. Such a change in the antistrand requires one replication to effect a change in the mRNA.

d) *Amber* codons can arise from codons for glutamine or tryptophan which in mRNA are CAG and UGG respectively. What is the base composition of mRNA *amber* and *ochre* triplets? Explain.

40. *Phosphate-negative* nonsense mutants in *E. coli* are incapable of synthesizing functional alkaline phosphatase. Only fragments of the protein are formed. The size of the fragment depends on the position of the nonsense codon in the gene in each mutant. Weigert, M.G. and A. Garen (Nature 206:992, 1965) established the base sequence in one of the nonsense codons by comparing the amino acid substitutions in 21 revertant strains arising from the nonsense mutant *H12*. The revertants were due to induced mutations in the nonsense codon responsible for *H12*. In each of these revertants, tryptophan at a specific site in peptide 1 in

wild-type alkaline phosphatase was replaced by 1 of the following 6 amino acids: glutamine, glutamic acid, tyrosine, serine, leucine, or lysine. The RNA codons for tryptophan and these 6 amino acids are shown below.

Amino acid	RNA codon
Tryptophan	UGG
Glutamine	CAA, CAG
Glutamic acid	GAA, GAG
Tryosine	UAU, UAC
Serine	UCU, UCC, UCA, UCG, AGU, AGC
Leucine	UUA, UUG, CUU, CUA, CUC, CUG
Lysine	AAA, AAG

Thus the nonsense triplet must have arisen from the tryptophan triplet by a single-base change and must also, by a single-base change, provide one of the possible codons for each of these amino acids. Only one triplet satisfies all the conditions. Which is it and why?

41. Weigert, M.G. and A. Garen (J. Mol. Biol. 12:448, 1965) showed that three nonsense mutants, 12, 45, and G-5, at different sites in the alkaline phosphatase structural gene in *E. coli* could arise from either a codon for tryptophan or a codon for glutamine. All 3 mutants were suppressed by the same suppressor (*Su1*), which replaced tryptophan with serine in suppressed mutant 12, whereas in the other 2, *Su1* caused the substitution of serine for glutamine. Account for the fact that the original codon in which a mutation occurs to produce a nonsense codon can vary but the amino acid inserted by a specific suppressor gene, e.g., *Su1* is in all cases the same, e.g., serine.

42. In *wild-type E. coli* glycine occurs at position 210 in the A polypeptide of tryptophan synthetase. In the mutant *A46*, which does not grow on minimal medium, glycine is replaced by glutamic acid. A revertant of *A46*, designated *A46 PR8*, grows on minimal medium but much more slowly than *wild-type*. When this revertant is crossed with *wild-type*, some of the recombinants are *A46* and others *PR8* (do not grow on minimal medium). The cross T^+ $PR8^-$ $A46^+$ x T^- $PR8^+$ $A46^-$ produced 101 $PR8^+$ $A46^+$, of which 81 were T^- and 20 were T^+, and 85 $PR8^-$ $A46^-$, of which 20 were T^- and 65 were T^+. Note that T is to the left of the A gene, in which the 2 mutant sites PR and $A46$ occur. The amino acid sequences at the relevant positions in the *wild-type* A protein and in the 3 mutant A chains were found to be as shown. At all other positions the proteins were identical and *wild-type* in constitution (Helinski, D.R. and C. Yanofsky, J. Biol. Chem., 238:1043, 1963).

	Amino acid positions	
Strain	174	210
Wild-type	Tyr	Gly
A46	Tyr	Glu
A46 PR8	Cys	Glu
Pr8	Cys	Gly

Map the mutations responsible for *A46* and *PR8* and discuss the significance of the fact that the mutant *A46 PR8* is phenotypically *pseudo wild-type*.

43. a) J.D. Smith et al. (Cold Spring Harbor Symp. Quant. Biol. 31:479, 1966), using a transducing phage φ80 carrying the *E. coli* nonsense suppressor $Su3^+$, found that when the phage and its suppressor replicate, an increased synthesis of tyrosine tRNA occurs. This tyrosine tRNA recognizes UAG and appears not to recognize the 2 normal tyrosine codons UAU and UAC. Replication of a φ80 strain carrying the *wild-type* $Su3^-$ allele leads to a similar increase of tyrosine tRNA, which recognizes the normal tyrosine codons UAU and UAC but not UAG. What do these results suggest regarding the function of the *Su* gene?
b) Andoh, T. and H. Ozeko (PNAS 59:792, 1968) found, using phage φ80 $Su3^+$, that the DNA from this phage hybridizes with the tyrosine tRNA from both the Su^+ and Su^- species of *E. coli*. One segment per DNA molecule hybridizes with one tRNA molecule. Do these results confirm your answer in (a)? Explain.
c) What part of the tRNA might mutation of $Su3^-$ to $Su3^+$ change? Explain.

44. Zipser, D. and A. Newton (J. Mol. Biol. 25:567, 1967) demonstrated that:

1) The nonsense mutant (*NG813*) in the z (β - galactosidase) gene of *E. coli* is due to a single-base change in a *wild-type* codon. This codon also mutates to form the *amber* (UAG) codon in the mutant *NG1012*.

2) *NG813* is convertible to UAA (*ochre*) by a single-base change. The possible codons in *NG813* that can be changed by single-mutational steps to the *ochre* and the *amber* codons are shown in the table below:

Possible mutant codons which are one mutational step away from *ochre* (UAA)	Possible *wild-type* codons which are one mutational step away from *amber* (UAG)
UAU (Tyr)	UAC (Tyr)
UAC (Tyr)	UAU (Tyr)
UUA (Leu)	UUG (Leu)
UCA (Ser)	UCG (Ser)
UGA (--)	UGG (Trp)
AAA (Lys)	AAG (Lys)
CAA (Gln)	CAG (Gln)
GAA (Glu)	GAG (Glu)

a) Which is the nonsense codon in *NG813*? Explain your decision.

b) Which codon in the *wild-type* z gene can mutate to form both nonsense mutants *NG813* and *NG1012*? Explain.

45. The head proteins produced by *wild-type* T4 phage and by the mutant H36 on 2 different *E. coli* strains have the N-terminal amino acid sequences shown (Stretton, A.O.W. and S. Brenner, J. Mol. Biol. 12:456, 1965).

Strain	N-terminal amino acid sequences
Wild-type	Ala-Gly-Val-Phe-Asp-Phe-<u>Gln</u>-Asp-Pro-Ile-Asp-Ile-Arg...
H36 on *E. coli* strain 1	Ala-Gly-Val-Phe-Asp-Phe
H36 on *E. coli* strain 2	Ala-Gly-Val-Phe-Asp-Phe and
	Ala-Gly-Val-Phe-Asp-Phe-<u>Ser</u>-Asp-Pro-Ile-Asp-Ile-Arg...

a) Was the mutation from *wild-type* to *H36* a substitution, deletion, or addition of a base pair?

b) Why does *H36* produce an incomplete protein on strain 1 of *E. coli* and both complete and incomplete types on strain 2?

c) Why is glutamine replaced by serine in the *H36* protein on strain 2?

d) Would you expect the phenotype of *H36* on strain 2 to be *mutant*, *wild-type* or *pseudo wild-type*? Explain.

46. a) Nonsense mutations cause premature termination of polypeptide chain growth. It is known that addition (+) or deletion (-) of a base pair in the B1 segment of the B gene in the rII region of phage T4 results in a mutant phenotype (no B activity). If both an addition and a deletion are present in this region, a *pseudo-wild-type* phenotype is restored. Brenner, S. and A.O.W. Stretton (J. Mol. Biol. 13:944, 1965) found that double and triple mutants had the phenotypes shown below, when grown on *E. coli* K12 Su^- (a strain on which nonsense mutants do not grow). Do nonsense mutants exert their effects at the level of translation or at the level of transcription? Explain.

Type	Mutant	Phenotype
double	(*amber*, +) or (*amber*, -)	no growth
double	(*ochre*, +) or (*ochre*, -)	no growth
triple	(+, *amber*, -)	near *wild-type* growth
triple	(+, *ochre*, -)	near *wild-type* growth

b) Although most double (+ -) mutants in the B1 segment grow on *E. coli* K12 Su^-, one (*FC73, FC23*) does not. It does, however, grow on *E. coli* K12. Offer an explanation for these results.

Chapter 23
Development and Regulation

1. Theoretically all cells in a multicellular organism have the same genotype. If so, why do different cells differentiate at different times?

   According to a certain hypothesis, the chromosomes of the egg contain determiners for the eyes, ears, and other body organs, and as cell division takes place, the determiners for each organ are sorted out. Explain how the chromosomes divide at cell division and why this method of division of the fertilized egg cannot account for this assortment.

2. a) List and discuss the criteria that might be used to distinguish V-type position effects from gene mutations.
   b) What is the *spreading effect* associated with V-type position effects? Advance an hypothesis that could explain it.

3. a) The *tortoise-shell* pattern appears only in female cats and rare males of chromosome constitution XXY. Why?
   b) Would you expect incompletely sex-linked genes like *bobbed, bb,* in *Drosophila* to show dosage compensation? Explain.
   c) What is the basis for the deduction that the heteropycnotic (condensed) X rather than the isopycnotic one is functionally inactive in *normal* (XX) female mammals?
   d) Why do heterozygous carriers of X-linked recessive disorders exhibit greater phenotypic variability than heterozygous carriers of autosomal recessive disorders?

4. Ohno, S. and B.M. Cattanach (Cytogenetics 1:129, 1962) showed that in mice, a segment of an autosome inserted into the X behaved cytologically and functionally like an integral part of the X. Would you expect a piece of an X translocated to an autosome to lose its ambivalent nature and behave like an integral part of the autosome? Explain.

5. In male mammals the single X in somatic cells always behaves euchromatically; in female mammals one X remains euchromatic, the other manifests positive heteropycnosis (is heterochromatic) along most of its length. Why does this difference in the cytological behavior of the X chromosomes exist between the two sexes? What is the functional consequence of this behavior?

6. Compare and contrast the dosage-compensation mechanism in mammals with that in *Drosophila*.

7. Histones suppress gene action (DNA-dependent RNA synthesis) in eukaryote chromosomes. Suggest a possible mechanism.

8. When lactose is added to a growing culture of *Escherichia coli*, the cells begin to make enzymes necessary for lactose utilization. In contrast, when tryptophan is added to the culture medium, synthesis of the enzymes of the tryptophan pathway ceases. Contrast the 2 systems, explaining their opposite modes of action.

9. a) Compare and contrast unregulated (constitutive) and regulated (inducible or repressible) enzyme synthesis.
   b) Why would one expect mutations in a promoter to alter the potential for operon expression? Can mutations in the operator affect the rate of operon expression?
   c) Gilbert, W. and B. Müller-Hill (PNAS 56:1891, 1966) isolated a protein that binds to isopropyl thiogalactoside (IPTG), which, like lactose, is an inducer of the *lac* operon. There was no binding between IPTG and the protein isolated from either i^- (deletion) or some i^- (revertible) mutants. Moreover, differences in binding of the protein isolated from *wild-type* and certain i^- mutants and IPTG were also detected. What is the significance of these facts?
   d) What role does feedback play in the molecular control of genetic activity?

10. *Tfm (testicular feminization)* vs. *tfm (normal sexual development)* is an X-linked allele pair in mice. Testosterone, when administered *in vivo* to *tfm/tfm* female or castrated *tfm*/Y male mice, elicits the following responses:

    i) An immediate translation of pre-existing kidney mRNAs.

ii) A hundredfold increase in kidney alcohol dehydrogenase and glucuronidase.
iii) A very marked increase in the transcription of kidney rRNA genes.
iv) An increase in an RNA polymerase, tentatively identified as RNA polymerase I, which is specific for the transcription of rRNA genes.

Tfm/Y mice showed neither response 2 nor 3 and gave no evidence of response 1 or 4 (Ohno, S. "Sex Chromosomes and Sex-linked Genes", Springer-Verlag, New York, 1971). Compare this type of control mechanism with that of the *lac* operon.

11. *Hen feathering* vs. *cock feathering* in fowl (males ZZ, females ZW) is determined by a single pair of autosomal alleles. HH and Hh males are *hen-feathered*; hh males are *cock-feathered*. Females of any genotype are *hen-feathered*. In Sebright Bantams both sexes are *hen-feathered*. In White Leghorns males are *cock-feathered* and females are *hen-feathered*. Work done in the 1920s and 1930s, significant in elucidating the determination of these traits, may be summarized as follow:

i) Removal of testes in *hen-feathered* males, H-, or of the ovary in females (regardless of genotype) results in *cock-feathering* (Roxas, H.A., J. Exp. Zool. 46:63, 1926; Eliot, T.S., Physiol. Zool. 1:286, 1928).
ii) A castrated Sebright Bantam cock, HH, that had developed *cock-feathering* became *hen-feathered* again after implantation of Leghorn testes, hh. In the reciprocal transplantation, the castrated Leghorn male continued to express *cock-feathering* (Roxas, H.A., J. Exp. Zool. 46:63, 1926).
iii) Danforth, C.H. (Biol. Gen. 6:99, 1930) showed that transplantation of potentially *cock-feathered* skin from an hh male to a female of any genotype, H- or hh, results in *hen-feathering*. Transplantation of skin from hh females to *cock-feathered*, hh males results in *cock-feathering*. Transplants of skin from *hen-feathered*, H-, males and females to *cock-feathered* males, hh, continue to develop *hen-feathers*. Finally, transplants of skin from females to males in breeds in which both sexes are *hen-feathered* do not alter feathering type. Note that within breeds like the Leghorn, transplants acquire the trait of their host.
a) Which is the organizing tissue and which the target tissue in the expression of feather type? Explain.
b) On what are the sexual differences in feathering based? Explain.
c) Is the initial effect of the H and h alleles in the organizing or target tissue? Explain.

12. In *Drosophila melanogaster*, the recessive X-linked allele *fu* for fused vein locus has pleiotropic effects causing partial sterility and ovarian tumors in *fufu* females and a reduced number of ocelli in both *fufu* females and *fu/Y* males. *Fufu* females produce offspring of the expected genotypes and in the expected proportions:

Fufu ♀ x *fu/Y* ♂ → 1 *fused* ♀ : 1 *nonfused* ♀ : 1 *fused* ♂ : 1 *nonfused* ♂

Fused (*fufu*) females x Fu/Y males produce females only; the male progeny die in early embryogenesis. Embryonic development is normal in all viable progeny. The cross *fufu* females x *fu/Y* males is sterile; no progeny are produced (Lynch, C.J., Genetics 4:501, 1919; Counce, S.J., Z. Vererbungsl. 87:462, 1956). Zygotes from the latter cross develop normally for 5 hours after fertilization. Thereafter development in the germ layers of the ovary is aberrant and leads to the pleiotropic effects mentioned above. These embryos do not hatch but remain alive far beyond the hatching time of *wild-type* embryos (Counce, S.J., Z. Vererbungsl. 87:462, 1956). When *fufu* ovaries are transplanted into a genetically normal environment, they behave autonomously, the eggs being as inviable as they are in *fufu* mothers (Clancy, C.W. and G.W. Beadle, Biol. Bull. 72:47, 1937).
a) Why do these results indicate that Fu and not *fu* is required for viability?
b) When does the Fu allele act (produce Fu substance)? Explain.
c) What is the function of the Fu substance, and where is it probably produced and stored?

13. In domestic fowl, the mutant allele F (*frizzle*) of an autosomal gene changes the structure and morphology of the feathers; the change being mild (curly feathers) in

heterozygotes, *Ff*, and extreme in homozygotes, *FF*. In *extreme-frizzle* birds the defective feathers are curly, brittle and easily broken, leading to near nakedness. Various abnormalities result from this primary aberration. Reduced insulation results in rapid heat loss, lowers body temperature, and impairs the ability to adjust to changing environmental temperatures. Consequently the metabolic system of these birds is disturbed causing many abnormalities. Krimm, S. (J. Mol. Biol. 2:247, 1960) showed that feather keratin (a protein) in *FF* birds is much more poorly organized than that of *normal, ff*, birds and that the amino acid content of the mutant protein differs from that of normal keratin. Outline in chronological sequence the cause-effect relationship between the DNA, keratin, and the deleterious pleiotropic effects that occur in *extreme-frizzle* fowl.

14. In *Drosophila melanogaster*, the recessive allele *lozenge-clawless*, lz^{cl}, at the lz^+ autosomal locus causes many abnormalities of the eyes, tarsals, and female genitalia (Anders, G., Rev. Suisse Zool. 54:269, 1947). *Lozenge-clawless* females lack sperm-storage organs and ovarian glands and are sterile. Mutant, $lz^{cl}lz^{cl}$, males have normal genitalia and are fertile. Anders, G., Z. Vererbungsl 87:113, 1955) found that XX flies homozygous for both lz^{cl} and *tra* (transforms XX flies into phenotypic males) were phenotypically male with normal male genitalia. Which of the genes, lz^{cl} or *tra*, acts first in development? Explain.

15. Somatic nuclei of *wild-type Chironomus* (2n) have nucleoli at corresponding positions on a pair of homologues. Beermann, W. (Chromosoma, 11:263, 1960) established a balanced lethal strain carrying only one nucleolus per somatic nucleus. This strain always produced about 25 percent progeny which lacked nucleoli. These flies develop normally to the gastrula stage, when aberrant features appear, followed by death. Offer an hypothesis, in molecular terms, which will account for these results.

16. In the squash (*Cucurbita pepo*) 2 pairs of alleles affect fruit shape. Plants that are *A-B-* are *disk* shaped (very wide), those *A-bb* or *aaB-* are *spherical* (width equals length), while *aabb* are *elongate*, (Sinnott, E.W., Am. Nat. 61:333, 1927). If growth is entirely dependent on cell division, suggest how these genes might determine shape.

17. Pea (*Pisum sativum*) cotyledons synthesize a specific pea-seed reserve globulin, which is not produced in other pea tissues, e.g., flowers and buds. Chromatin isolated from cotyledons and buds in an *in vitro* system produced the tabulated results (Bonner, J. et al., PNAS 50:893, 1963).

Source of chromatin	Pea-seed reserve globulin
Pea cotyledon	Synthesized
Pea buds	Not synthesized
Pea bud chromatin minus the histone	Synthesized

a) What substance is involved in repression of gene activity?
b) How may it function in gene-action suppression?

18. a) In the mouse (females XX, males XY) the *b* locus (*B* = *wild-type* and *b* = *brown* coat color) is on chromosome 8. A *variegated* mutant female arose among the offspring of irradiated *wild-type* males mated to *brown* females. Results of matings involving this female and her offspring are shown in the Table below (Russell, L.B. and J.W. Bangham, Genetics, 46:509, 1961):

Cross	Generation	Variegated ♂	Variegated ♀	Wild-type ♂	Wild-type ♀	Brown ♂	Brown ♀
1. *Variegated* mutant ♀ x *brown* ♂	F_1	0	9	8	1	5	3
2. *Brown* F_1 ♀ x *brown* F_1 ♂	F_2	0	0	0	0	1/2	1/2
3. *Wild-type* F_1 ♀ x *brown* F_1 ♂	F_2	0	0	1/4	1/4	1/4	1/4
4. *Variegated* F_1 ♀ x *brown* F_1 ♂	F_2	0	275	278*	45†	336*†	352§

*Most sterile. † Most fertile. *† Most fertile, few sterile. § Most ferile, few partially sterile

Note: (1) All *variegated* animals are *partially sterile* females. They have small *brown* areas irregularly interspersed with *wild-type* ones. The size of their litters ranges from 3.0 to 4.5; that of *fully fertile* females ranges from 5.7 to 11.0. (2) Matings among F_2's involving crosses identical to crosses 2, 3 and 4 among F_1's gave results similar to those in crosses 2, 3 and 4 respectively.

(1) Suggest an hypothesis to account for these facts.
(2) Indicate the cytological observations that would confirm your hypothesis.
b) Matings between *variegated* females and *brown* males produced some *partially sterile, brown* females which in crosses with *wild-type* males produced *partially sterile* females, all *wild-type*. Such females in crosses with *brown* males produced some *variegated* females, all *partially sterile*. Show whether these results agree with your hypothesis.
c) The allele pair *Ta ta* is X-linked. *TaTa* or *Ta/Y* show a *tabby* coat, *tata* or *ta/Y* show *wild-type* and *Tata* have a mosaic coat for the two patterns (called *heterozygous tabby*). Brown variegated, partially sterile, heterozygous-tabby females were produced from matings between the *brown variegated, partially sterile, wild-type* females and *bb/Ta/Y* males. When these females were test-mated with *brown, tabby* (*bb Ta/Y*) males, they produced the progeny shown below.

	Tabby genotypes			
	♀		♂	
Coat color	TaTa	Tata	Ta/Y	ta/Y
Brown	78	14	51	16
Variegated	0	68	0	0
Wild-type	11	0	16	57

Do these data permit a definite decision as to whether your hypothesis is correct or not. Explain.
d) Outline experiments to test your hypothesis. See Russell, L.B. and C.S. Montgomery (Genetics, 63:103, 1969) for the correct interpretation of these and similar results.

19. a) The alleles p (*pink* eye) and c^{ch} (*chinchilla* coat color) at the P and C loci in the mouse, *Mus musculus*, are located in the central segment of chromosome 1, an autosome. From the mating of $ppc^{ch}c^{ch}$ females with *wild-type*, $PPCC$, males treated with triethylenemelamine, Cattanach, B.M. (Z. Vererbungsl. 92:165, 1961) obtained one female with a *variegated* phenotype in which the central segment of chromosome 1 (about one-third of the autosome) carrying the *wild-type* alleles P and C was inserted into the X as shown:

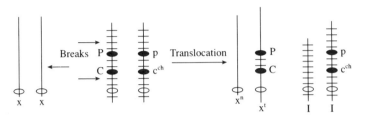

In animals with the two chromosome 1s after translocation:
i) X^tX^n females and X^tX^nY males expressed a *variegated* phenotype (patches of *wild-type*) and *white* (pc^{ch}) coat color.
ii) X^tY males and X^tO females express a *wild-type* coat color.

Explain whether these results are due to V-type position effect or dosage compensation.

b) Ohno, S. and B.M. Cattanach (Cytogenetics 1:129, 1962) showed that in *variegated* X^tX^n females and X^tX^nY males the somatic cells in *wild-type* patches contained a condensed (tightly coiled) X^n and a normal (isopycnotic) X^t (behaving in the same manner as the euchromatic autosomes). pc^{ch} patches contain cells with a condensed X^t and a normal X^n. No condensed chromosomes were found in somatic cells of X^tY and X^tO animals. The normal X always behaves isopycnotically in somatic cells.
Note: X^n = normal chromosome; X^t = translocated chromosome.
Are these results in accord with your explanation in (a)? Discuss briefly.

20. In *Drosophila melanogaster*, the genes Y (*gray* body), W (*dull red* eyes), and Spl (*normal* wing) are normally located near the tip of the euchromatic left arm of the X chromosome. Müller, H.J. (J. Genet. 22:299, 1930), and

Judd, B.H. (Genetics 40:739, 1955) obtained reciprocal translocations in which a segment containing these genes was moved to a heterochromatic region near the centromere of the right arm of chromosome 4, as shown below:

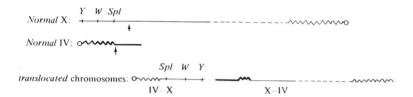

In translocation heterozygotes of the type shown in the following diagram a *variegated* phenotype was expressed for eye color (all flies had *white* eyes with *dull-red* speckles) and wing type (some flies had a *normal* and others a *split* wing):

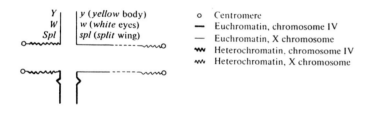

Matings between such females and *y w spl*/Y males produced *yy ww Spl Spl* females from which a homozygous translocation strain true-breeding for *y w Spl* was obtained. Such females, crossed with *y W Spl* males, produced *yy Ww Spl Spl* flies with *wild-type* eyes. These flies were all translocation heterozygotes with breaks at the same points as in the original 4-X translocation. The same results were obtained when w^a (*apricot* eyes) was used in place of *W*; *w* is an amorph (nonactive).

a) Explain why the *mottled* phenotype is not due to gene mutation and indicate how heterochromatin may be related to such *variegated* phenotypes.
b) Offer an explanation for the fact that in the first translocation heterozygote variegation was expressed for the traits controlled by *Splspl* and *Ww* but not for those specified by *Yy*.

21. Lyon, M.F. (Am. J. Hum. Genet. 14:135, 1962) reported on mosaic F_1 mice doubly heterozygous in the repulsion phase for mutant alleles of 2 X genes, one affecting coat color and the other hair structure. Patches of fur were either *mutant* for one character and *wild-type* for the other, or *vice versa*, and never *mutant* or *wild-type* for both. State whether or not this supports the Lyon hypothesis and why.

22. The Duchenne type of *muscular dystrophy*, a rare human trait with onset usually before the seventh year of life and death by age fifteen, can be detected by increased levels of creatine phosphokinase (CPK) and histological changes in the muscles. The test results of Pearson, C.M. et al. (PNAS 50:24, 1963) are presented in the pedigree.

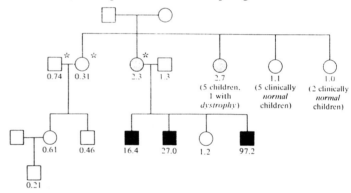

a) Why does this form of *muscular dystrophy* occur almost exclusively in boys?
b) On the basis of one gene with 2 alleles, suggest genotypes and the dominance relationship of the alleles in individuals indicated by an asterisk.
c) A few females have been reported to be clinically *affected*. Among subclinically *affected* females dystrophic muscle fibers range from few to many, and clinical symptoms appear only after about 50 percent of functional muscle is lost. What does this suggest regarding the

dynamic relationship between the two kinds of X's and gene inactivation according to the Lyon hypothesis?

23. In humans, the gene specifying the synthesis of glucose-6-phosphate dehydrogenase (G6PD) is on the X chromosome; the allele G specifies normal production and g no synthesis of the enzyme. Beutler, E. et al. (PNAS 48:9, 1962) found that individually tested red blood cells of *normal* males, $X^G Y$, and *normal* females, $X^G X^G$, show the same amount of G6PD activity. Some cells of heterozygous females, $X^G X^g$, were normal and others deficient for G6PD, but none showed intermediate activity. $X^g Y$, and $X^g X^g$ individuals, showed no enzyme activity. *In vitro* studies by Davidson, R.G. et al. (PNAS 50:481, 1963) of clones from single skin cells of $X^G X^g$ females also showed that some had *normal* and others little or no enzyme activity.
a) i) Why do erythrocytes of *normal* females not produce twice as much enzyme as those of *normal* males?
ii) Why do heterozygous, $X^G X^g$, females produce half the amount produced by *normal* males?
b) XO, XY, XX, XXX, XXXY and XXXX cells that are diploid for all autosomes and carry only the normal allele G all synthesize the same amount of G6PD. Relate this to your hypothesis in (a).

24. Epstein, C.J. (Science 163:1078, 1969) measured the activity of G6PD and LDH enzymes in the oocytes of XO and XX female mice. He found that LDH activity in both types of oocytes was the same but that the G6PD activity in XX oocytes was double that in XO oocytes. What is the probable cause of the difference in the activity of these two enzymes?

25. In kangaroos (females XX; males XY) one of the two X chromosomes in females is late-replicating, genetically inactive, and heterochromatic. The $X(X_e)$ of euros (*Macropus robustus erubescens*), which possess a *slow* moving electrophoretic form (S) of glucose-6-phosphate dehydrogenase (G6PD), is about 1 1/2 times larger than the $X(X_w)$ of wallaroos (*M. r. robustus*), which produce a *fast*-moving (F) G6PD. The karyotypes of the 2 subspecies, which interbreed freely and produce viable offspring, are otherwise identical. In F_1 females from euro females x wallaroo male, the X_w always replicates later than X_e. The reverse is true in F_1 females from reciprocal crosses (Sharman, G.B., Nature 230:231, 1971). The difference between *fast* and *slow* electrophoretic variants of G6PD is due to different alleles of an X-linked gene. Reciprocal crosses between euros and wallaroos produced the F_1 results shown (Richardson, B.J. et al., Nat. New Biol. 230:154, 1971).

Parents		F_1	
♀	♂	♀	♂
1. Euro (S)	Wallaroo (F)	S	S
2. Wallaroo (F)	Euro (S)	F	F

Is the mode of dosage compensation for X chromosomes in kangaroos the same as in eutherian mammals (random X inactivation)? If not, indicate how eutherian X inactivation may have evolved from the marsupial type of X inactivation.

26. The female mule has one $X(X_H)$ from her horse mother and the other $X(X_D)$ from her donkey father. The two X's are cytologically different. Giannelli, F. et al. (Heredity 24:175, 1969) found that the late-replicating X in 90 percent of somatic cells studied is X_D. However, electrophoretic studies of the G6PD enzyme (X-linked) in these females indicate that almost all of it is of the horse type. Is the X which replicates late and forms the sex-chromatin body the genetically inactive one?

27. Which of the following *E. coli* genotypes would be able to metabolize lactose? Do they do so constitutively or inducibly?

a) $i^+ p^+ o^+ z^+ / i^+ p^+ o^c z^+$
b) $i^- p^- o^+ z^+ / i^+ p^+ o^+ z^-$
c) $i^+ p^+ o^+ z^+ / i^+ p^- o^c z^+$
d) $i^+ p^+ o^+ z^+ / i^+ p^+ o^c z^-$
e) $i^- p^+ o^+ z^+ / i^+ p^- o^c z^-$
f) $i^+ p^+ o^+ z^- / i^+ p^- o^c z^+$

28. The lactose operon in *E. coli* consists of the regulator gene, i, the operator, o, and three structural genes z (makes β-galactosidase, an intracellular enzyme which hydrolyzes lactose into glucose and galactose), y (specifies permease, which aids in getting lactose into the cell), and a (specifies

acetylase, required for lactose utilization). The sequence of the genes in the lactose operon is shown.

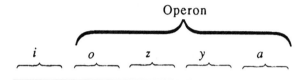

Jacob, F. et al. (C.R. Acad. Sci. 250:1727, 1960) and Jacob, F. and J. Monod (Cold Spring Harbor Symp. Quant. Biol. 26:193, 1961) found the functions of the alleles at the first 4 loci to be as shown in the table below:

Gene type		Function
Regulator		
	i^+	*Wild-type* allele for a diffusible repressor (protein) which inhibits synthesis of proteins specified by the structural genes by binding with o^+ operator DNA; lactose inactivates *wild-type* repressor
	i^-	Constitutive allele which results in no or an inactive form of repressor
	i^s	Superrepressor allele; results in a modified repressor protein which is unable to combine with the inducer lactose; the protein, being unbound, can bind to o^+ DNA and inactivate it
	i^d	Constitutive allele, inactivates i^+ product by aggregation
	i^q	Produces more repressor than i^+
Operator		
	o^+	*Wild-type* operator DNA, which turns on or permits synthesis of mRNA of structural genes; this operator is sensitive to, and inactivated by, i^+ and i^s repressors, permitting permanent synthesis of operon mRNA
	o^c	Constitutive allele which causes operator DNA to be insensitive to i^+ and i^s repressors, permitting permanent synthesis of operon mRNA
β-Galactosidase		
	z^+	Specifies synthesis of *wild-type* β-galactosidase
	z^-	Determines synthesis of no β-galactosidase or its mutant form, Cz (assume former)
Permease		
	y^+	Specifies *wild-type* permease
	y^-	Determines no or defective permease (assume former)

a) What is the expected order of dominance of the alleles at the regulator locus? At the operator locus? Explain.

b) Jacob, F. and J. Monod (Cold Spring Harbor Symp. Quant. Biol. 26:193, 1961) showed that the phenotypes of some of the haploid genotypes were those given in the following table:

	Inducer present		Inducer absent	
Genotype	β-galactosidase	Permease	β-galactosidase	Permease
1. $i^+o^+z^+y^+$	+	+	−	−
2. $i^+o^+z^-y^+$	−	+	−	−
3. $i^+o^+z^+y^-$	+	−	−	−
4. $i^-o^+z^+y^+$	+	+	+	+
5. $i^so^+z^+y^+$	−	−	−	−
6. $i^+o^cz^+y^+$	+	+	+	+
7. $i^so^cz^+y^+$	+	+	+	+
8. $i^+o^cz^+y^-$	+	−	+	−

Key: + = synthesis; − = no synthesis of enzyme

Do the phenotypes of these genotypes verify the functions of the alleles at the i and o loci? Indicate the genotypes that verify a particular function.

c) Phenotypes of various diploid genotypes were determined by the above authors to be those shown below.

	Inducer present		Inducer absent	
Genotype	β-galactosidase	Permease	β-galactosidase	Permease
1. $\dfrac{+\ +\ +\ +\ +}{i\ o\ z\ y}/\dfrac{+\ +\ -\ -}{i\ o\ z\ y}$	+	+	−	−
2. $\dfrac{-\ +\ +\ +}{i\ o\ z\ y}/\dfrac{+\ +\ -\ -}{i\ o\ z\ y}$	+	+	−	−
3. $\dfrac{-\ +\ -\ +}{i\ o\ z\ y}/\dfrac{+\ +\ +\ +}{i\ o\ z\ y}$	+	+	−	−
4. $\dfrac{+\ +\ +\ +}{i\ o\ z\ y}/\dfrac{+\ +\ -\ -}{i\ o\ z\ y}$	+	+	−	−
5. $\dfrac{s\ +\ +\ +}{i\ o\ z\ y}/\dfrac{+\ +\ +\ +}{i\ o\ z\ y}$	−	−	−	−
6. $\dfrac{s\ +\ +\ -}{i\ o\ z\ y}/\dfrac{+\ +\ +\ +}{i\ o\ z\ y}$	−	−	−	−
7. $\dfrac{+\ c\ -\ +}{i\ o\ z\ y}/\dfrac{+\ +\ +\ -}{i\ o\ z\ y}$	+	+	−	+
8. $\dfrac{+\ c\ +\ -}{i\ o\ z\ y}/\dfrac{+\ +\ -\ +}{i\ o\ z\ y}$	+	+	+	−
9. $\dfrac{+\ +\ -\ c\ -\ +}{i\ o\ z\ y}/\dfrac{+}{i\ o\ z\ y}$	+	+	−	+
10. $\dfrac{-\ c\ +\ -\ +\ -\ +}{i\ o\ z\ y}/\dfrac{+}{i\ o\ z\ y}$	+	+	+	−
11. $\dfrac{s\ +\ +\ +\ c\ +\ +}{i\ o\ z\ y}/\dfrac{+}{i\ o\ z\ y}$	+	+	+	+

i) Determine the dominance relationship between z^+ and z^-; y^+ and y^-.

ii) Are the allelic relationships at the i locus outlined in (a) verified by the phenotypes of genotypes 2 to 6? Answer the same question for the relationships at the operator locus by analyzing the results of genotypes 7 to 11.

iii) Does the cis-trans position of the alleles at the regulator locus affect their functions? Is the same true of the alleles at the operator locus? Explain.

29. The linked structural genes in the arabinose operon in *E. coli* are shown, along with a contiguously linked region marked X and a gene C. The enzymes specified by the three structural genes and the reactions the enzymes control are also indicated. The unlinked gene E specifies L-arabinose permease, which is concerned with active transport of arabinose.

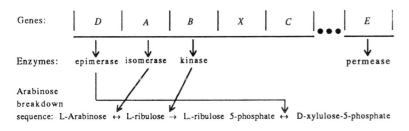

Englesberg, E. et al. (J. Bacteriol. 90:946, 1965) and Sheppard, D.E. and Englesberg, E. (J. Mol. Biol. 25:443, 1967) obtained the results given.

i)

Genotype	Arabinose present	Arabinose absent
$E^+D^+A^+B^+C^+$	All four enzymes produced to same extent	Deficient in all four enzymes
$E^+D^+A^+B^+C^c$		
$E^+D^+A^+B^+C^-$	Deficient in all four enzymes	

ii) Merodiploids are shown below for isomerase only. Essentially the same results have been obtained for enzymes specified by the other three genes.

		Enzyme activity, units	
Genotype	Arabinose	Diploid	Control*
A^-C^+/A^+C^-	Present	3.4	0.7
A^-C^c/A^+C^-	Absent	1.8	0.06
A^-C^c/A^+C^+	Absent	0.1	0.1

iii) Deletion of C results in no permease synthesis.

a) Is C an operator or regulator? Explain, indicating whether C functions negatively or positively.

b) What is the allelic relationship among C^+, C^c and C^-? Explain.

c) Indicate whether you expect both isomerase and kinase synthesis and activity in the following merodiploids when arabinose is and is not present:

i) $F'A^+B^+C^-/A^-B^+C^+$ ii) $F'A^-B^+C^-/A^+B^+C^-$
iii) $F'A^+B^+C^-/A^-B^+C^c$ iv) $F'A^+B^-C^+/A^-B^+C^c$

d) Deletion 709 includes gene C, the region between C and B, and part of B. Induced isomerase and kinase enzyme levels in this mutant in a haploid and merodiploid were as shown below (Sheppard, D.E. and E. Englesberg, J. Mol. Biol. 25:443, 1967):

	Isomerase	Kinase
Deletion 709 (D^-A^+)	≤ 1	0.1
$F'A^-B^+C^+/D^-A^+$ deletion	≤ 1	16.2

What is the probable function of the region X? Why?

30. In *E. coli*, the regulator gene is closely linked to an operon consisting of 2 structural genes (consider only one) and an operator. In the table the regulator, operator and structural genes are listed in correct sequence in each genotype. The phenotypes of the genotypes under induced and noninduced conditions are as shown. Which of the 3 genes is the regulator? The operator? The structural gene? Explain.

| | Phenotype | |
Genotype	Inducer absent	Inducer present
$a^-b^+c^+$	S	S
$a^+b^+c^-$	S	S
$a^+b^-c^-$	s	s
$a^+b^-c^+/a^-b^+c^-$	S	S
$a^+b^+c^+/a^-b^-c^-$	s	S
$a^+b^+c^-/a^-b^-c^+$	s	S
$a^-b^+c^+/a^+b^-c^-$	S	S

Key: S = enzyme synthesized in normal quantities; s = little or no synthesis; + = *normal* allele; - = *mutant* allele

31. The structural genes of the tryptophan operon of *Escherichia coli*, the enzymes they specify, and the reactions they control are as follows:

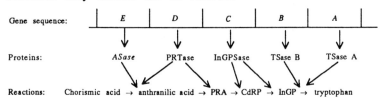

where: ASase = anthranilate synthetase
PRTase = phosphoribosyl anthranilate transferase
InGPSase = indoleglycerolphosphate synthetase
TSase = tryptophan synthetase
InGP = indole-3-glycerol phosphate
PRA = *N*-(5'-phosphoribosyl) anthranilate
CdRP = 1-(*O*-carboxyphenylamino)-1-deoxyribulose-5-phosphate

a) Deletions at the *E* end lead to nonrepressible transcription of *trp* genes. At which end of the operon is the operator located, and at which end of the operon would you expect mRNA synthesis to begin?
b) The formation of the tryptophan enzymes is coordinate. Nonsense mutations 1 and 2 occur in gene *E* (with 2 closer to *D* than 1) and nonsense mutation 3 in gene *B*.
i) On which side would you expect these mutations to reduce the relative rates of synthesis of the genes, the side proximal or distal to the operator?
ii) Which of the nonsense mutations, 1 or 2, would have a more drastic effect on enzyme synthesis? Answer the same question for mutations 1 and 3.

32. The *R1* and *R2* regulator genes control the formation of a repressor for alkaline phosphatase. Garen, A. and S. Garen (J. Mol. Biol. 6:433, 1963) found that certain constitutive mutations in these regulator genes respond to an external suppressor. Why do these results suggest that the products of the regulator genes (repressor for alkaline phosphatase) are protein molecules?

33. In *Salmonella typhimurium* the structural genes specifying the synthesis of the ten enzymes in the histidine biosynthetic pathway are clustered (in juxtaposition) in one region of the genome (Loper, J.C. et al., Brookhaven Symp. Biol. 17:15, 1964). Synthesis of all these enzymes is repressed simultaneously and to the same extent by excess histidine. Most mutations in this region result in the modification of structure and function of only one enzyme. However, mutations that occur at one end of this region result in nonsynthesis of all ten enzymes, although it is known that the structural genes for at least nine of them are *wild-type*. Why should the first kind of mutation cause structural alteration or loss of only one enzyme whereas the latter affects the synthesis of all enzymes?

34. Under a variety of conditions the synthesis of the tryptophan biosynthetic enzymes in *wild-type E. coli* is coordinate (see Problem 31) for enzymes specified by genes in the tryptophan operon. Yanofsky, C. and J. Ito (J. Mol. Biol. 21:313, 1966) found that enzyme production by 2 groups of mutants under repressed conditions were as shown (R^- is a regulator mutation producing a repressor which is only partially efficient and therefore permits elevated production of tryptophan enzymes under repression conditions). These workers also showed that within a gene, mutations in the second group proximal to the operator (*E-gene*) end had a more pronounced effect on protein synthesis specified by more distal genes in the operon than changes toward the other end of the gene.

		Percentage of wild-type value†				
Mutant	Class*	ASase	PRTase	InGPSase	TSase B	TSase A
R⁻ A9952	1	100	104	114	100	90
R⁻ B9763	1	100	-	108	0	103
R⁻ C3404	1	100	-	0	115	111
R⁻ D9885	1	100	0	91	117	107
R⁻ E9547	1	0	-	-	96	97
R⁻ A9796	2	100	95	100	49	0
R⁻ B40	2	100	111	83	0	3
R⁻ C9905	2	100	82	0	26	33
R⁻ D10242	2	0	0	8	9.3	10
R⁻ E9851	2	0	2	3	4	4

* Mutants in class 1 form CRM⁺; those in class 2 do not
† ASase value set at 100 percent in each case.

a) Mutants in group 2, but not 1, responded to certain suppressors. What is the probable nature of the mutation in each of the mutants in the 2 groups?
b) What is the difference in the effect of the mutation in each of the mutants in the 2 groups on the relative rates of polypeptide synthesis?
c) Offer an explanation to account for the tabulated results and the additional stated facts.

35. Regulation of the histidine operon in *Salmonella typhimurium* involves at least six genes (Anton, D.N., J. Mol. Biol. 33:533, 1968). Mutations at any one of these loci destroy the cell's ability to fully repress the histidine biosynthetic enzymes. Fink, G.R. and J.R. Roth (J. Mol. Biol. 33:547, 1968) studied dominance relationships between *mutant* and *wild-type* alleles at each of these loci using merodiploids containing an F' episome with the *wild-type* allele of the gene involved and the chromosome carrying the *mutant* allele. The levels of histidine biosynthetic enzymes, histidinol phosphate phosphatase (specified by *hisB*), and histidinol dehydrogenase (specified by *hisD*) were assayed as an index of derepression of the operon. Results for three genes (*hisI, W* and *D*) are presented. The relative specific activities are given first for each *mutant* and then for the diploid (same strain carrying the episome).

			Specific activity of*	
Line	Regulatory mutation	Genotype	B enzyme	D enzyme
	wild-type	T⁺W⁺	1	1
	hisT	T	26.3	
		F'T⁺/T⁻	3.5	
	hisW⁻	W	16.1	
		F'W⁺/W⁻	3.6	
1	wild-type	O⁺B⁺D⁺	1	1
2	hisO⁻	O⁻B⁻D⁺	0	12.4
3	hisO⁻	O⁻B⁺D⁻	12.0	0
4	hisO⁻	O⁻B⁺D⁺	11.2	15.1
5	hisO⁻	F'O⁺B⁺D⁺/O⁻B⁻D⁺	1.7	15.0
6	hisO⁻	F'O⁺B⁺D⁺/O⁻B⁺D⁻	16.5	1.2
7	hisO⁻	F'O⁺B⁺D⁺/O⁻B⁺D⁺	9.3	15.3

*All enzyme activities are expressed relative to *wild-type* taken as 1.

a) For each of these loci state whether the *mutant* or *wild-type* allele is dominant and why, and if dominant, whether in both cis and trans configurations.
b) Where would you expect *hisO* to map relative to the histidine operon structural genes and why?

36. Kiho, Y. and A. Rich (PNAS 54:1751, 1965) measured the relative size of polyribosomes showing β-galactosidase activity in *Escherichia coli*. Strains with deletions in the *y-a* (permease-acetylase) part of the lactose operon had smaller polyribosomes than *wild-type*. Amber mutants in the *z, y* and *a* genes also had smaller polyribosomes. *Wild-type* revertants of these had the larger *wild-type* polyribosomes. Do these results indicate that there is one mRNA for each gene or one messenger per operon? Explain.

37. The position of a nonsense mutation in a gene strongly influences the degree of polarity; operator-proximal mutations allow less efficient expression of the distal genes in the operon than operator-distal ones. Thus, there is a gradient of polarity along a gene (Fink, G.R. and R.G. Martin,, J. Mol. Biol. 30:97, 1967).

a) Suggest a mechanism to explain this phenomenon.
b) Why do nonsense but not missense mutations have polar effects?
c) Do nonsense mutations in the structural genes of an operon produce their effects at the transcription or translation level? Explain.

38. Spurway, H.S. (J. Genet. 49:126, 1948) carried out extensive studies on a phenotypically uniform line of *Drosphila subobscura*. A few females in matings with brothers, as well as unrelated males, produced *sterile* offspring of which all the females had rudimentary ovaries and the males did not have testes. Only *fertile* offspring were produced by other females from the abnormal line in matings with brothers as well as with related and unrelated males. Males from the abnormal line mated to sisters or unrelated females resulted in *fertile* offspring only. Results of some of the initial matings that helped clarify this unusual pattern of inheritance are tabulated below. Each cross involved one male and one female only.

♀ ♂	OFFSPRING
A x B	83 *fertile* ♂; when crossed with *fertile* sisters or unrelated females, these gave rise to *fertile* progeny only
	89 *fertile* ♀; when crossed with *fertile* brothers or unrelated males, these produced *fertile* progeny only
A x C	93 *fertile* ♂; breeding behavior the same as that of the 83 *fertile* ♂s from A x B
	97 *fertile* ♀; 24 of which when crossed with *fertile* related or unrelated males produced *sterile* offspring only; the other 73 females bred like the 89 females from A x B
D* x E†	58 *fertile* ♂; breeding behavior like that of fertile males from A x B
	61 *fertile* ♀; 29 of which when crossed with *fertile* related or unrelated males produced *sterile* offspring only (1/2 males: 1/2 females): the others produced *fertile* offspring

*Genotype identical to female A in second cross
†Brother of the 24 females in second cross

The mode of inheritance of this sex-limited trait in *Drosophila melanogaster* is identical to its inheritance in *Drosophila subobscura* (Niki, Y. and M.Okada, Wilm Roux's Arch. Dev. Biol. 190:1, 1981).

In *wild-type* females in these 2 and other *Drosophila* species the posterior tip of *young embryos* contains a normal region called the polar plasma which contains normal polar cells and polar granules composed of RNA and proteins (Counce, S.J., J. Morph. 112:129, 1963; Allis C.D. et al., Dev. Biol. 69:451, 1979). Embryos derived from mutant females that produce sterile adults only, either lack polar plasma and polar cells or these are defective (Niki, Y. and M. Okaka, Wilm Roux's Arch. Dev. Biol. 190:1, 1981; Mahowald, G.P. et al., Dev. Biol. 69:108, 1979; Niki, Y., Dev. Biol. 103:182, 1984).
a) Determine the mode of inheritance of this sex-limited trait (*grandchildless*).
b) Explain how and where the gene(s) may be acting to cause this effect.

Compare your answer to b) with the finding of Ilmensee, K. (Wilm Roux's Arch. 171:331, 1973), and Underwood, E.M. et al. (Dev. Biol. 77:303, 1980).

39. Attardi, G. et al. (Cold Spring Harbor Symp. Quant. Biol. 28:363, 1963) found that lactose operon mRNA synthesis in *E. coli* increased with addition of inducer. Do inducers act at the translation or transcription level? Explain.

40. *Drosophila melanogaster* embryos possess 14 segments: 3 head segments (Md, Mx, Lb), 3 thoracic segments (T1, T2, T3) and 8 abdominal segments (A1 through A8). All these segments are arranged in the aforementioned sequence from the anterior to the posterior regions of the fly. The genes in the bithorax (Bl) region determine the fate of the last 9 segments (T3 through A8). The research of Lewis, E.B. (Nature 276:565, 1978) indicates that probably 9 genes in the BX-C region determine the fate of the latter 9 segments. T2 development does not require the action of any of these genes. T3 formation requires the activity of the *wild-type* allele at one specific locus; the differentiation of A1 requires the action of 2 genes, the *wild-type* allele at the

first locus and the *wild-type* allele at a specific second locus. With each successive segment the activity of 1 more additional genes is required to effect differentiation.

Studies by Lewis, E.B. (Nature 276:565, 1978), Puro, J. and T. Nygren (Hereditas 81:237, 1975), Denell, R.E. and R.D. Frederick (Develop. Biol. 97:34, 1983) of the *Pc* (polycomb) locus have revealed the following:

i) Homozygotes and hemizygotes for *Pc*, Pc^3 and other such alleles have their thoracic and first 7 abdominal segments transformed to A8. Such flies die as late embryos.
ii) Gene dosage studies indicate that these mutant alleles represent the inactivated state of the Pc^+ gene.
iii) All the thoracic and abdominal segment of mutants homozygous for deletions of the *Pc* locus resemble the last abdominal segment.
iv) Viable and fertile adults heterozygous for the mutant allele, e.g., Pc^+Pc, express less extreme anterior to posterior transformations of thoracic and abdominal segments than do individuals homozygous or hemizygous for such mutant alleles.
v) The Pc^+ gene product is required for normal development throughout the larval period.

Zink, B. and R. Paro (Nature 337:468, 1989) using an immunostaining technique and antibodies raised against the 390 amino-acid long Polycomb protein showed that this protein binds to the BX-C region of polytene chromosomes of *wild-type* flies. No such staining was observed in Pc^-Pc^- mutant embryos.

a) Are the mutant alleles dominant or recessive to the *wild-type* allele at the *Pc* locus? Explain. What phenotypic ratio would you expect in the progeny of a mating between 2 heterozygotes, i.e., $Pc^+Pc^- \times Pc^+Pc^-$ and why?
b) The *Pc* locus interacts with the genes of the BX-C complex. Offer a plausible explanation based on the above facts as to how this locus regulates activity of the BX-C genes. Compare your explanation with that of Lewis, E.B. (1978), B. Zink and R. Paro (1989) and De Camillis, M., et al. (Genes and Dev. 6:223, 1992).

41. In *Drosophila melanogaster* a number of loci which affect segment pattern formation have maternal effects. The mutant alleles at 5 such loci produce the following deviations from normal segmental development:
i) *bcd* (*bicoid*), *exu* (*exuperentia*) and *swa* (*swallow*) all cause loss of head and thorax.
ii) *bic* (*bicaudal*) produces an abdomen at each end of the embryo. The polarity of the 2 abdomens is reversed.
iii) *osk* (*oskar*) prevents the abdomen from forming. Head and thorax develop in normal polarity.

The allele bcd^+ codes for a regulatory protein that forms a gradient by diffusion from the anterior region where maternal bcd^+ mRNA is localized by the exu^+ and swa^+ gene functions. In the following experiments, approximately 1% of cytoplasm and its cells were transferred from the anterior (*a*), middle (*m*) or posterior (*p*) portions of donor early embryos to (*a*) or (*p*) locations of recipient embryos. The results of these transfers are indicated below:

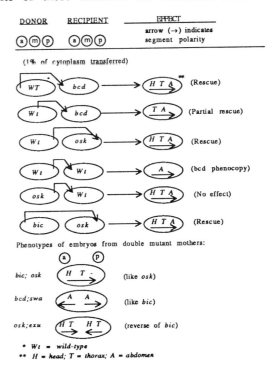

Explain what each of these results tells us about the specification of the normal segmental pattern in *D. melanogaster*?

42. The genes in the bithorax complex (BX-C) in *Drosophila melanogaster* determine the fate of development of the posterior portion of the body, specifically the last 2 thoracic segments T2 and T3 and the abdominal segments A1 through A8: these segments from anterior to posterior are arranged in the sequence T2, T3, A1, A2, A3, A4, A5, A6, A7 and A8. Each segment is also developmentally divided into 2 regions: an anterior and a posterior one. Many mutations in the BX-C region on the right arm of the third chromosome (bands 89E1 - 89E4) have been identified and studied primarily by Lewis, E.B. since 1947 (see Lewis, E.B. Nature 276:565, 1978; Lawrence, P.A. and G. Morata Cell 35:595, 1983). The phenotypic effects of 4 of these recessive mutations, typical of many others in this region, are as follows:

i) The first mutation, *bx* (for *bithorax*) causes the anterior segment of T3 to resemble the anterior segment of T2. Normally, only segment T2 develops a pair of wings; the third has a pair of halteres. In *bx* mutants the front portion of the halteres develops like the front half of wings.

ii) A second mutation, *pbx* (for posterior bithorax) produces an analogous transformation in the posterior half of T3. The double mutants *bx, pbx/bx, pbx* develop 2 pairs of wings.

iii) The mutation *bxd* (bithoraxoid) causes segment A1 to resemble segment T3. These mutants may have a fourth pair of legs on the transformed segment.

iv) The mutation which we symbolize *ab* changes A2 into A1.

v) Flies homozygous for a deletion of the entire BX-C region die as embryos. Inspection of these embryos reveals that T3 and each of the 8 abdominal segments resemble T2.

Flies have evolved from ancestors with 2 pairs of wings and a pair of legs on each of the segments of the body.
a) Are these mutations homeotic ones? Explain.
b) Which of the body segments can be considered to be the "developmental" ground state? Explain, stating your assumptions.
c) What is the expected number of genes in the BX-C complex? Explain. For ease of discussion, symbolize the genes as, a, b, c, etc.
d) Are the genes in this complex expressed successively from back to front or front to back?
e) What is the *wild-type* genotype? What would be the genotype of flies that convert A1 to T3, A2 to A1 and A3 to A2?

43. i) In *Drosophila melanogaster*, the majority of X-linked hypomorphic (leaky) alleles produce equivalent phenotypes in both sexes. For example the w^a (apricot) allele produces the same amount of pigment in both males and females (Muller, H.J., Harvey Lec. 43:165, 1950).
ii) *D. melanogaster* females heterozygous for *mutant* and *wild-type* alleles at X-linked loci do not exhibit phenotypic or biochemical mosaicism.
a) Does dosage compensation occur in *D. melanogaster*? Explain.
b) Is it achieved by X-chromosome inactivation as in mammals? Explain.

iii) The enzyme 6-PGD is a dimer specified by the X-linked gene *Pgd*. Using the technique of starch-gel electrophoresis, Kazazian, H. et al. (Science 150:1601, 1965) showed that strains true-breeding for the Pgd^a and Pgd^b alleles produced *fast* (F) bands and *slow* (S) bands of this enzyme. The cells of Pgd^aPgd^b females produced 3 different molecular forms of this enzyme and therefore 3 different bands: S, F and SF (intermediate mobility).
c) Explain how each cell that synthesizes 6-PGD can form 3 types of molecules and discuss the bearing of these results on the mechanism of dosage compensation in this species.
iv) Measurements of ^{3}H-uridine incorporation RNA transcripts on polytene X chromosomes by Mukherjee, A.S. and N. Beermann (Nature 207:785, 1965) was equal in the 2 sexes. Other studies indicate that the polytene X chromosome in males has half as much DNA as the paired polytene X chromosomes in females (Rudkin, G.T. in <u>The Nucleohistones</u>, Holden-Day, 1964; pp. 184-192). The polytene X chromosome in males is twice as wide as an autosome or each X chromosome in females (Offerman, C.A., J. Genet. 32:103, 1936).

Outline a plausible mechanism for dosage compensation in *D. melanogaster* based on this information.

44. a) The X-linked sex-lethal locus (Sxl) is involved in dosage compensation in *Drosophila melanogaster*. The *wild-type* allele Sxl^+ normally functions (is expressed) in females but not in males. The recessive loss-of-function allele Sxl^b has lethal effects in female ($Sxl^b Sxl^b$) embryos (Cline, T.W., Genetics 90:683, 1978). These females have much higher X-chromosome transcription levels than normal (Sxl^+) females (Lucchesi, J.C. and T. Skripsky, Chromosoma 82:217, 1981). Sxl^m is a dominant male-specific constitutive allele which is lethal to most male embryos during the first instar stage (Cline, 1978).

What role does the Sxl locus appear to play in dosage compensation? Explain.

b) At least 4 autosomal loci, *msl1*, *msl2*, *msl3* and *mle* control dosage compensation in *D. melanogaster*. The products of the *wild-type* alleles at these loci are essential for male, not female, viability. The recessive mutant alleles *msl1*, *msl2*, *msl3* and *mle* kill males throughout larval life but do not affect female viability (Baker, B.S. and J.M. Belote, Ann. Rev. Genet. 17:345, 1983). Belote, J.M. and J.C. Lucchesi (Nature 285:573, 1980) demonstrated that in all recessive *msl* mutant males the activities of enzymes, e.g., G6PD and GGPD, that are determined by X-linked genes were reduced by ~50% of their activities in *wild-type* individuals. The activities of autosomal enzymes, e.g., ADH and AO (alcohol oxidase) were altered in recessive *msl* males. Concomitant with results of enzyme assays was the finding that the rate of ^{3}H-uridine incorporation into chromosomal RNA in recessive *msl* males was ~60% of that in *wild-types* of both sexes. The same results were obtained with recessive *msl* males (Breen, T.R. and J.C. Lucchesi, Genetics 112:489, 1986).

What is or appears to be the function of the *msl* loci in dosage compensation in *Drosophila*? Explain.

45. a) Shih, C. et al. (Nature 290:261, 1981) extracted DNA from a human *bladder carcinoma* cell line EJIT24, and injected short fragments of this DNA into normal (noncancerous) mouse fibroblast cells (designated NIH3T3) *in vitro*. At very low frequency, some of the cells formed minitumors (a cancer-like mound of cells). Normal cells form a flat layer of cells. When the transformed cells were inoculated into normal mice, tumors grew. Cells from the minitumors were isolated and cultured. When DNA was extracted from this culture and applied once again to normal cells, minitumors of transformed cells appeared once again. The process was repeated 3 and more times with invariant results. DNA from these minitumors also caused cancer when inoculated into normal mice. Similar experiments with normal human cells does not induce tumor formation in mice.
i) Do these data indicate that carcinoma of the human bladder is inherited?
ii) Discuss at least 2 other genetic conclusions that can be drawn from this information.

b) Shih, C. and R.A. Weinberg (Cell 29:161, 1982) showed that the ability to transform normal cells into cancerous cells was contained within the *EcoR1* fragment of DNA from the cancerous mouse cells. Does this prove that the fragment is human in origin? That it contains only one gene? Explain.

c) The *normal* proto-oncogene (ras^+) protein (p21) which is 189 amino acids long binds to the nucleotide GTP at the inner surface of the cell membrane and hydrolyzes it to the less reactive form GDP. The mutant allele *ras* codes for a p21 protein which has glycine instead of valine at position 12. (Tabin, C.J. et al., Nature 300:143, 1982).

Speculate as to how the mutant allele (oncogene) product may lead to abnormal cell growth and tumor formation.

46. Propose a mechanism involving DNA methylation that would result in the inactivation of an X chromosome. Compare your explanation with that of Migeon, B. R. (Genet. Res. 56:91, 1990).

47. Homeo boxes, which were first discovered by Gehring, W.J. and his colleagues, are present in many higher eukaryotes (Gehring, W.J., Science 236:1245, 1987).
a) What are homeo boxes and what is their function?
b) What are homeo domains and what is their role in development?

48. The primary function of B cells of the immune system is to produce antibodies (immunoglobulins composed of heavy and light protein chains). The terminal regions of the q arms of normal chromosomes 8 and 14 carry the *c-myc* proto-oncogene and the *IgH* locus (for synthesis of the heavy chains of antibodies) at bands q24 and q32 respectively. *Burkitt's lymphoma* (*BL*) is a cancer of B cells. In *normal* B cells the homologues of both chromosome 8 and chromosome 14 are normal. In *BL* cells only one homologue of each of these 2 chromosomes is normal. The other chromosome 8 and chromosome 14 have been involved in a reciprocal translocation; the terminal segment of chromosome 8 with the *c-myc* gene is translocated to chromosome 14 next to the *IgH* locus. The segment of the latter chromosome distal to the *IgH* locus has been translocated to chromosome 8 (Dalla-Favera, R. et al., PNAS 79:7824, 1982).

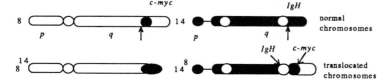

The protein produced by the *c-myc* gene in *BL* cells is identical to that produced in normal B cells and other somatic cells (Croce, C.M. and G. Klein, Sc. Am. 252(Mar):54, 1983).
a) Why are (i) gene mutation and (ii) position effect ruled out as possible causes of *BL*?
b) Propose a plausible mechanism for *BL*.
c) Outline an experiment which would permit you to test your proposal. See Nishikura, K. et al. (PNAS 80:4822, 1983) for an explanation of how *c-myc* becomes oncogenic in B cells with *BL*.

49. In *Drosophila melanogaster*, sex-lethal (*Sxl*) is the pivotal gene in sex determination. It functions in females (AAXX) and not in males (AAXY). Its activity state is chosen early in development in response to the X/A ratio, the primary sex determining signal. For discussion purposes assume females are Sxl^+Sxl^+ and males are Sxl^+/Y. Sxl^+ is transcribed in both sexes; it codes for an active full-length 354 amino acids long protein in females and a non-active one containing 48 amino acids in males (Bell, L.R. et al., Cell 55:1037, 1988).
a) Propose and illustrate a plausible mechanism whereby the same primary transcript in the 2 sexes results in a different protein in males and females.
b) The Sxl^+ protein shows sequence similarities with ribonucleoproteins. What does this information suggest regarding the possible role of the Sxl^+ protein with respect to the tra^+ and $tra-2^+$ downstream genes and products (primary transcripts and proteins) whose normal function is dependent on the normal action of the Sxl^+ allele?
c) Sxl^+ is a positive autoregulatory gene in females (Bell, L.R., Cell 65:229, 1991). Propose and illustrate a plausible mechanism that will account for its self-regulation in this sex. Is such a mechanism required in males? Explain.

50. a) The hereditary form of *retinoblastoma* (eye-cancer) shows a dominant mode of inheritance but a recessive phenotypic expression at the cellular level; that is *retinoblastoma* occurs only when the mutant allele Rb^- for this cancer or deletion of the *Rb* (*retinoblastoma*) locus is present in the homozygous or hemizygous state (Knudson, A.G., PNAS 68:820, 1971; Ann. Rev. Genet. 20:231, 1986). Is the normal allele (Rb^+) at the *retinoblastoma* locus a growth-promoter or a growth-suppressor? Explain.

b) In 1988 Huang, H.-J.S. et al. (Science 242:1563, 1988) cloned Rb^+ in a retrovirus which then transferred Rb^+ to a *retinoblastoma* cell line. The normal allele suppresses the tumor-forming capability of cultured *retinoblastoma* cells. Does this data confirm or refute your explanation in a)?

51. a) Studies in humans by Therman, E. et al. (Chromosoma 44:361, 1974; Hum. Genet. 50:59, 1979) have revealed the following:

i) Females with a normal X and an isochromosome for the short arm (1Xp) do not exist. In females with a normal X and an isochromosome for the long arm of the X(1Xq), the latter chromosome is inactivated and a certain proportion of Barr bodies is bipartite.

ii) In translocation heterozygotes in which whole arms have been reciprocally exchanged between an autosome and an X-chromosome, it is always the translocated chromosome with the q arm of the X that is inactivated.

iii) In females heterozygous for deletions of segments of the q arm of the X chromosome, all X's with deletions of segments distal to band q13 were capable of inactivation; longer deletions involving band q13 and proximal arm regions were not inactivated.

1) Does this information argue for a single or multiple centers of X-chromosome inactivation in mammals? Explain.
2) Where on the X is (are) this center(s) located? Explain.

b) Brown, C.J. et al. (Nature 349:38 and 82, 1991) cloned the gene *Xist* and hybridized it with RNA transcripts isolated from cells of both males and females as well as somatic-cell hybrids using the slot blot technique. Their results were as follows.

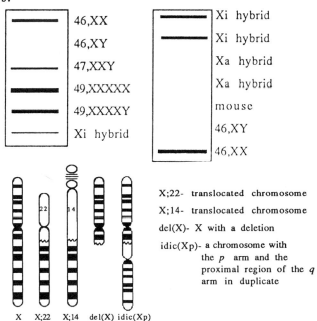

i) Where is *Xist* located on the X chromosome? Explain.
ii) Could it be associated with the X inactivation process? Explain how?

52. a) What are reciprocal-shift experiments?
b) Discuss the rationale used in these experiments to determine whether any 2 gene mediated steps in the cell cycle are dependent, interdependent or independent.
c) Hereford, L.M. and L.H. Hartwell (J. Mol. Biol. 84:445, 1974) grew 5 different cell-division-cycle (*cdc*) mutants under both permissive and restrictive conditions according to the reciprocal shift method. Their ability to complete the cell cycle was scored (+ for completed, and - for not completed). The results obtained were as follows:

cdc mutants	1st incubation: 2nd incubation:	restrict A, permit B permit A, restrict B	restrict B, permit A permit B, restrict A
cdc 14	*cdc* 4	+	-
cdc 2	*cdc* 18	+	+
cdc 8	*cdc* 14	-	+
cdc 8	*cdc* 4	-	+

i) Are the cell cycle steps governed by *cdc* 2 and *cdc* 18 independent, dependent or interdependent? Explain.
ii) What is the sequence of cell cycle steps determined by the mutant genes *cdc* 4, 8 and 14? Explain.

53. a) What is a tumor? What is a malignant tumor? What are its characteristics?
b) In humans the locus for G6PD synthesis is X-linked. The allele G^A codes for the A form of the enzyme and G^B specifies the B form. Linder, D. and S.M. Gartler (Science 150:67, 1965) studied $GA^A GA^B$ females. They showed that a single cell produces either the A or B isozyme. Each sample of normal uterine tissue consisting of many cells produced both forms of the enzyme. All cells of a uterine tumor (millions of cells) produced only the A or B form of G6PD. Some women with several separated tumors had some tumors with A isozyme and others with the B form but none of the tumors contained both A and B isozymes.

Does this data refute or support the single-cell origin of cancers? Explain.

54. a) What are proto-oncogenes? Oncogenes? Anti-oncogenes?
b) There are at least 5 different mechanisms by which proto-oncogenes can be activated or changed into oncogenes (Land, H. et al., Science 222:771, 1983). Discuss 3 of these mechanisms and name at least 1 form of cancer that is the consequence of each mechanism.
c) In all or most cancers, the pathway to malignancy is a multistep process (Land, H. et al., Science 222:771, 1983). In spite of this Shih, C. and R.A. Weinberg (Cell 29:161, 1982) were able to permanently transform normal mouse cells (N1H3T3) in culture with DNA from a human bladder carcinoma cell line EJ/T24 in one step.

Suggest an explanation for this apparent discrepancy.

55. a) *Src* (and some other) proto-oncogenes code for a tyrosine specific protein kinase (Collett, M.S. and R.L. Erikson, PNAS 75:2021, 1978).
i) What is a tyrosine specific protein kinase?
ii) Suggest how a mutant allele (oncogene) at the *src* locus might lead to cancer. (See Hunter, T., Sc. Am. 251 (Aug.):70, 1984.)

b) The human proto-oncogene *c-jun* codes for a DNA binding protein. All the data suggest that the protein binds to the enhancer sequences (Bohmann, D. et al., Science 238:1386, 1987). Suggest how this gene may function in gene regulation and how its mutant alleles (oncogenes) may be involved in inducing abnormal cell growth and cancer.

c) The proto-oncogene *erbB* codes for the receptor (a cell-membrane protein) for epidermal growth factor (EGF), a protein that stimulates cell division when joined to the receptor (Hunter, T., Sc. Am. 251 (Aug.):70, 1984). Suggest how a mutant allele (oncogene) at this locus might result in tumor-formation and cancer.

56. *Nasobemia* is an *antennapedia* (*ant*) mutant of *D. melanogaster* which in addition to 3 normal sets of legs, has legs growing on the head in place of antennae. (See Gehring, W.J. Sci. Am. (Oct.): 153, 1985.) What can you deduce about the effects of the normal allele ant^+ at this locus from this information? Where must the ant^+ be expressed in *wild-type* embryos?

b) Nüsslein-Volhard, C. and Wieschaus, E. (Nature 287:795, 1980) isolated *even-skipped* (*eve*) mutants in *D. melanogaster* in which all even-numbered segments of the thorax and abdomen are missing as well as *odd-shaped* (*odd*) ones in which the odd-numbered segments of the body are missing. Each of these is due to a mutant allele at a single locus. The normal alleles at these loci are expressed at the time of blastoderm formation.

i) What distribution of transcripts of each of these genes would you expect in normal embryos?
ii) How would you test your hypothesis if the genes have been cloned and antibodies against the proteins produced by these genes are available?

c) In *D. melanogaster,* mutant *dicephalic* (*dic*) embryos have 2 sets of anterior structures (heads, antennae, etc.) joined in the middle and lack posterior structures altogether. The ovarian structure called the follicle possesses 15 nurse cells and an oocyte. In the *dic* mutant, nurse cells are found at both ends of the embryo whereas in normal flies they occur only at the anterior end (Lohs-Schardin, M. Roux's Arch. Dev. Biol. 191:28, 1982).
i) Is the *dic* mutation a homeotic or maternal-effect one? Explain.
ii) Why do the results indicate the *dic* locus influences spatial polarity?
iii) The *dic* allele produces a maternal effect when homozygous. How would you provide evidence for this? Show the results expected.

Chapter 24
Population Genetics, Inbreeding, Outbreeding and Evolution

1. a) Some people believe that inbreeding *per se* leads to an increase in the frequency of recessive alleles in a population. Demonstrate that this is incorrect and that inbreeding affects only the distribution of the alleles among the genotypes.
   b) Describe the conditions under which inbreeding can cause harmful effects, in man and other animals.
   c) What effect does inbreeding have on (1) allele frequency and (2) heterozygosity?
   d) Self-fertilization results in a reduction in vigor in one species of plants but not in another. What conclusion could you draw regarding the form of natural breeding in the 2 species?
   e) Under what circumstances would inbreeding not have deleterious consequences?

2. a) Discuss the relatively different roles of inbreeding and outbreeding in the development of a well-adapted race of a cross-fertilizing species.
   b) Why is selection within a pure line futile?
   c) Can a population simultaneously mate at random with respect to one pair of alleles and assortatively with respect to others? Explain.

3. Demonstrate that the following rates of reduction in heterozygosity per generation are correct:
   a) One-half for self-fertilization.
   b) One-fourth for brother-sister matings.
   c) One-eighth for half-brother-half-sister matings.
   d) One-sixteenth for cousin matings.

4. A single *tall*, heterozygous Tt, pea plant and its progeny in subsequent generations are self-fertilized. How many generations would it take to attain approximately 94 percent homozygosity?

5. Three corn plants are selected, one heterozygous for *starchy* endosperm, $Wxwx$, and 2 homozygous for *waxy* endosperm, $wxwx$. If all plants produce the same number of progeny, what percent of the population will be $Wxwx$ after four generations of self-fertilization?

6. Demonstrate that offspring from a first-cousin marriage have 1 chance in 64 of being homozygous for an autosomal recessive allele for which one of the common great-grandparents is heterozygous.

7. The frequency of the recessive allele p causing *phenylketonuria* is about 1 in 200 in most North American populations.
   a) What is the risk of a child having the disease if the parents are unrelated and mating at random?
   b) How much is this risk enhanced if the parents are first cousins?
   c) If a *normal* man marries his niece, determine the probability of their first child being *affected* (assume that his sister's husband is homozygous *normal*). Would the chance of producing an *affected* child be greater or smaller if the man married his first cousin?

8. Sjögren, T. (Hereditas 14:197, 1931) showed that 1 percent of all marriages in Sweden are between first cousins. This 1 percent was responsible for 15 percent of all children with *juvenile amaurotic idiocy* (autosomal recessive). The frequency of this disease in the population as a whole was 1 in 30,000. What is the frequency of the allele responsible for this trait?

9. a) Two grandchildren of a woman with autosomal recessive *albinism* marry.
   i) What is the probability that both husband and wife carry the allele for *albinism*?
   ii) If both are carriers, what is the probability that a successful pregnancy will give an *affected* child?
   iii) Answer the above questions for the sex-linked recessive condition *ocular albinism*.

   b) A marriage involves first cousins. What is the probability that a child will be affected with *alcaptonuria* (rare recessive autosomal trait, incidence 1 in 40,000) if:

i) The common grandmother is *affected*?
ii) The wife's father is *affected*?
iii) None of the relatives are *affected*?

10. In the pedigree shown, in which the recessive alleles for *Duchenne muscular dystrophy* and *infantile amaurotic idiocy* of X-linked and autosomal genes respectively are segregating, a marriage occurs between first cousins, as indicated by the double line.

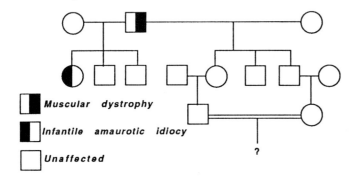

What is the probability that the offspring of the first-cousin marriage will be *affected*:
a) With *infantile amaurotic idiocy*?
b) With *Duchenne muscular dystrophy*?
c) With both conditions?
d) With neither?

11. Two clones of a wild plant species are allowed to reproduce annually for 10 years as follows:
i) One clone is propagated asexually.
ii) The other is reproduced by self-fertilization.

The numbers of offspring produced per generation are, on the average, the same in the 2 groups. State which of these 2 resulting lines is more likely to carry deleterious recessive alleles at a fairly large number of gene loci and why.

12. Morton, N.E. et al. (PNAS 42:855, 1956) discuss methods by which data from first-cousin marriages can be used to estimate the number of recessive lethals causing death before reproduction (adulthood) carried by phenotypically *normal* individuals in human populations. Basically, the method derives from the observation that for any heterozygous individual (X), the probability of any great grandchild (Y) by a cousin marriage being homozygous for a given gene is 1/16.

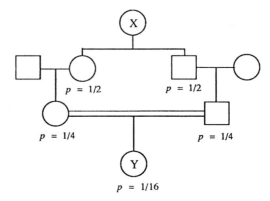

For n such genes (carried by X) the probability of death of such a great grandchild would then be $1/16 + 1/16 + \ldots$ to n terms $= n/16$. Sutter, J. (Biol. Med. 47:563, 1958) obtained the tabulated results from church records in rural France.

Deaths before adulthood (including stillbirths)	
Parents first cousins	0.25
Parents unrelated	0.12
Excess deaths from first-cousin marriages	0.13

Estimate the average number of recessive lethal equivalents per person (a recessive lethal equivalent is either one gene that causes death of homozygotes before reproduction or two genes each causing death of half the homozygotes before adulthood).

13. The frequency of autosomal alleles K and k in a large randomly mating population of snails is 0.6 and 0.4.
a) What will be the frequency of KK, Kk, kk genotypes after:
i) One generation of self-fertilization?
ii) Two generations of self-fertilization?
iii) One generation of a brother-sister mating?
iv) One generation of a first-cousin mating?

b) What genotypic frequencies would you expect if the inbreeding coefficient was 0.3?
c) If the snail population also carries the alleles M and m in frequencies $0.99M$ and $0.01m$, what frequencies of kk and

mm individuals are expected if only first cousins mate?
d) Compare these frequencies with those expected upon random mating.

14. Two different plant species are brought under cultivation. One is self-pollinating, the other cross-pollinating. Which one is more likely to respond to selection and why?

15. When animals from natural populations are inbred, various deleterious recessive traits appear in later generations. Moreover, the frequency of death *in utero* and from birth to reproductive age increases sharply.
a) Why have these alleles not been eliminated from the natural populations by natural selection?
b) Why do the traits show up more frequently upon inbreeding than in natural populations?

16. Define coefficient of inbreeding. Calculate the coefficients for the progeny of the following marriages:
a) Uncle and niece.
b) First cousins.
c) Second cousins.
d) Half-first cousins
e) Half-brother and half-sister.

17. Most pure breeds of dogs are more susceptible to diseases than mongrels. What genetic explanations can you offer for this?

18. Why may persistent inbreeding in a number of pure lines, followed later by crosses between them, result in more vigorous individuals than are produced by crosses among a similar number of lines that have not been inbred?

19. If 40 percent of the plants in a plot of alfalfa are heterozygous at a given locus, what proportion would be expected to be homozyous at this locus after two generations of self-fertilization (assume all plants produce the same number of seeds, all of which grow to maturity).

20. Determine the coefficient of inbreeding of the bull Domino from the pedigree given. Assume that Lamplighter has an inbreeding coefficient of 1/8.

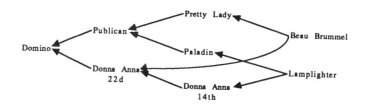

21. a) What is Hardy-Weinberg equilibrium?
b) What are the conditions necessary for the maintenance of this equilibrium in any population? Discuss.
c) What evidence is required before concluding that an allele pair is in Hardy-Weinberg equilibrium?
d) What kinds of sexually reproducing species are likely to show this type of equilibrium?
e) What frequencies of p and q give the greatest proportion of heterozygotes in Hardy-Weinberg equilibrium?

22. Show what is fundamentally wrong with the following statement: If *brown* eyes in man is due to a dominant allele at a single locus, the frequency of individuals showing the trait may be expected to increase in the population until a frequency of about 3 in 4 is obtained.

23. a) Do you think the frequency of dominant lethal alleles changes from generation to generation? Explain.
b) Explain why it becomes increasingly difficult to eliminate a recessive allele from a population as its frequency becomes lower.
c) What is one type of genetic change that contributes to evolutionary change in a nonadaptive fashion?
d) Which would you expect to be the more vulnerable to selection, a sex-linked recessive or an autosomal recessive? Why?
e) What are the conditions under which the frequency of individuals homozygous for an allele equals the rate of mutation to that allele?
f) Why is the frequency of a recessive allele that causes sterility in a population at equilibrium equal to the square root of the rate of mutation to it?
g) A dominant allele of one gene and a recessive allele of another have identical adaptive values. Against which type of allele will selection be more effective?
h) Does a disturbance of allelic equilibrium necessarily lead to a disturbance of genotypic equilibrium? Explain.

24. In the genus *Drosophila* the heaviest genetic loads appear to occur in common, ecologically versatile species, whereas the lightest loads appear to be found in rare, specialized species. Offer an explanation for this.

25. a) In sheep *white* vs. *yellow* fat is determined by a single pair of autosomal alleles (Mohr, O.L., J. Hered. 25:246, 1934). In a large randomly mating flock 23 percent of the sheep have *white* fat and 77 percent have *yellow* fat.
i) Can you determine from these data whether *white* is due to the recessive or dominant allele? Explain.
ii) Explain how you would determine whether the alleles are in Hardy-Weinberg equilibrium.

b) A sheep rancher in Iceland finds that the recessive allele *y* for *yellow* fat has become established in his flock of 1,024 and that about 1 out of every 256 sheep expresses the trait.
i) The rancher wishes to know how many of the *normal* sheep carry the recessive allele. Assuming the population is randomly mating for this gene and all genotypes have the same reproductive fitness, what is this proportion?
ii) How many of the 1,020 *white* individuals can be expected to be homozygous?
iii) Since only *white* individuals were selected for breeding, why is it that the recessive allele has not been completely eliminated from the population? How would you proceed to accomplish this? (Assume methods are now available for detecting heterozygotes.)

26. In 1958 Matsunaga, E. and S. Itoh (Ann. Hum. Genet. 22:111) reported the following *MN* blood-typing data (number observed) from the mining town of Ashibetsu in Hokkaido, Japan:

$L^M L^M$ 406 $L^M L^N$ 744 $L^N L^N$ 332

This population consisted of 741 married couples. The six types of matings and their frequencies were as follows:

$L^M L^M \times L^M L^M$ 58 $L^M L^M \times L^N L^N$ 88
$L^M L^M \times L^M L^N$ 202 $L^M L^N \times L^N L^N$ 162
$L^M L^N \times L^M L^N$ 190 $L^N L^N \times L^N L^N$ 41

a) Show whether this population is at Hardy-Weinberg equilibrium.
b) Derive expected frequencies for the various possible kinds of matings and state from inspection whether the data indicate that mating was at random with respect to the $L^M L^N$ pair of alleles.
c) You are studying an isolated community and find that the proportion of individuals with *N* blood type is approximately 4 percent.
i) What are the allele frequencies $p(L^M)$ and $q(L^N)$?
ii) What is the probable reason for this great difference in allele frequencies between this community and the general populace?

27. The human serum protein haptoglobins *haptoglobin-1*, *Hp-1* and *haptoglobin-3*, *Hp-3*, are specified by a single pair of autosomal codominant alleles, *Hp-1* and *Hp-3*, respectively (Smithies, O. and N.F. Walker, Nature 176:1265, 1955). Kamel, N.H. and E.I. Hammoud (J. Med. Genet. 3:279, 1966) tested 219 Egyptians for the presence of each haptoglobin. Of these,

9 showed only *haptoglobin-1* (presumed *Hp-1,Hp-1*)
75 showed only *haptoglobin-3* (presumed *Hp-3, Hp-3*)
135 showed both haptoglobins (presumed *Hp-1, Hp-3*)

a) What are the frequencies of the 2 alleles in the gene pool?
b) Are these results consistent with those expected of a large population mating at random?

28. You are permitted to take samples of a certain species once a year from a large natural park. The species you sample shows variation for a known pair of alleles lacking dominance.

Year	AA	AA^1	A^1A^1
1961	22	76	102
1962	24	72	84

What is the simplest interpretation of the data if the species you sample is:
a) A mammal?
b) A bisexual (monoecious) plant?
c) A unisexual (dioecious) plant?

29. The gamma globulin of human blood serum exists in two forms, $Gm(a^+)$ and $Gm(a^-)$, specified respectively by the dominant $Gm(a^+)$ and recessive $Gm(a^-)$ alleles of an autosomal gene. Broman, B. (Acta Genet. Stat. Med. 13:132, 1963) recorded the tabulated phenotypic frequencies in 3 Swedish populations. Assuming the populations were at Hardy-Weinberg equilibrium, calculate the frequency of heterozygotes in each population.

Region	No. tested	Phenotype, % $Gm(a^+)$	Phenotype, % $Gm(a^-)$
Norrbotten county	139	55.40	44.60
Stockholm city and rural district	509	57.76	42.24
Malmöhus and Kristianstad counties	293	54.95	45.05

30. *Phenylketonuria*, a lethal condition in early life, occurs in Caucasians with a frequency of about 1 in 40:000 individuals (Stern, C. "Principles of Human Genetics," Freeman, San Francisco, 1960).
a) What is the probability that a Canadian of Caucasian origin is heterozygous for the recessive autosomal allele causing this disease?
b) What are the frequencies p and q for the *normal* and *phenylketonuria* alleles respectively?

31. Large populations of a certain rodent are placed on each of two rodent-free islands for a population study. Each populations consists of 10 percent *coloured* animals and 90 percent *albinos*, the difference is due to different alleles of one gene and is uniform for a blood type Kk, due to the combined effect of a pair of codominant alleles.
a) The animals placed on the first island are all known to be homozygous for one or the other of the coat-type alleles.
i) Give the expected allelic, genotypic and phenotypic frequencies for each pair of alleles when this population reaches Hardy-Weinberg equilibrium if the traits are neutral with regard to selection.
ii) Show the expected frequencies of the six phenotypic classes representing the possible combinations of phenotypes for these loci at equilibrium.

b) Of the animals placed on the second island, only the phenotypes for coat type are known. If, at equilibrium, these phenotypes are found to be distributed as 31 percent *coloured* and 69 percent *albino*, what were the genotypic distribution frequencies of the original population?

32. On the basis of allele-frequency analysis of data from a randomly mating population Snyder, L.H. (Genetics 19:1, 1934) concluded that the *ability* vs. *inability* to *taste* phenylthiocarbamide (PTC) is determined by a single pair of autosomal alleles, of which T for *taster* is dominant to t for *nontaster*. Of the 3,643 Caucasians (whites) tested, 70 percent were *tasters* and 30 percent *nontasters*. Assume the population satisfies the conditions of Hardy-Weinberg equilibrium.
a) Calculate the frequencies of the alleles T and t and the frequencies of the genotypes TT, Tt and tt.
b) Determine the probability of a *nontaster* child from a *taster* x *taster* mating.
c) Determine the probability of a *taster* child from a *taster* x *nontaster* mating.

33. In man, the ability to taste phenylthiocarbamide (PTC) is determined by a pair of alleles, T for *taster* being dominant to t for *nontaster*.
a) In one large randomly mating population the frequency of *nontasters* is 0.04; in another it is 0.64. What is the frequency of heterozygotes, Tt, in each of these populations?
b) Synder, L.H. (Genetics 19:1, 1934) tested a random sample of 800 United States families for *ability* vs. *inability*

to taste PTC, taken from a population in which the frequency of t was 0.537. The results were as shown:

Parents	No. of couples	Offspring Tasters	Offspring Nontasters	Average family size
taster x taster	425	929	130	2.5
taster x nontaster	289	483	278	2.6
nontaster x nontaster	86	5	218	2.6

i) Assuming that the numbers of offspring per family do not deviate from the listed averages, calculate the frequency of parental genotypes as closely as possible and compare them with expected frequencies to derive an answer to the question: Are the genotypes for this pair of traits binomially distributed?
ii) Discuss possible explanations for the five *tasters* from *nontaster* x *nontaster* matings.

34. In a large randomly mating population the frequencies of the I^A, I^B and i alleles determining A, B and O blood group antigens are 0.6, 0.3 and 0.1, respectively. What are the expected frequencies for the blood groups A, B, AB and O?

35. a) Alleles S_1 and S_2 occur at a single locus on the X chromosome. The proportion of the S_1 allele is 0.60 in the female half of the population and 0.40 in the male half. What are the expected proportions among males and females in each of the two succeeding generations?
b) If *colour-blindness* is due to a recessive allele at a certain locus on the X chromosome and 18 women in 20,000 are *colour-blind* in a particular population, what is the expected frequency of *colour-blind* men in this population?
c) The frequency of a certain X-linked affliction in men is 1 in 20,000 and that of heterozygous women is 1 in 9,000. If *affected* individuals of a generation are prevented from mating, what are the expected frequencies of *affected* males and of heterozygous females in the next generation?

36. In cats, the genotypes BB and B/Y are *black*, bb and b/y, *yellow* and Bb *tortoiseshell*. In a sample of 281 Boston cats Todd, N.B. (Heredity 19:47, 1964) found the phenotypic distribution shown. Determine whether the population was at Hardy-Weinberg equilibrium.

Sex	Black	Tortoiseshell	Yellow
♀	102	48	4
♂	99	0	28

37. *Colour-blindness* in humans is determined by a recessive allele at an X-linked locus.
a) In a certain population, it occurs 20 times more frequently in males than in females.
i) What is the frequency of the allele for *colour-blindness*?
ii) What is the frequency of heterozygous females?

b) If mating is at random, what frequency of *colour-blindness* would you predict among women in a population at Hardy-Weinberg equilibrium in which 9% of the men are *colour-blind*?

38. It is difficult to arrive at a reliable estimate of mutation rates to recessive alleles in human populations. What are some of the reasons, and how might reliable estimates be obtained?

39. Mørch, E.T. (Opera Domo Biol. Hered. Hum. Univ. Hafniensis 3:1, 1941) has shown that the reproductive fitness of *chondrodystrophic dwarfs* is 20% that of *normals*. Ten of 94,075 births in a Copenhagen hospital were *dwarfs* of which 8 were born to *normal* parents and 2 to parents of whom one was a *dwarf*. From this information calculate:
a) The mutation rate of A_1 (*normal*) → A_2 (*dwarfs*).
b) The frequency of A_1 at genotypic equilibrium.

40. A certain large human population is at equilibrium for the recessive lethal allele a for *juvenile amaurotic idiocy* at an autosomal locus, which causes death before reproductive age. If the mutation rate $A → a$ in the population is 1 in 490,000 (1 gamete in 490,000 carries a instead of A):
a) What is the frequency of a?
b) What is the frequency of heterozygotes?

c) State the proportion of the population that would die because of this condition.

41. In a large randomly mating population at equilibrium, 1 child in 90,000 is born with *cystic fibrosis*, a condition caused by a recessive autosomal lethal, causing death before sexual maturity. Derive:
a) The rate of mutation, μ for C to c.
b) The proportion of the gene pool that contains c.
c) The proportion of heterozygotes in the population.

42. If the adaptive value of a recessive allele is 1.0, the homozygotes have a frequency of 0.01, and the mutation rate of the dominant to the recessive allele is 10^{-5} at equilibrium, what is the reverse mutation rate?

43. A recessive lethal allele causes the death of 1 person in every 20 homozygous for it before the age of reproduction ($s = 0.05$). The mutation rate from the dominant to the mutant allele is 1 in 200,000. The population is at equilibrium for this allele.
a) What proportion of homozygotes for the recessive allele die before reproductive age?
b) How much more frequent would homozygotes be if the recessive lethal allele were fully penetrant ($s = 1$)?
c) What would be the frequency of the recessive allele if $s = 1$?

44. a) In Europe, up to about 1848, in many species of moths the vast majority of individuals were *light-coloured*; e.g., in *Biston betularia* at least 99% of the population was estimated to be *light-coloured*. With industrialization the frequency of *melanic* (dark-coloured) variants in industrial areas increased until they now are the predominant forms, comprising 95 to 99% of many species. In most of the known cases *melanism* is caused by a dominant allele at a single autosomal locus (Kettlewell, H.B.D., Annu. Rev. Entomol. 6:245, 1961).
i) Propose an explanation for these changes and suggest how you might test your hypothesis experimentally.
ii) There is some evidence that the *melanic* alleles were originally recessive. How might such changes in dominance relationship have come about? What evolutionary significance could such changes have?

b) Equal numbers of *melanic* and *light* forms of *B. betularia* were released into an unpolluted wood in Dorset, England, a relatively unindustrialized area. Five species of birds were observed to eat 190 *betularia*, of which 164 were *melanic* and only 26 were *light*. In a polluted wood near Birmingham, an industrial area, the two types were again released in a 1:1 ratio of *melanic* to *light*. Redstarts ate 15 *melanics* and 43 *light*. A series of release and recapture experiments supported the visual-predation hypothesis. For example, near Birmingham 154 *melanics* and 73 *light* were marked and released; of 98 moths recaptured 82 were *melanic* and 16 *light* (Kettlewell, H.B.D., Heredity 12:51, 1958). Explain if these results support the hypothesis of a visual predation?

45. A corn plant heterozygous for a recessive allele, w, which causes *albinism*, is self-fertilized. The offspring consist of 275 *green* and 85 *white* seedlings. The latter die within about 3 weeks of their appearance.
a) The 275 *green* plants are permitted to cross-fertilize, and the seed is sown the next spring. The number of seeds is not determined, and the investigator, due to illness, is unable to look at the progeny until 6 weeks after seeding, at which time he finds 8,000 *green* plants. How many *albino* seedlings may be expected to have germinated and died? What is the frequency of w in this generation?
b) The population in the next generation consists of 5,000 plants. How many can be expected to be *green* and why?
c) How many generations would it take to reduce the frequency of w to 1 in 20?

46. Chung, C.S. et al. (Ann. Hum. Genet. 23:357, 1959) found that *deaf-mutism* in some families in Northern Ireland is due to a dominant allele at an autosomal locus and in other families it is due to a recessive allele at a different autosomal locus. The dominant alleles are the result of mutation occurring at a frequency of 5×10^{-5} and the recessives at a frequency of 3×10^{-5} per gamete. Assume that the reproductive fitness of dominant mutant homozygotes and heterozygotes is 25% that of *normal* homozygotes and the same is true for recessive mutant homozygotes in relation to *normal* homozygotes and

heterozygotes. What is the expected frequency of each of the two mutant alleles when equilibrium is attained?

47. a) What is meant by the terms *genetic load* and *genetic death*?
b) Explain why, in a population at equilibrium, different deleterious genes, although possessing different adaptive values and therefore different selection coefficients, e.g., the genes for *juvenile amaurotic idiocy* ($s = 1$), *retinoblastoma* ($s = 0.9$), and *chondrodystrophy* ($s = 0.8$), nevertheless may be expected to produce genetic deaths equal to their respective mutation rates.
c) In a population the proportion of individuals that carry a particular allele depends on both the mutation rate and on the degree of selection against the allele. State whether or not you agree with the statement and why.
d) Compare the effects of mutation and selection on genetic loads in Mendelian populations.

48. Individuals with dominant autosomal *retinoblastoma* and recessive autosomal *juvenile amaurotic idocy* almost never reach the age of reproduction. Assume the genes occur in separate populations.
a) If both defective alleles arise by mutation of the *normal* allele at the rate of 3×10^{-5} per gamete, what is the expected frequency of each when equilibrium is reached?
b) Which population would have a higher proportion of homozygous *normals*?

49. An interesting instance of a stable but disturbed equilibrium has been brilliantly investigated by Allison, A.C. (Br. Med. J. 1:290, 1954; Cold Spring Harbor Symp. Quant. Biol. 20:239, 1955). *Sickle-cell anemia* may be regarded as a recessive trait since the sickling process and its attendant severe anemia are suppressed in heterozygotes. These can be distinguished from *normals* by electrophoretic detection of sickle-cell hemoglobin in which the red blood cells become sickle-shaped when the sample is deoxygenated. Recessive homozygotes (*sickle-cell anemics*) do not usually survive long enough to reproduce. It is expected that with such a strong selection ($s = 1$) against the *sickle-cell* allele over many generations its frequency in any present-day population would be extremely low. This is true in most human populations. In some East African tribes, however, its frequency is abnormally high. This is entirely due to a selective advantage of heterozygotes over homozygous *normals*, occasioned by the superior ability of the former to recover from the severe malarial infections that occur in the region, so that the frequency of heterozygotes is actually higher than that of *normals*.
a) Assuming that the frequency $p(Hb^S) = 0.8$ and $q(Hb^A) = 0.2$, what would be the expected genotypic equilibrium frequencies for Hb^SHb^S, Hb^SHb^A and Hb^AHb^A?
b) After one generation, with complete selection against homozygotes, what would the gametic frequencies be? What would the frequency of heterozygotes be?
c) To offset this reduction in frequency of heterozygotes, what would the selective advantage for heterozygotes have to be?

50. In the highly inbred Amish community of Lancaster County, Pennsylvania, a form of dwarfism described as *Ellis-van Creveld syndrome* occurs. About 5 per 1,000 births and about 2 per 1,000 among the 8,000 living members of the group are characterized by the syndrome, which is absent from other Amish communities, such as those of Ohio and Indiana. It is probable that the ancestry of all *afflicted* persons in the Pennsylvania community trace back to a Samuel King and his wife, who immigrated to that area in 1774, although this couple are known not to have been *affected*. All *affected* individuals are known to have had *normal* parents (McKusick, V.A. et al., Bull. John Hopkins Hosp., 114-115:306, 1964).
a) State the most likely mode of inheritance of the syndrome and derive the frequency of the causal allele and of heterozygotes (with selection and under the assumption that the population is large and randomly mating).
b) Provide an explanation to account for the high frequency of the syndrome in the Pennsylvania Amish, together with its complete absence among the other Amish groups.

51. An isolated large human population is at equilibrium with regard to the recessive allele for *albinism*, which is present with a frequency of 0.1. Another isolated population of the same size has never produced any *albinos*. A prolonged period of extensive intermigration occurs between these two populations. If random mating is a continuous characteristic of both populations, what is the expected

frequency of *albinos* in each of these populations after intermigration?

52. Glass, B. (Am. J. Phys. Anthrop. 14:541, 1956) studied the *ABO* blood groups of a very small religious sect in Franklin County, Pennsylvania, known as the Dunkers. The ancestors of this sect consisted of 27 families who in the early eighteenth century came from the Rhineland region of Germany, near Krefeld. These people have remained relatively isolated, both sexually and culturally, from other people in the United Sates. Comparisons of present-day *ABO* blood-group proportions are given in the table.

Group	No. of people	Percent of *ABO* blood groups			
		O	A	B	AB
Dunkers	228	35.5	59.3	3.1	2.2
Rhineland Germans	3,036	40.7	44.6	10.0	4.7
United States	30,000	45.2	39.5	11.2	4.2

a) What is the most likely explanation for the shift in allele frequency in the Dunker isolate?
b) Suggest how this study might have been extended so as to confirm or contradict your answer.

53. In a survey of *albinos* among the Indian populations of Arizona and New Mexico, Woolf, C.M. (Am. J. Hum. Genet. 17:23, 1965) noted that in the majority of these populations *albinos* were absent. However, their frequency was high in three populations: 1 in 227 among Indians of Arizona, 1 in 140 among the Jemez Indians of New Mexico, and 1 in 247 among the Zuni of New Mexico. All 3 populations are culturally but not linguistically related. Other Indian tribes contain *albinos* at a much lower frequency (not as low, however, as that of the Caucasian population in the United States, which is 1 in 20,000). Explanations for the high frequency of *albinos* in these Indian populations are:
i) Cultural selection in past generations.
ii) Gene flow from one population to another.
iii) Genetic drift.
iv) Selection for the heterozygote.

a) Explain which of these possibilities seems to be the most likely cause?
b) Suggest how this survey might be extended to confirm or contradict your answer.

54. The water snake *Natrix sipedon* exhibits 4 major types of banding patterns: *A, no bands; B, slight banding; C, intermediate banding; D, complete banding*. All types of snakes occur on the islands in western Lake Erie. Almost all snakes from the mainland surrounding Lake Erie are of type *D*. Moreover, except in the area of western Lake Erie and one area in Tennessee, all known populations of this species are of type *D*. The distribution of the banding types in large samples of adult snakes and litters from pregnant females from the islands is illustrated (Canin, J.H. and P.R. Ehrlich, Evolution 12:504, 1958).

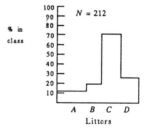
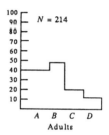

The percentage of relatively *unbanded* individuals is higher in adult than in young (litter) populations. Snakes in the laboratory show no evidence of pattern changes during development. The following processes might be responsible for the significant difference in pattern-type frequencies between adult and litter populations: a) Mutation; b) Genetic drift; c) Migration; d) Selection. Which of these factors seems most likely to be responsible and why?

55. The following represents a summary of data presented by Boyd, W.C. ("Genetics and the Races of Man," Little, Brown, Boston, 1950):
i) Among American Indians the tribal values for the frequency of *s (secretor)* vary from zero to 0.01; in Japanese, a nontribal people, the frequency is 0.24.
ii) In American Indians the *Rh* allele is absent; in different Asiatic populations it varies in frequency from 0.6 to 0.015.

iii) In the majority of American Indian tribes the *B* blood-group allele is absent; the remainder show values ranging from 0.16 to 0.024; in Asiatics the frequencies (tribal or otherwise) are never lower than 0.27.

iv) Among American Indians the tribal values for the *N* blood-group allele vary from 0.01 to 0.08; in Eastern Asiatics the frequencies are 0.18 for Chinese and 0.20 from Japanese.

It is a commonly accepted belief that the American Indians are the descendants of Asiatics who crossed the Bering Strait from Siberia. Assuming that at some place in Siberia these peoples "rested" for some considerable period, show how these observed differences in gene frequencies may be explained:
a) Between Asiatics and American Indians.
b) Between tribes of American Indians.

56. If a human trait is present with a very low frequency in the population, would institution of a program discouraging reproduction by *affected* persons be likely to cause significant changes in the frequency of the trait? Of the causal allele? Discuss fully.

57. Dobzhansky, T. and O. Pavlovsky (PNAS 41:289, 1955) studied a population of *Drosophila tropicalis* in the vicinity of Lancetilla, Honduras and found that it consisted chiefly of heterozygotes for an inversion in the second chromosome. Slightly more than 50% of the fertilized eggs developed to the adult stage. Most of the mortality occurred in the egg stage, but some larval and a small amount of pupal mortality also occurred. Offer an explanation of these observations.

58. In a certain population of beetles, *three-spot* vs. *four-spot* wings is determined by an autosomal pair of alleles. The allele *S* for *three-spot* has a frequency of 0.44 and *s* for *four-spot* a frequency of 0.55. What are the possible explanations for the intermediate frequencies of the two alleles?

59. In the South American species *Drosophila polymorpha*, body colour is monogenically determined, the body being *dark(EE)*, *intermediate(Ee)* or *light(ee)*. In natural populations, the frequencies of the three traits remain constant for many generations. Da Cunha, A.B. (Evolution 3:239, 1949) found that a laboratory population of F_2's from a cross between *dark(EE)* and *light(ee)*, consisted of 1,605 *dark*, 3,767 *intermediate* and 1,310 *light* individuals. Determine the relative adaptive values of the three genotypes and explain why the frequencies of the traits remain constant from generation to generation.

60. *Drosophila melanogaster* homozygous for the *Arrowhead* inversion are placed together with flies homozygous for the *Chiricuhua* inversion in a population cage. Sampling after one generation of mating reveals that the two kinds of inverted chromosomes are present at frequencies of 0.8 (*Arrowhead*) and 0.2 (*Chiricuhua*). After 15 generations, these frequencies are found to have changed to 0.6 and 0.4 respectively; the latter values remain unchanged in generations beyond the fifteenth. State the most probable explanation for these results and describe how the change and subsequent stabilization of ratios comes about.

61. a) What is meant by transient polymorphism?
b) Populations of *Biston betularia* in industrial areas of England are now almost entirely *melanic*. The transition took about 50 years. In other species the transition has taken place much more rapidly. For example, the change to *melanism* at Hamburg, Germany, in a species of the genus *Tethea* took place in the period 1904-1912. What factors can you think of that may affect the rate of displacement of one allele by another?

62. Different species of animals may in some instances be more closely similar in appearance than different breeds within a species are. For example, several species of bears appear more alike than certain breeds of dogs, e.g., the Labrador retriever and the Chihuahua. Why do we designate the former as belonging to different species while grouping all dogs in the same species?

63. a) What other mechanisms besides gene mutation may be considered as providing raw material for evolution?
b) State what basic features of evolution theory apply to sexually and asexually reproducing organisms alike and what basic differences may exist in the evolutionary process with respect to these two kinds of organisms.

64. a) Define *race* and discuss the genetic differences that exist between two such entities, illustrating your answer from humans.
b) Explain why genetic studies of populations are a prerequisite to classifying a species into races.
c) Is it easier to classify organisms into species than into races? Explain.
d) Although both species and races are natural biological entities, the lines of demarcation between the former are more distinct than between the latter. Why?
e) What is the important change that transforms races into species?
f) Do you think that the number of human races will increase in the future? Explain.
g) Why is it impossible to decide the race to which an individual belongs by comparing his (her) phenotype with a racial average?
h) People with *O, A, B* or *AB* blood groups do not belong to distant races. Why?
i) Some people contend that interracial marriages are biologically undesirable, others contend the reverse. Which do you believe and why?

65. Explain how species relationships can be measured at (1) the protein level and (2) DNA level through *in vitro* hybridization.

66. What are the mechanisms by which new species may arise suddenly? Illustrate your answer with examples.

67. What reasons can you give for including all human beings in the one species, *Homo sapiens*?

68. Domesticated species of plants and animals exhibit greater variability than the corresponding wild species. How do you account for this?

69. Is morphological specificity alone adequate for defining a species? Explain, citing examples. If not, what other criteria should be used? Which of these criteria can and which cannot be used to distinguish species among asexual organisms?

70. The increase in chromosome number, occurring in polyploids, is in itself not important in speciation. However, when it is combined with interspecific hybridization, it becomes an important factor in speciation. Discuss.

71. Explain how sympatric races may become reproductively isolated.

72. Certain species occur only in restricted geographic regions. What genetic explanation can you offer to account for this?

73. The Monterey and Bishop pines in California are different sympatric species which shed their pollen at different times. Do you think that hybrids between them would have a chance of becoming established in nature?

74. What are the different kinds of barriers causing reproductive isolation? Discuss each and cite an example where possible.

75. The three primary processes governing the rate and direction of evolution are mutation, recombination, and selection. Is any one of these more important than the others and why?

76. Using two parental species with different chromosome numbers, illustrate the process of speciation by amphidiploidy.

77. In allotetraploid organisms like cultivated tobacco and the macaroni wheats, a fairly large number of characters are determined by duplicate genes. In the allohexaploid bread wheat certain characters are determined by triplicate and others by duplicate genes. Account for the origin of these systems.

78. Discuss the role of heterochromatin in karyotype evolution involving changes in chromosome numbers, referring to specific examples where possible.

79. In the perennial ciquefoil plant species (*Potentilla glandulosa*) in California, three allopatric races inhabit the *coast* ranges, the *foothills* and *alpine* regions, respectively, of the Sierra Nevada. In the coastal habitat, where

temperatures rarely fall below the freezing point, plants of this species grow throughout the year; in contrast, the habitats of the other regions are characterized by cold winters with heavy snowfall. When reciprocally transplanted, the *coastal* and *foothills* races do not thrive as well as they do in their native habitats. Nevertheless, the *foothills* race shows vegetative growth during the winter on the coast, and the *coastal* race, becomes dormant in the winter in the *foothills*. Both these races are usually killed within a year when transplanted to the *alpine* zone. The *alpine* race, growing as a dwarf in its own environment, remains so when planted on the coast. Hybrids between the races are vigorous and fertile. The species was once very uniform genotypically and phenotypically (Clausen, J., D.D. Keck and W.M. Hiesey, Carnegie Inst. Wash., D.C. Publ. 520, p.1. 1940). Describe the probable manner in which the adaptation of geographic races to specific habitats arose.

80. Sturtevant, A.H. (Genetics 5:488, 1920; Carnegie Inst. Wash., D.C. Publ. 399, p.1. 1929) performed reciprocal crosses between *Drosophila melanogaster* and *D. simulans*, both of which have an X-Y method of sex determination, and obtained sterile progeny with rudimentary gonads. The results of his studies on the inheritance in these offspring of traits known to be X-linked in the parental species are shown.

Parents		Offspring	
♀	♂	♀	♂
melanogaster (XX) x simulans		regular	none
melanogaster (XXY) x simulans		regular*	exceptional*
simulans (XX) x melanogaster		regular	none
simulans (XXY) x melanogaster		regular	exceptional

* *Regular* offspring inherit the determiners of X-linked traits normally (viz. males inherit them only from female parent, and females from either parent); *exceptional* offspring inherit these determiners abnormally (viz. males inherit them from the male parent only, and females inherit them from the female parent only).

a) Suggest an hypothesis to account for this type of interspecific hybrid sterility.

b) Suggest experiments that might have been made to test this hypothesis had the offspring been fertile and outline the results expected.

81. Two species can give rise to a new species after interspecific hybridization by allopolyploidy (amphidiploidy), by introgression (natural selection of recombinant progeny) of the interspecific hybrid crossed with its parents, or by the occurrence of recombinants that are reproductively isolated from the parental species among its own progeny. The following is an account of the work of Lewis, H. and C. Epling (Am. J. Bot. 33:21s, 1946) and C. Epling (Am. Nat. 81:104, 1947) in the larkspur (*Delphinium* spp.). *D. gypsophilum* is intermediate in morphology between *D. hesperium* and *D. recurvatum* and occupies a new habitat which is intermediate between that of the other two species. All three species are diploid with 2n = 16. The F_1 hybrid between *D. recurvatum* and *D. hesperium* resembles *D. gypsophilum*. The progeny from the cross (F_1 hybrid x *D. gypsophilum*) are more fertile than those obtained by backcrossing the F_1 hybrid to either parent or by crossing *gypsophilum* with either of the other two species. What is the phylogenetic relationship among these three species, and which of the above methods of speciation is responsible for it? Give reasons for your decisions.

82. *Drosophila pachea*, found only in the Sonora Desert, breeds exclusively in the stems of the senita cactus (*Lophocereus schottii*), which synthesize the sterol Δ^7-stigmasten-3β-ol. It does not reproduce in the laboratory unless a piece of cactus stem or the sterol is added to the medium. No other *Drosophila* species utilizes the stem of this cactus for breeding although 2 other species in this area breed in the fruits of this cactus. The sterol and several other alkaloids of the cactus are lethal to these other species (Heed, W.B. and H.W. Kircher, Science 149:58, 1965). *D. pachea* is ecologically isolated from other sympatric *Drosophila* species. Does it necessarily follow that the species arose by divergence of sympatric races?

83. Moore, J.A. (Genetics 31:304, 1946) collected living samples of frogs (*Rana*) from 4 different geographic habitats along the east coast of the United States. These populations were intermated and the F_1 hybrids were studied for development and head size. The results of Moore, J.A. (Biol. Symp. 6:189, 1942) in modified form are shown below:

Rates of development and mature head size in hybrids

♀	♂ Vermont	New Jersey	Central Florida	Southern Florida
Vermont	Normal	Normal or very slight acceleration; head normal or very slightly enlarged	Moderate retardation; head considerably enlarged	Marked retardation; head extremely enlarged
New Jersey		Normal	Very slight retardation; head very enlarged	Slight retardation head moderately enlarged
Central Florida		Slight retardation; head slightly enlarged	Normal	Normal; head normal
Southern Florida	Marked retardation; head considerably reduced			Normal

The average seasonal temperatures for egg-to-adult development are Vermont, 45°F; New Jersey, 40°F; central Florida, 29°F; and southern Florida, 27°F.

Note: (1) Vermont-southern Florida hybrids are very inviable so few would survive in nature; all other hybrids develop into *normal* adults. (2) The greater the distance between populations the greater the retardation in rate of development, the expression of the defects, and the proportion of inviable offspring.

a) Would you classify these 4 populations as different races of the same species? Explain.
b) If so, how would you account for:
i) The gradual manner in which reproductive (genetic) isolation becomes established?
ii) Failure of these populations to reach the status of distinct species?

84. What requirement(s) must be satisfied for a single mutation to give rise to a new species?

85. Polyploidy has been relatively unimportant in the evolution of animals. With some exceptions, e.g., autotetraploid amphibians such as the South American frog *Odontophrynus americanus*, polyploid animals are restricted to certain of the lower forms (insects, crustacean, etc.). This cannot be said for chromosomal aberrations such as reciprocal translocations, deletions and inversions.
a) Discuss some of the probable reasons for the apparent unimportance of polyploidy in evolution in the animal kingdom.
b) Cite evidence that supports the latter statement.
c) Explain why aberrations have been more important than polyploidy in animal evolution.

86. Grant, V. (Genetics 54:1189, 1966) crossed *Gilia malio* (2n = 36) x *G. modocensis* (2n = 36) and obtained a highly sterile F_1 hybrid, intermediate in phenotype between the parents, with an average of six bivalents per meiocyte. By intercrossing and selecting for fertility, *fully fertile* (2n = 36) F_{10} plants were obtained with a new combination of morphological characters. These plants produced *sterile* hybrids in crosses with both parents.
a) How does this form of speciation differ from amphidiploidy?
b) Suggest how the F_{10}'s were derived and what their chromosome constitution might be like.

87. a) What is an *evolutionary tree*?
b) The human hemoglobin gene Hb^α is on chromosome 16 and codes for the α polypeptide (*pp*) chain which is 141 amino acids (AAs) long. The hemoglobin genes Hb^β, Hb^δ and Hb^γ are in juxtaposition on chromosome 11 and code for the β, δ and γ chains respectively, each 146 AAs long. The α pp

differs from the β, δ and γ pps by 84, 85 and 89 AAs respectively. The β pp differs from the δ and γ chains by 10 and 39 AAs respectively whereas the δ and γ pps differ by 41 AAs. (Data from McKusick, V.A., <u>Mendelian Inheritance in Man</u>, 6th Edition. The Johns Hopkins University Press, Baltimore, 1983.)

ii) Construct an evolutionary tree for these 4 pps, based on the information provided.
ii) Do these data indicate the direction of evolution of these 4 genes? Explain.
iii) Which chromosomal mechanism(s) were probably associated with the origin of these closely related genes? Explain.
iv) How would you account for the fact that gene Hb^α is on chromosome 16 and the other 3 genes are on chromosome 11 given that these genes came about through duplications?

88. a) Explain what is meant by the phrase "genetic distance between species".
b) What are some of the techniques that comprise the molecular approach and its utilization as a yardstick for the estimation of the genetic distance between two species?
c) The genetic distance between humans and chimpanzees is as small as that between 2 sibling species of fruit flies or of another group of organisms (King, M.C. and A.C. Wilson, Science 188:107, 1975). How would you account for the paradox that the genetic distance between these 2 species is very small and yet they differ significantly at the organismal level in terms of their anatomy, way of life, etc.?
d) Will molecular phylogenetic analysis provide the necessary insights into understanding the basis for and the pace of evolution? What meaningful function is such analysis likely to serve?

89. The rate of change in a protein molecule as determined from correlations between amino acid substitutions and the time of divergence of two organisms according to fossil records is *constant* for a given protein but *varies* widely for different proteins (Dickenson, R.E., J. Mol. Evol. 1:26, 1971). The time (in millions of years) that is required for organisms that have diversed from one another to reveal a 1% change in the amino acid sequences for 3 different classes of proteins is as follows: i) Class a - 1.1 million years; ii) Class b - 5.8 million years; and iii) Class c - 20 million years.
a) Why would the rate of change be different for each class of protein?
b) What conclusions can you make about the 3 classes of proteins from these values, i.e., which of the classes represents the hemoglobin, cytochrome oxidase and fibrinogen types of proteins? Explain.
c) Histones are chromosomal proteins in eukaryotic species. A modification in the amino acid sequence, anywhere along the length of a histone protein, would alter its binding capacity and the subsequent packaging of the DNA. How would the rate of change in these nucleoproteins compare with the rates for the 3 classes of proteins shown above? Explain your answers.

90. The amino acid sequence of *cytochrome c*, a protein involved in cellular respiration, varies from species to species. The data below (modified from Fitch, W.M. and E. Margoliash, Science 155:279, 1967) shows the number of amino acid changes in the sequence of this protein for five organisms, relative to that of humans:

Chimpanzee - 0; Rhesus monkey - 1; Donkey - 11; Horse - 12 and Yeast - 44

Which of the 6 organisms are: i) most closely related? ii) most distantly related? Explain your answers.

91. a) A large fraction (40-60%) of mammalian DNA comprises non-repeated DNA sequences. Most of a species genetic information is contained within this fraction of unique sequences. The other large fraction of mammalian DNA is made up of repetitive sequences (Britten, R.S. and D.E. Kohne, Science 161:529, 1968).
i) Which of the 2 kinds of DNA would you expect to renature very rapidly? Why?
ii) Which of the 2 fractions would be most suitable in determining the extent of nucleotide changes in 2 related species since their divergence? Why?

b) Assessment of genetic distance between 2 species can be accomplished through analysis of their DNA contents and/or the detection of changes in the amino acid sequence of a given protein. Are both of these molecular approaches likely to prove equally effective and reliable in phylogenetic analysis? Explain.

c) The table below (modified from Kohne, D.E. et al., J. Hum. Evol. 1:677, 1972) shows the percentage of nucleotide changes between the DNA of humans and that of various other primates:

Species	% of nucleotide change
Man - green monkey	9.5
Man - capuchin	15.8
Man - galago	42.0
Green Monkey - chimp	9.6
Green Monkey - gibbon	9.6
Green Monkey - capuchin	16.5
Green Monkey - galago	42.0

i) Which of these primates are: closely related? more distantly related? Explain.
ii) What additional information would you require to verify your answers.
iii) Construct an evolutionary tree from these data.

92. In 1979, Bruce, E.J. and F.J. Ayala (Evolution 33(4):1040) investigated the phylogenetic relationships among 9 hominoid species by studying allelic variations at 23 gene loci that code for red blood cell and plasma proteins. Their results for 16 loci in 5 of the 9 species are given in the table below: At each locus the most common allele found in humans is called 100; all other alleles at each locus in the 5 species are named *vis-a-vis* allele 100 by adding or subtracting the number of millimeters by which the protein specified by a specific allele differs in its migration in electrophoretic gels from that encoded by allele 100. Fox example, in the chimpanzee, the allele at the *Ak* locus migrates 4mm less than the allele in humans and is refered to as allele 96. Where more than one allele has been detected in a species, the frequency for each allele is given in parentheses.

Locus	Human (*Homo sapiens*)	Chimpanzee (*Pan troglodytes*)	Gorillaa (*Gorilla gorilla*)	Gilbon? (*Hylobates lar*)	Siamang (*Symphalanges syndactylus*)
Ada	100	103	-	-	-
Ak	100	96	98(.20) 100(.80)	92	96
Cat	100	100	100	-	-
Fum	100	-	100	100	100
G6pd	100	100	100	102	102
Got-s	100	100	100	100	98
Icd-s	100	96	100	100	100
Ldh-A	100	96	96	96	96
Mdh-s	100	100	100	93(.62) 100(.38)	102
Dia	100	100	85(.67) 95(.33)	100(.67) 108(.33)	100
Pgm-1	100	96(.62) 100(.88)	100	100	103
6-Pgd	100	97	97(.25) 100(.75)	100	100
Alb	100	100	100	100	102
Cer	100	100	98	98	98
Est-A	100	100	98	102	102
Hpt	100	105	107	107	107
Lap	100	100	100	100	100

a) Calculate the genetic identity, I and genetic distance, D, among the 5 species using the method of Nei, M. (Am. Nat. 106: 283, 1972).

b) Draw the most likely configuration of the phylogenetic relatedness.

Note: Additional questions on raciation and speciation are present in Chapters 14 (Euploidy: Haploidy and Polyploidy) and 16 (Chromosome Aberrations).

Chapter 14 - questions 11, 12, 13, 14, 16, 17, 18, 21, 16, 37 and 40
Chapter 16 - questions 16, 17, 34, 35, 49, 50, 51, 53 and 54

Chapter 25
Current Approaches to Genetic Analysis: Somatic Cell Hybrids, RFLPs and Recombinant DNA

1. a) What are: Cell strains? Permanent cell lines? Somatic cell hybrids?
   b) What properties do these 3 cell types exhibit? Which techniques and what properties would you use to distinguish somatic hybrid cells from their progenitor cells?
   c) What is somatic cell genetics? Why are interspecific somatic-cell hybrids which involve human cells of importance in human genetic and cytogenetic analysis?
   d) Interspecific somatic-cell hybrids can also be generated in plants, for example, between tobacco and petunia cells. However, such hybrid cells in plants are of limited importance in genetics. Why?
   e) What is the minimal number of different clones from human-rodent somatic cell hybrids that is necessary to determine which human chromosome carries a given human gene? Explain.

2. The structural-gene loci specifying human glucose-6-phosphate dehydrogenase (G6PD) and HGPRT are X-linked. Miller, D.J. et al. (PNAS 68:116, 1967) derived 6 interspecific somatic hybrid cell lines by fusing mouse cells able to synthesize G6PD but deficient in HGPRT with human diploid cells sufficient for both G6PD and HGPRT. Cell fusion was mediated using ultraviolet-inactivated Sendai virus. Hybrid cells and their clones were grown on HAT medium. Human HGPRT and mouse G6PD were present in all 6 hybrid lines and the 105 clones derived from them. In 2 of the 6 hybrids and in 47 of the 105 clones from the other 4 hybrid lines, human G6PD was absent. Note the following points:

(i) Reversion to HGPRT synthesis was not observed in "control" fusions within the mouse parental lines.
(ii) HGPRT produced by all mouse-human hybrid cells, with or without human G6PD, was always of the human type.
(iii) Inactivated Sendai virus can produce chromosome breaks.
(iv) After cloning, cells did not undergo phenotypic changes.

Explain which of the following hypotheses is the most plausible to account for the absence of human G6PD in 47 out of 105 clones:

(a) Use of human cell donors with a slightly leaky or nonleaky mutant allele at the G6PD locus.
(b) Somatic mutation of the *wild-type* to *mutant* allele at the G6PD locus.
(c) Repression of synthesis of human G6PD or inhibition of activity of the enzyme.
(d) Back mutation or virus-dependent induction at the mouse HGPRT locus, resulting in synthesis of HGPRT indistinguishable from human HGPRT.
(e) Loss of the G6PD structural locus.

3. In humans both glucose-6-phosphate dehydrogenase (G6PD) deficiency and hypoxanthine guanine phosphoribosyl transferase (HGPRT) deficiency are due to recessive alleles of different X-linked genes. Diploid fibroblasts from a G6PD-deficient male were hybridized by Siniscalco, M. et al. (PNAS 62: 793, 1969) with fibroblasts from a HGPRT-deficient male. The 4n=92 mononucleate hybrid cells synthesized both G6PD and HGPRT. Sex chromatin was not observed in these cells. Offer a plausible explanation for these surprising results.

4. The following data are based on the results of Creagan, R. et al. (Am. J. Hum. Genet. 26:604, 1974); Ruddle, F.H. et al. (Nature 227: 251, 1970); and Taylor, A.K. et al. (Genomics 10: 425, 1991). Cells from a karyotypically and enzymatically normal human fibroblast cell line (2n=46) were fused with cells from a permanent mouse cell line (2n+7 = 47) that possessed a mutant allele at the HGPRT locus and thus did not produce the enzyme. Six hybrid clones, selected on HAT medium, were assayed for the presence (+) or absence (-) of 9 human enzymes, each coded for by a single gene. The clones were also examined

cytologically for the presence (+) or absence (-) of human chromosomes. The results were as follows:

Human enzymes	Hybrid clones					
	A	B	C	D	E	F
Isocitrate dehydrogenase (IDH)	-	+	-	-	+	-
Hexosaminidase A (HEX-A)	+	-	-	+	+	-
Lactate dehydrogenase B (LDH-B)	+	-	+	-	-	+
B-galactosidase (B-GAL)	-	+	-	+	-	+
Enolase-1 (ENOL-1)	-	+	+	+	-	-
Carboxyl ester lipase (CEL)	-	-	-	+	+	+
Peptidase B (PEP-B)	+	-	+	-	-	+
Aspartyl glucosaminidase (AGA)	-	-	-	-	-	-
Malate dehydrogenase (MDH)	-	+	-	-	+	-
Human chromosomes						
1	-	+	+	+	-	-
2	-	+	-	-	+	-
9	-	-	-	+	+	+
12	+	-	+	-	-	+
15	+	-	-	+	+	-
19	+	-	-	-	-	+
22	-	+	-	+	-	+
X	+	+	+	+	+	+
Y	+	+	-	-	-	+

a) Which of the 9 genes are syntenic and which are asyntenic? Explain.
b) Explain whether it is possible to determine if the syntenic genes are linked or not.
c) Does chromosome 9 carry any of these genes? Explain.
d) Which of these genes is located on chromosome 22? Which chromosomes carry the other genes? Explain.
e) Give 2 possible explanations for the results when enzyme AGA is assayed for.
f) On which of the chromosomes is the gene for HGPRT located? Why?
g) Are any of the 9 genes located on the Y chromosome? Explain.

5. Dalla-Favera, R. et al. (PNAS 79: 7824, 1982) fused TK⁻ mouse fibroblast cells with human lymphocyte and Burkitt lymphoma cells. Hybrid cells and clones were selected on HAT medium. By using a radioactively labeled probe for the human c-myc proto-oncogene, 16 different clones were analyzed by the Southern blotting technique. Their results for 10 clones were as follows:

Human chromosomes	Hybrid Clones								
	A*	B	C	D	E	F	G	H	J	K
1								+		+
2				+						
3	+							+		+
4					+		+			
5					+			+		+
6			+							
7	+	+								
8	+			+	+					+
8q⁻							+			
9										+
10										+
11							+			
12				+			+			
13										+
14	+	+						+		+
14q⁺									+	
15		+			+					
16	+						+			
17	+	+			+					
18		+						+		+
19				+						
20						+	+			+
21		+								
22	+	+			+		+	+		+
X		+								
Human c-myc	+			+	+				+	+

* The identity of the clones has been simplified for ease of presentation. With respect to a given human chromosome and the human c-myc gene, + = presence and lack of + = absence in a particular clone.

Note: Chromosomes 8q⁻ and 14q⁺ are products of a reciprocal translocation between the short terminal regions of the q arms of chromosomes 8 and 14.

a) Give a complete cytogenetic explanation of these results.
b) *Burkitt's lymphoma* (*BL*) is a malignancy of the B cells in the human immune system. In normal B cells all chromosomes are normal. In B cells of individuals with *BL* all chromosomes except 8q⁻ and 14q⁺ are normal. In 14q⁺ chromosomes the gene coding for the heavy protein chains of antibody molecules is juxtaposed next to the breakpoint in 14q⁺ chromosomes.

In view of this information and your answer to (a), propose a plausible mechanism to explain why B cells in *BL* patients become malignant. Compare your explanation with that of Croce, C.M. and Klein, G. (Sc. Am. 252(3): 54, 1985).

6. Ricciuti, F.C. and F.H. Ruddle (Genetics 74: 661, 1973) fused cells from the human fibroblast line KOP-2 with hypoxanthine-guanine phosphoribosyltransferase (*HGPRT*) deficient cells from the mouse RAG line derived from a renal adenocarcinoma strain. Hybrid clones were selected on HAT medium and analyzed for the expression of 3 X-linked human genes *HGPRT*, *G6PD* (glucose-6-phosphate dehydrogenase) and *PGK* (phosphoglycerate kinase) and the autosomal gene NP (nucleoside phosphorylase). The normal human chromosomes X and 14 have the morphologies shown in A.

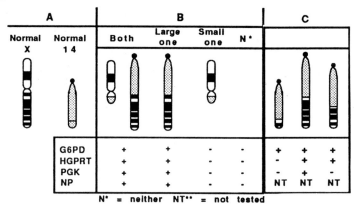

Clones with mutant chromosomes were tested for the presence (+) or absence (-) of the enzymes coded for by these 4 genes. The results are shown in B. Gerald, P.S. and J.A. Brown (Cytogenet. Cell Genet. 13: 29, 1974) showed that clones with other mutant X and 14 chromosomes expressed the phenotypes shown in C.
a) What types of mutant chromosomes were present in the original hybrid clones? Explain.
b) Does either of the chromosomes, X or 14, carry the NP gene? Explain.
c) Where specifically on the X chromosome are the 3 X-linked genes located? What is their sequence and why?

7. Jones, C. and F.T. Kao (Hum. Genet. 45: 1, 1978) used a hybrid clone JI which contains the entire chromosome complement of Chinese hamster ovary cells plus human chromosome 11, to isolate 5 subclones, each of which displayed a different terminal deletion of chromosome 11 diagrammed below:

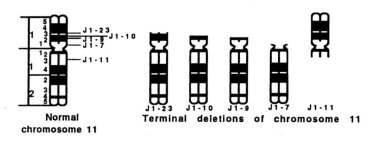

—indicates the breakpoints at which the deletions occurred in chromosome 11 in the 5 subclones

Gusella, J. et al. (PNAS 76: 5239, 1979) conducted cytogenetic studies using these 6 hybrid clones and the human genes coding for the β-globin protein, cell surface antigens *SA11*-1, -2, and -3 and the enzymes lactate dehydrogenase A (*LDH-A*) and acid phosphatase (*ACP-2*). The results of their studies using cloned β-globin genes and Southern hybridization, antisera that detect the 3 antigens and biochemical assays for the *LDH-A* and *ACP-2* are shown on the next page:

Clone	Presence of specific human markers					Assay for human Hbβ gene
	SA11-1	SA11-3	LDH-A	ACP-2	SA11-2	
J1	+	+	+	+	+	+
J1-23	−	−	+	+	+	+
J1-10	−	−	−	+	+	+
J1-9	−	−	−	+	+	−
J1-7	−	−	−	−	+	−
J-11	+	+	+	+	−	+

a) Which segment of chromosome 11 is deleted in each of the mutant chromosomes in the 5 subclones?
b) With reasons for your answers, give a complete cytogenetic explanation of these results, indicating which of these genes is on chromosome 11 and the specific region of this chromosome that carries each of these genes.
c) Briefly outline the important practical and theoretical implications of these kinds of studies.

8. For regional mapping of the human carboxyl ester lipase (*CEL*) gene on chromosome 9, Taylor, A.K. et al. (Genomics 10: 425, 1991) utilized 3 hybrid clones, each of which contained all the Chinese hamster chromosomes and a specific human chromosome with a X/9 reciprocal translocation. Clone CF11-4 possessed the 9 pter → 9q34::Xq13 → Xqter translocation. Clones CF57-14 and CF57-1 contained the 9pter → 9q24::Xq12 → Xqter chromosome. Subclone CF57-IR did not possess the translocated chromosome. The 9q34 → 9qter region occupies ~ the distal 1/4 of the q arm. Total human genomic DNA and DNA of the human chromosomes from the clones was digested with *PuvII* and hybridized with DNA from the *CEL* gene. The results of Southern blot analysis are shown in the top portion of the illustration below. The DNA from clones CF11-4 and CF57-14 was probed with DNA from the 4 markers *gelsolin, MCO A12, AKI* and *Lamp 92*. The results of Southern blot analysis with these markers are shown in the bottom portion of the illustration.

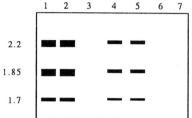

Lanes 1&2, human genomic DNA; lane 3, DNA from clone CF11-4; lane 4, DNA from clone CF57-14; lane 5, DNA from clone CF57-1; lane 6 DNA from clone 57-1R; lane 7, hamster DNA

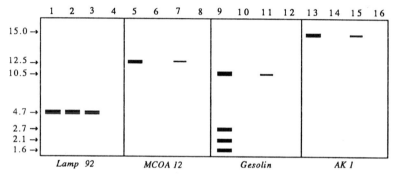

Lanes 1,5,9,&13 contain human genomic DNA. Lanes 2,6,10,&14 contain clone CF11-4 DNA. Lanes 3,7,11,&15 contain CF57-14 DNA. Lanes 4,8,12,&16 contain hamster DNA.

a) With reasons for your answer indicate the arm of chromosome 9 that carries the CEL gene and the specific region of the arm that this marker resides in.
b) Which of the latter 4 markers is closest to the centromere? Explain. Why is it not possible to determine the sequence of the other 3 markers from these data?

9. a) What is a palindromic DNA sequence? Why are such sequences important in molecular genetics studies?
b) What are restriction endonucleases? Why are these enzymes important in genetic studies?
c) What are restriction fragment length polymorphisms (RFLPs)? Are allelic RFLPs usually dominant-recessive, overdominant or codominant? Explain.
d) Assume that A,C,G and T are present in equal proportions and randomly arranged in a long DNA molecule.

You cleave the DNA with *BamHI* which recognizes the 6-bp sequence:

GGATCC
CCTAGG

i) On the average how long would the restriction fragments of DNA be? Explain.
ii) What would your answer be if the enzyme was *HpaI*, which recognizes the 4-bp sequence:

GCGC
CGCG

e) How many restriction enzyme recognition sequences are present in the DNA duplex shown below?

5'- C A G A G A A A T C C G G G C A T T T C G A C G A T C G C A G A A T T C G A -3'
3'- G T C T C T T T A G G C C C G T A A A G C T G C T A G C G T C T T A A G C T -5'

f) What types of mutations are likely to produce differences in restriction fragment lengths? Illustrate.
g) Bacterial species produce restriction endonucleases. Why do these enzymes fail to fragment the DNA of the species that produces them?

10. A 34 kb long DNA duplex has 2 *BamHI* recognition sequences (B) and 1 sequence recognized by *TaqI* (T) as shown below:

$$\underset{\text{B}}{\underline{\text{5 kb}}}\underset{\text{T}}{\underline{\text{9 kb}}}\underset{\text{B}}{\underline{\text{8 kb}}}\underline{\text{12 kb}}$$

How many bands and of what fragment lengths would you expect in an electrophoretic gel when this DNA molecule is digested with:
i) *BamHI*? ii) *TaqI*? iii) both enzymes?

11. A large sample of identical 32 kb long DNA duplexes is cut with either *EcoRI* or *BamHI* or with both restriction enzymes. In each case, the fragments are separated from each other in an electrophoretic gel and are of the following sizes:

EcoRI lane: 3 bands with fragment lengths of 7, 9 and 16 kb;
Bam HI lane: 2 bands with fragment lengths of 12 and 20 kb;
EcoRI and *Bam HI* lane: 4 bands with fragment lengths of 3, 4, 9 and 16 kb

Draw a restriction map of this DNA duplex.

12. One possible overlay of the restriction maps of a 600 bp segment of DNA digested separately with the restriction enzymes *EcoRI* and *HindI* is shown below to the left. The electrophoretic bands of the double digest are also presented.

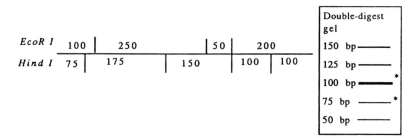

The end segments are denoted by asterisks. Is the overlay shown above correct? Explain.

13. A linear DNA duplex upon cutting with *EcoRI* produced 4 fragments E1 to E4 which were 8, 5, 3 and 2 kilobase pairs (kb) long, respectively. When the same molecule was cut with *BamHI*, again 4 fragments, designated B1 to B4 were obtained which were 7, 5, 3.5 and 2.5 kb long, respectively. *EcoRI* fragments were isolated and digested with *BamHI* with the following results:

E1 gave subfragments of 5, 2 and 1 kb
E2 was not cut further
E3 gave subfragments of 2.5 and 0.5 kb
E4 was not cut further

BamHI fragments were isolated and cut with *EcoRI* with the following results:

B1 gave subfragments of 5 and 2 kb
B2 was not cut by *EcoRI*
B3 gave subfragments of 2.5 and 1 kb
B4 gave subfragments of 2 and 0.5 kb

a) Draw a restriction map of the DNA molecule, showing the positions of all restriction sites.
b) Which *EcoRI* fragments would hybridize to fragment B3 by Southern blotting?

14. a) Morimoto, R. et al. (Nucl. Acids Res. 4: 2331, 1977) cleaved the mitochondrial DNA (mtDNA) from the yeast *Saccharomyces cerevisiae* with the restriction endonucleases *EcoRI* and *HpaI*. Fragments were separated by agarose gel electrophoresis and photographed under UV light. The molecular sizes (kb) of the cleavage products produced by digestion with these enzymes were as follows:

Digestion

Fragment Number	EcoRI	HpaI	EcoRI 23.7 fragment with HpaI	EcoRI 8.3 fragment with HpaI	HpaI 20.6 fragment with EcoRI
1	23.7	23.7	10.0	8.3	10.0
2	17.3	20.6	5.6		8.3
3	10.0	14.7	5.4		3.5
4	8.3	7.1	3.2		
5	7.8	5.6			
6	3.5	3.2			
7	2.4	2.3			
8	1.7	77.2			
9	0.9				
10	0.2				
	75.8				

With reasons for your answer, draw a restriction map (position of restriction enzyme sites) of the mtDNA of yeast as accurately as possible.

b) The *EcoRI* 23.7 kb fragment was digested with *SalI* and produced 2 fragments; 14.4 and 9.6 kb in length. Digestion of the latter fragment produced 5.4 and 4.2 kb fragments. Does this additional information help resolve any ambiguities in the map? Explain.

15. Restriction fragment length polymorphisms (RFLPs) can be used as genetic markers in prenatal diagnosis. The following table gives the frequency of associations of *HpaI* fragment sizes and hemoglobin A and S genes in a sample of American Blacks (Kan Y.W. and A.M. Dozy, PNAS 75: 5631, 1978)

HpaI Fragment Size	β^A Gene	β^S Gene
7.6 kb	0.88	0.31
7.0 kb	0.09	0.01
13.0 kb	0.03	0.68

A couple is concerned about their second pregnancy. Their first child was diagnosed with *sickle-cell anemia*. Both parents had great grandparents in Ghana and Nigeria.

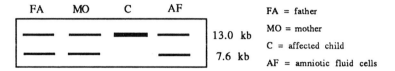

FA = father
MO = mother
C = affected child
AF = amniotic fluid cells

Diagram of an autoradiograph of *HpaI* digested DNA hybridized with a β-globin cDNA probe:

i) Diagram the DNA fragments containing the β-globin gene and the *HpaI* restriction sites for each member of the family.
ii) What information can you give the parents with regard to the genotype of the fetus? What assumptions are you making?
iii) What criteria must be met before this method can be useful for prenatal diagnosis?

16. Kan, Y.W. and A.M. Dozy (PNAS 75: 5631, 1978) carried out restriction mapping of normal ($\alpha^A \alpha^A \delta^A \delta^A \beta^A \beta^A$) individuals and those affected with *homozygous a thalassemia* (deletion for both closely linked α-globin genes on chromosome 4), *hereditary persistent fetal hemoglobin* (HPFH) (deletion for the closely linked δ- and β-globin genes on chromosome 11), *homozygous hemoglobin Lepore* (δ- and β-globin fusion gene), *sickle-cell trait* ($\beta^A \beta^S$) and *sickel-cell anemia* ($\beta^S \beta^S$). Autoradiographs of normal human globin DNA digested with *EcoRI* and *HpaI* and probed with a mixture of α- and β- globin cDNA is shown in Figure 1. The sequence of the δ– and β-genes is 5' - δ - β -3'.

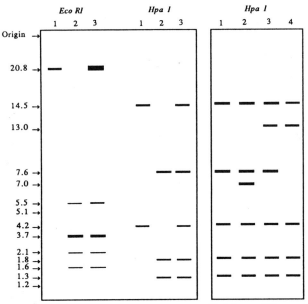

Fig.1. Lanes: 1- *HPFH*; 2-*homozygous α-thalassemia*; 3-*normal*

Fig.2. Lanes: 1&2-*normal*; 3-*sickle-cell trait*; 4-*sickle-cell anemia*

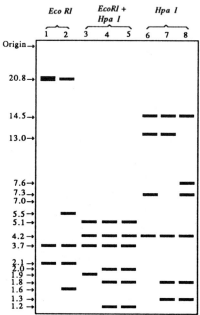

Fig.3. Lanes: 1,3&6 *Hemoglobin Lepore*; 2,5&8 *normal*; 4&7 *sickle-cell anemia*

a) With reasons for your answers indicate which *EcoRI* and *HpaI* bands contain α and which possess the β- and δ-globin genes.

b) Explain whether it is possible to determine which fragment(s) contained the β-globin gene and which possess the δ-globin gene.

Among all *black* individuals, the *EcoRI* patterns were identical to that of all *normal* individuals (Fig. 1, lane 3) whereas *HpaI* produced 7.0 and 13.0 variants of the 7.6 kb fragment (Fig. 2). All 3 of these fragments were found to be allelic.

Autoradiographs of *EcoRI*, *HpaI* and the double restriction (*EcoRI + HpaI*) digestion patterns of DNAs from a *normal* individual with the 7.6/7.6 kb pattern, a *sickle-cell anemia* individual with the 13.0/13.0 kb pattern, and a patient with *hemoglobin Lepore* are shown in Fig. 3. α- and β-globin DNAs were used as probes. *EcoRI* digestion of *hemoglobin Lepore* yielded 2.1 kb and 3.7 kb-bands. Only the 2.1 fragment hybridized with the 5' probe. The 5' β-probe also hybridized with the 5.5 kb *EcoRI* band of *normal* DNA. Digestion of *normal* DNA with *HpaI* produced 3 fragments of β and δ origin which were 7.6, 1.8 and 1.3 kb in length. Double digestion of *normal* DNA yielded 3.7, 2.0, 1.8 and 1.2 kb fragments. The 5' probe hybridized with the 7.6, 1.8 and 2.0 kb fragments. Double digestion of DNA from a *sickle-cell anemia* individual and a *normal* one erased the differences seen in the *HpaI* digestion clone (compare lanes 4 and 5 and lanes 7 and 8 in Fig. 3).

Analyse the autoradiographs carefully, and take the additional information into consideration.

a) With reasons for your answers indicate which fragment(s) carry the β-globin gene and which one(s) carry the δ-globin genes.

b) Map the *EcoRI* and *HpaI* sites in the region of the δ- and β-globin genes in both *normal* and *hemoglobin Lepore* DNA.

17. A normal DNA duplex segment 950 base pairs (bp) long is cleaved by the restriction endonuclease *XhoI* at the 4 sites indicated by the arrows.

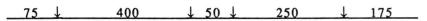

Three mutant forms of this DNA segment were isolated and cleaved with *XhoI*. Electrophoresis of total digests gave the following banding patterns:

Mutants		
A	B	C
	475 bp ———	
		425 bp ———
275 bp ———		
250 bp ———	250 bp ———	250 bp ———
175 bp ———*	175 bp ———*	175 bp ———*
125 bp ▬▬▬		
75 bp ———*	75 bp ———*	
50 bp ———	50 bp ———	50 bp ———

Asterisks indicate end-labeled fragments. With reasons for your answers, indicate the nature of the mutation in each of the 3 mutant forms and the location of each change.

18. A DNA duplex 750 base-pairs long contains 4 *BamHI* restriction sites as indicated by the arrows on the restriction map below:

```
 75  ↓    200    ↓ 50 ↓    175    ↓    250
————————————————————————————————————————————
```

Indicate the number and size of fragments that you would expect after total and partial *BamHI* digestions of this DNA. Denote the end-labeled segments with asterisks.

19. The pedigree below illustrates the segregation of a recessive disorder. RFLP analysis was carried out on this family to determine the carrier status of females at risk. Southern blots containing the appropriate restriction digests were probed with 3 unrelated polymorphic DNA fragments (1, 2, 3) which are located close to the disease locus (order unknown). The resulting autoradiograph is shown below the pedigree.

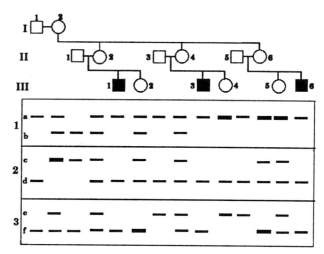

a) Is the disorder determined by a recessive allele of an X-linked or autosomal gene? Explain.

b) Examine the RFLP data carefully and predict the carrier status for all females in generation III. Explain your answers.

c) What do these data tell you about the genotypes of the grandparents? Explain.

20. A strain of *Neurospora crassa* possesses a mutant allele of a gene involved in synthesizing tryptophan. How would you proceed to isolate the *wild-type* allele (DNA) of this gene?

21. The recognition sites and positions of cleavage in duplex DNA by the restriction enzymes *BamHI* and *BglII* are indicated by the arrows.

```
            ↓                           ↓
BamHI  5' G G A T C C 3'     BglII  5' A G A T C T 3'
       3' C C T A G G 5'            3' T C T A G A 5'
                   ↑                           ↑
```

a) Will these enzymes produce complementary sticky ends? Explain.

You ligate 2 DNA duplexes, one cleaved by *BamHI* and the other by *BglII*.

b) Explain whether it would be possible to separate the 2 DNA molecules with *BamHI* or *BglII*.

22. Since RNA cannot be cloned directly how would you proceed to clone the mRNA coded for by a given structural gene, e.g., β-globin gene in humans?

23. The cells of sea urchin embryos produce large amounts of histone mRNA and histones. How would you detect a cDNA corresponding to this protein?

24. What is a cloned DNA fragment? A cDNA library?

25. The DNA duplex of an aster virus has 4 *BamHI* recognition sequences. Large amounts of the viral coat protein are synthesized in aster cells infected with this virus. The cDNA corresponding to the mRNA of the gene (*Cp*) coding for this protein has been cloned in a *E. coli* plasmid. Cleavage of the DNA by *BamHI* produces 5 fragments of different size.
a) Is the viral DNA circular or linear? Explain.
b) How would you use Southern blotting to determine which of the 5 fragments contains the gene *Cp*?

26. *Normal* mouse fibroblast cells transformed with human bladder carcinoma cells form minitumors *in vitro*. DNA from transformed cells injected into normal mice causes tumor development (Shih, E. et al., Nature 290: 261, 1981). The ability to transform normal mouse cells into cancerous ones is due to a *mutant* allele (oncogene) ras^- of the *normal* allele (proto-oncogene) ras^+ (Shih, C. and R.A. Weinberg, Cell 29: 161, 1982)

Neither the DNA nor RNA probes of this gene were initially available. Moreover, it should be noted that ~ 300,000 short base-pair sequences called Alu are scattered throughout the human genome. These sequences do not occur in the mouse genome. Alu-specific radioactively labeled sequences have been generated.
a) Using the bacteriophage lambda (λ) how would you establish a genomic library of the transformed cells containing ras^-?
b) How would you proceed to identify the few λ clones that carry the ras^- oncogene? How would you confirm that these clones contain this cancer-causing gene?

c) The *ras* locus is ~ 5,000 base-pairs long. How would you proceed to determine the nature of the difference between ras^+ and ras^-? Compare your procedure with that used by Tabin, C.J. et al. (Nature 300: 143, 1982) and Reddy, E.P. et al. (Nature 300: 149, 1982).

27. a) Indicate the rationale behind the dideoxy method developed by Sanger, F. et al. (PNAS 74: 5463, 1977) for determining the sequence of bases in a DNA chain. Outline the protocol for DNA sequencing.
b) The chromosome of the single-stranded DNA virus M13 was sequenced by Williams, S.A. et al (BioTechniques 4: 138, 1986) using the dideoxy method. A segment of an autoradiograph of one of their dideoxy electrophoresis gels is shown below. The letters A, C, G, T refer to the ddATP, ddCTP, ddGTP and ddTTP reaction mixtures.

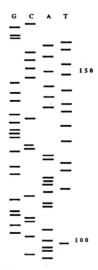

i) What is the sequence of the bases from position 100 through position 150?
ii) Which is the 5' end of the DNA molecule? Explain.

28. *Familial hypercholesterolemia* (*FH*) is a hereditary human disease caused by mutations in the LDL receptor gene which regulates plasma cholesterol (LDL-C). A typical pedigree of an inbred Druze family which was studied by Landsberger, D. et al. (Am. J. Hum. Genet. 50: 427, 1992) shows the plasma LDL-C levels for 8 of the individuals.

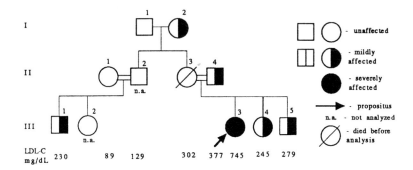

a) Is the mutant allele determining *FH* dominant or recessive? Explain.

These investigators sequenced the coding strand of the gene in the propositus (*mutant*) as well as that in a normal individual using the dideoxy chain-termination method. Both the *normal* and *mutant* alleles (symbolize them *L* and *l* respectively) revealed identical sequences for all the exons and introns except exon 4 whose partial sequences are shown below:

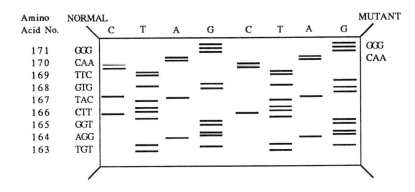

b) Identify the partial sequence of bases in exon 4 of *normal* and *mutant* individuals.
c) What is the nature of the mutation? In which codon does it occur and what is its consequence with respect to the protein specified by the receptor gene?
d) Why is the propositus more severely affected than the other members of this family?

29. The human gene that codes for *factor VIII*, is ~ 2,400 base-pairs long. It is cloned into the plasmid pBR322 which has only 1 *EcoRI* recognition sequence. Subsequently you are provided with a test tube that has a very high concentration of *E. coli* with the recombinant DNA molecule.
a) How would you separate the gene from the recombinant plasmid and how would you isolate it?
b) How would you proceed to determine its base pair sequence?

30. Giles, R.E. et al (PNAS 77: 6715, 1980) cleaved human mitochondrial DNA (*mt* DNA) with the restriction endonuclease *HaeII* and found *typical* (O, □) and *atypical* (●, ■) cleavage patterns in 33 members of a 3-generation family shown below.

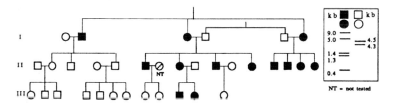

Account for the mode of inheritance of human *mt*DNA.

31. In humans the gene for the receptor of the haemopoietic regulator, *granulocyte-macrophage colony stimulating factor* (*GM-CSF*) is determined by a single gene *A*. Gough, N.M. et al (Nature 345: 734, 1990) studied the mode of inheritance of *HindIII* RFLPs of the *GM-CSF* receptor locus in a 3-generation family presented below. *A1* and *A2* are different alleles of fragment *A* of this gene.

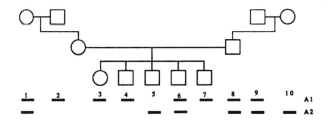

a) What is the chromosomal location of the A locus? Explain.

b) Could an aneuploid female (XO) be used to show that the mode of inheritance indicated in (a) is correct? Explain.

32. a) *Retinoblastoma*, Rb (eye-cancer) is a rare human condition which occurs in ~ 1 in 20,000 individuals. The following typical pedigree of the hereditary form of the cancer is from Weinberg, R.A. (Sc. Am. 259 (Sept.): 44, 1988):

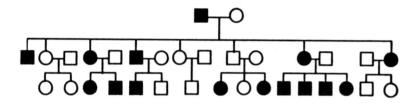

What is the mode of inheritance of the predisposition to *retinoblastoma*? Explain.

b) i) In hereditary *Rb*, usually both eyes are affected early in life. In sporadic forms usually only one is affected late in life.

ii) The *Rb* and Est D (esterase enzyme) loci are located in band q14 of chromosome 13 (13q14) (Sparkes, R.S. et al., Science 219: 971, 1983).

iii) A patient with hereditary bilateral *Rb* has a q14 band deletion in one homologue of chromosome 13 in all body cells. The tumor clones from this individual are monosomic for chromosome 13; the single chromosome 13 had a q14 band deletion (Benedict, W.F. et al., Science 219: 973, 1983).

iv) Using RFLP analysis, Cavanee, W. et al (Science 228: 501, 1985) showed that tumor cells contained 2 copies of the mutant allele for *Rb* (Rb^-Rb^-) whereas other cells were heterozygous (Rb^+Rb^-).

v) Fusion of *retinoblastoma* tumor cells and *normal* cells yields hybrid cells that possess all the chromosomes from 2 such cells which grow normally (are nontumorigenic).

Why are these data with respect to phenotypic expression at variance with the mode of inheritance of *Rb*? Outline a plausible explanation of these results which is also congruent with the mode of inheritance.

c) Speculate as to how the Rb^+ and Rb^- alleles may function in regulating cell growth and division. See Weinberg, R.A. (Sc. Am. 259 (Sept.): 44, 1988) and Whyte, P. et al. (Nature 334: 124, 1988) for confirmation or refutation of your explanation.

33. In humans the *KEL* gene codes for a specific red blood cell antigen. The *PIP* locus on the q arm of chromosome 7, specifies the synthesis of prolactin-inducible protein. Zelinski, T. et al. (Ann. Hum. Genet. 55: 137, 1991) studied the mode of inheritance of a specific pair of alleles at each of these loci in a 3-generation family shown below. The *KEL1* and *KEL2* alleles specify the synthesis of the *KEL1* and *KEL2* antigens respectively. The *PIP1* and *PIP2* alleles were determined using a genomic probe which defines a *TaqI* polymorphism where *PIP1* = 5 kb and *PIP2* = 4 kb.

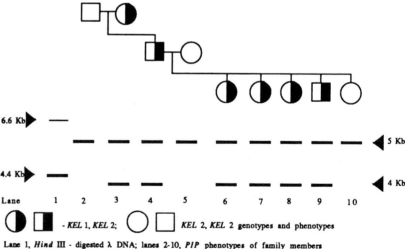

Lane 1, *Hind* III - digested λ DNA; lanes 2-10, *PIP* phenotypes of family members

Is the *KEL* locus on chromosome 7? Explain giving genotypes of all 9 individuals and indicating for each whether it is a parental or recombinant genotype.

34. *Wilm's tumor* (cancer of the kidney) occurs in ~ 1 in 10,000 young children. In 1987 Weissman, B.E. et al. (Science 236: 175) obtained the following results:

i) When cells from a *Wilm's tumor* (homozygous for a deletion of the p13 region of chromosome 11) were injected into *nude* (*nunu*) mice that are prone to malignancies, these animals developed tumors.
ii) A normal chromosome 11 was inserted into Wilm's tumor cells by micro-cell transfer. When these modified Wilm's tumor cells were injected into *nude* mice the tumors did not develop.
iii) When *nude* mice were injected with Wilm's tumor cells that carried normal chromosomes other than chromosome 11 the tumorigenicity of the Wilm's tumor cells was unaffected.

A single gene (*WT*) is involved in determining *normal* vs. *cancerous* cell growth and division. What conclusions can you draw from these data regarding the functions of the *normal* (WT^+) and oncogene (WT^-) alleles at this locus?

35. a) What is a telomere? What are its functions?
b) Szostak, J.W. and E.H. Blackburn (Cell 29: 245, 1982) constructed a linear plasmid vector, from a circular plasmid which is capable of replicating in yeast, and possessed the terminal regions of the linear rDNA plasmid of *Tetrahymena*. They found that this construct was taken up by yeast cells and was capable of accurate and stable replication.

What does this finding indicate with respect to telomere structure and its mechanism of replication in these and other species?

c) Szostak and Blackburn removed one *Tetrahymena* telomere from this linear plasmid vector by restriction digestion and replaced it with a yeast telomere which possessed the repeat sequence:

5' - T T G G G G - 3'
3' - A A C C C C - 5'

Maintaining this linear vector in yeast cells, they discovered 2 important findings: (i) the telomeres of different plasmids showed size heterogeneity; (ii) Only yeast telomeric repeats were added to the *Tetrahymena* telomeres.

Propose 2 explanations that can account for these findings.

d) In 1989 Greider, C.W. and E.H. Blackburn (Nature 337: 331) isolated a DNA polymerase that had a RNA component with the 5' - CAAACCCAA - 3' sequence in the coding region.

i) Which mechanism proposed in c) does this discovery support and why?
ii) How would you proceed to determine whether this RNA component was essential for replication of telomeres? Indicate the results expected in your experiment(s)?

36. a) Regardless of the specific base-pair sequence, the telomeres of all eukaryotic chromosomes have one feature in common. What is it?
b) Why are DNA polymerases unable to replicate telomeres? What are telomerases and why are they able to replicate telomeres?
c) Telomerases, unlike other enzymes involved in DNA replication, are ribonucleoproteins. Why is this so? How would you isolate a mutant defective in telomerase activity in a species like yeast?

37. a) Distinguish between a centromere and a kinetochere.
b) What is the function of a centromere? Outline one experiment that would provide results to confirm the stated role of this region of a chromosome.
c) The following observations in the yeast, *Saccharomyces cerevisiae*, have provided an insight into yeast centromeres and their function:

i) Insertion of discrete regions of the centromere into high copy number plasmids (autonomously replicating and randomly segregating) causes these plasmids to behave stably (replicate and segregate normally) during mitosis and meiosis and to be maintained in 1 or 2 copies per cell (Tschumper, G. and J. Carbon, Gene 23:221, 1983).
ii) In a functional region (element II) of less than 200 base-pairs (bp) there are sequences in all *S. cerevisiae*

chromosomes that have a greater than 90% AT content. As long as a high AT content is maintained in this region, the centromere will still function normally despite point mutations and small deletions in this element (Fitzgerald-Hayes, M. et al., Cell 29: 235, 1982).

iii) Extending for 2.0 to 3.5 kilobases on either side of the centromere, is an extremely ordered array of 160 bp nucleosomes present one after another for long stretches, which is not observed in the bulk of the chromatin (Bloom, K.S. and J. Carbon, Cell 29: 305, 1982).

iv) The centromere regions have no open reading frames (ORFs) of greater than 30 to 40 bp, and have neither start nor stop consensus sequences for RNA polymerase III (Fitzgerald-Hayes, M. et al., Cell 29: 235, 1982).

What conclusion(s) can you make from each of the facts presented? How do you have to modify or supplement your answer to (b) in view of these observations?

d) Microtubules are approximately 200 Å wide. How many of them would you expect to see bound to: (i) each yeast centromere? (ii) each centromere of a chromosome in mammals? Explain.

e) At least 5 different centromere associated proteins have been discovered in mammals. One of these, CENP-E, first appears in late prometaphase, and remains associated with the centromere throughout metaphase. It dissociates from the centromere during anaphase after which it remains at the cell's equator and is eventually discarded (Yen, T.J. et al., EMBO 10: 1245, 1991). Propose a plausible role for this protein. Injection of antibody mAB177 into late prometaphase cells prevents cell transition from metaphase to anaphase. Does this finding support your hypothesis? Explain.

f) Why are centromeres of higher eukaryotes likely to be more complex than those of yeast? Compare your comments with those of L. Clarke (TIG 6: 150, 1990) and Willard, H.F. (TIG 72: 410, 1990).

38. Adachi, Y. et al (Cell 64: 137, 1991) found that in chicken erythrocyte nuclei, which have a very low content of topoisomerase II, chromosomes are not condensed. The chromosomes in these nuclei condense normally when this enzyme is added. What conclusion can you make from these simple observations?

39. *Wild-type* strains of the fission yeast *Schizosaccharomyces pombe,* a haploid eukaryote, replicate their DNA and proceed through all stages of the cell cycle in a normal manner under both permissive (25 °C) and restrictive (35°C) temperatures.

a) A cell-division cycle gene $cdc2^+$ codes for a 34 kb protein kinase which is required for initiation of DNA replication and entry into mitosis.

b) Another gene $wee1^+$ codes for an inhibitor that delays initiation of mitosis until cells have attained a certain size.

In 1989 Molz, L. et al. (Genetics 122: 773) observed that:

i) The double mutant strain ($cdc2$-3W(=c), $wee1$-50(=w)) expresses a *wild-type* phenotype at the permissive temperature (25°C) whereas at the restrictive temperature (35°C) it results in a lethal phenotype termed mitotic catastrophe (mc). This phenotype includes cells which possess a 2C amount of DNA, with less than 1% of the daughter cells being able to go through a complete cell cycle. Often one of the daughter cells is anucleate or has only a minute nucleus.

Speculate as to the possible nature and location of the defects in the cell cycle in this double mutant (c, w).

ii) Gene mutations were induced by chemical mutagenesis in this double mutant strain (c, w). The strains containing these induced mutations grew, divided, and formed colonies (expressed a *wild-type* phenotype) at the restrictive as well as the permissive temperatures. Each of these induced (triple) mutants was crossed with a *wild-type* strain. Some of the segregants in each cross displayed the mc phenotype.

Why were the induced mutants able to express a *wild-type* phenotype under restrictive conditions?

iii) Diploids heterozygous at each of the induced mutant gene loci but homozygous for c and w were produced at 25°C. When assayed at 35°C they expressed the lethal phenotype.

Is each mutant allele dominant or recessive to its *wild-type* allele in its ability to suppress *mc*?

40. Uemura, T. et al. (Cell 50: 917, 1987) obtained cold-sensitive (*cs*) and temperature-sensitive (*ts*) mutant alleles at the *top2* and *nda3* loci in the fungus *Schizosaccharomyces pombe*; *top2* codes for topoisomerase II and *nda3* codes for β-tubulin.

a) The *cs* mutants express a *wild-type* mitotic phenotype at 36°C (permissive temperature) and a *mutant* mitotic phenotype at 20°C (nonpermissive-restricted temperature). The reverse is true for *ts* mutants.
b) The *cs top2* mutant forms spindles, however the "structurally aberrant chromosomes are transiently pulled apart but fail to separate" and it fails to form colonies at 20°C. At 36°C the somatic cells of this mutant proceed through mitosis and grow normally.
c) In cells of *cs nda3* at 20°C chromosomes condense normally, but spindles do not form, mitosis does not occur and cells fail to grow. At 36°C these mutant cells possess normally condensed chromosomes, spindles form, mitosis occurs and cultures grow normally.
d) The double mutant *cs top2-cs nda3* forms long entangled chromosomes; spindles are not produced and mitosis does not occur at 20°C. When the double mutant cells are shifted to the permissive temperature, chromosomes condense, spindles form and mitosis proceeds normally. If cells at the permissive temperature are incubated in the presence of nocodazole (tubulin polymerization inhibitor), spindles do not form, chromosomes condense but chromatids do not separate.
e) In pulse-shift experiments, the cultures of *cs nda3* and *cs top2 - cs nda3* were first arrested at 20°C for 12 hours, transferred to 36°C for a series of short periods (2-8 minutes), then transferred back to 20°C.

i) Chromatids of chromosomes in the *cs nda3* mutant separated if cells were exposed to 36°C for 4 minutes even though no spindles were formed at 4 minutes after shift to 36°C.

ii) Separation of chromatids of chromosomes in the double mutant *cs top2-cs nda3* was partially impaired after a 4-minute pulse.

f) Temperature - shift experiments with the *ts top2 - cs nda3* double mutant strain gave the following results:

i) Cells were first arrested at 20°C for 7-10 hours (*cs* β-tubulin inactivated, topoisomerase II is active). Chromosomes in these cells condense as in the *cs nda3* mutant.

ii) Cells were then transferred to 36°C for 30 minutes. β-tubulin was reactivated, spindles formed, but chromatid separation was partly blocked. The chromosomes had a "streaked" appearance. They were arranged along the mitotic spindle and looked like a string of nonseparating chromosomes. Note: at 36°C topoisomerase is inactive, β-tubulin is active, normal spindles form, but chromosomes are "streaked".

The enzyme topoisomerase II specified by the *wild-type* allele at the *top2* locus is required for mitosis.

a) What specific mitotic functions are (i) the *top2* locus and its enzyme (ii) the *nda3* locus and its enzyme, required for? Justify your answer.
b) How might the *ts top2 - cs nda3* double mutant be maintained and propagated if the permissive temperature of one mutation is the restrictive temperature of the other?

APPENDIX

Table 1. "Chi-square" Values

Degrees of Freedom			Probability values				LEVEL OF SIGNIFICANCE ↓	
	0.95	0.75	0.50	0.30	0.20	0.10	**0.05**	0.01
1	0.0^2393	0.102	0.455	1.074	1.642	2.705	3.841	6.635
2	0.103	0.575	1.386	2.408	3.219	4.605	5.991	9.210
3	0.352	1.21	2.366	3.665	4.642	6.251	7.816	11.345
4	0.711	1.92	3.357	4.878	5.989	7.779	9.488	13.277
5	1.15	2.67	4.351	6.064	7.289	9.236	11.070	15.086
10	3.94	6.74	9.342	11.781	13.442	15.987	18.307	23.209

From the table one can generate the **probability (p)** of obtaining a deviation (chi-square value) as great as or greater than the calculated value. For example, in an experiment with 3 degrees of freedom, a chi-square value equal to or greater than 9.837 will be found to occur by **CHANCE** alone only two out of every hundred times (i.e., $p = 0.02$). Since the accepted level of significance is arbitrarily set at $p = 0.05$, the deviation observed in the experiment is said to be significant, thus necessitating the rejection of the null hypothesis proposed to account for the experimental observations.

NOTE: The closer a generated p-value is to the level of significance, the less certain one has to be of the proposed null hypothesis.

Table 2. Coding dictionary
mRNA (5' → 3') for the 20 amino acids

First letter	Second letter				Third letter
	U	C	A	G	
U	UUU } Phe UUC UUA } Leu UUG	UCU UCC } Ser UCA UCG	UAU } Tyr UAC UAA } Nonsense* UAG	UGU } Cys UGC UGA Nonsense* UGG Trp	U C A G
C	CUU CUC } Leu CUA CUG	CCU CCC } Pro CCA CCG	CAU } His CAC CAA } Gln CAG	CGU CGC } Arg CGA CGG	U C A G
A	AUU AUC } Ileu AUA AUG Met	ACU ACC } Thr ACA ACG	AAU } Asn AAC AAA } Lys AAG	AGU } Ser AGC AGA } Arg AGG	U C A G
G	GUU GUC } Val GUA GUG	GCU GCC } Ala GCA GCG	GAU } Asp GAC GAA } Glu GAG	GGU GGC } Gly GGA GGG	U C A G

*Chain-terminating codons.

Table 3. Metric equivalents

Metric prefixes					
Prefix	Abbreviation	Meaning	Prefix	Abbreviation	Meaning
deci	d	10^{-1}	micro	μ	10^{-6}
centi	c	10^{-2}	nano	n	10^{-9}
milli	m	10^{-3}	pico	p	10^{-12}

Conversion factors for length

	Meter (m)	Micrometer* (μm)	Angstrom (Å)
1 meter (m)	1	10^{6}	10^{10}
1 centimeter (cm)	10^{-2}	10^{4}	10^{8}
1 millimeter (mm)	10^{-3}	10	10^{7}
1 micrometer* (μm)	10^{-6}	1	10^{4}
1 nanometer* (nm)	10^{-9}	10^{-3}	10
1 picometer* (pm)	10^{-12}	10^{-6}	10^{-2}
1 inch	2.54×10^{-2}	2.54×10^{4}	2.54×10^{8}

* The unit formerly called the micron (μ), i.e. 1×10^{-6} m, has been renamed the micrometer, and the prefixes millimicro- and micromicro- have been replaced as shown.

Some constants

Average molecular weight * of nucleotide	= 330
Average molecular weight of amino acid	= 110
Distance between nucleotides	= 3.4 Å
Distance between two amino acids on a polypeptide chain	= 3.8 Å
Diameter of DNA helix	= 20 Å

* Sum of atomic weights of all atoms in molecule

Answers to Selected Questions and Problems

Chapter 1: Physical Basis of Heredity

1. a) Major function of mitosis - Production of daughter nuclei, and if cytokinesis accompanies karyokinesis, daughter cells identical to each other and the parent cell with respect to (i) both chromosome number and kinds of chromosomes and (ii) genotype. Function is accomplished because for each chromosome replication and division into chromatids there is a single nuclear division (equational separation of the chromatids of each replicated chromosome).
   (i) DNA synthesis during S period of interphase and chromosome division into chromatids.
   (ii) prophase - unwinding of chromatids, shortening of chromosomes.
   (iii) metaphase - auto-orientation of the chromosomes on the metaphase plate.
   (iv) anaphase - longitudinal division of centromere, movement of chromatids (now chromosomes) of each chromosome to opposite poles (equational division).
   (v) telophase - nuclear membranes form around each anaphase group, cell divides in two, chromosomes become long and thin, nucleoli reappear. Daughter nuclei and cells identical to each other and parent.

   b) (i) Animals - furrowing; Plants - cell-plate formation
      (ii) Centrosomes (centrioles) occur in all animals but only in a few (lower) plant species - responsible for aster formation; may play subsidiary role in determining position of poles of spindle and in forming the spindle.

   c) Same in (i) diploid and haploid cells of same organism and (ii) different organisms of same and different species. Since the major function of mitosis is to produce daughter cells identical to each other and the parent cell and since this involves only one replication and division of chromosomes into chromatids followed by a single equational division during which all chromosomes behave autonomously, it is basically the same in all cells and eukaryotes.

   d) Because homologous chromosomes do not normally synapse and then segregate.

5. a) The first one is definitely haploid (none of the chromosomes have homologues). The second cell is at anaphase and is also haploid for the same reason.
   b) First - interphase, anaphase, telophase
      Second - interphase, prophase, metaphase, telophase.

6. a) To bring about a reduction in the number of chromosomes to exactly one-half that of somatic cells and zygotes. Specifically, to produce meiotic products and, directly or indirectly, gametes, each with a haploid (n) chromosome number which contain one or the other parental homologue of each pair. This is accomplished in almost all eukaryotes by having two divisions: one reductional and the other equational. In most species reductional precedes equational. Specifically, reduction in chromosome number occurs because homologous chromosomes (i) synapse (at zygotene), (ii) co-orient (at metaphase I) and (iii) segregate (at anaphase I).
   Meiosis is necessary to maintain constancy of the chromosome number of a sexually reproducing species from generation to generation. If reductional division did not occur the chromosome number would increase (double) after each fertilization in sexually reproducing species.

b) Genetic functions:
  (i) Permits crossing over which involves the reciprocal exchange of homologous segments between non-sister chromatids.
  (ii) Permits random segregation of homologues of each chromosome pair (physical basis for Mendel's law of segregation).
  (iii) Allows any two or more different (non-homologous) chromosome pairs to segregate independently of each other - recombination at the interchromosomal level. Physical basis for Mendel's law of independent segregation (independent assortment).

All these events produce genotypic variability at the haploid and diploid levels.

8. a) Differences between first division of meiosis and mitosis. The following occur during first meiotic division but not during mitosis:
(i) pairing of homologous chromosomes to form bivalents
(ii) crossing over and chiasma formation
(iii) random and independent co-orientation of chromosome pairs on the metaphase I plate
(iv) segregation of homologues (reductional division) at anaphase I to produce haploid (n) secondary meiocytes

b) The second meiotic division is essentially a mitotic one. Differs from mitosis in that (i) chromosomes present in haploid (n) number, (ii) chromatids do not exhibit relational coiling and (iii) each chromatid may be quite different genetically from its sister because of crossing over (first division).

9. i) False. A chromosome may pair only with its homologue.
ii) True. Each chromosome duplication is accompanied by a nuclear division.
iii) True. Because during the first meiotic division homologues paired, and then segregated to opposite poles.
iv) False. If a primary spermatocyte had 18 chromosomes, since each individual obtains one chromosome of each pair from each parent, the maximum number of paternal chromosomes possible is 9.
v) True. Proof is derived from studies of meiosis in individuals with 2 or more heteromorphic chromosome pairs.
vi) False. In the somatic cells one member of each pair is paternal. The gamete can only contain one member of each pair, therefore it cannot contain more paternal chromosomes than the somatic cells.

10. Meiosis cannot occur in haploid individuals since chromosomes do not possess homologues necessary for synapsis, crossing over, co-orientation and segregation. Meiosis may occur in a haploid species only if a diploid zygote (= meiocyte) is part of its life cycle as for example, in *Neurospora crassa*.

11. Because chromosomes do not segregate at random. Instead, homologues segregate to opposite poles and as a consequence eggs possess one homologue of each pair (either paternal or maternal member of each pair).

13. AA^1BB^1 because (i) zygote receives chromosomes A and B from the male and the homologues A^1 and B^1 from the female parent and (ii) somatic cells arise from a zygote by mitosis.

14. b) Yes. If meiosis is not preceeded by replication and division of chromosomes into chromatids, single chromosomes could then synapse (at zygotene) and segregate (at anaphase I). If this happened, the second division could be eliminated since it occurs to separate chromatids to opposite poles (and therefore nuclei and cells). Illustrate this for yourself.

21. None of the progeny will have only rod-shaped chromosomes after four or any number of generations of self-fertilization. Barring chromosome mutations, all progeny will possess the original karyotype.
Reason - During meiosis in both male and female meiocytes, homologues segregate at anaphase I and their chromatids separate equationally at anaphase II with the result that gametes are formed each with one chromosome of each chromosome pair. Therefore, the chromosomes contributed by the female gamete are morphologically equivalent to those contributed by the male gamete. As a consequence the progeny, like the parents will possess *each of the different kinds* of chromosomes in pairs.

22. a) The four types of meiotic products on the male side (2_s, 3_s; 2_s, 3; 2, 3_s; and 2, 3) occur in equal proportions because of independent segregation of the 2 chromosome pairs.
b) Yes, because meiosis is basically identical in both sexes.
c) Four gamete types (male and female), unite at random

23. a) All four offspring in a litter are phenotypically identical (quadruplets), because all four are genotypically identical. The offspring are genotypically identical because a single zygote divides mitotically to produce two daughter cells, each of which in turn undergoes mitosis, giving rise to four genetically identical cells. These four identical cells separate and develop into phenotypically identical offspring.
b) The simplest explanation is that at least one parent is heterozygous at a locus determining banding pattern, e.g., ♀ Aa x ♂ aa → 1/2 Aa: 1/2 aa where A determines one banding pattern and a determines a different banding pattern.
c) No -- because all four offspring in a litter are derived from the same zygote. For example, if the species has the XX (♀), XY (♂) method of sex determination and if the zygote was XX, all four offspring would be female (XX).

24. If equational division preceded reductional division and the other meiotic events remained the same as in the general pattern, this difference could account for all similarities and differences between the 2 patterns.

25. a)   (1) 0   (2) 0   (3) 0   (4) 100 %
Reason - Homologues synapse and then segregate to produce haploid (n) meiotic products each with one or the other chromosome of each pair. That is, each meiotic product will possess a long and a short metacentric, a long and a short acrocentric and a long and a short telocentric (Note: answer to (3) is also 100 % if one interprets this to mean a long and a short (1) metacentric, (2) acrocentric, (3) telocentric.
b) 3 long pairs - 1 metacentric, 1 acrocentric, 1 telocentric
3 short pairs - 1 metacentric, 1 acrocentric, 1 telocentric.

29. Expect hybrids to be sterile. None of the chromosomes have homologues; they will be present as univalents. Since chromosomes move independently of each other, each has a 50 : 50 chance of going to a given pole, and there are 63 chromosomes, all or most meiotic products (and gametes) will possess unbalanced chromosome complements due to unequal distribution and chromosome loss. Low probability of gametes possessing (1) all 63 chromosomes, (2) 32 horse chromosomes, (3) all donkey chromosomes.

31. a) From union of polar nuclei and one sperm. Divides by mitosis to form triploid endosperm (exception in lily family in which endosperm is pentaploid). In flowering plants (angiosperms), food for developing embryo.

b) (i)

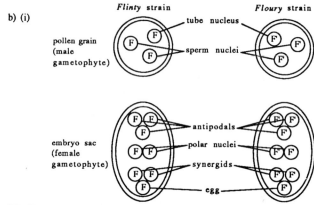

(ii) Reciprocal crosses give different results because (i) endosperms have different genotypes (*flinty* male x *floury* female - F x F^1F^1 - FF^1F^1; *flinty* female x *floury* male - FF x F^1 - FFF^1) and (ii) neither allele is dominant over the other.

32. (1) In plants, there are two sperm nuclei; one fertilizes egg → zygote, the other fertilizes polar nuclei → triploid endosperm nucleus.
(2) The tube nucleus contains digestive enzymes to digest pathway through the tissues to the megaspore mother cell.

34. The contributions of sperms and eggs to the zygote are as follows:
(i) *nuclear contributions are equal* - nuclei are of the same size and both contain the same number (n = 23) and kinds of chromosomes (one of each pair).
(ii) *cytoplasmic contributions are unequal* - the egg contains and contributes a large amount of cytoplasm to the zygote; whereas the sperm contains and contributes very little or none.
Since a child inherits equally from both parents the above facts suggest, but do not prove, that the hereditary determiners are located in the nucleus, specifically on the chromosomes. The difference in size between sperms and eggs is due to non-genetic (non-chromosomal) material, cytoplasm, of which the egg contains a large amount and the sperm very little.

37. a) The species is haplontic. The *green* and *blue* parents, being true-breeding, each produce only one kind of gamete, e.g., G and G^1. If the species was diploid all zygotes would be 2n (GG^1). They would divide mitotically and all plants would be genotypically and therefore phenotypically identical (*green* or *blue* or *intermediate*). If the species is haploid the G and G^1 gametes would unite to form a diploid GG^1 zygote as in diploid species, but the zygote would function as a meiocyte and produce G and G^1 gametes in a 1 : 1 ratio. Each would then divide mitotically to produce its own genotype and phenotype.

42. a) Syngamy involves fusion of a female gamete and a male gamete but it occurs in both diploid and haploid species.
A gamete + *a* gamete → zygote → meiosis → spore → haploid plant
b) True - both female and male individuals develop.
c) False - although hermaphroditic organisms usually self-fertilize, they can cross-fertilize, e.g. corn, nematodes.
d) False - many organisms, e.g. bacteria, *Paramecium* reproduce asexually by fission.
e) False - in certain tissues cytokinesis is not followed by karyokinesis resulting in the doubling of the chromosome number in the tissue.

f) True - isogamous gametes contain equivalent amounts of cytoplasm thus allowing the male to develop parthenogenetically.
g) False - e.g., Hymenoptera - unfertilized eggs develop parthenogenetically into haploid (*n*) males.

46.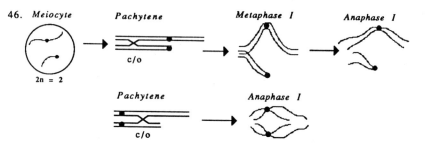

The centromere is located at the end of the short chromosome (which is therefore telocentric); the long chromosome is either metacentric (as illustrated) or acrocentric.
Note: If the centromere in the short chromosome was not at the end of the chromosome and crossing over occurred in the short arm, a long and a short chromatid would go to each pole at anaphase I.

48. (i) The plants are likely members of different, closely related species.
Reasons: (a) All 4 chromosomes morphologically different - none of the chromosomes has homologues.
(b) If plants were members of same species, only bivalents would have been present at metaphase I instead of univalents.
(ii) If each univalent moves randomly to one or the other pole, at anaphase I (which is the usual mode of behaviour of univalents) and the chromatids subsequently separate normally (equationally) at anaphase II we expect $(1/2)^{n=5}$ = 1/32 of the meiotic products of the hybrid to have the same chromosome number as its somatic cells. (1/2) = the probability of each chromosome moving to a given pole at anaphase I and n = the number of chromosomes in the hybrid.
If the univalents behave in any other way the proportion of meiotic products with the same chromosome number as in the somatic cells will be less than $(1/2)^{n=5}$.
(iii) Six. Each parent contributes a haploid number of chromosomes to each offspring.

Chapter 2
Monohybrid Inheritance

1. a) No. Phenotype is the result of genotype-environment interaction (i.e. P = G + E). The genotype of one generation may affect the phenotype of the next generation regardless of the latter's own genotype (e.g., *dextral - sinistal* coiling in snails), but not *vice versa* - a phenotype *per se* cannot be inherited.

   b) True. In polygenic (multigenic) inheritance different combinations of genes (duplicate, cumulative will give the same phenotype, e.g., medium height. Same phenotype may also be due to specific types of gene interaction, e.g., epistasis and complementary genes or specific kinds of allelic interaction, e.g., dominance, e.g., *Aa = AA*. Individuals of identical genotype may be phenotypically different because of different environments. For example, treatment of certain genetic disorders, e.g., *phenylketonuria*, will alleviate symptoms (change phenotype) while the genotype remains the same as in untreated individuals.

3. a) Interaction between alleles, e.g., dominance, recessiveness, etc.
   b) (i) No problems with respect to determining genotype at any locus since no dominance, codominance, etc.
   (ii) In many haploid species can study all products of a given meiocyte separately from products of other meiocytes in an unordered or ordered manner (unordered and ordered spore analysis respectively). This provides more genetic information than analysis of a random sample of meiotic products of many meiocytes as in diploids. For example, such analysis can permit determination of gene - centromere distances and whether genes are on the same or different arms of a chromosome. Random spore analyses do not permit such information to be obtained.

4. a) Yes, these results are expected. Transplantation of the ovary does not change the genotype of its cells. The implanted ovary (*BB* = *black*) in the *albino* female would produce *B* gametes which upon fertilization would produce *black* progeny only regardless of the genotype of the male. Since the male was *albino* (*bb*) all the progeny would be heterozygous (*Bb*) and *black*.
   albino ♀ with ovary (*BB*) x *albino* ♂ (*bb*) → F$_1$ all *Bb*

   b) The bearing of the results on the Lamarckian doctrine (acquired characters are inherited) is indirect. They nevertheless indicate that an acquired character is not inherited.

7. a) (i) Mutation of recessive *e* to dominant *E* (*earless*) at one of the loci for ear development.
   (ii) Male is a homozygous recessive segregant from a rare mating of two heterozygotes (*Ee* x *Ee* → *ee earless* male).
   (iii) Phenocopy. *Earless* males of *normal* genotype but *earless* phenotype because of environmental influence, e.g., mal-development.

   b) Cross 1 - Cross *earless* male with *normal* females.
   Cross 2 - Mate *normal* parents producing *earless* male and study their progeny.
   (i) If the *earless* male is due to mutation, since all individuals in the previous generation were *normal* (true-breeding line), the trait is dominant to *normal* as it is very highly unlikely that a mutation will affect two alleles at the same time to produce a homozygous dominant; a recessive mutant allele would not express its effects because its partner would be dominant.
   Expectations:
   Cross 1 - ♂ (*Ee*) x ♀ (*ee*) → 1 *Ee* (*earless*) : 1 *ee* (*normal*)
   Cross 2 - ♂ (*ee*) x ♀ (*ee*) → all *ee* (*normal*).

   (ii) If *earless* male is a recessive segregant expect:
   Cross 1 - ♂ (*ee*) x ♀ (*EE*, rarely *Ee*) → all *Ee* (*normal*) or rarely 1 *Ee* (*normal*) : 1 *ee* (*earless*)
   Cross 2 - ♂ (*ee*) x ♀ (*EE*) → F$_2$: 3 *E_* (*normal*) : 1 *ee* (*earless*)

   (iii) Phenocopy. *Earless* male same genotype as all other mice, therefore all progeny *normal* in all crosses.
   Note: *Earless* male could arise as a result of a somatic mutation. Same genotype as all other mice except for ear tissue, therefore all its progeny would be *normal*.

8. a) *Kandiyohi* is due to a dominant allele because all F$_1$'s were *kandiyohi* and three quarters of the F$_2$ expressed the trait.
   b) Two-thirds (approximately 37) of the *kandiyohi* F$_2$ frogs would be expected to be heterozygous.
   c) All, since they are determined by a recessive allele which cannot produce its effect unless it is present in the homozygous condition.
   d) By performing a progeny test or testcross. The *kandiyohi* individuals would be crossed with the *wild-type* ones (homozygous recessive). If a *kandiyohi* frog was homozygous all its progeny would be *kandiyohi*. If heterozygous, half of its progeny would be *kandiyohi* and half *wild-type*.
   e) Because meiosis is basically identical in both sexes and results in the same kinds and proportions of gametes in both males and females. The gametes of the two sexes also unite at random.

9. a) A single pair of alleles is involved. The allele, *P*, for *pubescent* is dominat to *p* for *glabrous*.
   b)
   (1) *pubescent* (*Pp*) x *glabrous* (*pp*)
   (2) *pubescent* (*PP*) x *pubescent* (*PP*)
   (3) *glabrous* (*pp*) x *glabrous* (*pp*)
   (4) *pubescent* (*PP*) x *glabrous* (*pp*)
   (5) *pubescent* (*Pp*) x *pubescent* (*Pp*)

   c) Cross 2. Since all 63 progeny are *pubescent* it is highly likely that at least one of the parents is *PP*.
   (i) If both parents *PP*, all progeny *PP* - none would produce *glabrous* progeny upon self-fertilization.
   (ii) If parents *PP* and *Pp*, 1/2 the *pubescent* progeny would be *Pp* and these plants would be expected to produce *glabrous* plants upon self-fertilization.

   Work out the results for crosses 4 and 5 in the same manner.

10. *Peroneal muscular atrophy* is determined by a dominant allele (at a specific locus). Barring mutations, if a child expresses a trait that occurs in at least one of the parents it must also possess the allele for the trait. It is true that in some cases the trait could be due to a recessive in the homozygous condition. This would require that the *unaffected* parent be heterozygous. If the trait, however, was always due to a recessive allele it should, at least occasionally, and in reality frequently if rare, occur when both parents are *unaffected*. Since it occurs in a child only when at least one parent is *affected*, it is a dominant.

12. a) A single autosomal allele pair appears to be involved. Allele *S* (*short* hair) is dominant to the allele *s* (*long* hair).

Cross 1 - *short* (SS) x *long* (ss) → all *short* (Ss)
Cross 2 - *short* (Ss) x *long* (ss) → 1 *short* (Ss) : 1 *long* (ss)
Cross 3 - *long* (ss) x *long* (ss) → all *long* (ss)

b) One way to test the hypothesis is to intercross the *short*-haired progeny from cross 1. If the progeny size is large expect 3 *short* : 1 *long*-haired rabbits.

short (Ss) x *short* (Ss) → 3 *short* (S_) : 1 *long* (ss)

15. The question is concerned with matings within a breed which with few exceptions are true-breeding. Also *aniridia* is an undesireable trait which would be selected against.
Since the trait occurred in approximately 1/2 of the foals in matings of "Godvan" (*affected*) with the 124 *normal* (*unaffected*) mares it must be determined by a dominant allele. Since Godvan's parents were *normal* (aa), he expressed the trait because of a mutation of $a \to A$ in the meiocytes or gametes of one of the parents.

Parents   Godvan x Mares   Progeny
aa x aa (a→A) →   A a   aa   Aniridia Aa : normal aa

If the trait was due to a recessive allele not only would "Godvan" have to be homozygous (aa) but all (or almost all) the mares would have to be heterozygous (Aa) since 50% of the offspring are affected. Since *aniridia* is an undesireable trait it is highly improbable that the causative recessive allele would be present in such a high frequency in a domesticated breed.

17. a) I would tell Burch that Fraser's bull is only partly the cause of the *hairless* calves.
b) (i) Results of crosses made by other people (e.g., Hutt and Saunders) showing that it was recessive.
(ii) Since *hairlessness* occurs among the progeny of *normal* parents it must be due to a recessive allele. Each *hairless* individual must receive a recessive allele from each of the parents.

Mating to confirm stand
Cross *hairless* male offspring with their mothers (some of Burch's *normal* heifers). If some of the progeny are *hairless*, some of Burch's heifers must be Hh.

c) Numbers of *normal* and *hairless* calves expected
Fraser's bull - Hh; Burch's cows - HH (never any *hairless* calves until these 2 series of matings).
Therefore expect ≈ 1/2 of the heifers to be HH, the other half to be Hh.

Burch's bull	Heifers	Calves	Phenotypes and theoretical ratio	Expect among 46 calves
Hh	x 1/2 (HH)	2HH : 2Hh	7 normal :	40.25
Hh	x 1/2 (Hh)	1HH : 2Hh : 1hh	1 hairless	5.75

d) Twenty-two (4/7)

20. a) Haploid. If they were diploid the progeny in the F_1 would all be genotypically identical and therefore phenotypically uniform (identical or very similar).

Since the organism is haploid the traits are those expressed in the haplont generation. When these haploids mate they produce a zygote which undergoes meiosis to produce haploid spores, the initial cells of the haplont generation. Since the haploid strains carried different alleles of one gene the meiotic products (spores), and the cells they give rise to, would therefore be of 2 types (both parental) in equal proportions.

b) Law of segregation.

22. a) These tetrads would be recognized in *Neurospora* by doing an ordered-tetrad analysis. All 21 of the 31 tetrads analysed would show a segregation pattern different from 4 : 4 (first division segregation).

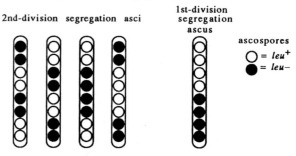

b) Crossing-over between the gene locus and centromere. Thus first meiotic division is equational at the *leu* locus and second division is reductional. The reverse is true when no exchange occurs in the gene - centromere region. The specific second-division segregation pattern depends on the manner of orientation of the chromatids of each chromosome in each secondary meiocyte. Each chromatid has a 50 : 50 chance of facing either AII pole.

25. a) Either dominant or recessive. If one assumes *alkaptonuria* is due to a dominant allele then *affected* individuals must be heterozygous since matings wtih *normals* (homozygous recessive) produce both *affected* and *unaffected* progeny. If one assumes that *alkaptonuria* is due to a recessive allele then the *normal* phenotype must be due to a dominant allele and all *normals* (disregarding the lower left hand portion of the pedigree) must be heterozygous. Work this out for yourself.

b) The trait is likely due to a dominant allele as it is highly unlikely (but not improbable) that *normal* heterozygotes (allele is rare in population) would enter the family so frequently. For other reasons see answer to question 10.

c) It neither confirms nor rejects, but is consistent with the hypothesis in (b) (note that consanguinity in this case adds no new information). However, since consanguineous matings are involved, *alkaptonuria* can also be explained by recessive inheritance (determined by a recessive allele). Critical evidence indicated below is missing:
(i) If parents of the first generation *affected* individuals were *unaffected* this would lead to rejection of the dominant allele hypothesis and would indicate the trait is recessive.
(ii) Pedigree(s) or portion(s) of pedigrees involving matings between *unaffected* individuals collectively producing a 3 *unaffected* : 1 *affected* ratio. This would support the dominant mode of inheritance.

Chapter 3
Dihybrid and Multihybrid Inheritance

1. a) 1/4 (1/16 of each of the four types: *AABB*, *AAbb*, *aaBB*, *aabb*)
   b) 0   c) 0

2. Segregation and independent assortment permit the formation of recombinant genotypic combinations in gametes which unite at random to form new (recombinant) diploid genotypes in addition to parental ones. This genotypic variability in a species permits nature to "try out" all these combinations. Favourable ones are selected and increased in frequency, unfavourable genotypes are eliminated or maintained at a low frequency - genotypic variability is the basis for evolution. Barring the infrequent gene and chromosome mutations (most detrimental), almost all the progeny of asexually reproducing species are genotypically identical to the parent; if the environment changes, placing the progeny at an adaptive disadvantage, they may be less viable, reproduce at a lower rate, or even die.

4. *O* - one pod, *o* - three pod; *N* - normal leaf, *n* - wrinkled leaf
   Cross 1
   Since the progeny consists of *one-pod* and *three-pod* individuals in approximately equal proportions and the allele for *three-pod* is recessive, the *one-pod, normal* parent must be heterozygous (*Oo*) and the *three-pod, normal* homozygous recessive (*oo*). Since 3/4 of the progeny are *normal* and 1/4 *wrinkled*, both parents must be heterozygous *Nn*. Therefore the genotype of the 1st parent = *OoNn* and that of the 2nd parent = *ooNn*.
   A similar type of reasoning can be applied to derive the genotypes of the parents in the other crosses.
   Thus for:
   Cross 2: 1st parent = *OoNn*   2nd parent = *Oonn*
   Cross 3: 1st parent = *OONn*   2nd parent = *oonn*
   Cross 4: 1st parent = *OoNN*   2nd parent = *oonn*
   Cross 5: 1st parent = *Oonn*   2nd parent = *ooNn*

5. Corn is monoecious, has no sex-chromosomes and therefore no sex-linked genes. Reciprocal crosses give different results because:
   (i) Endosperm is 3n (if it was 2n such differences would be impossible since all genes autosomal).
   (ii) Female contributes two haploid polar nuclei, male one gamete nucleus to the primary endosperm nucleus (3n).
   (iii) The allele (*F*) for *floury* is incompletely dominant to (*F'*) for *flinty* (or vice versa), whereas the allele (*S*) for *starchy* is dominant to *s* for *sugary*. Assume that a single allele pair determines each pair of traits.

Parents		F_1
♀ (2 polar nuclei, each n)	♂ (n gametes)	Endosperm (3n)
Sugary, flinty (*ssF'F'*)	Starchy, floury (*SF*)	Starchy (*Sss*), flinty (*F'F'F*)
Starchy, foury (*SSFF*)	Sugary, flinty (*sF'*)	Starchy (*SSs*), floury (*FFF'*)

7. a) Two allele pairs are segregating. In the first cross segregation occurs for both pairs of alleles. In the second cross segregation occurs for the pair of alleles that determines plumage.

   P_1 rose, blue-splashed *RRBB*   x   single, black *rrB'B'*

   F_1 rose, blue *RrBB'*

   F_1 (*RrBB'*) x single, black (*rrB'B'*) F_1 (*RrBB'*) x rose, blue-splashed (*RRBB*)

RrBB' (rose, blue)	64	*RRBB* and *RrBB* (rose, blue-splashed)	108
RrB'B' (rose, black)	57	*RRBB'* and *RrBB'* (rose, blue)	113
rrBB' (single, blue)	61		
rrB'B' (single, black)	59		

   b) The allele *R* (*rose*) is dominant to *r* (*single*) because the F_1 phenotype resembles that of the *rose* parent. The allele *B* (*blue-splashed*) is incompletely dominant to *B'* (*black*) because the F_1 phenotype (*blue*) is *intermediate*, relative to the two parental phenotypes *blue-splashed* and *black*.

   c) Inheritance is independent, since the ratio in the first cross is 1 : 1 : 1 : 1.

8. a) *Pituitary dwarfism* is determined by a recessive allele *b*; *normal* by the dominant *B*.
   (i) Two *unaffected* parents (III-1 and -2) give rise to an *affected* individual
   (ii) Trait skips generation.

   Cataract is determined by a dominant allele *A*; *normal* by the recessive allele *a*.
   (i) Trait does not skip generations.
   (ii) Cannot be recessive since both III-5 and -6 are *affected*, yet they have *normal* (*unaffected*) children.

   b) IV-1 has *pituitary dwarfism*, IV-6 has *cataracts* and *pituitary dwarfism*. Since all the children are *dwarfs* but some have *cataracts* and others do not verifies the hypothesis in (a). Therefore their genotypes are IV-1 = *aabb*; IV-6 = *Aabb*

   c) 1.

Marriage	Progeny
III-5 x IV-1	1/4 *AaBb* - cataract, normal
AaBb x *aabb*	1/4 *aaBb* - normal, normal
	1/4 *Aabb* - cataract, dwarf
	1/4 *aabb* - normal, dwarf

   2. III-2 x IV-5
   aaBb x *A_B_*   (assume *AABB*) → *AaBB* and *AaBb*   all cataract, normal

   d) (1) 1 (100%)   2) 0
   (3) Note: Both IV-2 and IV-4 are *aa*. Since both these individuals have siblings that are *pituitary dwarfs* their parents must both be heterozygous (*Bb*). Therefore, probability that IV-2 is heterozygous is 2/3 and probability that IV-4 is heterozygous is also 2/3. Probability of both IV-2 and IV-4 being heterozygous is 2/3 x 2/3 = 4/9. Probability of a *dwarf* child from such a mating

is 1/4. Therefore, the chance that the first child of IV-2 x IV-4 is a *dwarf* is: 4/9 x 1/4 = 4/36 = 1/9

12. a) Two genes (2 pairs of alleles) since 4 phenotypes occurred in the testcross progeny.
    b) No. One must compare the phenotypes of the parents and the F_1 or note the phenotypic ratio in the F_2 progeny.
    c) They show that (i) the F_1 parent was heterozygous for two pairs of alleles, (ii) the allele pairs are inherited independently and (iii) 4 types of gametes are produced by the F_1 parents in equal proportions.

14. a) Two allele pairs are involved. They segregate independently since the $3^{n=2} = 9$ F_2 phenotypes and genotypes appear in the proportions (1 : 2 : 2 : 4 : 1 : 2 : 1 : 2 : 1) as expected with independent assortment and incomplete dominance within both allele pairs.
    LL - long; *LL'* - oval; *L'L'* - round //*RR* - white; *RR'* - purple; *R'R'* - red.

    (1) F_1 *oval, purple* (*LL'RR'*) x *long, white* (*LLRR*) parent → 1 *long, white* : 1 *long, purple*: 1 *oval, white* : 1 *oval purple*

    (2) F_1 *oval, purple* (*LL'RR'*) x *round, red* (*L'L'R'R'*) parent → 1 *oval, purple* : 1 *round, purple*: 1 *round, purple* : 1 *round, red*

    b) (1) *Long, puple* (*LLRR'*) x *oval, purple* (*LL'RR'*)

    Genotypic and phenotypic ratio

    | 1 *LLRR* - long, white | 1 *LL'RR* - oval, white |
    | 2 *LLRR'* - long, purple | 2 *LL'RR'* - oval, purple |
    | 1 *LLR'R'* - long, red | 1 *LL'R'R'* - oval, red |

    (2) *Oval, puple* (*LL'RR'*) x *round, white* (*L'L'RR*)

    Genotypic and phenotypic ratio

    | 1 *LL'RR* - oval, white | 1 *LL'RR'* - oval, purple |
    | 1 *L'L'RR'* - round, purple | 1 *L'L'RR* - round, white |

    (c) Genotypic and phenotypic ratio

    | 1 *LLRR* - long, white | 1 *L'L'R'R'* - round, red |
    | 1 *LLR'R'* - long, red | 1 *L'L'RR* - round, white |

    (d) Lines for producing *oval, purple* (*LL'RR'*)
    (i) *long, white* (*LLRR*) and *round, red* (*L'L'R'R'*)
    or
    (ii) *long, red* (*LLR'R'*) and *round, white* (*L'L'RR*)

15. Two independently inherited allele pairs are involved; *S* (*stippled*) is dominant to *s* (*non-stippled*) and *P* (*patternless*) is incompletely dominant to P^l (*twin spot*).

    P_1 *Stippled, patternless* (*SSPP*) x *non-stippled, twin spot* (*ssP^{l}P^l*)

    F_1 *Stippled, crescent* (*SsPP^l*)

F_2 Genotypes and ratios	F_2 Phenotypes and ratio	
1 *SSPP* + 2 *SsPP*	3 *stippled, patternless*	(35)
4 *SsPP^l* + 2 *SSPP^l*	6 *stippled, crescent*	(83)
1 *SSP^{l}P^l* + 2 *SsP^{l}P^l*	3 *stippled, twin spot*	(29)
1 *ssPP*	1 *non-stippled, patternless*	(12)
2 *ssPP^l*	2 *non-stippled, crescent*	(27)
1 *ssP^{l}P^l*	1 *non-stippled, twin spot*	(9)

Note: Six phenotyes because of incomplete dominance at the *P* locus and complete dominance at the *S* locus.

True-breeding *crescent* phenotypes cannot be developed since they are due to the heterozygous genotype (*Pp*). The best way to get the most *crescents* is to cross *patternless* with *twin spot*.

16. Two possible explanations:
    a) One allele pair with incomplete dominance so that each genotype results in a different phenotype: *AA* (*black*); *Aa* (*intermediate*) and *aa* (*pink*).

    b) Two allele pairs - *black* is due to a dominant allele at the first locus and homozygous for the recessive at the second locus (*A_bb*); *pink* is due to homozygosity for the recessive allele of the first gene and a dominant allele of the second gene (*aaB_*). *Intermediates* would have both dominants (*A_B_*). The double recessive genotype is probably inviable. To verify, cross 2 *intermediates*. If 3 phenotypes in approximately a 1 *black* : 2 *intermediate* : 1 *pink* are obtained this would verify that the character is determined by 1 pair of alleles. If the cross always gives rise to reduced litters, then it may be suspected that the character is determined by two allele pairs and that the genotype *aabb* is inviable.

20. a) The data reveal that the allele pairs are segregating independently for the following reasons:
    (i) In the first 2 tetrad classes, both allele pairs show first division segregation only. Moreover, the frequency (19/109) of PD tetrads classes 1 - 5 with f^+ and mt^+ in coupling) is approximately equal to the frequency (21/109) NPD classes 2 and 6 tetrads (with f^+ and mt^+ in repulsion).

    (ii) Per cent recombination =
    $$\frac{\text{Recombinant spore pairs}}{\text{Total spore pairs}} \times 100 = \frac{80+12+120+4+6}{109 \text{ tetrads} \times 4} \times 100 = 50.9$$

    In addition to showing that the allele pairs segregate independently the data also indicate that the allele pairs are on different chromosome pairs. Basis for latter conclusion - It is possible that 2 loci may be so far apart on a chromosome that the number of recombinant spore pairs would be 50 percent. Such cases, however, are not ambiguous unless one or both of the loci shows free recombination with the centromere, i.e., 67 percent second division segregation. The cross f^+mt^+ x fmt^- is not of this kind; f shows 30.7 percent (1/2 second-division segregation; 67/109 x 100 = 61.4 x 1/2) and *mt* shows 5.9 percent (1/2 second-division segregation, 13/109 x 100 = 11.8 x 1/2) recombination with its centromere.

    b) Illustrate this for yourself.

    c) Four genotypic and phenotypic classes of spores are expected in equal proportions since the 2 genes assort independently.

f^+mt^+ - *nonfluffy*, "plus" mating type; f^+mt^- - *nonfluffy*, "minus" mating type

fmt^+ - *fluffy*, "plus" mating type; fmt^- - *-fluffy*, "minus" mating type

21. The following kinds and proportions of tetrads are expected in the cross thi^+ad^+ x thi^-ad^- since: (i) the 2 genes are on different (non-homologous) chromosomes, (ii) different chromosome pairs and therefore allele pairs orient at MI and segregate at AI independently of each other, and (iii) no crossing over occurs between the gene loci and their centromeres. As a consequence only two basic alignments (each with two variations) are possible in a large number of meiocytes.

Tetrads

(i)	thi^+ad^+		thi^-ad^-	(ii)	thi^+ad^-		thi^-ad^+
	thi^+ad^+	or	thi^-ad^-		thi^+ad^-	or	thi^-ad^+
	thi^-ad^-		thi^+ad^+		thi^-ad^+		thi^+ad^-
	thi^-ad^-		thi^+ad^+		thi^-ad^+		thi^+ad^-
	1/4		1/4		1/4		1/4

The same results (tetrads) would be obtained in the cross thi^+ad^- x thi^-ad^+ for reasons (i) and (ii) stated above.

Chapter 4
Probability and Chi-Square

2. The probability of III-1 and III-2 having *short*-haired or *long*-haired offspring depends on (i) the possible genotypes of these individuals; (ii) the probability of these individuals possessing specific genotypes:
a) Given that II-2 is *SS* and since II-3 is *ss* (has *long* hair) III-1 must be *Ss* (probability 100%).
b) II-5 is *SS*.
Since the parents of II-6 both have *short* hair, there are 4 possible combinations with respect to their genotypes, each with an equal chance of occurrence.

	I-5	I-6	Probability of occurrence	Chance of a heterozygous (*Ss*) offspring	Overall probability of heterozygous offspring
i)	SS	SS	1/4	0	0
ii)	Ss	SS	1/4	1/2	1/8
iii)	SS	Ss	1/4	1/2	1/8
iv)	Ss	Ss	1/4	2/3	1/6

Total probability of II-6 being *Ss* regardless of the genotypes of its parents = 1/8 + 1/8 + 2/12 = 5/12

p of III-2 being *Ss* = 5/12 x 1/2 = 5/24; p of III-1 being *Ss* =1
Therefore p of their offspring being *long*-haired = 1 x 5/24 x 1/4 = 5/96
p of their offspring being *short*-haired = 1 - 5/96 = 91/96

3. a) Fifteen. **Reason** - since ram is *Bb*, for ewes to produce *affected* (*bb*) progeny they must also be heterozygous.
b) Solution: The probability that a heterozygous ewe will not have a *white* lamb in 3 matings to a heterozygous male is determined as follows: each lamb has a 3/4 chance of being *black*; therefore, the probability that we will fail to detect a heterozygous ewe with 3 lambs is 27/64. Therefore, the probability that we will detect a heterozygous ewe having only 3 lambs is 37/64. If x = number of heterozygous ewes in the flock, then 37/64 x = 15, i.e., x = 26. Twenty-six heterozygous ewes probably exist in the flock; since 15 have already been detected, there should be 11 more such ewes.
c) $(3/4)^6$ = 729/4096.

4. Parents $Gg \times Gg$; G = *normal*; g = *galactosemia*

a) $(1/4)^2$ = 1/16
b) Let a = chance child will be *normal* = 3/4; b = chance child will be *galactosemic* = 1/4. Chance of 3 *normals* and 1 *galactosemic* can be derived from the binomial: $(a + b)^{n=4} = a^4 + 4a^3b + 6a^2b^2 + 4ab^3 + b^4$. Select appropriate term - $4a^3b$. Substitute known probabilities and multiply, i.e., $4 \times (3/4)^3 \times (1/4)^1$ = 108/256 = 27/64 chance of 3 *normals* and 1 *galactosemic*.
c) 1
d) Insufficient information to determine probability of being *Gg*.
e) 1/2
f) Assume *normal* sister's husband is *GG*, since he is *normal* and not related to the family.
Chance of *normal* sister being heterozygous = 2/3;
chance of her transmitting *g* to any child = 1/2;
chance of any child being heterozygous = 2/3 x 1/2 = 1/3
g) 1/4
h) $(3/4)^3$ = 27/64
i) $3ab^2 = 3 \times 1/4 \times (1/2)^2$ = 3/16
j) Chance of any child being a *boy* = 1/2;
chance of any child being a *girl* = 1/2.
Probability of any child being:
(1) a *galactosemic*, *girl* = 1/4 x 1/2 = 1/8;
(2) a *normal*, *boy* = 3/4 x 1/2 = 3/8
k) 3/8 x 1/8 = 3/64
l) Probability of at least 2 *normal* children is the sum of the probabilities represented by the terms $6a^2b^2 + 4a^3b + a^4$; i.e., 243/256.
a = chance of *galactosemic*, *girl* = 1/8; b = chance of *galactosemic*, *boy* = 1/8
$6a^2b^2$ = chance of 2 *galactosemic*, *girls* and 2 *galactosemic*, *boys*
$6 \times (1/8)^2 \times (1/8)^2 = 6 \times 1/64 \times 1/64$ = 6/4096
m) Chance first 2 children will be *galactosemic* = 1/4 x 1/4 = 1/16. Chance first 2 children will be *normal* = 3/4 x 3/4 = 9/16. Chance both will be *galactosemic* or both *normal* = 1/16 + 9/16 = 10/16 = 5/8

5. a) (i) 1/2; (ii) 1/2 x 1/2 = 1/4; (iii) 1/2 + 1/2 = 1
b) (i) 1/2 x 1/2 = 1/4; (ii) 1/2 x 1/2 = 1/4;
(iii) 1/4 x 1/4 = 1/16; (iv) 1/2 x 1/2 = 1/4

8. n = 6; sex ratio 1:1, a = chance child will be a *boy* = 1/2;
b = chance child will be a *girl* = 1/2
a) $(a + b)^6 = a^6 + 6a^5b + 15a^4b^2 + 20a^3b^3 + 15a^2b^4 + 6ab^5 + b^6$
(i) Chance - 4 *boys*, 2 *girls* - $15a^4b^2 = 15(1/2)^4 (1/2)^2$ = 15/64
(ii) *All same sex* (all *boys* or all *girls*), i.e., $a^6 + b^6 = (1/2)^6 + (1/2)^6$ = 1/32
(iii) At least 3 *girls*, i.e., $b^6 + 6ab^5 + 15a^2b^4 + 20a^3b^3$ = 1/64 + 6/64 + 15/64 + 20/64 = 21/32

(iv) *No fewer than 2 girls and 2 boys* - i.e. $15a^4b^2 + 20a^3b^3 + 15a^2b^4 = 15/64 + 20/64 + 15/64 = 50/64$

(v) *3 or more girls*. Answer the same as for at least 3 *girls* = 21/32 and is calculated in the same way.

(vi) Chance of 4 *boys* and 2 *girls* = $15a^4b^2 = 15 \times (1/2)^4 \times (1/2)^2 = 15/64$.

Chance of 2 *boys* and 4 *girls* = $15 a^2b^4 = 15 \times (1/2)^2 \times (1/2)^4 = 15/64$.

Chance of 4 *boys* and 2 *girls* or 2 *boys* and 4 *girls* = $15/64 + 15/64 = 15/32$.

(vii) Chance eldest a *boy* = 1/2; chance youngest a *girl* = 1/2. Chance of both (eldest a *boy*, youngest a *girl*) = $1/2 \times 1/2 = 1/4$.

b) (i) $20 a^3b^3$ - 3 *boys* and 3 *girls* = 20/64. (ii) % of families of 6 having 3 *boys* and 3 *girls* = 31.25.

10. a) Let *h* be the recessive allele for *hypotrichosis*, *H* the allele for *normal*. (1) Both parents must be heterozygous, *Hh*, since some of their children are *hh* (*hypotrichotic*). (2) Each *normal* has a 2/3 probability of being *Hh*, *p* of all three *normals* being heterozygous is $(2/3)^3 = 8/27$.

b) (1) The man's genotype is *Hh* because he is *normal* and has some *affected* children. (2) Probability of *normal* children being *Hh* in this marriage is = 1.
(3) Probability of first child being *hypotrichotic* = $1 \times 2/3 \times 1/4 = 1/6$. Probability of four *normal* children from this marriage is 418/768.
Explanation:
The *normal* individual from (2) must be *Hh*. The *normal* individual from (a) may be either *Hh* or *HH* with probabilities of 2/3 and 1/3 respectively. Thus, the marriage may be either: (i) $Hh \times 2/3 Hh$, or (ii) $Hh \times 1/3 HH$.
If it is (i), the probability will be $2/3 \times (3/4)^4 = 162/768$. If it is (ii), the probability is $1 \times 1/3 = 256/768$. The combined probability of the marriage giving *all 4 normal* offspring is = $162/768 + 256/768 = 418/768$.

11. If the F_1 is heterozygous for a single autosomal allele pair and produces 360 progeny, expect 90 (1/4 x 360) of the progeny to be homozygous recessive (*compound*). Subjecting the results to a χ^2 test reveals a value of 3.33. The probability of a χ^2 of 3.333 (1 *df*) is between 0.10 and 0.05. Therefore, it is likely that a single autosomal allele pair is involved in producing the results given.

16. a) Single allele pair; *W* (*white*) dominant over *w* (*coloured*). $\chi^2 = 0.533$ which for 1 degree of freedom has a probability value between 0.75 and 0.50. Hypothesis is consistent with the data. Two independently inherited allele pairs interacting in a dominant-recessive epistatic manner; *A-B-*, *A-bb* and *aabb* produce *white* and *aaB-* produce *coloured*. F_1's - *AaBb* and F_2's - 13 *white* (9 *A-B-*, 3*A-bb*, 1 *aabb*): 3 *coloured* (3 *aaB-*).

$\chi^2 = 1.50$ which for 1 degree of freedom has a probability value between 0.50 and 0.30. This hypothesis is also consistent with these data. On the basis of χ^2 tests alone, the former hypothesis is more acceptable.

b) Take the 36 *coloured* birds and intercross.
(1) If colour is determined by a single pair of alleles (*Ww*) all the progeny should be *coloured* (*ww* x *ww* → all *ww*).
(2) If two allele pairs with dominant-recessive epistasis interaction responsible for the results, the progeny of *some* crosses should consist of both *coloured* and *white* birds, e.g., *aaBb* x *aaBb* → 1 *aaBB* : 2 *aaBb* : 1 *aabb*, e.g., 3 *coloured* : 1 *white*.

17. Hypothesis - the 2 sexes occur in equal proportions (1:1 ratio).
a = chance of a *girl* = 1/2; b = chance of a *boy* = 1/2.
The terms of the expanded binomial $(a + b)^{n=4}$ indicate both the type of sibship, i.e., combination of *girls* and *boys* in a family of 4 (disregarding the sequence in which they are born) and the probability of occurrence of each combination.

Term/sibship	a^4	$4a^3b$	$6a^2b^2$	$4ab^3$	b^4
	(4 girls, 0 boys)	(3 girls, 1 boy)	(2 girls, 2 boys)	(1 girl, 3 boys)	(0 girls, 4 boys)
Probability	1/16	4/16	6/16	4/16	1/16

Sibship	No. of families observed (O)	expected (C)	$(O - C)^2 / C$	χ^2
0 boys, 4 girls	12	15	9/15	0.60
1 boy, 3 girls	69	60	81/60	1.35
2 boys, 2 girls	84	90	36/90	0.40
3 boys, 1 girl	57	60	9/60	0.15
4 boys, 0 girls	18	15	9/15	0.60
			Total χ^2 =	3.10

Probability for χ^2 of 3.10 (4 *df*) is between 0.50 and 0.30. The data are consistent with the hypothesis.

Chapter 5
Gene Interaction and Lethal Genes

2. a) No. Segregation of alleles located at corresponding positions on homologous chromosomes is dependent on their behavior during meiosis.
   b) No. Kinds of gametes depend on an individual's genotype, and chromosome behavior at meiosis.
   c) No. F_2 genotypes depend on genotype and chromosome behavior in F_1's plus the fact that gametes unite at random.
   d) and e) Yes. In all cases of gene interaction, the activity (expression) of one gene (the epistatic locus) influences that of the hypostatic locus (loci). The kinds of phenotypes that will occur and their proportions will depend on the type of gene interaction.

3. No. To detect a gene via breeding procedures, it must exist in at least two allelic forms.

5. *Cross 1.* P_1 *Grey (AARRCC)* x *albino* line 1 (___cc) F_1 *Grey*

   a) The distribution of F_2 phenotypes - 174 *Grey* : 65 *black* : 80 *albino* approximates a 9:3:4 dihybrid ratio; therefore the true-breeding strains carry different alleles of two genes only. One of these genes is that determining pigment development (*grey, C-; albino, cc*). Since *grey* and *black* differ in genotype with respect to alleles of the first gene and since the *grey* parent is *AA*, the *albino* parent must be *aa*. Therefore, genotype of *albino* line is *aaRRcc*.

   P_1 *Grey (AARRCC)* x *albino* line 1 (*aaRRcc*) - F_1 *AaRRCc* (*grey*)
   F_2 9 *A_RRC_* - *grey*; 3 *aaRRC_* - *black*: 4 (3A - RRcc + 1aaRRcc) *albino*

   b) *Recessive epistasis. The allele c, for albinism, is epistatic to the colour genes;* thus the F_2 ratio is 9 : 3 : 4.

   Cross 2.
   a) Since the F_2 distribution (48 *grey* : 16 *albino*) is in a 3 : 1 ratio, the two strains must carry different alleles of the *C* gene. The *albino* strain must have been *AARR*; otherwise other coloured types would have appeared in the progeny. Therefore genotype of *albino* line 2 is *AARRcc*.
   b) Interallelic-interaction; allele *C* for *pigment formation* is dominant over *c* for *albinism*.

   Cross 3.
   a) Since the coloured progeny appear in a ratio of ~ 3 *grey (A_R_)* : 1 *yellow* (*A_rr*), the *albino* line must have been homozygous for the dominant allele *A* (necessary for both phenotypes) and the recessive allele *r* (which in the homozygous condition causes *yellow*). Genotype of *albino* line 3 is *AArrcc*.
   b) Recessive epistasis. Allele *c* for *albinism* is epistatic to the alleles of the colour genes (*Aa* and *Rr*).

   Cross 4.
   a) Since the coloured F_2 phenotypes (292 *grey* : 87 *yellow* : 88 *black* : 32 *cream*) appear in ~ a 9 : 3 : 3 : 1 ratio, the *albino* line must have been homozygous for the recessive alleles *a* and *r* in addition to *c*. That the *albino* line is homozygous for all three recessives (*aarrcc*) is also evidenced by the fact that the overall F_2 phenotypes are in a ratio of 27 *grey* : 9 *yellow* : 9 *black* : 3 *cream* : 16 *albino*.
   b) The *A* and *R* genes interact in a complementary manner in the presence of allele *C* to produce the various colours. The dominant *A* results in *yellow*, the dominant *R* in *black*, the homozygous double recessive *aarr* in *cream*, and the double dominant *A_R_* in *grey*. The recessive allele *c* (*albinism*) is epistatic to the alleles at the colour loci *A* and *R*.

6. a) 12 *tri* : 3 *tetra* : 1 *penta*. *Reason* - two independently inherited allele pairs *Tt* and *Cc*, interact in determining these phenotypes.
   b) Because of dominant epistasis. The dominant allele *T* for *trimolting* is epistatic to the alleles *C* (*tetramolting*) and *c* (*pentamolting*) at the *C* locus. Therefore true-breeding *trimolting* strains can be of 2 genotypes - *TTCC* and *TTcc*.
   c)

Cross	Parental genotypes	F_1	F_2
1. *tri x penta*	*TTcc x ttcc*	*Ttcc*	3 *T_cc* (*tri*) : 1 *ttcc* (*penta*)
2. *tetra x penta*	*ttCC x ttcc*	*ttCc*	3 *ttC_* (*tetra*) : 1 *ttcc* (*penta*)
3. *tri x tetra*	*TTcc x ttCC*	*TtCc*	9 *T_C_* + 3 *T_cc* = 12 *tri*
			3 *ttC_* = 3 *tetra*
			1 *ttcc* = 1 *penta*

14. a) Since 1/16 of the F_2 express the recessive (*winter*) phenotype, 2 pairs of alleles (*Ss* and *Ww*) must be involved. They are interacting in a duplicate dominant epistatic manner. Either *S* or *W* is sufficient to produce the dominant (*spring*) phenotype.
    b) *Spring*, since F_1's are *SsWw* and either dominant is sufficient to produce the trait.
    c) Homozygous for recessives at both loci (*ssww*).

17. a) Two pairs of alleles determine capsule shape. *Interaction* - dominant phenotype (*triangular*) is due to duplicate non-complementary dominant alleles of different genes and the recessive phenotype (*top-shaped*) is due to complementary recessives at both loci. An alternative explanation is duplicate

dominant epistasis, i.e. the dominant allele at either locus is sufficient for expression of *triangular* phenotype.
b) Gene symbols.

Gene 1	Gene 2
T - triangular, t - top-shaped	T^1 - triangular, t^1 - top-shaped

(1) Ttt^1t^1 x Ttt^1t^1 or ttT^1t^1 x ttT^1t^1

(2) TTT^1T^1 x ttt^1t^1 or ttT^1T^1 x ttt^1t^1 or any other combination of genotypes as long as one locus is homozygous dominant in at least one of the parents.

(3) TtT^1t^1 x Ttt^1t^1 or TtT^1t^1 x ttT^1t^1

(4) Ttt^1t^1 x ttt^1t^1 or ttT^1t^1 x ttt^1t^1

25. Since only two phenotypes are observed in the F_2, regardless of the number of allele pairs involved, there is complete dominance of one allele over the other in all pairs.
a) If only *one* allele pair was involved *(R (red)* vs. *r (white))* the F_2 distribution should approach 3 *red* : 1 *white*. The actual results deviate significantly from expected; therefore more than one allele pair is involved.
b) If two pairs of alleles are involved, since only two phenotypes occur in the F_2, they could be interacting in either (i) a non-complementary dominant or (ii) complementary dominant manner.
If the former, the F_2 ratio should be 15 *red* : 1 *white*. This is significantly different from the actual results. Therefore b) i) does not explain the distribution.
If due to complementary dominant alleles the F_2's should consist of more *reds* than *whites* and in a 9 : 7 ratio. The actual F_2 distributions are reversed - more recessive than dominant types are present. Therefore, two pairs of alleles acting in a dominant complementary manner also do not explain these results. **Therefore, at least 3 pairs of alleles must be involved.**

c) If three allele pairs are involved the dominant alleles may be acting in a *non-complementary* or *complementary* manner.
i) If *non-complementary* (and F_1 *AaBbCc*), the F_2 phenotypic ratio would be 63 *red* : 1 *white* (this can be derived using the branching method). The actual results deviate significantly from this expected ratio. Therefore can reject this explanation.
ii) *If due to complementary dominant alleles* red is due to the presence of the dominant alleles at all 3 loci. Absence of the dominant allele at any of the 3 loci will produce *white*. This type of interaction predicts an F_2 ratio of 27 *red* (*A_B_C_*) : 37 *white* (includes all other genotypes which can be determined by using the branching method). The actual results of 285 *red* and 378 *white* are in close approximation to this modified expected ratio of 27 *red*: 37 *white*.

33. The results cannot be explained by a single allele pair with the allele for *deaf-mutism* being either dominant or recessive. If the trait was due to a single dominant, *unaffected* parents (I-1 and I-2) could not produce an *affected* offspring (I 1x2 → II - 4,6,8,10) and if it was due to an autosomal recessive, 2 *affected* (III-7 and III-8) could not produce *unaffected* children (IV - 1,2,3,4,5,6). Therefore, 2 allele pairs are involved which interact in a duplicate recessive epistatic or duplicate complementary dominant manner. Individuals with both dominants (A_B_) would be *unaffected* and all others (A_bb, aaB_, aabb) *affected*. This accounts completely for the results. For example: - III - 7 *affected* (*AAbb*) x III - 9 *affected* (*aaBB*) produce *unaffected* (*AaBb*) children because of complementation.

38. Since it is impossible to develop a true-breeding *montezuma* variety and when two *montezumas* are mated some of the progeny are *wild-type*, this indicates all *montezumas* must be heterozygous (*Mm*). A 2 : 1 ratio in the offspring of a cross between two *montezumas* (2/3 *montezumas* : 1/3 *wild-type*) suggests that *montezumas* must be heterozygous for an incompletely dominant lethal allele, the *wild type* being homozygous (*mm*) for the non-lethal recessive allele. The homozygous genotype (*MM*) must therefore be lethal.

39. a) (1) Using the symbols *N* for the *normal* allele and *n* for the lethal allele the genotypes are as follows:

(i) Father (*Nn*) x Mother (*NN*) → Daughters: 1/2 *NN* : 1/2 *Nn*

(ii)
Father	Daughters	Progeny
Nn x 1/2 *NN* →		1/4 *NN* - normal
		1/4 *Nn* - normal
Nn x 1/2 *Nn* →		3/8 *N* - normal
		1/8 *nn* - lethal

(2) Recessive lethal.

b) Probably not. To detect early acting lethals the organism would have to be fairly prolific. In early abortions the foetuses are often reabsorbed. Large progeny sizes from matings between two *normal* (heterozygous) individuals are needed to be able to detect reductions in progeny sizes.

41. The reasons why *droopy-winged* individuals occur among females only, are: i) In bees, females are diploid, males haploid. ii) The allele *D* for *droopy* is lethal when homozygous or hemizygous (when present in single dose, as in haploid males). The allele *d* for *normal* does not result in lethality.

Since males with *D* die before sexual maturity they cannot transmit the allele. Therefore *droopy* females cannot be homozygous for *D*. Only possible matings are:

i) *dd (normal)* ♀'s x *d (normal)* ♂'s → ♀ + ♂ offspring all *normal*
ii) *Droopy Dd* ♀'s x *normal d* ♂'s → 1/2 *Dd droopy* ♀'s; 1/2 *dd normal* ♀'s: 1/2 *d normal* ♂'s; 1/2 *D* (lethal)

46. a) The cross involves two independently inherited allele pairs (*Pp* and *Ss*) with the *plum* and *stubble* traits being determined by the dominant alleles *P* and *S* respectively. *Red* and *normal* are due to the recessive alleles *p* and *s* respectively. A modified F_2 ratio with a 2 : 1 segregation for each pair of traits indicates that an incompletely dominant lethal is involved at each locus. Thus:

P_1 *plum, stubble (PpSs)*   x   *plum, stubble (PpSs)*
Viable offspring: 4 *PpSs* :   2 *Ppss* :   2 *ppSs* :   1 *ppss*
                   plum, stubble   *plum, normal*   *red, stubble*   *red, normal*

All other genotypes (*PPSS*, *PPS_*, *P_SS*) are lethal because of the homozygous presence of one or both incompletely dominant lethals, *P* and *S*.

b) The only cross that will breed true is *red, normal* x *red, normal*. Crosses involving the other three strains will never breed true. In addition, they will always give rise to reduced progeny sizes. The cross *plum, stubble (PpSs)* x *red, normal (ppss)* which would produce 4 phenotypes: *plum stubble, plum normal, red stubble* and *red normal* would verify the hypothesis.

47. The registered tom turkey is heterozygous for a Z-linked recessive lethal allele (Z^l). The cross is as follows:

$Z^L Z^l$ x $Z^L W$ -   Males: 1/2 $Z^L Z^L$ and 1/2 $Z^L Z^l$
                              Females: 1/2 $Z^L W$ and 1/2 $Z^l W$ (lethal)
Therefore the overall ratio of 2/3 males: 1/3 females

50. a) *Brachydactyly* is probably determined by an incompletely dominant lethal allele of an autosomal gene. **Reasons**: The condition is equally frequent in both males and females. The trait is not observed to skip generations. Two *affected* parents have both *affected* and *unaffected* progeny. If it were sex-linked two *affected* parents could not have *unaffected* daughters.
c) Each *affected* individual is probably heterozygous, since the homozygous dominant genotype is lethal.

Chapter 6
Multiple Alleles

3. a) One gene and 3 alleles are involved. **Reason**: F_2's of all possible crosses between the truebreeding species show a monohybrid (3 : 1) ratio.
   b) Allele A for *striped* is dominant to both the allele a for *plain* and a^m for *moricaud*. The allele a^m for *moricaud* is dominant to that for *plain* a. **Basis for conclusion**: F_1 phenotypes, conclusion verified by F_2 results. The genotypes of the different phenotypes therefore are: *striped* (AA, Aa^m or Aa); *moricaud* ($a^m a^m$, $a^m a$) and *plain* (aa).

4. The most plausible explanation is that the traits *silver*, *platinum* and *white-face* are determined by 3 different (p, P and P^W) alleles of one gene - allele p for *silver* is non-lethal; allele P for *platinum* is lethal when homozygous and allele P^W for *white-face* is lethal when homozygous.

   Reasons: i) Crosses that show segregation give monohybrid ratios. ii) Since the ratio in crosses 3 and 4 is 2 : 1 and the litter sizes are smaller than in the previous two crosses, the alleles for *platinum* and *white-face* must be lethal when homozygous. This conclusion is verified by the results indicating that one genotype PP^W is lethal. Otherwise (PP^W) should have expressed a fourth phenotype, either *platinum* or *white-face*.

   Note: The results can also be explained by 2 allele pairs with *silver* due to $aabb$, *platinum* caused by $Aabb$, *white-face* determined by $aaBb$ with $AAbb$, $aaBB$ and $A\_B\_$ genotypes should produce fatal effects.

6. a) All ratios, including the ones in the critical crosses *superdouble* x *double* and *superdouble* x *single* are monohybrid, i.e., 3 : 1 or 1 : 1.
   b) *Allelic relationships*: S^1 (*superdouble*) is dominant over S (*single*) and s (*double*); S is dominant over s, i.e., $S^1 > S > s$.
   Bases for conclusions:
   (1) The results of cross 1 indicate that the difference between *single* and *double* is due to the allele (S) for *single* being dominant to s for *double*.
   (2) The results of cross 2 indicate that (i) *superdoubles* are heterozygous, carrying an allele for *superdouble* and one for *double*, and (ii) the allele (S^1) for *superdouble* is dominant to s for *double*.
   (3) That S^1 is dominant to S for *single* is indicated by the results of the third cross. Since *superdoubles* arose in the *double* variety they must be S^1s. These results also verify the conclusion of cross 1 that S is dominant to s.
   (4) The data in crosses 4 and 5 verify the above conclusions and also indicate that *superdoubles* can be $S^1 S$ or $S^1 s$.

8. a) One gene, 3 alleles involved. B (*black*) is dominant to b^1 and b; b^1(*black*) is incompletely dominant to b (*sapphire*) so that $b^1 b^1$ are *black*, $b^1 b$ *platinum* and bb *sapphire*. Therefore A *black* is BB, B *black* is Bb, C *black* is $b^1 b^1$. *Platinum* is $b^1 b$ while *sapphire* and domesticated animals are due to the genotype bb.
   b) A (BB) x   E ($b^1 b$)   → all *black* (1/2 Bb^1 + 1/2 Bb)
      B (Bb) x   E ($b^1 b$)   → 2 *black* (1/2 Bb^1 + 1/2 Bb) : 1 *platinum* ($b^1 b$) : 1 *sapphire* (bb)
      C ($b^1 b^1$) x   D (bb)   → all *platinum* ($b^1 b$)

9. Three alleles of one gene are involved.
   Reason: If only two alleles B and B' with incomplete dominance were involved, each genotype would produce its own phenotype with *black* or *tortoise-shell* or *red* being heterozygous (BB^1). Working from the left to the right side of the pedigree since *black* x *tortoise-shell* (1 x 2) produce *black* (5) and *tortoise-shell* (4) progeny, one of these phenotypes must be due to heterozygosity (BB^1); the other must be due to one of the 2 homozygous genotypes (BB or $B^1 B^1$) and therefore *red* must be due to the other homozygous genotype. Regardless whether the *black* (1) parent is homozygous and the *tortoise-shell* (2) heterozygous or vice-versa, it would be impossible for a *black* (1) x *tortoise-shell* (4) mating to produce a *red* (8), if only 2 alleles were involved.
   i) *black* (BB^1) x *tortoise-shell* (BB) → *black* (BB^1) and *tortoise-shell* (BB)
   ii) *black* (BB) x *tortoise-shell* (BB^1) → *black* (BB) and *tortoise-shell* (BB^1)

   Neither cross will produce *red* which would have to be due to $B^1 B^1$. Since *tortoise-shell* (2) x *red* (3) produce *red* (6) and *tortoise-shell* (7) progeny, *tortoise-shell* would be due to the heterozygous genotype had this explanation been correct. Therefore 3 alleles must be involved.

   B - *black*, dominant to b^1 and b.
   b^1 - *tortoise-shell*, incompletely dominant or dominant to b
   b - *red*, recessive to B and b^1.
   Individuals 1 to 9 are of the following genotypes:
   1 - Bb; 2 - $b^1 b$; 3 - bb; 4 - $b^1 b$; 5 - Bb^1; 6 - bb; 7 - $b^1 b$; 8 - bb; 9 - $b^1 b$

10. a) The 5 mutations occurred in 3 different genes. l and o are allelic since they do not complement each other; m and n are allelic since they do not complement each other; the l/o mutations are at a separate locus from the m/n mutations since lm, ln, om and on crosses complement each other to produce *wild-type* progeny. p is at the third locus since complementation occurs in the females progeny from crossing of p in the crosses with l, m, n and o.
    b) The genes containing the mutations l/o and m/n are on autosomes since males and females have the same phenotype in reciprocal crosses. Genes carrying mutation p is on the X chromosome since there is a difference in

phenotype between males and females when p is used as the female parent and l, m, n or o are each used as the male parent.

c) P_1 l strain ($llMM$) x m strain ($LLmm$) - F_1 $LlMm$ (wild-type)

F_2 9 wild-type (9 $L\_M\_$) : 7 brown (3 L_mm + 3 $llM\_$ + 1 $llmm$)

18. a) In each cross the difference in expression of pigmentary pattern between the parents is due to alleles of one gene: the allele for *kandiyohi* is dominant to the allele for *wild-type* as is the allele for *burnsi* dominant to the allele for *wild-type*. Impossible from this information alone to determine whether *kandiyohi*, *burnsi* and *wild-type* are due to alleles (3) of one gene or specific allele pair combinations of two or more genes.

b) (1) *Multiple allelic series*. Let C^K represent allele for *kandiyohi*, C^B allele for *burnsi* and c allele for *wild-type*.

F_1 *kandiyohi* (in cross 1) would be $C^K c$; F_1 *burnsi* (cross 2) would be $C^B c$.

F_1 *kandiyohi* ($C^K c$) x F_1 *burnsi* ($C^B c$) → $C^K C^B$ (mottled burnsi)

                                                                   $C^K c$ (*kandiyohi*)

                                                                     $C^B c$ (*burnsi*)

                                                                     cc (*wild-type*)

(2) Independent assortment involving two allele pairs where allele K of one pair determines *kandiyohi* and allele B of the other determines *burnsi*.

Thus: *Kandiyohi* = K-bb, *burnsi* = kkB-; *wild-type* = $kkbb$

F_1 *kandiyohi* ($Kkbb$) x F_1 *burnsi* ($kkBb$). Progeny will be of four types:

$KkBb$ (mottled burnsi); $Kkbb$ (kandiyohi); $kkBb$ (burnsi); $kkbb$ (wild-type)

c) Cross *mottled burnsi* x *wild-type*.

Results for (b) (1) - $C^K C^B$ x cc → $C^K c$ (*kandiyohi*) : 1 $C^B c$ (*burnsi*)

Results for (b) (2) - $KkBb$ x $kkbb$ 1 $KkBb$ mottled burnsi; 1 $Kkbb$ kandiyohi; 1 $kkBb$ burnsi; 1 $kkbb$ wild-type

d) The actual results support the 2-gene hypothesis. Since 4 types of progeny occur this indicates that the *mottled burnsi* must have been heterozygous for 2 allele pairs and therefore produced 4 types of gametes. Had the *mottled burnsi* parent been heterozygous for one allele pair it could have only produced 2 gamete types and therefore only 2 kinds of progeny. P for χ^2 of 1.934 (3 d.f.) is between 0.50 and 0.75. The χ^2 substantiates the conclusion drawn.

20. The calf resulted from transplantation of a fertilized egg from another animal into the cow.

Reason: The alleles for blood antigens are dominant. Therefore at least one of the parents should be of the same blood type as the calf for each antigen. This is not so for blood types A, W_3 and S. All 3 antigens are present in the calf but in neither of the parents. Had the calf possessed only one antigen that the parents did not synthesize one could have argued that a mutation occurred in one or the other parent. That the calf possessed 3 antigens which neither parent possessed, precludes mutation as a possibility and supports the proposed hypothesis.

22. The *ABO* blood groups are determined by a single multiple allelic series in which the I^A and I^B alleles are codominant and both are dominant to i. Since the father belongs to group AB and the mother to group O they can have children belonging to groups A and B only ($I^A I^B$ x ii → $I^A i$ and $I^B i$). Thus the children of blood types O and AB are from other matings. It is possible for the mother to have a group O child (ii x $I^A i$ or $I^B i$) but not an AB child because the alleles for these antigens are codominant and she would have to carry at least one of them, i.e., she herself would have to be of blood type A ($I^A i$) or B ($I^B i$). Therefore the O child is hers from a previous marriage and the AB child is adopted.

24. a) C is fully morphic since it produces *full* (100%) effect in single dose (increase in C dosage does not alter the phenotype). The alleles c^k, c^d and c^r are hypomorphs; each produces c (*wild-type*) effect but less efficiently and an increase in dosage alters the phenotype in the direction of *full* produced by C. c^a is an amorph since it fails to produce a phenotypic effect in single or double dose.

25. The genes for *Hb-S* and *Hb-C* are allelic as indicated by the pedigrees in which these genes are seen only to segregate and never to recombine. Since they can only segregate, they must occupy corresponding positions in homologous chromosomes. If they were non-allelic, some of the gametes would carry both genes and some would carry neither and therefore in matings between *Hb-A* (○, □) and *Hb-S*, *Hb-C* (●, ■) individuals some *normal* (*Hb-A*) and *doubly abnormal* (*Hb-S*, *Hb-C*) individuals would occur.

The genes for *Hb-HO-2* and *Hb-S* (and therefore *Hb-C*) are non-allelic. In 2 definitive portions of the pedigree, individuals with both traits (*HB-S, Hb-HO-2*) married to *normal* (*Hb-A*) ones produce some children with both *Hb-S* and *Hb-HO-2* and some others with neither of these mutant hemoglobins. This indicates that the genes for *Hb-S* and *Hb-HO-2* can recombine as well as segregate (from their normal alternatives) and must therefore occupy non-corresponding positions in the genome.

Chapter 7
Polygenic Inheritance

1. a) (i) It would show a more or less continuous variation in expression from one extreme phenotype through various intermediates to the other extreme phenotype.
   (ii) Since differences are one of degree, the phenotypes for such a character cannot be classified into discrete classes. Measurement is necessary for classification. Individuals are described numerically in such terms as inches and ounces and then arbitrarily classified.
   (iii) Expression of such a character is usually easily influenced by the environment.
   (iv) Such a character often shows transgressive segregation, i.e., the appearance of individuals that are more extreme for an expression of the character than either parent. In human populations such individuals can occur in any generation.

   b) The statement is true. Multiple genes are located on chromosomes and transmitted from one generation to the next in exactly the same way as major genes. The difference between multiple genes and major genes that interact is in the way they express their effects. The contributing alleles at polygene loci are non-dominant and their effects are minute and cumulative (additive or multiplicative). Although the alleles at major duplicate gene loci may also be non-dominant, their individual effects are fairly large permitting easy identification of these genes. Moreover, the effects of the dominant or non-dominant alleles at these loci are not cumulative. Thus in the one case phenotypic variation is more or less continuous whereas in the other case it is not.

   c) There are 2 main reasons why it is more difficult to study the inheritance of quantitative characters:
   (i) Quantitative characters show a continuous variation in phenotypic expression whereas qualitative characters show a discontinuous variation in expression. Also see answer to 1 a).
   (ii) The expression of quantitative characters is more easily influenced by the environment than that of qualitative characters.

   d) A polygene is one of many genes which affect the expression of a quantitative character. One allele of each such gene has minute effects on the phenotype which is identical or very similar to the phenotypic effects of contributors at other polygene loci in the system. These non-dominant contributors interact in a cumulative manner which may be additive or multiplicative. A modifying gene is any gene that by interaction affects the phenotypic expression of major or multiple genes at other loci. Many modifiers can be detected only by their effects on the expression of other nonallelic genes; others may have phenotypic effects of their own.

2. a) In the F_2, approximately 1 in 270 is like one parent (*large*) and approximately 1 in 270 is like the other parent (*small*). The expected proportion of F_2's with either extreme phenotype is $(1/4)^{n=4}$. Thus 1/256 should be *large* and 1/256 should be *small*. Since the actual proportions of extreme F_2 phenotypes approximate the expected fairly closely, 4 polygenes (allele pairs) are involved.

   b) Since the F_1's show an *intermediate* phenotype and since there is no evidence of transgressive segregation:

   (i) The contributors at each of the 4 loci act arithmetically.
   (ii) the *small* strain is homozygous for all neutral alleles (e.g., *aabbccdd*) and the *large* strain is homozygous for all 4 contributors (*AABBCCDD*). If the phenotypic difference between the 2 strains is x, each contributing allele is responsible for a phenotypic expression equal to x/8. For example, if the *small* strain animals weigh 32 ounces and those in the *large* strain weigh 64 ounces, each contributing allele in the genotype would add (64-32/8) = 4 ounces to the weight of a rabbit.

3. The variation in expression of *brachydactyly* may be due to multiple gene interaction. These genes act as modifiers of the incompletely dominant allele for this trait. For example if B = brachydactyly, b = normal, in $B\_$ individuals *aaccdd* - normal, *Aaccdd* - slighly short, *AACcdd* - slightly shorter etc. to *AACCDD* - extremely short.

8. a) Expect $(1/4)^n$ of the F_2 to show one extreme expression of the phenotype (36 grams) and $(1/4)^n$ to show the other extreme (20 grams). Since 1/255 show each of the extreme expressions and $(1/4)^n$ = 1/256, 4 polygene loci are involved.

   b) strain A-*aabbccdd* (24 grams) x strain B-*AABBCCDD* (32 grams)
   F_1 - *AaBbCcDd* (28 grams)

   c) Since distribution in the F_2 is normal and the F_1 average is arithmetic ($m_A + m_B$ /12 = 24 + 32/2 = 28), gene action is arithmetic (additive).

   d) Two grams. **Explanation** - Residual genotype *aabbccdd* (20 grams) has no contributors. Maximum phenotype (36 grams) has 8 contributors and is *AABBCCDD*. Difference between residual and maximum phenotypes is 16 grams; each of the 8 contributors is responsible for 16/8 = 2 grams.

9. The basic weight = lowest weight shown = 3 lbs. Since weight range is 3 to 6 lbs. and the difference between weight classes always involves a 1/2 lb., each contributing allele adds 1/2 lb. to weight.

   a) The genotypes of plants A and B.
   Since the cross between plant A x plant B shows a range in the F_2 of 3 to 5 lbs., this indicates that four contributing alleles, each adding 1/2 lb. to the base weight of 3 lbs. are segregating. The genotypes of A and B must therefore each contain a pair of contributors. The fact that when plant A is crossed with plant C a 6 lb. individual can be obtained indicates that a third allele pair must be present; both plants A and B must be homozygous for the neutral allele of this 3rd gene. Hence:
   Plant A: *AAbbcc* (3+1=4lbs.); Plant B: *aaBBcc* (3+1=4 lbs.)
   F_2 progeny *AABBcc* (3+2=5 lbs.) → *aabbcc* (3 lbs.)

   b) The genotype of plant C.
   A range of 3 to 5 lbs. in its immediate progeny indicates this plant is heterozygous for two allele pairs and has 2 contributing alleles. It could be *AabbCc*, *aaBbCc* or *AaBbcc*.

	Plant A	x	Plant C
	AAbbcc		?

Offspring range: 3.5 to 4.5 lbs.
3.5 lbs. - These offspring have only one contributor (supplied by parent A). Therefore plant C must have at least one neutral allele at two of the three loci.

4.5 lbs. - These offspring have 3 contributory alleles. This is only possible if there are contributing alleles at only two of the loci, thus: A and B, or A and C, or B and C.

Selection in the offspring produces 6 lb. types, i.e., plants with 6 contributors. These would be $AABBCC$. This is only possible if plant C has contributing alleles B and C. Its genotype therefore must be $aaBbCc$.

Plant A		Plant C		Offspring
$AAbbcc$	X	$aaBbCc$	→	$Aabbcc$ (3.5 lbs.) to $AaBbCc$ (4.5 lbs.)

Selection among progeny of $AaBbCc$ will produce some $AABBCC$ (6 lb.) offspring.

Plant B		Plant C		Offsspring
$aaBBcc$	X	$aaBbCc$	→	$aaBbcc$ (3.5 lbs.) to $aaBBCc$ (4.5 lbs.)

Selection among progeny of $aaBBCc$ will produce some $aaBBCC$ (5 lb.) offspring.

12. a)

Parent and generation	Mean (arithmetic)	Standard deviation
P19	10.3	1.12
P782	18.0	2.27
F_1	15.9	1.28
F_2	14.1	1.57

b) If the environmental conditions were the same, the standard deviation of the F_2 should be greater than that of the P_1 and F_1 generations because of greater genotypic variability due to segregation, recombination and random gamete union in the F_1. The P_1's and F_1's would be genotypically uniform. The P_1's would be homozygous and the F_1's would possess the same heterozygous genotype.

c) If the polygene hypothesis is correct the F_1 and F_2 means should be midway (14.1) between the parental means. Although the F_2 mean very closely approximates this value, there is a difference of 1.7 between the F_1 and expected value. This difference may be due to (i) chance, (ii) F_1 average being based on significantly fewer plants than the F_2 mean, (iii) all the genes not acting in an arithmetic manner and (iv) P782 not being completely homozygous or its genotype has a wider norm of reaction than that of P19.

d) The data suggest that gene action is arithmetic. With geometric gene action the F_1 and F_2 means would approximate 14.2 and 13.5 respectively. The actual means are much closer to those expected with additive action (14.2).

(e) Some reasons why it is not possible to determine the actual number of genes involved from the data given:
(i) F_2 population size is not large enough since the number of genes involved is large.
(ii) Environmental conditions were not uniform for all individuals in all generations. This not only contributed to a broad range of phenotypes in the parental strains but also created a problem in separating phenotypic similarities and differences due to environmental factors from phenotypic similarities and differences due to genotypic differences. This leads to the impossibility of estimating the actual number of F_2 phenotypic classes and therefore the number of genes involved. It is possible to estimate the number of genes involved by using the formula:

$$N\text{(no. of genes)} = \frac{D^2\text{(difference between parental means)}}{8(F_2 \text{ variance} - F_1 \text{ variance})}$$

$$= \frac{(18.0 - 10.3)}{8((1.57)^2 - (1.28))^2} = 10$$

Variance = standard deviation squared.

13. a) The strain is homozygous at all polygene loci. Self-fertilization will always produce offspring of the same uniform genotype as that in the parents. Any phenotypic variation will be ineffective since acquired traits are not heritable.

b) Since the F_1 yield is the same as that of the two parental strains, all 3 populations must contain the same number of contributors. Those in the parents are in the homozygous condition. Since F_2 and F_3 plants are obtained which yield significantly more than the F_1 and parents, some of the contributors possessed by the second strain are not at the same loci as those found in the first strain. For example, one parent could be of genotype $AABBCCddeeff$ and the other $aabbccDDEEFF$. The F_1's ($AaBbCcDdEeFf$) would have the same number of contributors as the parents and therefore the same yield. Segregation and recombination in the F_1 can produce some F_2's that have more contributing alleles (7 to 12) than the parents. Some would have fewer than 6 contributors. If the total number of genes is large, individuals homozygous for all the contributors would probably not be found in a limited F_2. Selection in further generations would be required to obtain them. This is a case of transgressive segregation.

Chapter 8
Sex Determination and Sex Differentiation

2. *Klinefelter's* and *Turner's* syndromes both arise as a consequence of non-disjunction during the first and/or second meiotic division in males and/or females.

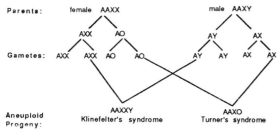

Turner's may also arise because of chromosome loss during gamete formation in either parent to produce an (n-1) gamete.
Illustrate the other situations for yourself.

b) Sex chromosome loss in either sex during meiosis (because of lagging or some other reason) will produce n-1 (AO) gametes which, upon union with n (AX) gametes, produce monosomic AAXO (*Turner* individuals). For example, loss of Y in the male would lead to an (AO) sperm, which upon fusion with an ootid (AX), will result in a *Turner's* syndrome (AAXO).

Possible explanations for the difference in frequency of the 2 syndromes are:
(i) *Klinefelter's* is the result of more than one kind of abnormal chromosome constitution (XXY, XXYY, etc.).
(ii) Approximately 95% of AAXO zygotes (or embryos) abort; monosomics are less viable than trisomics.
(iii) Higher rate of non-disjunction in females than in males. If non-disjunction occurred only in females, only AAXXY and AAYO zygotes would be formed and all of the latter type would abort since the Y carries few, if any, essential genes.

4. a) Four hypotheses may be advanced for the location of male and female determining genes:
(i) female determining genes on A (autosomes); male determining ones on the Y.
(ii) female determiners on the X, male determiners on the Y.
(iii) female determiners on the X, male determiners on A.
(iv) female determining genes on A and X, male ones on Y.

b) (i) The X chromosome carries female-determining genes. **Evidence** - Comparison of males and females with the same number of sets of autosomes and different numbers of X chromosomes shows that as the number of X chromosomes increases expression of female traits increases, e.g.,
AAXY - Male; AAXX - female; AAXXX - superfemale; AAAXX - intersex; AAAXXX - female

(ii) Autosomes carry male-determining genes. **Evidence** - Comparison of individuals with the same number of X chromosomes and different numbers of sets of autosomes indicates that the expression of male traits increases with an increase in the number of autosomal sets, e.g., AAXX - female; AAAXX - intersex; AAAAXX - male

c) The primary mechanism is the ratio :

$$\frac{\text{number of X chromosomes (female determining genes)}}{\text{number of sets of autosomes (male determining genes)}}$$

Subsequent to the action of the primary mechanism, many other genes function to determine direction and extent of sexual differentiation.

5. a) Progeny:
XX *Tratra* - normal (fertile) females; XX*tratra* - sterile males.
XY *tratra* - normal (fertile) males; XY *Tratra* - *normal* fertile males.
Overall ratio - 2 *normal* males : 1 *sterile* male : 1 *normal* female or 3 males : 1 female.

b) The *tra* allele when homozygous is epistatic to the female determiners on the X. *tra* is very rare in nature since it is selected against. When flies are *Tra*-, sex is determined by X/A balance.

c) One may hypothesize that the *wild-type* allele *Tra* directly or indirectly leads to an increased activity of those X chromosomal genes which produce female-determining substances, or it suppresses the autosomal genes which produce male-determining chemical compounds. The *tra* allele would be inactive or very slightly active in this respect. Therefore in *normal* females (XX*Tra*_), the masculinizing action of the autosomes would be suppressed by the female determining action of the X-chromosome genes only in the presence of *Tra*.

12. The gene or genes involved are epistatic to the male sex-determining mechanism in AAXY males and cause(s) the individuals to develop in the female direction. The genotype may be said to be sex-limited in its expression.
Modes of inheritance that are definitely eliminated are:
(i) Holandric - since male-to-male transmission is not characteristic of the pedigree.
(ii) Autosomal recessive with expression limited to males - since both parents would have to be carriers of a rare recessive.
(iii) Both sex-influenced modes of inheritance - since only the male sex is affected.

a) Possible modes of inheritance:
(i) Sex-linked dominant or recessive with expression limited to males. Either type of X-linked gene transmission and action can account completely for the results. *Unaffected* female carriers transmit the gene to 1/2 their male offspring. For example, if it is a recessive: carrier females are $X^A X^a$, non-carrier females are $X^A X^A$ and *affected* males are $X^a Y$.
(ii) Autosomal dominant allele, limited in expression to males. Since the trait occurs in successive generations which arise from a common female it could be due to this type of autosomal gene. For example, carrier females would be Aa, non-carrier females would be aa, *unaffected* males would be aa and *affected* ones, Aa.

b) The most feasible method, to distinguish between the two possible modes would be a linkage analysis study, for which reliable autosomal and X-linked markers would have to be available. Linkage with either an X-linked or an autosomal gene would eliminate the other possibility. Alternatively, if a biochemical test for heterozygosity were available, then both males and females could be detected as heterozygotes in the case of autosomal sex-limited modes of inheritance whereas only females would be detected as heterozygotes in the case of X-linked inheritance. This test would not be feasible if the limitation in gene expression occurs at the biochemical level.

Answer c) and d) yourself.

16. a) (i) They suggest that sex is determined by alleles at one locus. Allele *M* for *maleness* is dominant to the allele *m* for *femaleness*. Self-fertilization of an *M m* (male) plant would give the results obtained by Rick and Hanna.
(ii) The male is the heterogametic ($X^m Y^M$) sex. *M* is on the Y chromosome and *m* on the X chromosome.

$X^m Y^M \times X^m Y^M \rightarrow$ $1 X^m X^m \female : 2 X^m Y^M \male : 1 Y^M Y^M \male$
                                         1 female   :   3 male

b) (i) Yes. In the progeny of self-fertilized males one would expect $X^m Y^M$ and $Y^M Y^M$ males in a ratio of 2 : 1. This is shown by the fact that when a random sample of these 155 males were mated with females $X^m X^m$, 8 produced only males (indicating that the males were $Y^M Y^M$) and 17 produced males and females (indicating the males were $X^m Y^M$).
(ii) The allele for *maleness* is dominant. Evidence discussed in (a) and (b).
(iii) Female plants are homogametic $X^m X^m$; therefore self-fertilization would produce only females.
(iv) Both *pistillate* and *staminate* flowers on male plants are $X^m Y^M$.

c) One could obtain seed producing only *staminate* plants by crossing $Y^M Y^M$ males with *normal* ($X^m X^m$) females.
By self-fertilizing males ($X^m Y^M$), obtain $Y^M Y^M$ individuals. These can be distinguished from $X^m Y^M$ in crosses with females. Propagate the $Y^M Y^M$ plants asexually, i.e., by cuttings.
Cross: $X^m X^m \times Y^M Y^M \rightarrow X^m Y^M$ (all males)
These males will yield 25% more than females. Necessary to repeat cross every generation. This is done with commercially important crops, e.g., onions, corn.

17. Sex in *Lebistes reticulatus* is determined by the XY mechanism; females are homogametic (XX) and males are heterogametic (XY). **Reason:** Since *black-spot* occurs only in males and does not occur in this sex unless the father expresses it, the trait must be determined by a holandric gene which can reside only in the differential portion of the Y chromosome, transmitted by males to males only. This carries with it the corollary that the X chromosome does not possess and therefore cannot transmit the hereditary determiner for *black-spot*. For example:

XX ♀ x XY ♂ (*black-spot*) → F_1 XX ♀ ; XY ♂ (*black spot*)

$F_2 \rightarrow$ ♀ XX; ♂ XY (*black-spot*)

If the gene for *black-spot* is on the differential portion of the X, since this region is present twice in homogametic (female) individuals and only once in heterogametic (male) individuals, the gene will be carried and transmitted by both sexes and reciprocal crosses between true-breeding lines will give different results related to sex.

21. a) The female is heterogametic (ZW) since:
(i) Crosses between 2 genetic females produce progeny of both sexes.
(ii) Z/AA individuals are females and ZZ/AA males.
(iii) ZZZ/AA individuals are supermales.

b) Male-determining genes are on the autosomes since ZZ/AA individuals are males and ZZ/AAA animals are intersexes. The addition of a set of autosomes to the male karyotype alters the male phenotype in the female direction. Since Z/AA animals are sterile, the female fertility genes must reside on the W chromosome.

c) The mechanism of sex determination is the ratio of the number of Z chromosomes to the number of sets of autosomes. This conclusion is based on the sex phenotypes of the monosomic (AAZO) and trisomic (AAZZZ) and triploid-disomic (AAAZZ) individuals. The male determiners on one Z are stronger than female-determiners on one set of autosomes. Thus:
If Z/A = 1.0 as in ZZ/AA, the individual is a male. If Z/A = 0.5 as in ZW/AA or ZO/AA, the individual is a female. If Z/A > 1.0, animal is a supermale. If Z/A < 0.5, animal is a superfemale. If Z/A is between 0.5 and 1.0 as in ZZ/AAA (Z/A = 0.67), the individual is an intersex.

d) Body colour is a sex-limited character. Sex differentiation is autonomous as evidenced by the occurrence of gynanders with *chocolate* on one side of the body and *gray* on the other.

e) *C* for *chocolate* is dominant to *c* for *gray* (limited in expression to females). The *chocolate-gray* individuals start out as males AAZZ which are homozygous for *c*. They arise because of loss of one of the Z chromosomes during the first cleavage division of the zygote.

24. a) The data indicate that sex is determined by a one-gene multiple allele system, e.g., $X^a, X^b, \ldots X^z$: hemizygous and homozygous individuals develop into males; heterozygous ones develop into females.
Sample genotypes of parents and offspring of first four crosses:
(i) ♀ $X^a X^b$ x ♂ $X^c \rightarrow n \male (1/2 X^a, 1/2 X^b); 2n \female (1/2 X^a X^c; 1/2 X^b X^c)$
(ii) F_1 ♀ $X^a X^c$ x F_1 ♂ $X^b \rightarrow F_2$ group A $\rightarrow n \male (1/2 X^a, 1/2 X^c); 2n \female (1/2 X^a X^b, 1/2 X^c X^b)$
(iii) F_1 ♀ $X^a X^c$ x F_1 ♂ $X^a \rightarrow F_2$ group B $\rightarrow n \male (1/2 X^a, 1/2 X^c); 2n \male (X^a X^a); 2n$ ♀ $(X^a X^c)$
(iv) F_2 ♀ $X^a X^c$ x F_2 group B n ♂ $(1/2 X^a; 1/2 X^c) \rightarrow n \male (1/2 X^a, 1/2 X^c); 2n \male (1/2 X^a X^a; 1/2 X^c X^c); 2n$ ♀ $(X^a X^c)$

b) Since diploid females produce haploid eggs and such females produce triploid males and female offspring in matings with diploid males, the latter individuals must produce unreduced diploid sperm to account for the large number of triploid offspring.

c) Since diploid females are fertile and produce haploid eggs the absence of 2n and 3n progeny in the last cross is due to sterility of triploid males which may have one or more of a number of causes, some of which are: (i) defective gonads and no meiocytes, (ii) meiocytes are produced but do not initiate or complete meiosis, (iii) meiosis proceeds normally but aside from a very minute percentage of meiotic products being balanced (n and 2n), the majority are defective in morphology or size and incapable of effecting fertilization.

26. a) In the *Mexican axolotl*, sex can be explained by one gene and two alleles. The female is the heterogametic sex ($Z^m W^M$)
- W carries the dominant allele *M* for *femaleness*.
- Z carries the recessive allele *m* for *maleness*.

female Z^mW^M x neomale Z^mW^M → 1 Z^mZ^m (males) : 2 Z^mW^M : 1 W^MW^M (females).

6 F_1 females (W^MW^M) x *normal* male (Z^mZ^m) → all females (Z^mW^M).

11 F_1 females (Z^mW^M) x *normal* male (Z^mZ^m) → 1 female (Z^mW^M) : 1 male (Z^mZ^m).

b) Hormones play no role in sex determination, only in sex differentiation. If hormones determined sex, neomales and neofemales would not have the sex determining genotype of the zygote (embryo) from which they are derived, e.g., in the *Mexican axolotl*, neomales would be ZZ and not ZW if hormones determined sex. These results indicate that embryos in these organisms are bipotential with respect to sex and that the direction of sex differentiation is dependent on which of the two kinds of hormones, male or female, is in excess. The synthesis of these hormones depends on the sex-determining genotypes of the gonads. Whether male or female hormones are in excess normally depends on the sex determining genotype of the gonads. Experimentally one can reverse sex differentiation by administering hormones of the opposite sex during the sexually neutral stage of embryonic development or at an appropriate time thereafter. In summary, sex is determined genetically. The sex-determining genotype is responsible for synthesis of hormones which regulate sexual differentiation.

28. a) These results suggest that WW individuals which would comprise 25% of the total progeny, are viable and are of the female sex.

ZW x ZW → 1 ZZ : 2 ZW : 1 WW, i.e., 1 ♂ : 3 ♀

If the WW animals were inviable the progeny would have been produced in a 1 : 2 ratio of males : females.

b) By crossing the female progeny to normal males. About 1/3 of these crosses should produce only female progeny (WW x ZZ → ZW). The remainder of the crosses (2/3) should produce males and females in equal proportions (ZW x ZZ → 1 ZW : 1 ZZ).

Chapter 9
Sex Linkage, Sex-Influenced and Sex-Limited Characters

1. a) Because sex-linked recessive traits are readily expressed and observed in the hemizygous condition of the heterogametic sex while autosomal dominants are readily expressed and observed in either sex in either the homozygous or heterozygous condition.

   b) Yes. Males are heterogametic (XY) and possess only one dose of any completely sex-linked gene (are hemizygous for the causative allele). If they possess a recessive allele its trait will be expressed. For a recessive trait to be expressed in the homogametic sex such individuals must be homozygous for the recessive allele. The chances of this occurring are lower than those for the heterozygous genotype and also less than those for the hemizygous one in males. Thus, while 1/2 the male progeny of a heterozygous female, and all of a homozygous recessive one will express the recessive trait, none of the female progeny will unless the male parent also carries the recessive allele.

   c) That the trait occurs in all the sons of an *affected* father and not in any of his daughters. Moreover, the trait could never occur in the sons unless the father expressed it. This would indicate that the gene is present on the differential segment of the Y chromosome which is carried only by males and transmitted only to their sons.

4. *Short*-bristles is determined by an X-linked incompletely dominant lethal allele, S. *Long*-bristles is due to the recessive non-lethal allele, s.

   (i) Virgin female - *short* $X^S X^s$ x male - *long* $X^s Y$
   F_1 ♀ 1/3 *short* $X^S X^s$; 1/3 *long* $X^s X^s$
   F_1 ♂ 1/3 *long* $X^s Y$; $X^S Y$ (die)

   (ii) F_1 female - *long* (ss) x F_1 male - *long* (sY)
   F_2 $X^s X^s$ *long* ♀ ; $X^s Y$ *long* ♂

   (iii) F_1 female - *short* $X^S X^s$ x F_1 male - *long* $X^s Y$
   F_2 ♀ 1/3 *short* $X^S X^s$, 1/3 *long* $X^s X^s$; ♂ 1/3 *long* $X^s Y$; $X^S Y$ (die)

5. a) Reciprocal crosses give different results because the pair of traits is determined by a sex (Z)-linked pair of alleles. The allele (L) for *wild-type* is dominant to l for *lacticolour* since the F_1's in cross 1 are *wild-type* and the F_2 phenotypic ratio is 3 *wild-type* : 1 *lacticolour*.

   Cytogenetic explanation of the crosses:
   Cross 1
   P_1 ♀ *lacticolour* ($Z^l W$) x ♂ *wild-type* ($Z^L Z^L$) → F_1 ♂ ($Z^L Z^l$); ♀ ($Z^L W$) → F_2 ♂ (1/4 $Z^L Z^L$, 1/4 $Z^L Z^l$); ♀ (1/4 $Z^L W$, 1/4 $Z^l W$) - 3 *wild-type* : 1 *lacticolour*

   Cross 2
   P_1 ♀ *wild-type* ($Z^L W$) x ♂ *lacticolour* ($Z^l Z^l$) → F_1 ♂ ($Z^L Z^l$); ♀ ($Z^l W$) → F_2 ♂ (1/4 $Z^L Z^l$, 1/4 $Z^l Z^l$); ♀ (1/4 $Z^L W$, 1/4 $Z^l W$) - 1 *wild-type* : 1 *lacticolour*

   b) The female is the heterogametic sex. If it were homogametic, the F_1 and F_2 results of the reciprocal crosses would be reversed, i.e., in cross 1 the F_1 males would be *lacticolour* and the F_1 females would be *wild-type* while in cross 2 both sexes in the F_1 would be *wild-type*.

7. a) *Ocular albinism* is a recessive X-linked trait. Therefore, an *affected* mother must be homozygous recessive. Since the sons normally receive their X only from their mother, they will only have the recessive allele and all will be similarly *affected*. An *affected* father usually trasmits his X with the recessive allele only to his daughters. Since the trait is rare he will likely marry a homozygous *normal* woman (rarely a heterozygous one and still less frequently a homozygous recessive one). Therefore all his sons and daughters will be *normal* but the daughters will be carriers (heterozygous). The sons of an *affected* father will be *affected* only if the mother is a carrier (in which case approximately 1/2 the sons will be *affected*) or if she is *affected* (in which case all the sons will be *affected*).

   b) Illustrate for yourself.

15. a) *Black* vs. *yellow* vs. *tortoise-shell* is determined by a single X-linked pair of alleles, Bb. The phenotype *tortoise-shell* is due only to a heterozygous (Bb) genotype. Barring aberrations, only females can possess such a genotype and phenotype since only they contain 2 X chromosomes in their complements. Males normally have one X and therefore can only produce *black* or *yellow* fur. If the gene was on an autosome, the Bb genotype and phenotype would appear equally frequently in both sexes. Thus the *tortoise-shell* cat must have been a female. *Tortoise-shell* is a mosaic phenotype, consequence of the fact that in certain regions of the coat one allele of the pair is active whereas in other regions of the coat the other allele is functional.

    b) Theoretically, one would expect an equal number of *black* and *yellow* males (3 : 3). However, in a litter of 6, the phenotypic proportions could vary from 6 *black*, 0 *yellow* to 0 *black* and 6 *yellow*.

    c) Probably male ($X^B Y$) since it would receive the X chromosome from the *yellow* (female) parent ($X^B X^B$).

    d) Father probably was *yellow* ($X^B Y$), since there are *yellow* ($X^B X^B$) females in the litter of the *tortoise-shell* ($X^B X^b$) female.

    e) Parents were: mother - *black* ($X^b X^b$); father - *yellow* ($X^B Y$) since litter contains *tortoise-shell* females. The *black* female in the litter is probably the result of non-disjunction in the male parent. Mutation would be a less likely possibility.

    f) Should there not be any *black* females in the cat's litter, the *yellow* tom may be incriminated as the father. Should there fail to be some *yellow* females in the litter, then this would exonerate the *yellow* tom from being the father.

16. a) *Incontinentia pigmenti* is most likely caused by an X-linked incompletely dominant lethal allele *Ip*. Individuals hetertozygous ($Ipip$) for the allele express the trait. The hemizygous ($X^{Ip} Y$) dominant genotype in males is inviable; homozygous ($X^{ip} X^{ip}$) and hemizygous ($X^{ip} Y$) recessive genotypes are *unaffected* (*normal*).

Reasons for conclusions:
(i) Expression of the condition in every generation is indicative of a dominant trait, as is the observation that an *affected* individual always has at least one *affected* parent and the fact that *affected* females produce *affected* and *unaffected* offspring as well as abortions.
(ii) The ratio of approximately 2 females : 1 male among the offspring of *affected* females strongly suggests lethality; specifically an X-linked lethal. That the allele has lethal effects is supported by the high frequency of infant deaths (approximately 25%) among the children of *affected* females.
(iii) Only females are *affected* in direct descent through several generations.
(iv) If the allele is not sex-limited in expression and is chromosomally located, since only females are *affected* the transmission pattern of the trait is best explained by an X-linked gene. Otherwise one would expect both sexes to be *affected* and in about equal numbers.

b) Other possible modes are: (i) sex-influenced dominant in females and recessive in males; (ii) female limited, autosomal dominant; (iii) cytoplasmic inheritance with lethality in males.

Criteria to distinguish the different modes of inheritance:
(i) The sex of the abortuses would be helpful in establishing a definitive mode of inheritance. Should the abortuses all be males, this would strongly suggest an X-linked incompletely dominant lethal mode of inheritance.
(ii) Analysis of linkage data. Demonstration of linkage with an established X-linked gene would be conclusive proof that the gene *Ip* is on the X chromosome. A similar demonstration with an autosomal gene would prove that the gene is on one of the 22 autosomes. Either would eliminate cytoplasmic inheritance as a possible mode of transmission.

20. Since reciprocal crosses produce different results, *slow* vs *rapid* metamorphosis is determined by a single pair of alleles (Sr) that is completely X-linked with females carrying 2 alleles of the gene and males possessing only 1 (are hemizygous). This indicates that the chromosome carrying the locus is present twice in females and once in males. Since all other pairs of traits which are determined by pairs of alleles on all 5 chromosomes in the genome show the same transmission patterns as *slow* vs. *rapid*, all 5 chromosomes, e.g., 1, 2, 3, 4 and 5 are present in double dose (twice) in females and once in males. The most logical conclusion is that in *Pseudococcus nepae*, females are diploid (2n=10) and males are haploid (n=5). An alternative explanation is that both sexes are diploid but the genome in males that is derived from their father becomes heterochromatic and genetically inactive and therefore functionally males are haploid.

26. a) Since in each family *RP* (i) can occur in an offspring when neither parent is *affected* and (ii) it occurs much more frequently in males than in females, a recessive allele at a completely X-linked locus is responsible for the condition. Analysis of the data in each family does not indicate whether the recessives in the different families are allelic or not. The *affected* female offspring from the mating between III-12 and III-13 indicates that the recessives in families B and C are allelic e.g., r and therefore *affected* males in these families are X^rY and *affected* females are X^rX^r

b) Because the recessive gene i.e., p in family A is not allelic to the recessive r in families B and C.

affected ♂ X^{pR} x *affected* ♀ $X^{Pr}X^{Pr}$

All daughters $X^{pR}X^{Pr}$ are *unaffected* because of complementation. All sons are $X^{Pr}Y$ and *affected*.

c) Yes. Even though dosage compensation occurs, since the females are homozygous X^rX^r each cell will possess an active X with the recessive allele and therefore will express its potential (same is true for X^rY males).

32. a) (i) *Ectodermal dysplasia* is probably determined by an X-linked recessive allele. **Reasons:**
(1) X-linked, since only males are *affected*; therefore females would be carriers of the recessive allele. Moreover, females with *affected* offspring have approximately 1/2 their sons *affected*.
(2) Recessive, because the trait is observed to skip generations and also because two *unaffected* parents have an *affected* child.

(iv) *Xg blood antigen* is due to an X-linked dominant allele.
Reasons:
(1) X-linked since more females *affected* than males. Furthermore, an *affected* father always has all his daughters *affected*.
(2) Dominant because an *affected* individual always has at least one *affected* parent. As a corollary, two *unaffected* parents always have *unaffected* offspring.

(v) *Reifensteins's syndrome* - Two explanations are plausible: X-linked recessive or sex-limited in males, caused by an autosomal recessive.
(1) That the gene is recessive is evident from two observations 1. The trait skips generations. 2. Two *unaffected* parents can give rise to *affected* progeny.
(2) The observation that only males are *affected* is indicative of either sex-linkage or limitation of expression in one sex only.
It would be difficult to distinguish between the two possibilities on the basis of pedigrees alone.

37. a) The modes of inheritance that can be eliminated are:
(i) Sex-limited, since the trait can occur in both sexes.
(ii) Autosomal dominant, since 2 *normal* parents can have an *affected* child. With a dominant mode of inheritance at least one parent must be *affected*. Moreover, such traits occur equally frequently in both sexes.
(iii) Holandric, since the trait occurs in both sexes. Holandric traits can occur in males only.
(iv) Autosomal recessive, since the *affected* persons are usually of one sex. With a recessive mode of inheritance, both sexes should be *affected* with equal probability.
(v) Sex-linked dominant, since 2 *normals* have an *affected* child. If the trait were of this type, for a child to be *affected* at least one parent must also express the trait.
(vi) Sex-linked recessive, only if the *affected* child of 2 *normal* parents is always a female.

b) The trait is determined by a sex-influenced allele, dominant in males and recessive in females, since a son is *affected* when neither of the parents expresses the trait (both *normal*). The *affected* son must have received the determining allele from his mother who must have been heterozygous since she was not *affected*.

38. a) Since *white* occurs only in females and can occur when neither parent expresses it, *white* is a sex-limited trait. Had the trait been hologynic, all daughters of a *white* mother would be *affected*. Moreover, daughters would not be *affected* unless the mother also expressed the trait.

b) Since the ratio in the last cross is 3 : 1 a single allele pair is involved.

c) Since the ratio in cross (5) is 3 *white* : 1 *yellow* and only in females, the allele *W* for *white* is dominant to *w* for *yellow*.

d) The allele pair is autosomal. **Reasons:** (1) The phenotypic ratio in the last cross is 3 : 1 and only in females. (2) Were it Z-linked, all the female progeny in cross (5) would have been *white*.

39. c) *Transverse vaginal septum* is due to a recessive allele of an autosomal gene with expression limited to females rather than a recessive allele of an X-linked gene for the following reasons:
(i) Occurrence of *affected* girls in each of 3 sibships.
(ii) All 6 parents of the *affected* girls share a common ancestral couple.
(iii) The trait is expressed in every consanguineous marriage in the pedigree.
(iv) The incidence of expression in females is characteristic of an autosomal recessive (approximately 25% are *affected*). Were it due to an X-linked recessive with expression limited to females, 50% of the females from marriages between 2 *unaffected* parents would be expected to be *affected*.
(v) If the trait was due to an X-linked recessive allele, e.g., *a*, *affected* daughters would have to be $X^a X^a$. Since the mothers of each of the families with *affected* girls are *unaffected*, they would be $X^A X^a$ and the fathers of these families would have to be $X^a Y$. The father of the family with *affected* girls on the extreme right is X^a since his mother was from the same family as his wife, who has to be a carrier, regardless of whether the gene is autosomal or X-linked. However, the fathers of the left and centre families with *affected* girls have *unaffected*, unrelated mothers who would highly likely be $X^A X^A$. This conclusion is also supported by the fact that the trait is highly likely to be rare and therefore few individuals would possess its determining allele. If this assumption is correct, these fathers would be *unaffected* and $X^A Y$. Therefore it would be impossible for their daughters to express a sex-limited trait determined by an X-linked recessive allele since they could not be of the genotype $X^a X^a$.

Chapter 10
Linkage, Crossing-over and Genetic Mapping

4. a) Testcross Phenotypes

smooth, round ($PR//pr$) - recombinant		12
smooth, long ($Pr//pR$) - parental		123
peach, round ($pR//pr$) - parental		133
peach, long ($pr//pr$) - recombinant		12

   percent recombination = $\dfrac{\text{number of recombinants}}{\text{total number of progeny}} \times 100 = 8.6$

   b) The significant excess of parental phenotypes and significant deficiency of recombinant phenotypes compared with the 1:1:1:1 ratio expected if the genes segregate independently is evidence that the allele pairs are linked.
   Since the genes are close together and exchange occurs at the 4-strand stage there is a 2:1 relationship between percent crossing-over and percent recombination.
   Map distance between P and R = percent recombination between P and R = 8.6 map units. Percent crossing-over between the genes = 17.2.

5. a) The 2 autosomal allele paris [T (*trembling*), t (*normal*) and R (*rex*) r (*long*)] are on the same chromosome and less than 50 map units apart.
   Basis for conclusion: Considered separately each pair of traits is present in the expected 1:1 ratio among the testcross progeny. When the 2 pairs of traits are considered simultaneously the 4 phenotypic classes are not present in equal proportions. Recombinant types significantly less frequent and parental types are significantly more frequent than expected with independent segregation. This indicates that the 2 genes are on the same chromosome.

   b) The *trembling, rex* females were heterozygous in repulsion ($Tr//tR$). The conclusion is based on the fact that among the testcross progeny the complementary single dominant phenotypic classes, e.g. *trembling, long* (dominant, recessive) and *normal, rex* (recessive, dominant) were the most common (parental). Had the females been heterozygous in coupling ($TR//tr$) the double dominant (*trembling, rex*) and double recessive (*normal, long*) testcross progeny would have been the parental types and, therefore, the most frequent. The cross can be represented as follows:

   Trembling, rex ♀ ($Tr//tR$) x *normal, long* ♂ ($tr//tr$)

   Testcross progeny:
52	$Tr//tr$	*trembling, long*	P
21	$TR//tr$	*trembling, rex*	R
22	$tr//tr$	*normal, long*	R
54	$tR//tr$	*normal, rex*	P

   c) Percentage recombination = $43/149 \times 100 = 28.85$

   d) Since the loci in question are autosomal and provided crossing-over occurs with the same frequency in both sexes, the results of the reciprocal cross should not have been any different.

   e) For simplicity of calculation assume percent recombination to be 30 percent. The frequencies of the 4 types of gametes from heterozygous ($tR//Tr$) males and females would be as follows:
   $P = (0.70) = (0.35\ Tr) + (0.35\ tR)$
   $R = (0.30) = (0.15\ tr) + (0.15\ TR)$

   If these male and female gametes united at random, the phenotypic frequencies among the F_2 progeny would be as follows:

trembling, rex	0.5225 - 52%
trembling, normal	0.2275 - 23%
normal, rex	0.2275 - 23%
normal, long	0.0225 - 2%

   Illustrate the Punnett square for yourself.

7. E (*gray*) is dominant to e (*ebony*) and C (*normal*) is dominant to c (*curled*) as indicated by the F_2 ratios of 3 *gray* : 1 *ebony* and 3 *normal* : 1 *curled*. Since the F_2 phenotypic ratio deviates significantly from a 9:3:3:1 distribution, the 2 allele pairs Ee and Cc on chromosome 3 must be incompletely linked (less than 50 map units apart). Since the F_1 males and females are heterozygous in coupling ($EC//ec$) and no crossing-over occurs in *Drosophila* males, the F_1 males will form only 2 types of gametes (those with both dominant alleles EC, and those with both recessive alleles, ec) in equal proportions:

   P_1 *ebony, curled* ♀ ($ec//ec$) x *gray, normal* ♂ ($EC//EC$)
   F_1 *gray, normal* ($EC//ec$)
   F_2

female gametes \ male gametes	$E\ C$		$e\ c$	
non c/o $E\ C$	$\dfrac{E\ C}{E\ C}$ gray, normal 88		$\dfrac{E\ C}{e\ c}$ gray, normal 88	
$e\ c$	$\dfrac{E\ C}{e\ c}$ gray, normal 88		$\dfrac{e\ c}{e\ c}$ ebony, curled 88	
c/o $E\ c$	$\dfrac{E\ c}{E\ C}$ gray, normal 14		$\dfrac{E\ c}{e\ c}$ gray, curled 14	
$e\ C$	$\dfrac{E\ C}{e\ C}$ gray, normal 10		$\dfrac{e\ C}{e\ c}$ ebony, normal 10	

   One half of the F_2 genotypes will result from F_1 male gametes containing both dominant alleles and they will express the double dominant phenotype *gray, normal* regardless of the kind of gamete they receive from the F_1 female. Since the total number of F_2's is 400, 200 of the 288 *gray, normals* would constitute this half of the F_2 genotypes and can be excluded from calculations of map distance. The other half of the F_2 genotypes all receive ec from the F_1 male parent; this portion of the F_2 constitutes a testcross type of progeny with respect to these 2 genes.
   Their genotypes (= phenotypes) resulting from a union of F_1 male (ec) gametes and F_1 female gametes will indicate the kinds and proportions of gametes that the F_1 females produced. The $EC//ec$ (*gray, normal*) and $ec//ec$ (*ebony, curled*) individuals are parental types; the $Ec//ec$ (*gray, curled*) and $eC//ec$ (*ebony, normal*) ones are recombinants. The 4 occur in a distribution of 88:88:14:10.

Total number of F_2's that should be used therefore in calculating the percent recombination = 200.
Total number of recombinants in this portion of the F_2 = 14+10=24.
Percent recombination = 24/200 x 100 = 12.0
Therefore, the map distance between *E* and *C* is 12 map units.

15. The trihybrid F_1's receive *Cy* (chromosome 2), *Ubx* (chromosome 3) and Adh^F from the balanced lethal stock and the corresponding alleles *cy*, *ubx* and Adh^S from the *wild-type* strain. Since there is no crossing-over in males and F_1 males are crossed with *wild-type* females that are $Adh^S Adh^S$, to determine whether chromosome 2 or chromosome 3, if either, carries the *Adh* gene, one simply has to determine whether $Adh^F Adh^S$ are inherited independently of *Cy* or *Ubx* or in complete linkage with one or the other.

The results are as follows:

	Wings	halteres	Adh
1	curly	enlarged	intermediate
2	curly	normal	intermediate
3	straight	enlarged	slow
4	straight	normal	slow

The gene for isozyme type is inherited independently of that for haltere type but is completely linked with that for wing form. Therefore, *Adh* is carried on chromosome 2.

17. a) Analyse the data taking characters 2 at a time.
Since the 4 phenotypic classes are not present in equal proportions expected with independent segregation and 2 classes (recombinants) are significantly less frequent than the 2 others (parentals), the allele pairs *Tt* and *Gg* for *tunicate* vs. *non-tunicate* and *glossy* vs. *non-glossy* respectively, are linked.
(i) *Tunicate* vs. *non-tunicate* and *glossy* vs. *non-glossy*

Tunicate, glossy	103	P
Tunicate, non-glossy	18	R
Non-tunicate, glossy	20	R
Non-tunicate, non-glossy	102	P

Percent recombination = 38/243 x 100 = 15.6

(ii) *Tunicate* vs. *non-tunicate* and *liguleless* vs. *non-liguleless*
Since the 4 phenotypes are present in equal proportions (recombinants are as common as parentals), the allele pairs *Tt* (*tunicate* vs. *non-tunicate*) and *Ll* (*liguled* vs. *liguleless*) are not linked.

Tunicate, liguled	58	Non-tunicate, liguled	59
Tunicate, liguleless	63	Non-tunicate, liguleless	61

(iii) *Glossy* vs. *non-glossy* and *liguled* vs. *liguleless*
Similar reasoning indicates that *Gg* and *Ll* are not linked.

Glossy, liguled	61	Non-glossy, liguled	56
Glossy, liguleless	62	Non-glossy, liguleless	64

b) The F_1 genotype is :

Note: (i) Since *Ll* is inherited independently of *Tt* and *Gg*, one cannot indicate which of the alleles of the *Ll* pair entered the F_1 with *TG* and which with *tg*.

(ii) The unlinked genes are either on different (nonhomologous) chromosomes or the same one but 50 or map units apart. Therefore, the 3 genes are either on the same chromosome (in the sequence *T-G-L* and, therefore, *T* and *L* are at least 65.6 map units apart) or *T* and *G* are on one chromosome and *L* is on another.

c) Percent recombination between *T* and *G* = 38/243 x 100 = 15.6. Therefore, *T* and *G* are 15.6 map units apart.

18. The cross may be diagrammed as follows:

brown (*CCbb*) x albino (*ccBB*) → F_1 black (*CcBb*)
F_1 black (*CcBb*) x albino (*ccbb*) →
F_2 black (*CcBb*)   102;   brown (*Ccbb*)   198
    albino (*ccBb*) and (*ccbb*)   300

a) On the basis of the phenotypes observed in the testcross progeny and knowing that *c* prevents pigment formation, the type of gene interaction involved is recessive epistasis, when *c* is present in the homozygous condition it masks the expression of the alleles at the *B* locus.

b) With independent assortment of the 2 allele pairs and random union of gametes the 4 testcross genotypes should have been produced in approximately equal proportions (150 of each). With recessive epistasis, the number of *albinos* should have been twice that of either *blacks* or *browns*, i.e., one should have observed either 204 or 396 *albinos*. Subjecting the results to a χ^2 test and testing for independent assortment of the 2 genes reveals a χ^2 value of 32.06. The probability value for this χ^2 is less than 1 percent; the actual results deviate significantly from those expected with independent segregation. The allele pairs are, therefore, linked.
The F_1 *blacks* are heterozygous in repulsion (*Cb//cB*). For every *C B* recombinant gamete (*black*) that was produced one *cb* recombinant (*albino*) was formed. Since 102 of the testcross progeny were *black* an equal number of the *albinos* must be recombinants.
Percent recombination = (102 + 102)/600 x 100 = 34

c) The genotype of the *brown* testcross progeny rabbits is *Cb//cb*. If these individuals were intercrossed, 3/4 of their progeny would be *brown* (*C_bb*) and 1/4 *albino* (*ccbb*) because segregation would occur only at the *C* locus.

22. a) (i) To calculate percent recombination from F_2 data of the cross in coupling, one should determine which F_2 phenotype is determined by a specific single F_2 genotype. The only F_2 phenotype that qualifies is the *white, glabrous* (double recessive) one. Sixteen percent of the F_2 are *white, glabrous* (*bp//bp*). Since each such individual is the product of 2 *bp* gametes, the frequency of the F_1 gametes that are *bp* can be determined by taking the square root of the frequency of the F_2 (*bp//bp*) genotype, (i.e., square root of 0.16 = 0.4 = 40%). Since the two types of parental gametes are formed with equal frequency, the *BP* F_1 gametes must also occur with the same frequency, 40%. Therefore parental gametes (*BP* and *bp*) comprise 80% (or 0.8) of all the F_1 gametes produced and 20% (or 0.2) constitute recombinant gametes. Since the 2 types of recombinant gametes are expected to be equally frequent, 10% should be *Bp* and 10% should be *bP*. Therefore F_1 gamete types and proportions are *BP* 40%, *bp* 40%, *Bp* 10%, *bP* 10%.

%R between B and P is 20. The loci are 20 map units apart.
(ii) Calculation of %R from F_2 of the repulsion cross can be made the same way. 1% of the F_2 progeny are *white, glabrous* ($bp//bp$). Therefore, square-root of $0.01 = 0.1$ or 10% of the gametes are bp. The same frequency of gametes should be of the reciprocal type (BP). Therfore toatal percentage of recombinant gametes ($BP + bp$) = 10 + 10 = 20%. Percent recombination between B and P is 20%.

b) Cross 1: heterozygous in coupling
   Cross 2: heterozygous in repulsion.
   Basis for answer - information in (a)

24. Let H vs h determine *no hemophilia* vs. *hemophilia* and C vs. c determine *normal colour vision* vs. *red-green colour blindness*. Since the second-generation female is *normal*, produces *colourblind* sons and daughters and her father was *hemophilic*, she must be heterozygous in repulsion ($X^{H,c}//X^{h,C}$) receiving $X^{h,C}$ from her father and $X^{H,c}$ from her mother. Her husband is $X^{H,c}/Y$. The third-generation offspring and their classification is as follows:

a) 1 and 4 are non-recombinants with genotypes $X^{H,c}/Y$ and $X^{h,C}/Y$ respectively.
b) 2 and 6 are recombinants with genotypes $X^{H,C}/Y$ and $X^{h,c}/Y$ respectively.
c) 3, 5 and 7 cannot be classified because they may be recombinants or non-recombinants. 3 - $X^{H,c}//X^{hC}$ (non-recombinant) or $X^{H,C}//X^{h,c}$ (recombinant); 5 and 7 $X^{H,c}//X^{H,c}$ (non-recombinant) or $X^{H,c}//X^{h,c}$ (recombinant).

26. The gene sequence is H-T-C.
Reasoning: In pedigree 1, two recombinant chromosomes are identified. Both crossing-over events occurred in the male parent, giving rise to chromosomes recombinant for H and T and simultaneously for H and C, but not for T and C. Therefore H is one outside marker. In pedigree 2, the crossing-over event again occurred in the male parent, resulting in a crossover chromosome with a recombinant genotype for C and T and simultaneously for C and H, but non-recombinant for T and H. This indicates that C is the other outside or flanking marker. Therefore, the information from the 2 pedigrees leads to the conclusion that the gene sequence is H-T-C.

32. a) Determine the distance by dividing the percent second-division segregations by 2.

f - centromere = (67/ 109x2) x 100 = 30.7 map units
mt - centromere (4 crosses) = 6.5 map units
c - centromere (2 crosses) = 7.7 map units
p - centromere (3 crosses) = 13.1 map units

b) The percent recombination between any 2 genes can be calculated in 2 ways:

(i) (R) = $\frac{1/2 TT + NPD}{PD + NPD + TT}$ x 100

or (ii) $\frac{\text{recombinant spore pairs}}{\text{total spore pairs}}$ x 100

(i) f - mt percent recombination (R) = $\frac{1/2(69) + 21}{109}$ x100 = 50.9

Percent R = 222/436 x 100 = 50.9 map units

Similarly percent recombination for the other genes, using either method is:
p - mt = 20.6 map units; c-p = 12.2 map units; c-mt = 14.7 map units.

c) (i) $f^+ mt^+$ x $f mt^-$. Since percent recombination between the 2 loci is 50.9 and segregation classes 1 and 2 are insignificantly different (in approximately a 1:1 ratio), the 2 genes are on separate chromosomes.
(ii) If the sum of the distances of 2 linked markers from a centromere is equal to the distance between the 2 markers, the 2 markers must reside on opposite sides of the centromere of a particular chromosome. Based on this reasoning and in light of the various calculations shown in (a) and (b), the locations of the markers are as follows:

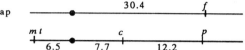

d) Yes. Tetratypes (second-division segregations) would not occur if crossing-over occurred at the 2-strand stage. For example, in the cross $c^+ mt^+$ x $c^- mt^-$ segregation classes 3, 4 and 7 would not have appeared if c/o occurred at the 2-strand stage.

37. There appear to be 3 separate linkage groups:
(i) y; (ii) nic-8 - cho and (iii) phe-2 - pan - sm
Basis for decision:
(i) Among the selected *yellow* segregants the 4 phenotypic classes shown occur in approximately a 1:1:1:1 ratio.
(ii) Among the *yellows* there is a 1:1 ratio of *nic-8, cho. Nic-8, Cho* and *phe-2, pan, sm : Phe-2, Pan, Sm*. Had either of these 2 linked groups been on the same chromosome as y they would have been inherited *en bloc*, e.g., y, nic-8, cho and Y, Nic-8, Cho if the cross was in coupling and the number of phenotypic classes would have been reduced to 2.
(iii) The allele pairs (Nic-8 nic-8, Cho cho) are inherited independently of the Phe-2 phe-2, Pan pan, Sm sm ones.

43. Some of the reasons why map distances do not represent actual spatial distances are:
(i) The frequencies of crossing-over and recombination are affected by environmental factors, e.g., temperature, age, genetic conditions, sex, etc.
(ii) The chromosome is composed of two kinds of chromatin, (a) heterochromatin, in which little or no exchange occurs and (b) euchromatin, in which crossing-over occurs. Therefore different chromosomes and different regions of the same chromosome may show differences in frequencies of crossing-over. As a consequence, the same physical distances may be represented by different exchange frequencies and *vice versa*. Since the distance between any two genes on a genetic map (in map units) is based on the frequency of crossing-over in the region between these markers, and since the procedure used to construct genetic maps does not permit one to distinguish between heterochromatin and euchromatin, a cytological map will often be different from a genetic one.

(iii) Percent recombination for loosely linked genes may not accurately estimate map distance because some of the chromosomes would possess parental genotypes as a result of an even number of exchanges. Such chromosomes would lead to an underestimation of map distance.

50. a) Since the testcross phenotypic ratio for each pair of traits is 1:1, each pair of traits is determined by a single pair of alleles. Let Bb determine *black* vs. *white*, Ww determine *wavy* vs. *straight* and Ll determine *long* vs. *short* respectively.

b) Yes. This is indicated by the (i) recombination values calculated from the testcross data, all of which are significantly less than 50 as shown below and (ii) by the fact that two of the eight expected phenotypes are missing.

Black, straight (R) = 49   Black, long (R) = 42   Straight, long (P) = 398
Back, wavy (P) = 348   Black, short (P) = 355   Straight, short (R) = 7
White, wavy (R) = 47   White, long (P) = 365   Wavy, long (R) = 9
White, straight (P) = 356   White, short (R) = 38   Wavy, short (P) = 386

%R=(96/800)100=12   %R=(80/800)100=10   %R=(16/800)100=2

c) The F_1's were heterozygous in the repulsion phase since their phenotype is different from that of both parents.
Genotypes P_1 Black, wavy, short   BWl x White, straight, long   b w L

                       BWl   b w L

       F_1 Black, wavy, long   BWl

                                                      bwL

d) Two. *Black, wavy, long* and *white, straight, short* because the three genes are closely linked as indicated in b) with the maximum recombination value being 12 between *B* and *W*. Since double crossing over does not occur within such short regions, and is necessary to generate the double crossover phenotypes and genotypes, these two classes of individuals are missing.

e) One can derive the sequence of the three genes in two ways:
(i) Compare the three recombination values shown in b). The largest percent recombination is that between the genes *B* and *W*: these genes must be the outside markers with gene *L* being in between them.
(ii) Compare double crossover phenotypes and genotypes with parental ones:

      Parental   Double crossover
Black, wavy, short BWl   *Black, wavy, long* BWL
White, straight, long bwL   *White, straight, short* bwl

The alleles *L* and *l* for the traits *long* and *short* have been transposed to produce the double crossover phenotypes and genotypes. Since this requires two reciprocal exchanges, one on each side of the *Ll* pair, the *L* locus is in the middle, with *B* and *W* being the outside markers. Therefore the sequence of the genes on the linkage map is *B - L - W*.

Based on the genetic (=map) distances given in b), the linkage map is as follows:

    B   L   W
    +------------------------------+---+
            12 mu   2 mu

f) The coefficient of coincidence is zero (0) because double crossover genotypes and phenotypes were not produced. Therefore interference will be 1.0 (complete).

52. a) The parental genotypes are *cmcm yy FF* and *CmCm YY ff*. If the gene order was *Cm Y F* the double crossover phenotypes should be *cm Y F* and *Cm y f*. Since the observed double crossover progeny classes are *Cm y F* and *cm Y f*, the gene order cannot be *Cm Y F*.

b) Since the *Cmcm* allele pair was transposed to generate double crossover types from parental chromosomes, this pair must be located between the other two. Therefore the correct gene sequence is *Y Cm F*.

c)

Parents   y   cm   F   X   Y   Cm   f
             y   cm   F   Y   Cm   f

F_1 Y   Cm   f
     y   cm   F

Chapter 11
Extranuclear Inheritance and Related Phenomena

3. The cell divisions in *Spirogyra* are mitotic and therefore all the cells should have an identical genotype. If plastid type was determined by nuclear genes, all the plastids in each cell should have been of one type. If it was determined by cytoplasmic genes not located in plastids, these should affect all plastids in a cell in the same way with only one plastid type being present. Thus, the perpetuated differences between the plastids must be due to genes located within the plastids.

6. a) The traits *green* vs. *white* vs. *variegated* leaves are determined by plasmagenes of two types: one determining *coloured* and the other *colourless* plastids.

   b) **Critical observation:** Flowers on *variegated* branches, when self-fertilized or pollinated with pollen from any source, e.g., flowers on *green*, *white* or *variegated* branches, produce progeny with *green*, *white* or *variegated* leaves.

7. a) (i) Cytoplasmic inheritance - cytoplasm occasionally transmitted by male gamete. (ii) Maternal effect (influence).
   (iii) Some entity, e.g., a virus in the milk of females of the *high cancer* strain. The entity is not present in the milk of females of the *low cancer* strain.
   Note: Not a Dauer modification since lines were found to breed true for such a change.

   b) Explanation (iii) in (a) appears to be the most plausible for these results. Reason: If offspring of females of a *high cancer* strain are suckled by their mothers they all develop cancer. However if suckled by females of a *low cancer* strain they fail to develop cancers. Similarly the offspring of *low cancer* strain females when suckled by their mothers fail to develop cancer but when suckled by females of a *high cancer* strain they do develop cancer.

8. The most plausible explanation is that *streptomycin-resistance* vs. *streptomycin-sensitivity* is determined by a cytoplasmic gene which exists in two allelic forms. Although the zygote receives the alleles of this cytoplasmic gene from both mating types, only the alleles from one (mt^+) mating type are transmitted to all the meiotic products and therefore to all the progeny. The alleles from the mt^- parent are concurrently eliminated and not included in the gametes. There is uniparental transmission of the alleles of this cytoplasmic gene; this is an example of maternal inheritance (all progeny express the phenotype of the female (mt^+) parent). The preferential transmission of cytoplasmic genes from the female (mt^+) parent and loss of the corresponding male (mt^-) genome may be due to (i) only mt^+ contributes cytoplasm and its genes to the zygote or (ii) alleles from the mt^- parent are somehow excluded from the gametes.

11. a) Only maternal (cytoplasmic) transmission can account for the inheritance of *myoclonic epilepsy*. **Basis for conclusion:** (i) The hereditary determiner (gene) is transmitted only by the female parents in each generation. (ii) Offspring of both sexes are *affected*. An autosomal mode of inheritance can be eliminated because (a) the trait should occur equally frequently in the 2 sexes. (b) the gene should be transmissible by males as well as females. It is not. (c) Since the disorder is rare, it is highly likely that *affected* individuals should be heterozygous, e.g., Mm and *unaffected* ones be homozygous recessive (mm); therefore 1/2 the children of an *affected* x *unaffected* mating should be *affected*. The actual results do not conform to these expectations. (iv) The probability of all 8 individuals in generation II being *affected* is $(1/2)^8 = 0.004$. The inheritance pattern is also inconsistent with an autosomal recessive mode of inheritance. Individuals II-3, II-4 and II-7 have *affected* children and *affected* grandchildren. This mode of inheritance would require that their spouses and spouses of their children be heterozygous (Mm). This is highly unlikely since the disorder is rare.

An X-linked recessive mode of inheritance would require I-1 to be X^m/Y and I-2 X^MX^m (highly unlikely). Similarly, the daughters of II-3 and II-7 are *affected* as are those of III-20 and III-21. This would require the husbands of all these women to be X^m/Y and therefore *affected*. This is not the case. The pedigree is also inconsistent with an X-linked dominant mode of inheritance because all the daughters of an *affected* male are not *affected*.

b) Yes. Since the gene determining this disorder is transmitted only by the female parent and human mitochondrial (mt) DNA is transmitted by a mother to all her children, the finding of a mutant allele in the mtDNA coding for $tRNA^{LYS}$ is consistent with cytoplasmic inheritance and the disorder being determined by the mutant allele at the $tRNA^{LYS}$ locus. The variations in proportions of *wild-type* vs. *mutant mt* DNA are also consistent with cytoplasmic inheritance; each cell contains thousands of mitochondria and *mt* DNA. These are randomly distributed to daughter cells during both mitosis and meiosis since mechanisms (spindles) do not exist for equal distribution of these organelles.

c) The variable expressivity of the disease is due to (i) the fact that the thousands of mt are randomly distributed during cell division and (ii) the intracellular ratio of the *mutant* and *wild-type mt* DNAs. The mutant phenotype is expressed only when the proportion of mt DNA reaches a certain threshhold. The severity of the symptoms is dependent on the proportion of the 2 types of mtDNAs in the zygote and cells derived from it - the higher the proportion of mutant mtDNAs the more severe the phenotype.

13. Jenkins (1924) showed that 1/4 of the F_2 of a cross between *green* ♀ and *iojap* ♂ had *iojap* progeny. Chlorophyll development is therefore influenced by a recessive allele *ij* of an autosomal gene. Since the F_1 generations of reciprocal crosses give different results and in all crosses the progeny phenotype resembles that of the maternal parent it is highly likely that a cytoplasmic determinant is also involved in the determination of the *iojap* trait.

b) On the basis of Rhoades' (1943) data the following hypothesis may be proposed: The allele *ij* of the nuclear (chromosomal) gene, when homozygous, acts as a mutagen causing some of the cytoplasmic determinants determining chlorophyll development to mutate to an allelic form which is unable to direct the synthesis of chlorophyll. Thus, plastids determined by the mutant alleles are *white* and those determined by the *wild-type* allele are *green*. The cytoplasmic determinants affected by *ij* are inherited through the cytoplasm of the egg and not by the sperm (pollen); the *iojap* condition shows maternal inheritance.

c) The results imply that the cytoplasmic genes are in the chloroplasts but do not prove that they are so located.

d) (i) Yes, since the mutant cytoplasmic genes produce their effects in the absence of the recessive autosomal allele *ij* or when it is heterozygous (*Ijij*).
(ii) Yes. Reason same as in (i).

14. a) *Male-sterility* is caused by a cytoplasmic determinant (gene or genes, *f*) transmitted directly by the female parent. The situation can be explained as follows:
Assume that the *male-sterile* strain and the *male-fertile* lines are identical in nuclear genotype *msms*. The cytoplasmic determinant is indicated within parentheses.

♀ (*male-sterile*) msms (*f*) x ♂ (*male-fertile*) msms (*F*) → F_1 msms (*f*) *male-sterile* because of cytoplasmic factor (*f*), obtained from the female parent. Subsequent backcrossing, will not restore fertility because of the cytoplasmic determinant *f*.
Reciprocal cross:
♀ (*male-sterile*) msms (*F*) x ♂ (*male-sterile*) rare pollen msms (*F*) → F_1 msms (*F*) *male-fertile*. True-breeding, since nuclear genotype for both parental strains is homozygous *msms*.

15. When V4 is used as the female parent in crosses the F_1 is *runner* and the F_2 shows a phenotypic ratio of 9 *runner* : 7 *bunch*. In reciprocal crosses when V4 is the male parent, the F_1 is *bunch* and the F_2 phenotypic ratio is 5 *runner* : 11 *bunch*.
Conclusions:
(i) Since reciprocal crosses give different results in the F_1, involvement of a cytoplasmic component in the expression of growth habit is highly likely.
(ii) The nuclear genotype is also important since the F_2 generations from reciprocal crosses produce segregation ratios (9:7 and 5:11) that are characteristic of two independently inherited allele pairs. Growth habit is determined by the interaction of 2 independently inherited pairs of alleles with each other and the *V* and *v* cytoplasms. The type of interaction between the cytoplasm and nuclear genotypes may be demonstrated using the reciprocal crosses involving the true-breeding varieties V4 and NC2.

V4 aaBB (V) ♀ x NC2 AAbb (v) ♂ → F_1 AaBb (V) s.f. → F_2
V4 aaBB (V) ♂ x NC2 AAbb (v) ♀ → F_1 AaBb (v) s.f. → F_2

F_2 phenotypes in *V* cytoplasm	Genotypes	F_2 phenotypes in *v* cytoplasm
Runner	1 AABB	Runner
Runner	2 AaBB	Runner
Runner	2 AABb	Runner
Runner	4 AaBb	Bunch
Bunch	2 Aabb	Bunch
Bunch	2 aaBb	Bunch
Bunch	1 AAbb	Bunch
Bunch	1 aaBB	Bunch
Bunch	1 aabb	Bunch

(i) In the presence of *V* (V4) cytoplasm, *runner* growth habit is due to the complementary interaction of the dominant alleles of 2 independently inherited loci. The *bunch* phenotype is due to the absence of a dominant allele at at least one of the loci.

(ii) In the presence of *v* cytoplasm 3 or more dominant alleles are required for the expression of the *runner* phenotype whereas the *bunch* trait is caused by the presence of two or fewer dominants.

The other crosses can be explained in the same way.

20. The *wild-type* allele, *Gs*, at the *grandchildless* locus may be responsible for the production of a substance which is transmitted through the cytoplasm of ovarian eggs and is essential for the proper development of the gonads during embryogenesis. Another possibility is that the *grandchildless* allele *gs* (when homozygous) causes a biochemical block in females which results in an accumulation of precursor materials in the egg cytoplasm that prevent the subsequent normal development of the gonads.

21. Reciprocal crosses give the same results in all generations, except the F_1, and the ratio of *dextral* and *sinistral* F_3 broods (each from a single F_2 snail) is 3:1. Therefore a single autosomal pair of alleles determines the direction of coiling; the allele *D* for *dextral* is dominant to *d* for *sinisistral*. Thus:
(i) the *dextral* and *sinistral* parents are *DD* and *dd* respectively,
(ii) F_1's in reciprocal crosses are genotypically identical (*Dd*), and
(iii) F_2's in both crosses are genotypically identical - each consists of 3 genotypes - 1 *DD* : 2 *Dd* : 1 *dd*.

The results in the F_2 and F_3 generations are those that are normally obtained in the F_1 and F_2 generations respectively. This suggests that the phenotype of an individual depends on its mother's genotype and not on its own genotype (maternal influence).

The direction of coiling of the shell depends upon the orientation of the spindles during the first and subsequent early cleavage divisions of the zygote. The orientation of the spindles is determined by the nature of the egg cytoplasm which in turn depends upon the mother's genotype. The $D\_$ genotype in the mother produces a substance(s) that is (are) stored in the egg cytoplasm, causing the spindles during mitosis to orient in a right-left direction relative to the long axis of the organism and cells, resulting in coiling to the right (*dextral*). The *dd* genotype in the mother produces a different form of the substance(s) causing orientation of the spindles in a left-right direction (opposite to the orientation specified by $D\_$ genotypes) and resulting in *sinistral* coiling. The maternal effect in this case is permanent, lasting throughout the life of the individual.

22. a) Reciprocal crosses yield different results indicating that an autosomal pair of alleles alone cannot explain the observations
(i) A sex-linked mode of inheritance is not feasible since the results differ significantly from those expected with sex-linkage.
(ii) Y-linked inheritance is eliminated since the hereditary determiners are carried and transmitted by both sexes.

b) These results are an example of maternal influence. The genotype *Aa* has a transitory effect. In the presence of the dominant allele *A* at a particular locus tryptophan is converted to kynurenine. This diffusible substance is stored in the egg cytoplasm and affects the phenotype of the larvae (causes the larvae to be pigmented). Kynurenine lasts only a few days; after that the individual's own genoypte takes over and determines the phenotype of the individual.
In the first cross the mother is *Aa*, therefore kynurenine is present in the cytoplasm of her eggs. All larvae but only *Aa* adults are *pigmented* (*brown eyes*). In the second cross the mother is *aa*, therefore only the larvae and adults that receive *A* from the father are *pigmented* and *brown-eyed* respectively.

c) One of the parents had *dark brown* eyes and the other *red* eyes as an adult.

23. a) *L. angustifolius* is a bisexual species. It does not possess sex chromosomes and therefore none of its genes can display sex-linkage. Possible explanations for the results are:
(i) Cytoplasmic inheritance (ii) Dauer modification (iii) Maternal inheritance (influence)

b) Since the leaves of F_1's more than a month old, regardless of the direction of the cross, are *bitter*, this eliminates explanations (i) and (ii). The expression of *sweet* vs. *bitter* appears to be due to the genotype of the mother that has a transitory effect (month or less) in the progeny. The proportions of phenotypes in the F_2 generations suggest that a single gene is involved.

c) (i) The genotype of the mother affects the expression of the character during the first 5 to 6 weeks of growth. After that, each individual's phenotype is determined by its own genotype.
(ii) First Cross:
P_1 ♀ *sweet* (bb) x ♂ *bitter* (BB) → F_1 Bb - leaves *sweet* first month because of their mother's genotype (bb); *bitter* later because of their own genotypes (Bb). Approximately 3/4 of the F_2 have leaves that are *bitter* throughout the life of the plant and all ripe seed are *bitter*. 1/4 of the F_2 have leaves that are *bitter* the first month because of their mother's genotype. However, all ripe seed and mature plants are *sweet* because of their own genotypes. Explain the results of the second cross, using the same reasoning.

27. Yes. The evidence is indirect and as follows:
a) Since α is the main DNA band, it is highly likely that it is derived solely or mainly from nuclear DNA, where most of the genetic material is located.
(i) In DNA from purified chloroplasts, the relative quantity of β satellite DNA increases significantly over its proportion in whole cell DNA. This implies that β DNA is in the chloroplasts and constitutes at least a portion of all chloroplast DNA. This conclusion is supported by the finding that in normal and mutant strains with chloroplasts, the β satellite band is present as well as the α one.

b) Since mutant strains of *Euglena*, which lack chloroplasts, do not show the β satellite band although the α band is present as in normal strains, suggests that chloroplasts contain DNA (specifically β band DNA).

c) The fact that α and β bands have different base ratios, indicates they come from different sites or organelles in the cell. It does not indicate by itself that they are derived from the DNA in the nucleus and chloroplast respectively.

Chapter 12
Genetics of Bacteria and Viruses

1. a) Conjugation requires physical contact between donor and recipient cells which in the U-tube experiment, is prohibited due to the presence of a filter between the 2 strains. Therefore conjugation can be ruled out as a mechanism.

   Barring mutation, 2 other mechanisms can transfer genes from one bacterium to another: transduction and transformation. Add DNase to the reaction mixture and/or test for the presence of phage.

   b) Transformation involves the transfer of DNA fragments into a recipient cell and their subsequent integration into the recipient's chromosome. Treating the mixture with DNase will degrade the DNA fragments and thus prevent transformation. Therefore if no recombinants are formed after the addition of DNase to the mixture one could conclude that transformation is responsible for the transfer. However, if recombinants are observed, transformation can be eliminated, leaving transduction as the probable mechanism of change. If transduction is responsible one should be able to detect the presence of phage.

9. a) The prototrophs in this case could arise by mutation, conjugation, transduction or transformation.

   Mutation as an explanation can be ruled out because (i) it would require the highly improbable coincidental occurrence of reverse mutations at 3 specific loci in one or the other or both parent strains and (ii) the frequency of reversion to prototrophy (at 3 loci) was higher than the reversion to single- and double-prototrophy. Conjugation appears to be the most plausible mechanism because the region covering the 3 loci in either parent strain is rather large. In both transduction and transformation only short fragments of donor DNA are transferred. Furthermore, neither viruses nor DNA extracts were fed into the system, which are necessary characteristics of transduction and transformation respectively.

   b) Perform a U-tube experiment. Place one auxotrophic strain in one arm of the U-tube and the second strain in the other arm of this structure. If conjugation is responsible for the genetic transfer, no recombinants will be observed because the U-tube prevents physical contact between the 2 strains. Either transduction or transformation will have occurred if recombinants are formed in the U-tube. If transformation is responsible, the addition of DNase, which degrades naked DNA, will prevent the transfer of genetic material. If transduction is responsible, one should be able to detect the presence of viruses in the reaction mixture and DNase will have no effect on the frequency of occurrence of the prototrophs (recombinants).

10. a) (i) Clone no. 1 must be doubly auxotrophic $thr^-leu^-his^+pro^+pan^+bio^+$ since it grows only on the plate supplemented with thr and leu.
    (ii) Clones 2 and 4 possess the same genotype. Since the grow in the presence of thr and leu or thr and his but not in the presence of leu and his, they must be singly auxotrophic for thr ($thr^-leu^+his^+pro^+pan^+bio^+$).

    (iii) Clone no. 3 grows in the presence of pan and bio or pro and bio but not in the presence of pro and pan. This clone is therefore auxotrophic for bio only ($thr^+leu^+his^+pro^+pan^+bio^-$).
    (iv) Clone no. 5 grows only on medium supplemented with pan and bio and must therefore be a double auxotroph of genotype $thr^+leu^+his^+pro^+pan^-bio^-$.
    (v) Clone no. 6 cannot be auxotrophic since it grows on all the plates. It is prototrophic for all markers ($thr^+leu^+his^+pro^+pan^+bio^+$).

    b) Clone no. 7 could be any triple, quadruple, or multiple auxotroph since it grows on complete medium. If three, four and five nutrients in various combinations were added and the clone does not grow, one would then know that the clone is a complete auxotroph. Similarly if addition of a specific combination of three, four or five growth factors permitted growth, one could conclude that the clone is auxotrophic for those three, four or five nutrients.

12. a) The results support Hayes' hyypothesis that only a portion of the male's genetic material is transferred. **Reason:** Conjugation interruption breaks the donor chromosome and recombinants can only occur for those donor alleles that enter the F^- parent; recombinants for the donor alleles at these loci arise in one specific sequence, at precise times and not in a random manner. This indicates that only a specific portion of the donor genome is transferred, always with the same end first. If the entire chromosome was transferred and integration of the donor segment was at random, different kinds of recombinants could arise at various times of conjugation interruption.

    b) (i) Since the donor genes enter F^- in a specific order, e.g. thr^+ first, leu^+ second, azi^s third, etc. and at precise times:
    E. coli has only one linkage group and therefore chromosome in which the genes are arranged in a linear sequence as they are in eukaryotes.
    (ii) A particular Hfr strain always transfer the same part (end) of the Hfr chromosome (called origin, O) first to the recipient.
    (iii) The genes on the donor chromosome enter in a sequence which must correspond to their sequence on the Hfr chromosome.
    (iv) The times at which different donor genes enter the recipient are directly proportional to the distance between them. Therefore if is possible to construct a linkage map which not only gives the correct gene sequence but also the correct distance between the genes in relation to time (1 time unit = 1 minute). Thus the linkage map of the H genes given is:

Sequence	O	thr	leu	azi	T1	lac	gal	λ
Time units		8	8 1/2	9	11	18	25	26

(v) The amount of genetic material transferred during conjugation varies with time.
(vi) Integration of donor chromosome segment(s) into the recipient chromosome involves an even number of exchanges.

17. An even number of crossovers is required to integrate the exogenote into the endogenote (host chromosome).

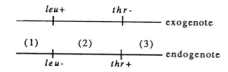

Crossing-over in regions (1) and (2) will produce prototrophic recombinants ($leu^+ thr^+$) which can grow on minimal medium. On minimal medium supplemented with *threonine*, $leu^+ thr^-$ recombinants arising by crossing-over in regions (1) and (3) will grow in addition to the prototrophs. Therefore

percent recombination = $\dfrac{\text{number of prototrophs}}{\text{number of recombinants}} = \dfrac{25}{125} \times 100 = 20$

20. a) Donor chromosome tends to break randomly during transfer. The probability that genes located at the distal end of the chromosome will enter and form merozygotes, diminishes as their distance from the origin increases. Genes located between the point of origin and the selected marker should have equal or almost equal frequencies of recombination. Those distal to the selected genes exhibit a gradient of frequencies which reflects their sequence and distance apart on the chromosome. When a selected marker is transferred, all genes proximal to it, with respect to the point of origin are also transferred. The frequencies of recombination for genes distal to the selected marker depend primarily on the chance of their being transferred, which is a function of their distance from the selected marker. Since a gradient of recombination frequencies is obtained for all unselected donor traits and their genes among the $thr^+ leu^+$ selected recombinants, all the unselected markers are distal to O with respect to $thr^+ leu^+$.

b) The gradient of recombination frequencies for the unselected markers reflects the order of these markers on the chromosome. Since thr^+ and leu^+ are on the proximal part of the chromosome they permit one to determine the sequence of the other genes on the basis of their decreasing frequencies of recombination. The gene sequence therefore is:

O - (thr - leu) - axi - T1 - lac - gal .... $\begin{matrix} mal \\ xyl \\ man \end{matrix}^+$

*The data do not permit the exact sequence of these 3 markers.

22. a) Conjugation and transformation can be eliminated as the mechanisms of genetic transfer. **Reasons:**
(i) Conjugation cannot be responsible for genetic recombination because prototrophs were observed in the U-tube experiment.
(ii) That transformation is not the mechanism is indicated by: (a) addition of donor DNA to the recipient culture does not result in recombinants, (b) treatment of the U-tube culture with DNase does not affect the formation of recombinants.

b) That transduction is the mechanism of genetic transfer is supported by all the observations, e.g., viruses in the donor strain cultures, absence of any effect of DNase, presence of prototrophs in U-tube experiments, etc. Transduction is mediated by a vector which is a virus and, therefore, cell contact is not necessary to obtain genetic recombinants. During lytic infection of nonlysogenic donor cells, as well as after the induction of lysogenic donor cells, short bacterial chromosome segments (less than 1 to 2 percent of the total chromosome length) are incorporated into some (1 in 10^6) of the progeny phage particles which become vectors. Vector phage are biologically inactive because their chromosome is incomplete or totally missing. Therefore, transduced cells are usually nonlysogenic. After the bacterial chromosome segment has been transferred to a recipient, it may pair with the homologous region of the recipient chromosome and a portion or it may be integrated into the recipient chromosome through an even number of crossovers. In Zinder and Lederberg's experiments, a few lysogenic recipients ($met^- thr^+$) were induced. The released phage infected the donor strain ($met^+ thr^-$) cells where they went through the vegetative state producing progeny viruses, causing lysis and death of the infected cells. Some of the released phage particles carrying the donor gene met^+, infected and transferred the donor chromosome segment into the recipient cells. The met^+ segment of the donor was integrated (replaced the allele met^-) into the recipient genome by an even number of exchanges producing recombinant prototrophs ($met^+ thr^+$).
With this as a background illustrate the process yourself.

24. a) The *large* colonies are the result of complete transduction. The segment of the donor fragment (exogenote) with the *wild-type* allele or site is substituted for the mutant allele or site in the recipient chromosome (endogenote) by an even number of crossovers. When completely transduced cells divide, all the cells in the colony receive a replica of the *wild-type* allele. All such cells can grow and divide on minimal or any other medium and, therefore, colonies will be *large*.
The frequency of *large* (*wild-type*, prototrophic) colonies in a given cross is a linear function of the distance between the mutant sites. If the mutant sites are far apart the probability is high that crossing-over will occur between them, resulting in the substitution of the *wild-type* site for the mutant site. Thus the frequency of *large* colonies will be high. However, if the mutant sites are close together the frequency of *large* colonies will be low. The situation with respect to the 3 mutants is as follows:
(i) In crosses ath-1 x ath-6 and ath-4 x ath-6 the mutant sites exist in different genes (cistrons) with the consequence that crossing-over can readily occur between them producing many large colonies.
(ii) In the cross ath-1 x ath-4 the mutant sites occur in the same gene (cistron) and, therefore, are close together. Exchange between the sites will be rare and, therefore, the number of *large* colonies will be small.
(iii) In each of the crosses ath-1 x ath-1, ath-4 x ath-4 and ath-6 x ath-6 mutant sites at corresponding positions are involved. Exchange cannot occur between such mutants and therefore, no *large* colonies can arise.

b) *Minute* colonies arise as a result of abortive transduction. Their presence in a given cross indicates that the mutant genes in the donor and recipient mutant strains are functionally non-allelic (in different genes or cistrons). On the other hand, their absence in a cross indicates that the mutant genes in the parental strains are functionally allelic (in the same gene or cistron).
Since *minute* colonies are not formed in crosses between ath-1 and ath-4, the mutants ath-1 and ath-4 must be functionally allelic (possess mutant sites in the same gene). Since *minute* colonies always appear in crosses between ath-1 and ath-6 as well as between ath-4 and ath-6, the mutant ath-6 must not be allelic to ath-1 and ath-4 (possesses a mutant site in a different gene).

c) At least 2 genes are involved in adenine-thymine formation; ath-1 and ath-4 are the result of mutations in one gene and ath-6 is mutant for another.

25. a) Cotransduction indicates close linkage between the loci involved. Since $cysB$ is common to all 6 sets of reciprocal crosses, and the trp loci of the donor are cotransduced with $cysB$ in each of the 6 experiments, it can be concluded that each of the trp genes A, B, C and D is linked to $cysB$; all 4 trp loci must be linked to each other.
Results and conclusions:
In each of the 6 sets of reciprocal transductions the first cross involves a double mutant recipient and a single mutant donor while the second cross has a single mutant as a recipient and the double mutants as a donor. Therefore, the expectations are as follows in all 6 crosses, using the first set of reciprocal transductions as an example:
(1) If the gene sequence is $trpD$ - $trpB$ - $cysB$, 4 crossovers are required to produce *wild-type* recombinants in the first cross and 2 in the second one and,

therefore, *wild-type* recombinants should arise in significantly different frequencies in reciprocal crosses.
(2) If the gene sequence was *trpB - trpD - cysB*, 2 exchanges would be required to produce *wild-type* recombinants in both crosses and, therefore, *wild-type* recombinants would have occurred in insignificantly different frequencies in both crosses.
Since the frequencies of *wild-type* recombinants are significantly different for the first set of reciprocal transductions, one would conclude that the gene sequence for the first cross is *trpD - trpB - cysB*. On the basis of the above reasoning the sequence of loci in each of the 6 sets of crosses is as follows:

Cross	Sequence
1	*trpD - trpB - cysB*
2	*trpD - trpA - cysB*
3	*trpB - trpA - cysB*
4	*trpD - trpC - cysB*
5	*trpC - trpA - cysB*
6	*trpC - trpB - cysB*

From (1) and (3) it is evident that the order of the 3 *trp* loci is *trpD - trpB - trpA* and from (4) and (6) that *trpC* is between *trpB* and *trpD*. Thus the order of the 4 *trp* genes and the *cysB* locus is *trpD - trpC - trpB - cysB*.

b) The sequence of the closely linked *trp* genes suggests that the genes are clustered into one operon and are under the control of one operator and one regulator. Thus, the enzymes controlling the various steps in the pathway would, or might be, sequentially transcribed. It also suggests that the close association in a definite sequence of the 4 *trp* loci is of evolutionary advantage to *Salmonella*. Having the genes linked in the same sequence as the biochemical reactions they determine, may lead to increased efficiency of their performance and thus confer a selective advantage on the organisms in which such an arrangement is present.

33. a) Only one strand (either one) of transforming DNA is integrated into the recipient genome. The inference is that the renatured (hybrid) DNA should not be able to yield double transformants because:
(i) the genes for *cathomycin-resistance* and *streptomycin-resistance* should be in the trans-configuration in the DNA duplex, that is, the gene for *cathomycin-resistance* should be in one strand and the gene for *streptomycin-resistance* should be in the other and,
(ii) if one of the 2 strands is always destroyed on uptake, only one of each hybrid DNA should be transferred to any one transformant. Each transformant should express either *cathomycin-resistance* or *streptomycin-resistance*, but not both. The occurrence of doubly-resistant transformant clones indicates that the hybrid transforming DNA carries both the gene for *cathomycin-resistance* and the gene for *streptomycin-resistance* on the same strand. If the 2 genes were on different (complementary) strands they would segregate at the first replication of the tranformant DNA and produce mixed but not pure clones. Such unexpected results may be accounted for as follows: Heating may cause single strand breaks. On cooling, 2 single stranded fragments, each carrying a different marker, could unite and pair with either complementary regions of an intact strand derived from either DNA. Upon replication a true-breeding double-resistant mutant would be formed.

b) Because of the high degree of homology, DNAs of the same species are more complementary than DNAs of different species. Reannealing denatured DNAs from different organisms and comparing their degrees of homology can be used to clarify or aid in their taxonomic relationships.

43. The data are only compatible with the second hypothesis that, the linkage map of T4 is circular. **Explanation:**
If the linkage map was linear, equal numbers of the $h42$ and $ac41$ recombinants should bear $r67^+$ and $r67$ alleles. Since this is not obtained it indicates that the $r67$ locus not only fails to segregate independently of loci $h42$ and $ac41$ but appears to be liked more closely to the distal $h42$ locus than to the proximal $ac41$ locus. The latter conclusion is arrived at as follows: Since $r67$ is linked to $h42ac41$ it could be either to the right of $ac41$ or to the left of $h42$. If it was to the right of $ac41$ the majority of the $h42$ $ac41$ recombinants are expected to be $r67^+$ and not $r67$ since only one exchange is required to produce the former and the latter two. If $r67$ was to the left of $h42$ the majority of the $h42$ $ac41$ recombinants are expected to be $r67$ and a minority $r67^+$ since fewer exchanges are required to produce $h42$ $ac41$ $r67$ than $h42$ $ac41$ $r67^+$ genotypes. This latter expectation was always obtained.
The only way to resolve the paradox is to conclude that the linkage map of T4 is circular, having neither a beginning nor an end. Only in this way can $r67$ simultaneously be both more and less closely linked to $h42$ than $ac41$.
The reason why the linkage map of T4 is circular, whereas the chromosomes are linear is that the chromosomes of this species are terminally repetitious and circularly permuted.

44. a) T2 must be haploid since one observes phenotypes arising from segregation in the first generation of a cross and this is atypical of diploid organisms.

b) At least three genes specify plaque type ($r1$, $r7$ and $r13$) since different recombination values of 24%, 12.3% and 1.7% are observed in the three crosses $hr^+ \times h^+r1$, $hr^+ \times h^+r7$ and $hr^+ \times h^+r13$ respectively.

c) The 3 genes are linked to the h locus as indicated by their recombination values. There is 24% recombination between h and $r1$, 12.3% between h and $r7$ and 1.3% between h and $r13$. Thus $r13$ is most closely linked whereas $r1$ is least closely linked to the host-range gene, h.

d) Four different arrangements of the four loci can satisfy the condition that $r13$ is closest to h and that $r1$ is most distant from h. Illustrate this for yourself.

e) Complementary gene action occurs among the r mutants. If the genes function in the same biochemical pathway.

$$\text{precursor} \xrightarrow{} A \xrightarrow{r1^+} B \xrightarrow{r7^+} C \xrightarrow{r13^+} \text{wild-type plaques}$$

A block in any one of the 3 steps will result in a *mutant* plaque.

46. a) The mutants map in at least three different genes.
Explanation: Assume that extremely low percent recombination indicates mutants are non-allelic (in different genes). Since each of the three subgroups among the 74 mutants when crossed to the three testers, gave few *wild-type* recombinants with one of the testers and many with the other two, indicates that these mutants occur in three different genes. Specifically, the 65 r mutants that gave few recombinants with $r48$ and several with r ED b 50 and $r67$ occur in one gene (symbolize it rI); those (seven) that gave very few *wild-types* with r ED b 50 and many with the other two testers occur in another gene (symbolize it rII); and those (the latter two) that produced extremely few *wild-types* with

$r67$ and a high percentage with $r48$ and $r\,ED\,b\,50$ reside in a third gene (symbolize it $rIII$).

Since 64 of the initial 194 mutants are closely linked, as indicated by the very low percent recombination among them in two-factor crosses, and since 63 of these mutants and 7 of the 74 that produced few *wild-types* with $r\,ED\,b\,50$ gave no recombinants with $r\,ED\,d\,f\,41$ (a deletion mutant), indicates that these 70 mutants are in a segment of the chromosome absent from this deletion and, therefore, in juxtaposition. Seven of these (those that produced few *wild-types* with $r\,ED\,b\,50$) are in gene rII. Although the other 63 are closely linked to these seven mutants there is no information on whether they are in gene rII or another closely linked gene. Therefore, at least three r loci are involved in these results.

b) (i) $r\,ED\,d\,f\,41$ is a deletion (multisite) mutant since it does not revert and does not produce *wild-type* recombinants with at least two point-mutants. Specifically, it gives no such recombinants with 70 mutants.

(ii) All other mutants (137) are of the point (single base-pair substitution) type since each reverts to r^+ at low frequency.

48. a) Since each amino acid is specified by a specific sequence(s) of three consecutive bases (a codon) and an average polypeptide contains 200 amino acids, each kind of polypeptide must be coded for by a gene that is $200 \times 3 = 600$ bases long. Therefore, the DNA of $\phi X174$ would possess $5500/600 = 9$ different genes which specify 9 different polypeptides.

b) (i) The T4 chromosome consists of double-stranded DNA. It must contain $1.2 \times 10^8/720 = 1.66 \times 10^5$ nucleotide pairs. Since any two nucleotides are 3.4Å apart, the T4 chromosome is $3.4 \times 1.66 \times 10^5 = 5.644 \times 10^5$Å $= 56.4\,\mu$ in length.

(ii) One recombination unit $= 1.66 \times 10^5/2500 = 620$ nucleotide pairs. If two h mutants with mutations in adjacent bases are crossed, the probability of obtaining an h^+ recombinant is $1/619 \times 100 = 0.16\%$.

49. a) The seven mutations occured in two different genes (cistrons).

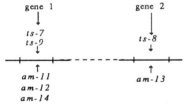

Basis for conclusion: Lack of complementation (no or few progeny) in crosses between mutants within each group and complementation (*wild-type* or near *wild-type* number of progeny) in crosses between mutants from the different allelic groups.

b) Yes. Both genes can undergo either the amber or temperature-sensitive type of mutation as indicated by the fact that crosses between certain ts and certain am mutants (the allelic ones) do not complement, e.g., no complementation in crosses between ts-8 and am-13. The amber and temperature-sensitive mutations arise by the generation of nonsense and missense codons respectively.

c) The results of the cross ts-7 x ts-9 suggest that the protein specified by gene 1 may be a dimer. The protein resulting from the union of polypeptides specified by the 2 mutant alleles would have some catalytic activity unlike those formed from polypeptides specified by the one of the other mutant allele. As a result some interallelic complementation would occur and some progeny would be produced. This is corroborated by actual data. The data do not provide any clues as to the possible number of polypeptide chains in the protein produced by gene 2.

50. Only *wild-types* ($sus_1^+\;sus_2^+$) grow (form plaques) on the non-permissive host. These arise because of recombination. The order of the markers cannot be c sus_1 sus_2; otherwise the majority of the *wild-types* would be c^+ and a minority c in the cross $c\;sus_1\;sus_2^+$ x $c^+\;sus_1^+\;sus_2$. Theoretically, the sequence of the three markers could be $sus_1\;sus_2\;c$ or $sus_1\;c\;sus_2$. Each of these sequences could produce more than 50 percent *clear* and less than 50 percent *turbid* plaques. For the first of these two sequences this could occur if c was linked to at least its closest marker sus_2. With the second sequence this could occur if c and sus_2 were closely linked and c and sus_1 were loosely linked. Since the three markers are closely linked and in three-factor crosses the unselected marker (in this case c) is usually to the left of the two selected ones or to the right of both of these markers, the order of the markers is $sus_1\;sus_2\;c$.

53. a) Percent recombination (R) = $\dfrac{\text{sum of two recombinant types}}{\text{sum of all types}} \times 100$

(i) %R for the cross $s\;mi^+$ x $s^+\;mi$ is:

$$\dfrac{65+46}{647 + 502 + 65 + 46} \times 100 = 9.5$$

(ii) The recombination values for the other 4 crosses are as follows:

Cross $s\;c^+$ x $s^+\;c = 2.8$ Cross $s\;co_1$ x $s^+\;co^+_1 = 2.8$
Cross $co_1\;mi^+$ x $co^+_1\;mi = 4.7$ Cross $c\;mi^+$ x $c^+\;mi = 6.2$

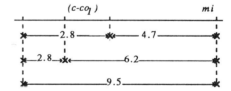

Since (i) both c and co_1 lie between s and mi and are closer to s than mi and (ii) both c and co_1 show 2.8 percent recombination with s but 6.2 and 4.7 % with mi respectively indicates that c and co_1 are closely linked. The positioning of genes with c and co_1 with respect to the outside markers cannot be determined unequivocally from these data. To determine the gene sequence unequivocally, three-gene crosses involving c and co_1 and one of the outside markers should be made.

b) The sequence is c -co_1- mi. Basis for conclusion in (a).

62. a) So as to reduce the number of crosses required to map the numerous mutations and to assign the mutations to genes affecting independent functions. Mutations in the same gene do not complement or only partially, whereas those in different genes do. Since each gene occupies a specific continuous short chromosomeal region, mutations that do not complement should be in the same short region (gene) and, therefore, closely linked. Initial mapping experiments can be done with representatives (one mutant) from each functional group or gene.

b) One would (i) make 2-factor crosses between mutants representing the various genes in all possible combinations, (ii) determine the percent recombination for each cross (2-particular genes/cross) and (iii) compare the various recombination values (iv) determine the correct gene sequence and their distances apart as accurately as can be determined from such data.

The crosses and recombination percent determinations are carried out as follows:

(i) Add equal numbers of phage of two type, e.g., *ts1* and *ts2* under permissive conditions (25°C) to a liquid culture of a bacterial host, e.g., *E. coli* B, which phage of both types can infect. The multiplicity of infection is adjusted to have approximately 10^9 phage and 10^8 bacteria present.

(ii) Allow phage multiplication to proceed until lysis occurs.

(iii) The progeny phage in the lysate are diluted to appropriate concentrations, e.g., from 10^{10}/ml to 10^3/ml, for plating on bacterial lawns so that no more than approximtely 100 plaques are formed per assay plate.

(iv) Samples of the dilute lysate are plated at 25°C to assay total progeny (parentals and recombinants). Other samples are plated at 42°C to detect and measure *wild-type* (ts^+) progeny. Percent recombination is calculated as twice the percent of *wild-type* (ts^+) progeny since the complementary recombinants which occur as frequently as ts^+ types cannot be detected.

65. **Detection of nonsense mutant in gene 23.**

This is done by infecting a nonpermissive host bacteria, e.g., *E. coli* B, with T4 phage carrying this nonsense mutant allele. Although such mutants, being *wild-type* at all other loci, will replicate their DNA and synthesize their tail and other proteins, they will not synthesize any head proteins (specifically the proteins would be incomplete) and, therefore, none would be seen in electron microscope observations of lysates of such infected bacteria. Such mutants in the nonpermissive host, therefore, fail to grow; they do not produce any progeny.

The reason why incomplete polypeptides are synthesized in nonpermissive *E. coli* strains is that the nonsense codon, e.g., UAG, cannot be translated as sense because the host does not possess a nonsense suppressor. Suppressors are mutant alleles (su^+) of genes, each of which specifies the synthesis of one of the minor types of transfer RNA (tRNA). Non-suppressor strains, like *E. coli* B, possess *wild-type* alleles (su^-) of tRNA genes which are transcribed into tRNA molecules with the *wild-type* nucleotide sequence. Such a tRNA would have normal affinity for the amino acid it carries and would recognize (pair=hydrogen bond) the codon or codons for that amino acid but would not recognize the nonsense codon. Consider a gene that specifies a tRNA that carries leucine (tRNA^{leu}), with a 3'- AAC -5' anticodon. This tRNA would pair with the leucine codon 5'- UUG -3' but not with the nonsense codon, e.g. UAG, which would be read by release factors resulting in an incomplete protein.

Maintenance of nonsense mutant in gene 23.

This is done by infecting a permissive host, e.g., *E. coli* CR63. In such hosts the nonsense mutant can synthesize the gene 23 product, the head protein and, therefore, can grow (form progeny phage) and thus maintain the mutant genotype.

The reason why the mutant can produce complete head proteins in the permissive host is that the nonsense codon is translated as sense because the host possesses a nonsense suppressor (su^+). Such a mutant allele would possess a mutation in the anticodon specifying region of the particular minor tRNA gene, e.g., the one that carries leucine. This mutation has no effect on the affinity of the tRNA molecule, e.g., tRNA^{leu}, for its amino acid leucine, and thus leucine tRNA complexes form normally. It does, however, change the anticodon of the tRNA^{leu} so that, instead of being 3'- AAC -5' and pairing with a 5'- UUG -3' leucine codon, it is 3'- AUC -5'. This mutant tRNA recognizes a 5'- UAG -3' nonsense codon and inserts leucine into the polypeptide chain in the position corresponding to the nonsense codon. Thus a complete functional polypeptide chain and protein is produced. Thus the nonsense codon is suppressed. The mutant suppressor molecules permit a mutant codon to be translated incorrectly but efficiently. Therefore, in permissive hosts all the phage components are synthesized and complete progeny phage are produced. These can infect other permissive hosts and thus perpetuate the nonsense genotype.

Chapter 13
Genotype-Environment Interactions; Expressivity, Penetrance, Phenocopies and Pleiotropism

1. Weismann's experiment demonstrated that characters are not acquired; instead the potential to express a character was inherited and transmitted from generation to generation. The experiment was significant in that it disproved the Lamarckian theory of acquired characters. The following conclusions, using modern-day genetics parlance, can be made:
   (i) Tail length in mice is genetically determined.
   (ii) The homozygous genotypes of the pure-breeding strain will express the *normal* phenotype in all generations provided the interaction between genotype and environmeent is constant from generation to generation.

3. a) The *susceptibility* to disease would be hereditary. **Reasoning:** If the condition was determined solely by environmental factors, all the rats in the group should have expressed rickets since they were all fed a vitamin D-deficient diet. Since 75 percent of the rats were *affected*, this suggests that the group is genotypically heterogeneous, i.e., it comprises rats some of which possess the potential for *susceptibility* and others that for *resistance* with the potential for *susceptibility* being expressed only in a vitamin D-deficient diet environment. However, although *susceptibility* to disease is attributed to heredity it would still be due to the interaction of a specific genotype (potential for expression of the trait) and a specific environment (a diet deficient in vitamin D).
   Note: If all the animals were genetically *resistant* they would produce the same phenotype (no rickets) with or without a vitamin D-deficient diet.

   b) Expression of the disease would be attributed to environment. However, as in the case in (a) it is still due to an interaction between genotype (G) and environment (E).
   Explanation: The line is a true-breeding *susceptible* one, therefore all the rats must be genotypically identical for *susceptibility*. That some (45%) of the individuals are *affected* and others (55%) are not is attributable to their different environments; those on a vitamin D-deficient diet are *affected* and those on vitamin D-sufficient diet *unaffected*.

7. In a large population the probability of an individual being male (p) or female (q) is 1/2. The probability of 2 fertilizations (non-identical twins) is given by (p+q) (p+q) = $(p+q)^2 = p^2 + 2pq + q^2$. Thus the frequencies of the three types of dizygotic twins will be: $p^2 = 1/4$ = both male; $q^2 = 1/4$ = both females; $2pq = 1/2$ – male and female.
   a) Concordance refers to the agreement of members of a twin pair with respect to the expression of a particular trait (i.e., individuals exhibit the same expression). Discordance refers to their disagreement with respect to its expression.

   b) (i) An identical concordance value for both MZ and DZ twins implies that the gentic basis for the character or trait in question is fairly uniform throughout the population.
   (ii) The character is probably determined by one or a few genes that have a very narrow norm of reaction (highly heritable) and, therefore, the genotype will have the same, or almost the same expression in many different environments. The differences between DZ twin members are due solely or mainly to genotypic differences between them. In this case, 55% of the DZ twin pairs have different genotypes for the character under study.
   (iii) Less than 100% concordance among MZ twins implies that the genotype in question has a wide norm of reaction; its expression is influenced by differences in the environment.

9. a) Any phenotype that is expressed always involves an interaction between a gene-produced effect (enzyme) and the environment. Occasionally, the environment can cause a change in the phenotype of mutant individuals so that they mimic a trait known to be determined by the alternate allele of the gene, i.e., the *galactosemics* (*gg*) that become *normal* upon being fed galactose-free diets are phenocopies of *normals*.

   b) Yes, since the change is not a permanent and heritable one, i.e., it is not a mutation. The environment (galactose-free diet) merely modifies the expression of the allele *g*. Each of the parents will still transmit the causative allele to their offspring who will in turn express the trait provided the expression of the allele is not modified by their environment.

11. The fact that many lethal and semi-lethal hereditary diseases can be "treated and cured" through diet manipulation or some other means is indeed gratifying. However, phenocopies are capable of transmitting the causative gene(s) to future generations, thereby adding to the genetic load of human populations. This ability of phenocopies to reproduce, transmit and add to the genetic load raises moral and genetic issues. Should such individuals refrain from perpetuating the deleterious gene(s) in the population, thereby reducing the genetic load and fulfilling their moral obligations to both future generations and society? Or should they exercise their right to procreate and leave society to cope with the consequences?
    Other medical advances which further confound the moral and genetic issues implicit in the "treatment and cure" of deleterious conditions are the techniques such as amniocentesis and chorionic villi sampling which permit the monitoring of pregnancies, thereby facilitating *in utero* detection of a condition. Should these reveal that a foetus is *afflicted*, then a therapeutic abortion is suggested. And the question of abortions in itself is not without controversy.
    The above statements are not intended to serve as an exhaustive discussion of the subject. There are many other facets to the moral and genetic issues arising from the "treatment and cure" of deleterious hereditary diseases.

13. The importance of genotype in the final expression of limb anomalies in the given sample is minimal since only 5% (6 out of 130) of reported cases reveal the phenotype without a history of intake of thalidomide. A strong involvement of the environment (intake of the drug during pregnancy) is indicated by the fact that approximately 73% of the reported cases reveal the phenotype with a history of intake of the drug. A mean coefficient of correlation (0.98) between drug intake duing pregnancy and expression of limb anomalies suggests that most of the *phocomelia* like progeny were induced by thalidomide; these individuals are phenocopies of those who possess the trait because of a mutant genotype. Their genotype remains unaffected by the drug. In cases where thalidomide was not taken by the mother, the condition is either heritable or caused by some other environmental agent.

16. a) Phenocopies usually arise in experimental organisms through manipulation of their environment which leads to the expression of a phenotype that mimics a particular genetic trait, even though the allele(s) ordinarily responsible for the trait is (are) known to be absent. The non-genetic form of the trait is said to be a

phenocopy. Clinicians may therefore attempt to obtain these so-called "mimics" for human anomalies known to have a hereditary basis, by manipulating the environments of their parents.

b) The advice given by medical practitioners to patients who are "*normal*" phenocopies should pertain to:
(i) Recurrence of the condition in the patient itself if controlled diets are ignored.
(ii) Probability of occurrence of the condition in the offspring of these "*normal*" patients. The allele(s) for the latent trait whether recessive or dominant will always be transmitted to the offspring. However, expression of the condition would depend upon the genotype of the conceived offspring which in turn would depend upon the genotype for the condition in question (dominant or recessive) of the other parent.

17. a) In order that valid conclusions be possible with regard to gene transmission and expression, it is imperative that the experimenter be able to control one or the other variable (genotype or environment) in genetic experiments. This experimental requisite is innate in genes with complete penetrance and constant expressivity since such genes express the same genotypic potential in any environment. Environmental factors can now be varied and an attempt made at studying gene properties, i.e., ascertaining the number of genes involved and their inter-relationships, the mode of transmission of these genes and their nature of biochemical expression.

b) Implicit in the study of gene action is an assessment of the roles of genotype and environment. If either component can be manipulated so as to produce a specific phenotype, undoubtedly the study of gene action will be facilitated, since an attempt can then be made at ascertaining the nature of interaction of the two components for a given set of conditions. Phenocopies are the ideal tool for such investigations in that they are environmentally induced mimics of a specific genotype, thus enabling an assessment of the nature of gene action.

19. a) No. The c^h allele has only one primary phenotypic effect. It produces a specific enzyme which transforms a colourless substance into *black* pigment in all cells in all parts of the body.
b) At the secondary and/or further levels of phenotypic expression (e.g., whether the fur is *black* or *white*) the effects of the c^h allele differ in different parts of the body.
c) The *Himalayan* genotype $c^h c^h$ produces the same enzyme in all the cells of the different parts of the body. However its activity is modified by differences in the environment (in this case temperature) in different parts of the body to produce alternate phenotypic effects (*black* or *white*), thus resulting in a variegated phenotype. Specifically, the enzyme is active at temperatures below $34°C$ and is inactive at $34°C$ and higher temperatures. Since the temperature of the body proper of maturing and adult animals is $34°C$ and that of the extremities is $27°C$, $c^h c^h$ (*himalayan*) rabbits will express a mosaic phenotype: Their bodies will be *white* and their extremities *black*. At birth, the temperature of all body parts will be $34°C$ or above, so that very young $c^h c^h$ individuals will have a *white* phenotype throughout.
d) The allele c^h cannot be termed pleiotropic since its primary biochemical effect (the enzyme) does not lead to a syndrome of effects at the secondary and tertiary levels.

22. a) Since *hemolytic jaundice* is caused by a dominant allele, one would expect individuals heterozygous for the causative allele to express the trait. However, since the allele is incompletely penetrant (10% penetrance), the chance that the first child (heterozygous) will express the trait is reduced to 10%.

b) The probability of a child being heterozygous and expressing the trait = 10% x 50% = 5%. In such a mating 50% are expected to be Aa and if percent penetrance is 10, probability that 1 of the 10 children will be affected = 10% x 50% x 10 = 50%.

24. a) The results of the first two crosses indicate that, the allele T for *normal* is dominant to t for *tremor* and is completely penetrant.
The results of the cross are indicative of the t allele being incompletely penetrant. In the latter 4 crosses there is a dearth of *tremor* types. In cross 5 all the progeny should be *tremors* if t was completely penetrant. Since some were *normal* they possess a non-penetrant tt genotype and therefore are phenocopies of $T\_$ normals. The allele t also shows variable expressivity as evidenced by the fact that the degree of *tremor* ranges from *violent* to *barely perceptible*.

b) Yes. This would involve breeding for homozygosity at all loci for positive modifiers to achieve 100% penetrance and for negative modifiers to reduce penetrance to 0%.

25. a) An allele's primary function at the biochemical level is to produce an enzyme; since the genotype of all the cells of a plant should be identical, the same type of enzyme can be expected to be produced in all flowers of the plant. The observed difference in flower colour can be attributed to the nature of interaction of the gene product with the environment. Specifically, at temperatures above $86°F$ during a critical period of floral development, the enzyme is rendered inactive; pigment is not produced and the resultant flower colour is *white*. At temperatures below $86°F$ enzyme activity and pigment production are in effect, thus resulting in *red* colour.

b) The allele is not pleiotropic in that it has only one primary biochemical function, i.e., only one enzyme is produced. The allele is fully penetrant at temperatures below $86°F$ and non-penetrant at temperatures in excess of $86°F$. However, since expression of the allele varies with temperature, it shows variable expressivity.

c) No. In heterozygotes, one allele may express its potential whereas the other does not (dominance); both may express their effects (codominance) or only one of the alleles partially expresses its potential whereas the other allele's potential is not expressed (incomplete dominance). These relationships are affected by environmental factors, including temperature. In plants which are heterozygous for the alleles at the flower colour locus, the dominance relationship depends on whether the temperature during the critical period of floral development is above or below $86°F$.

27. a) The allele dw is pleiotropic, since a syndrome of effects (traits) is present including those mentioned. The initial effect of the gene is probably isolated to the pituitary, since a pituitary gland injected from a $Dw\_$ rat into a *pituitary dwarf* induces a phenocopy of *normal*, and removal of the pituitary from a *normal* rat produces a phenocopy of *dwarf*. The effects on the thyroid, thymus and adrenal glands are secondarily related to an initial effect on the anterior lobe of the pituitary.

b) This animal should be crossed with a homozygous recessive ($dwdw$) animal. If it was $dwdw$, all the offspring would be *dwarfs*. If it was a phenocopy it should produce only *normal* if its genotype is $DwDw$; *normals* and *dwarfs* in a 1:1 ratio can be expected if its genotype is $Dwdw$.

Chapter 14
Euploidy: Haploidy and Polyploidy

1. a) Autotriploids, if they produce meiocytes which undergo meiosis, are highly sterile for the following reasons:
   (i) Meiotic products (and gametes) are unbalanced, because of irregular, independent distribution of the three homologues of each kind to the poles at AI.
   (ii) In many unisexual species autotriploidy unbalances the sex-determining mechanism. In most cases either the gonads do not differentiate or meiocytes are not formed.

   b) By asexual means of reproduction, e.g., by bulb formation and grafting in the tulip and apple respectively.

9. a) The triploid genotype was $GlGlglWs_3ws_3ws_3$.
   Both testcrosses indicate directly the kinds and proportions of gametes formed by the triploid heterozygotes.
   (i) The cross 3n x $glgl$ produced 89 Gl and 20 gl progeny. This distribution of phenotypes approximates a ratio of 5 dominant to 1 recessive and indicates that the triploid was $GlGlgl$ in which segregation of the chromosomes carrying these alleles was at random at anaphase I; n and $n+1$ gametes for alleles at this locus were formed in the ratio of 2 Gl : 1 gl : 2 $Glgl$: 1 $GlGl$. Had the triploid been $Glglgl$ its gametic ratio and the testcross phenotypic ratio would have been different. Determine this for yourself.
   (ii) The ws_3ws_3 x 3n produced Ws_3 and ws_3 progeny in a ratio approximating 1 : 2; this indicates that the triploid was $Ws_3ws_3ws_3$. The chromosomes carrying these alleles segregated at random at anaphase I such that the n and $n+1$ gametic ratio was 1 Ws_3 : 2 ws_3 : 2 Ws_3ws_3 : 1 ws_3ws_3. Since the triploid was used as the male parent in the cross, certation occurred with only the haploid (n) gametes effecting fertilization.

   b) The $GlGlws_3ws_3$ parent supplied the 2n gamete to form the triploid. A $GlGlws_3ws_3$ unreduced gamete from the diploid uniting with a $glWs_3$ gamete would produce the triploid of genotype $GlGlglWs_3ws_3ws_3$. If the origins of the gametes were reversed the triploid would have possessed the $GlglglWs_3Ws_3ws_3$ genotype and would have given results significantly different from those that Rhoades obtained.

11. a) *Gossypium hirsutum* has 13 chromosomes that are the same or sufficiently similar to the 13 chromosomes of *G. thurberi* so that they pair to form 13 bivalents at meiosis in the hybrid. It also has 13 chromosomes that show a similar relationship with the 13 chromosomes of *G. herbaceus*. Since *G. thurberi* x *G. herbaceus* hybrids are sterile and contain 26 univalents, it is logical to conclude that *G. hirsutum* is an amphidiploid derived from the two 26-chromosome species *thurberi* and *herbaceus*.

    b) Produce an experimental amphidiploid by crossing *G. thurberi* with *G. herbaceus* and treat the hybrid seed with colchicine. Compare the experimental amphidiploid with *G. hirsutum*. If it is morphologically similar to *G. hirsutum* and has 26 bivalents, and if in crosses with *G. hirsutum* it gives rise to completely fertile offspring (with 26 bivalents), then *G. hirsutum* must be an amphidiploid, derived in the same manner as the experimental amphidiploid.

13. a) The haploid chromosome numbers of the three species are: *B. campestris* n=10; *B. nigra* n=8; *B. oleracea* n=9.
    (i) The evidence that n=10 for *campestris* comes from the crosses (i) *napus* x *campestris* and (ii) *carinata* x *campestris*. Hybrids from the first cross possess 29 chromosomes and form 10 bivalents and 9 univalents at meiosis. *Napus* has a 2n chromosome number of 38 and thus would contribute n=19 chromosomes to the hybrid. The number of chromosomes that must be contributed by a gamete from *campestris* must therefore be n=10. Further evidence can be obtained from the second cross.
    (ii) Similar analysis of the other crosses will verify the haploid chromosome numbers of *B. nigra* and *B. oleracea*.

    b) *Juncea, carinata* and *napus* are allopolyploids, specifically allotetraploids.
    (i) That *juncea* is an allotetraploid is indicated by the following facts:
    Hybrids from crosses between *juncea* (2n = 36 = 18^{II}) and *nigra* (2n = AA = 16 = 8^{II}) form $8^{II} + 10^{I}$. This indicates that *juncea* possesses 8 chromosomes that are homologous to the chromosomes in the *nigra* genome (n = A = 8). It also has a genome with 10 chromosomes derived from a 2n=20 species. It is not certain that *campestris* (2n=AA=20) provided *juncea* with the genome with 10 chromosomes because the results of the cross between *juncea* and *campestris* are not given. However, assuming that *campestris, nigra* and *oleracea* are the only diploid species involved in the formation of these polyploids and since *campestris* has an n=10 chromosome number, it is highly likely that *campestris* contributed the n=10 genome to *juncea*.

    B. campestris; (AA, 2n=20) x *B. nigra* (BB, 2n=16) → F_1 hybrid (AB, 10 + 8) → doubling of chromosome number → 4n=36 (*B. juncea*); a fertile allotetraploid.

    b) That *carinata* is an allotetraploid derived from hybrids of crosses between *B. oleracea* (CC) and *B. nigra* (BB) and *napus* is an allotetraploid derived from hybrids of crosses between *B. campestris* and *B. oleracea* can be deduced by the same kind of reasoning.

    The genome constitutions are as follows:
    B. juncea - AABB (4n = 36 = 18^{II}); genome A from B. *campestris* and genome B from *nigra*.

    B. carinata - BBCC (4n = 34 = 17^{II}); genome B from *nigra* and genome C from *oleracea*.

    B. napus - AACC (4n = 38 = 19^{II}); genome A from *campestris* and genome C from *oleracea*.

18. a) Alcohol dehydrogenase (*Adh*)
    (i) For a dimeric enzyme homozygotes will produce one band and heterozygotes three bands in electrophoretic gels; therefore parent P_2 is homozygous and parent P_1 is heterozygous.
    If disomic inheritance was responsible for these results the following results could have been expected:
    1) If one allele pair (*Ff*) was involved the progeny should consist of 2 genotypes and phenotypes in equal proportions 1 *FF* (*fast*) : 1 *Ff* (*slow*) and not three.
    2) If two independently inherited allele pairs (*FfSs*) were involved with codominance at both loci, the cross *FFSS* (*fast*) x *FfSs* (*intermediate*) should have produced three progeny phenotypes and 4 progeny genotypes. *Fast* (*FFSS*), *intermediate fast* (*FFSs, FfSS*) and *intermediate* (*FfSs*) in a distribution of 24, 48 and 24 plants respectively in the three phenotypic classes. The actual results deviate significantly; the P value for this expectation is < 0.05.
    The results are characteristic of tetrasomic inheritance. Using the symbols *F* and *f* for the alleles specifying the *fast* and *slow* electrophoretic forms of the

enzyme respectively, P_2 is $FFFF$ and P_1 is $FFff$. Twenty, 16 and 60 of the progeny express the phenotypes 1 (P_1), 2(=P_2) and 3. This is insignificantly different from the expected numbers of progeny (16; 64; 16). The P-value for this difference with 2 d.f. is > 0.50.

Aconitase (Aco)
(i) Since the enzyme is a monomer homozygotes will reveal one band and heterozygotes will produce 2 or more bands depending on the mode of inheritance and number of allele pairs that they are heterozygous for. P_1 is homozygous and P_2 is heterozygous.
(ii) Since the cross P_1 x P_2 produced six phenotypic classes with 13, 14, 11, 19, 14 and 14 plants in the phenotypic classes 1, 2, 3, 4, 5 and 6 respectively, (insignificantly different from a 1:1:1:1:1:1 ratio, P-value of > 0.75), only tetrasomic inheritance can account for these results. If P_1 is $a^b a^b a^b a^b$ and P_2 is $a^a a^b a^c a^d$ the progeny phenotypes possess the genotypes 1 - $a^a a^b a^b a^b$; 2 - $a^b a^b a^b a^d$; 3 - $a^b a^b a^b a^c$; 4 - $a^b a^b a^c a^d$; 5 - $a^b a^b a^b a^a$ and 6 - $a^b a^b a^a a^c$.

b) Since the mode of inheritance for both enzymes is tetrasomic the two varieties must be tetraploids. Tetrasomic inheritance occurs only in autotetraploids; therefore both *elegans* and *intermedia* are autotetraploids. Allotetraploids produce disomic phenotypic ratios in backcrosses, testcrosses, F_2's, etc.

c) One can confirm the answer to (b) cytologically using C-, Q- and other banding pattern techniques. Somatic karyotypes of *T. ulmifolia* should reveal the presence of 4 chromosomes of each kind. Observations of meiosis would reveal the presence of multivalents (quadrivalents only or quadrivalents and trivalents and univalents) or both multivalents and bivalents at MI.
Tetrasomic inheritance for other characters (morphological, biochemical, molecular or serological) will confirm the autotetraploid nature of *T. ulmifolia*.

20. a) P (*purple*) dominant to p (*white*) and S (*spiny*) dominant to s (*smooth*).

Cross	Parents	Probable Phenotypic ratio	Parental genotypes
1	*purple* x *white*	all dominant	$PPPp$ x $pppp$
2	*purple* x *purple*	all dominant	$PPP\_$ x $PPP\_$ *
3	*purple* x *white*	5 : 1	$PPpp$ x $pppp$
4	*purple* x *white*	1 : 1	$Pppp$ x $pppp$
5	*purple* x *purple*	3 : 1	$Pppp$ x $Pppp$
6	*spiny* x *smooth*	27 : 1	$SSSs$ x $ssss$
7	*spiny* x *smooth*	11 : 3	$SSss$ x $ssss$
8	*spiny* x *smooth*	27 : 1	$SSSs$ x $ssss$

* Different specific parental genotypes can give rise to progeny with the dominant phenotype only.

b) The gene responsible for flower colour shows the best fit to chromosome segregation. The expected 5:1, 1:1 and 3:1 ratios for crosses 3, 4 and 5 are almost in perfect accord with the actual data. Confirm this yourself using χ^2 tests.

23. a) To determine whether meiosis in *Planococcus citri* is standard or inverse one should induce autotriploids and study meiosis in these polyploids which will have $2n = 3x = 15$ chromosomes.

Expectations:
(i) If meiosis is standard (first division reductional, second division equational), since there are three homologues of each kind of chromosome, trivalents or bivalents and univalents would form and co-orient on the metaphase I plate. During anaphase I, segregation for each group of homologues would be 2:1 or 1:1 (1 lost) and independent of the segregation for other groups of three homologues. As a consequence secondary meiocytes and meiotic products produced by different meiocytes would possess different chromosome numbers ranging from n through various intermediate numbers (n+1, n+2, etc.) to 2n.
(ii) If meiosis is inverse all the chromosomes will orient independently on the metaphase I plate (in the same way as during mitosis - auto-orientation); therefore chromatids and not chromosomes will segregate at AI; one chromatid of each chromosome will migrate to each AI pole; all the secondary meiocytes will possess the same chromosome number $2n = 3x = 15$ as the primary meiocyte. Co-orientation at MII and reductional division at AII will produce meiotic products and gametes with different chromosome numbers ranging from $n = x = 5$ to $2n = 2x = 10$.

b) Chandra (1962) performed the above kind of experiment and found that meiosis in the mealybug is of the inverse type - first division equational and second reductional.

29. Assume these results are typical of all others for the species.
The results obtained with M-9 can be interpreted in two ways: (i) Alfalfa is a diploid and M-9 is heterozygous (Aa) at a single locus segregating in a disomic manner. (ii) This species is an autotetraploid and M-9 is a tetrasomic simplex of genotype $Aaaa$.
Other data are necessary to determine which interpretation is correct. If one self-fertilizes the *resistant* and *susceptible* S_1 plants, the types of S_2 (second generation self-fertilized) families will indicate which explanation is correct since the expectations for disomic and tetrasomic inheritance will be significantly different as shown below:

Type of S_2 family	disomic inheritance	Expected ratio if random chromosome segregation	random chromatid segregation
Homozygous *susceptible*	1/4	1/4	225/784
Segregating 1 *resistant* : 1 *susceptible*	1/2	1/2	360/784
Segregating 21 *resistant* : 1 *susceptible*	0	0	174/784
Segregating 35 *resistant* : 1 *susceptible*	0	1/4	0
Segregating 783 *resistant* : 1 *susceptible*	0	0	24/784
Homozygous *resistant*	1/4	0	0

The S_2 families from 36 S_1 plants derived from the *resistant* clone M-9 and the goodness of fit to expectations under disomic inheritance and two tetrasomic hypotheses are as follows:

Type of S$_2$ family	observed	disomic inheritance	Expected random chromosome segregation	random chromatid segregation
Nonsegregating *resistant*	1	9	0	1.15
Segregating 21:1 to 35:1	3	0	9	7.99
Segregating 3:1	20	9	9	16.53
homozygous *susceptible*	12	9	9	10.53
Total	36	36	36	36.00

Interpretation: The distribution of the 36 S$_2$ families definitely indicates that alfalfa is an autotetraploid. The results in these 36 families conform a little more closely to those expected with random chromatid than with random chromosome segregation (χ^2 for random chromatid segregation = 4.146, p = 0.20-0.30; χ^2 for random chromosome segregation = 4.00, p = 0.10-0.20). The fact that one S$_2$ family was nonsegregating *resistant* and was apparently *AAAa* (triplex) suggests that double reduction occurs which means that at least some quadrivalents occur.

31. a) The haploid number of *N. tomentosa* is 12.
Explanation: *N. tabacum* (2n=4x) is an allotetraploid originating from the F$_1$ hybrid between *N. tomentosa* and *N. sylvestris* by a doubling of the F$_1$ chromosome number. Therefore the F$_1$ hybrid must have had 24 chromosomes. Since 12 were contributed by *N. sylvestris* the remainder (24 - 12 = 12) must have come from the gametic contribution of *N. tomentosa*.

b) Twenty four bivalents are expected:
(i) If the chromosomes in *tomentosa* are not homologous with those in *sylvestris* or
(ii) If there is only slight homology between the chromosomes in the 2 genomes or
(iii) If the homology is extensive but diploidizing genes permit only completely homologous chromosomes to pair.
If these conditions hold only in part, fewer bivalents and more multivalents and univalents can be expected.

c) *N. tabacum* is an allopolyploid since it is sterile and originated from a hybrid from the cross *N. tomentosa* x *N. sylvestris*.

32. a) The somatic chromosome number of the second parent is 2n = 6. Since one parent has 2n = 4 chromosomes its gametic contribution must be n = 2 chromosomes to the hybrid. The remaining three chromosomes must originate from the gametic contribution of the other parent, whose somatic chromosome number must therefore be 2n = 6.

b) Parents of the hybrid are most likely members of different species for the following reasons:
(i) The hybrid and its progenitors do not possess the same chromosome number; they would if they were members of the same species. The hybrid has 5 chromosomes. One parent has a somatic chromosome number of 2n = 4, the other possessed a 2n = 6 somatic chromosome number.

(ii) The chromosomes are completely unpaired at MI in cells of the hybrid indicating that they are not homologous or even partially homologous. If they were from the same or closely related species, bivalents would have been observed.

c) If the hybrid behaves like most haploids in diploid species during meiosis, in most if not all meiocytes, the chromosomes would segregate at random to the poles at AI. The chance of each chromosome migrating to a given pole is 1/2; the chance of all chromosomes migrating to the same pole and therefore the chance of the hybrid producing a meiotic product with the same chromosome number as present in its somatic cells is $(1/2)^n$, where n = the number of chromosomes in the hybrid. Therefore $(1/2)^{n=5}$ = 1/32 = 3.125 percent of the meiotic products should possess 5 chromosomes.

33. The answer requires that one determine the kind of polyploid the doubled hybrids are. Since there is no chromosome pairing in the *purple sterile* F$_1$ hybrids but the doubled hybrids are *fertile*, it is most likely that the double hybrids are allotetraploids.

P$_1$ Species A (AA)   x   Species B (BB)   →   F$_1$ hybrids (A+B);*sterile*
        white (PPww)   *white* (ppWW)   PpWw - *purple*

F$_1$ hybrids; chromosome number doubled → Allotetraploid 4n = AABB *fertile*
                                                                    PPwwppWW *purple*

Conclusion: Since the chromosomes pair autosyndetically, the four chromosomes carrying the duplicate loci from the two species will form bivalents between completely homologous chromosomes only. At anaphase I segregation from each bivalent will be normal (1:1) with the result that all gametes will be balanced and possess a complete A and a complete B genome. The doubled hybrids will therefore form only one kind of gamete *PpWw*. Upon self-fertilization all the progeny will be genotypically identical *PPppWWww* (like the parent) and *purple*.

38. a) On the basis of the results (15:1 F$_2$ ratio of *pubescent* : *glabrous*), *Galeopsis tetrahit* can be classified as either a diploid or an allopolyploid (specifically an allotetraploid). The latter situation is more likely. There are very few duplications in diploid organisms. Therefore, it is unlikely that the plant is diploid, even though the F$_2$ ratio is characteristic of dihybrid duplicate dominant epistatic interaction. Therefore *G. tetrahit* is an allotetraploid.

b) (i) Cross the two species *G. speciosa* and *G. pubescens* and produce the experimental allotetraploids by doubling the chromosome number of the interspecific F$_1$ hybrids.
(ii) Determine the chromosome numbers of the three species and the hybrids from crosses between *G. speciosa* and *G. pubescens*. The interspecific F$_1$ hybrid should possess a chromosome number which is the sum of the n number of the parental species. If *G. tetrahit* and the other two species are diploids they will possess the same chromosome number or ones that differ by one or two chromosomes.
(iii) Cross *G. tetrahit* with the experimental allotetraploid derived from crossing *G. speciosa* with *G. pubescens*. The progeny should form bivalents only at meiosis if *G. tetrahit* is an allotetraploid obtained from a doubling of the chromosome number of the interspecific hybrid of *G. speciosa* and *G. pubescens*. If *G. tetrahit* is an allotetraploid its somatic cells should possess a chromosome

number which is the sum of the somatic chromosome complements of the other two species.

(iv) Cross *G. tetrahit* with each of the other two species.

1) If *G. tetrahit* was derived from *speciosa* x *pubescens* hybrids, both bivalents and univalents should be observed during the first division of meiosis in each F_1 hybrid.

2) If *G. tetrahit* was not derived from hybrids resulting from *speciosa* x *pubescens* crosses only univalents should be observed during MI in the hybrids.

(v) Cytophotometric and other molecular approaches could also be used to verify or refute the initial decision.

Chapter 15
Aneuploidy

1. a) An additional chromosome constitutes a much larger proportion of the total complement of diploids than it does of tetraploids. The addition of a single chromosome to the complement of a diploid organism will cause a greater imbalance than if it is added to a tetraploid.

5. b) The circle configuration (fourth from the left) is not possible in a primary trisomic. Reasons: One of the three chromosome would have to have two identical ends (homologues to one of the ends of each of the other two chromosomes) to achieve this kind of pairing association, i.e., a ring of three will be formed if the extra chromosome had two a or two b ends.

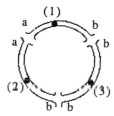

8. Insufficient information. One must know the *normal* chromosome number of the species to answer the question. If *normal* plants in the species have 2n=8 chromosomes, the plant is a monosomic. However, if the *normal* plants have 2n=6 chromosomes, it is a trisomic for a short chromosome in the genome. It is much more probable that the plant is a monosomic since meiocytes in these aneuploids regularly form univalents. If the plant were a trisomic for a short chromosome the meiocytes would have revealed a trivalent at metaphase I.

9. *Nystagmus* is determined by a recessive allele (X^n) of an X-linked gene since the parents of the *affected* female, who has an *affected* brother, are both *unaffected*; the father (AAXY) cannot carry the causative allele and her mother must be heterozygous AAX^NX^n. Thus the *affected* daughter must be missing the paternal X. Non-disjunction during the first or second division of spermatogenesis in her father would render her *affected* for *nystagmus* and an aneuploid AAX^nO.

10. a) The data from the ovarian-transplant offspring permit the refutation of the first three hypotheses.
    Reasons:
    (i) A high mutation rate of *Sf* to *sf* can be ruled out since all the daughters of the ovarian-transplant progeny were not capable of transmitting *sf* to their progeny. Had the original *scurfy* daughters whose ovaries were used in transplants been used the consequence of a mutation of *Sf* to *sf* (in the male parent), they would have been homozygous for the *sf* allele. In crosses with *normal* males all their daughters should have been heterozygous and capable of transmitting *scurfy*. Instead, only half the daughters were able to do so. Thus, mutation of *Sf* to *sf* is not plausible.
    (ii) Assuming that neo-females (sex-reversed males) are fully fertile, the ratio of males : females in the ovarian-transplant offspring should have been 2:1. Futhermore half the males should have been *normal* and all the females should have been heterozygous and capable of transmitting *scurfy*. Instead more females than males were obtained; none of the males were *normal* and only half the females were capable of transmitting *scurfy*.
    (iii) Basically a lack of *normal* males, their reduced frequency (expect 3 times as many males as females if *scurfy* females are $X^{sf}X^{sf}Y$) and the inability of all female progeny to transmit the allele for *scurfy*, can be presented as arguments against this third alternative.

    b) Monosomy ($X^{sf}O$) and spontaneous deletion ($X^{sf}X^o$) are both equally plausible. The deletion hypothesis explains all the breeding results if it is assumed that X^oY males die prenatally rather than being of the *scurfy* phenotype. According to the monosomy hypothesis *scurfy* females are $X^{sf}O$ and their non-transmitting daughters were $X^{Sf}O$. If this explanation is correct it means that the sex-determining mechanism is different from that in *Drosophila* where XO's are males but is similar to that in humans in which XO types are sterile females. The fifth hypothesis should be favoured since deletions of chromosomal segments in animals are in all likelihood lethal.

    c) A cytological analysis of the original *scurfy* female and some of its *normal* female progeny should help resolve the issue. Should the findings reveal that the X chromosome is missing, then the explanation of monosomy is favoured. However, should the analysis reveal a heteromorphic pair of sex chromosomes, then the alternative explanation is the causal one. The genetic expectations for monosomy and deletion are the same. Such studies will not permit one to determine which of the two explanations is the correct one.

11. a) The *n* gametes of *Nicotiana tabacum* will possess 24 chromosomes: 12 of the *sylvestris* genome and 12 of the *tomentosa* chromosome set. Monosomic plants in this species will form *n* and *n-1* gametes with 24 and 23 chromosomes respectively; *n-1* gametes will be lacking either a *sylvestris* or a *tomentosa* genome chromosome. The 35-chromosome hybrids received an *n-1* gamete (23 chromosomes) from *N. tabacum* and an *n* gamete (12 chromosomes) from *N. sylvestris*. If the *n-1* gamete was lacking a *N. sylvestris* chromosome, it would have contained 11 *sylvestris* and 12 *tomentosa* chromosomes. The 11 *sylvestris* chromosomes would pair with their 11 homologues from *N. sylvestris* leaving the twelfth chromosome from the *sylvestris* parent and the 12 *tomentosa* chromosomes unpaired. Therefore the 35-chromosome hybrids should have formed 11 bivalents and 13 univalents at metaphase I. However, since 12 bivalents and 11 univalents were consistently observed at metaphase I, the n-1 gamete must have contained 12 *sylvestris* chromosomes and only 11 *tomentosa* ones with the monosomic missing a T (*tomentosa*) genome chromosome. The 12 *sylvestris* chromosomes pair with their homologues from *N. sylvestris* to form 12 bivalents; the 11 *tomentosa* ones without pairing partners form the univalents.

    b) Eleven bivalents and 13 univalents, based on the reasoning above. Illustrate this for yourself.

16.

a) If crossing-over does not occur between the R and M loci - *red* males and *white* females will be the expected progeny:
(i) ♀ X^rY^R x ♂ X^rY^R → 1/4 ♀ X^rX^r (*white*); 1/4 ♂ X^rY^R (*red*); 1/4 ♂ X^rY^R (*red*); 1/4 ♂ Y^RY^R (*red*).

(ii) The unexpected progeny are 2 *white* $X^{rm}Y^{rM}$ ♂ and 2 *red* $X^{Rm}X^{rm}$ ♀

b) **Explanations:**
(i) Crossing-over between the colour (Rr) and sex-determining (Mm) loci is the most plausible explanation for the unexpected progeny.

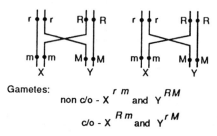

Gametes:
non c/o - X^{rm} and Y^{RM}
c/o - X^{Rm} and Y^{rM}

	X^{rM}	X^{Rm}	Y^{rM}	Y^{RM}
X^{rm}	$X^{rm}X^{rm}$ ♀ *white*	$X^{rm}X^{Rm}$* ♀ *red*	$X^{rm}Y^{rM}$* ♂ *white*	$X^{rm}Y^{RM}$ ♂ *red*
Y^{RM}	$X^{rm}Y^{RM}$ ♂ *red*	$X^{Rm}Y^{RM}$ ♂ *red*	$Y^{RM}Y^{rM}$ ♂ *red*	$Y^{RM}Y^{RM}$ ♂ *red*

*Unexpected progeny

(ii) No diagram necessary (iii) No diagram necessary
(iv) Non-disjunction - c/o between X and Y, followed by non-disjunction should give rise to exceptional progeny: *white* ♂ (X^rY^r) and *red* ♀ (X^RX-).

c) Progeny tests
♀ X^rX^r x ♂ X^rY^r → 1/2 ♀ X^rX^r (*white*); 1/2 ♂ X^rY^r (*white*)
Note: there are no *red* progeny. Males and females in approximately a 1:1 ratio.

♀ X^rX^R x ♂ X^rY^R → 1/4 ♀ X^rX^r (*white*); 1/4 ♀ X^rX^R (*red*); 1/4 ♂ X^rY^R (*red*); 1/4 ♂ X^RY^R (*red*)
Note: There are no *white* males. Because of the small population we do not get recombinants.
Ratio of males : females, 1:1. Ratio of *red* and *white* progeny, 3:1.

Explanations eliminated:
(i) Mutation - unexpected progenies although possible with mutation, not very likely because the frequency of unexpected progeny is very high (4/90).
(ii) Spontaneous - sex reversal - This is not plausible because the males and females are in a ratio of 1:1 which is not possible assuming spontaneous sex reversal.
(iii) Non-disjunction - if non-disjunction occurs in females, the unexpected male progeny will not be possible and *vice versa*.

21. Since sex is determined by a single pair of alleles located on one of the 6 pairs of chromosomes the expected sex ratio among the progeny should be 1 male : 1 female.
a) Chromosome 5 (*Reflex*) carries the sex-determining gene. **Basis for conclusion.** If a phenotypic sex ratio among the progeny of a given trisomic is significantly different from the normal diploid sex ratio of 1 male : 1 female this indicates that the chromosome present in triplicate in the particular trisomic carries the sex-determining gene. When the male parent is trisomic for chromosome 5, the phenotypic ratio (1 male : 2 female) is significantly different from the expected diploid one. Therefore chromosome 5 must be the carrier of the sex-determining gene. This conclusion is supported by the results of a Chi-square test for the progeny of the chromosome 5 (*Reflex*) trisomic. The results do not fit a 1:1 ratio (χ^2=218, P< 0.001) but do fit a 1:2 ratio (χ^2 = 0.6, P=0.96).
χ^2 for (i) *Savoy* is 0.012 (P = 0.95); (ii) *Oxtongue* is 0.046 (P = 0.85); (iii) *Star* is 16.5 (P = 0.15); (iv) *Curled* is 0.144 (P = 0.60); (v) *Wild* is 0.460 (P = 0.60).

b) If X^m and X^M represent the pair of sex determining alleles on the 2 chromosome 5 homologues, the *staminate* (male) plants in the *Reflex* (chromosome 5) trisomic line (2n + 1 =13) could theoretically have one of two chromosome constitutions and genotypes: (5^{II} + $X^mY^MY^M$) or (5^{II} + $X^mX^mY^M$). Since only n gametes function on the male side, the sex ratio among the progeny of the *Reflex* trisomic line indicates the genotype of these males. They must be 5^{II} + $X^mX^mY^M$, producing X^m and Y^M gametes in a 2:1 ratio since the progeny sex ratio is 2 female : 1 male.
i.e., $X^mX^mY^M$ x X^mX^m → 2 X^mX^m female : 1 X^mY^M male
If males in the *Reflex* trisomic line were $X^mY^MY^M$ they would produce n gametes in the ratio 1 X^m : 2 Y^M in which case the sex ratio among the progeny would have been the reverse (1 female : 2 male) of the one obtained.

c) The female parent of the *staminate* trisomics in (b) must have been trisomic ($X^mX^mX^m$) for chromosome 5. n+1 gametes are functional on the female side but such male gametes do not effect fertilization because of certation with their n counterparts.

23. **Trisomic A**

A distribution of 71 dominant : 44 recessive phenotypes was observed among the F$_2$ diploid offspring.
a) Since this distribution is significantly different from the one expected if the gene is on a chromosome in the disomic condition, the Dd allele pair must be on the chromosome present in the trisomic condition.

b) Genotype of the trisomic parent is Ddd since the F$_2$ distribution of 71 dominant : 44 recessive conforms very closely to that expected, i.e., Ddd x Ddd → 5 $D\_\_$: 4 dd.
Expect 64 dominant : 51 recessive
Obtained 71 dominant : 44 recessives
If the genotype of the trisomic was DDd one would have expected a ratio of 8 (102) dominant : 1 (13) recessive among the F$_2$ diploid progeny. This expectation is significantly different from the results obtained.

c) Since trisomic progeny were not produced among the F$_2$ progeny of the trisomic plants only (n) gametes must function on the female side (n - 100%, n+1 - 0%).

Trisomic B

a) The allele pairs Dd, Rr, Cc and Aa each exhibit 3:1 ratios among the F_2 diploid progeny; they are not located on the chromosome present in the trisomic condition. Allele pair Ll, however, gives a ratio that is significantly different from that expected with disomic inheritance; it must be on the chromosome present in triplicate.

b) Genotype of the trisomic for chromosome B parent is Dd, Rr, Cc, Aa and LLl. If the trisomic plant was Lll one would expect the following distributions of F_2 phenotypes:
Among trisomics - 7 (28) dominant : 2 (8) recessive
Among disomics - 5 (21) dominant : 1 (4) recessive

c) Since 25 and 36 of the 61 F_2 progeny are diploid and trisomic respectively the percent transmission of n gametes is 41 and that of n+1 gametes 59%.

Trisomic I

a) Reasoning similar to that shown for trisomics A and B will reveal that allele pair Rr is on chromosome I whereas Yy is not.

b) Genotype of trisomic I parent is Yy, RRr.
Evidence that the parent was RRr and not Rrr is found in the distribution of dominant and recessive phenotypes among the trisomic and diploid F_2 progeny. The actual results conform closely with those expected.

c) Since 102 and 7 of the 109 progeny are diploid and trisomic respectively the percent transmission of n gametes is 94 and that of n+1 is 6 percent.

25. Let R and S represent the duplicate genes governing reaction to stem rust races 11, 17 and 56. C.I. 12633 $RRSS$ Chinese Spring $rrss$ $R\_S\_$, R_ss and $rrS\_$ are *resistant* and $rrss$ are *susceptible*. Had the allele pairs been inherited independently, the F_2 ratios would have been 15 *resistant* : 1 *susceptible* or 93.75% *resistant* : 6.25% *susceptible*. In an F_2 totalling 1672 plants, 1567.5 are expected to be *resistant* and 104.5 *susceptible*.
Excluding the F_2 data from monosomic F_1 plants in line 13, the F_2 ratios of *resistance* : *susceptibility* in each of the other F_2's and the overall F_2 ratio (1412 : 260 or 5.43 *resistant* : 1 *susceptible*) are significantly different from the expected 15:1 ratio. This verifies the statement in the question that the duplicate genes are linked. The F_2 ratio from monosomic 13 F_1 plants is also significantly different from a 15:1 distribution.

a) The results obtained in all F_2 populations, except the one from F_1's monosomic for chromosome 13, were insignificantly different (see calculated Chi-square values) from the overall 5.43:1 F_2 ratio. Therefore all these F_2's are non-critical ones. The F_2 ratio among the plants from monosomic 13 F_1 plants was significantly different from the expected 5.43 : 1 ratio (χ^2 = 17.084, P < 0.01) indicating that chromosome 13 carries the duplicate genes for reaction to stem rust races 11, 17 and 56. Since n and n-1 gametes function in a 1:3 ratio on the female side and in a 24 : 1 ratio on the male side, disomic, monosomic and nullisomic F_2's are expected in a 24:73:3 ratio. The 131:3 distribution of *resistance* to *susceptibility* can be explained as follows: The 131 *resistant* plants consist of disomics and monosomics only which are homozygous and hemizygous respectively for the dominant alleles at the two loci. The 3 *susceptible* plants are nullisomics and therefore do not carry chromosome 13 and its genes including R and S. Had nullisomics not been produced, all F_2's from monosomic 13 would have been *resistant*.

Note: In non-critical F_2's, since R and S are not on the univalent in the monosomics but on a chromosome present in the disomic condition, these genes will be inherited in a linked manner, independent of the univalent in these aneuploids. Therefore not only will the overall F_2 ratio be 5.43 *resistant* : 1 *susceptible*, but it will also occur within the disomic, monosomic and nullisomic F_2 progenies.

b) The F_1's are heterozygous in the coupling phase. The results of all F_2 populations, except the one from F_1's monosomic for chromosome 13, are used in calculating the map distance because they represent the normal (disomic) pattern of inheritance for R and S.
1412 of the F_2's used in determining map distance were *resistant* and 260 *susceptible*. This indicates that about 84 percent of the F_2's expressed the dominant trait and about 16 percent the recessive phenotype. The homozygous recessive F_2 genotype $rrss$ is the product of the union of parental rs gametes which are formed with the same frequency on both the male and female sides. Since 16 percent of the F_2's are *susceptible*, 0.4 (the square root of 16) or 40% of the F_1 gametes produced by both male and female gametes must be rs. Since rs gametes are parental and the two F_1 parental gamete types are equally frequent, the RS ones will also be produced with a frequency of 0.40 (40%). The remaining 20 percent of the F_1 gametes are the recombinant ones (Rs and rS), each produced with a frequency of 10%. Therefore the two genes are 20 map units apart.

26. a) The basis for determining which chromosome carries a given gene is as follows:
All the non-critical F_2 progenies, those derived from monosomic F_1's whose univalents do not carry the gene under study, will reveal a 3:1 ratio of *resistant* to *susceptible* plants. The critical F_2 progeny derived from monosomic F_1s whose univalents carry the dominant allele of the gene under investigation will produce significantly different results. Specifically, all plants in the critical F_2 progeny will express the dominant phenotype (*resistance*) if nullisomics are not produced or a vast majority of plants will express the dominant trait and one or a few the recessive phenotype. The latter plants would be nullisomics derived from the rare union of $n-1$ male and female gametes because almost all (96%) male gametes are n.

b) An F_2 phenotypic ratio of 3 *resistant* : 1 *susceptible* observed in 16 of the F_2 progenies from monosmic *resistant* F_1s indicates that a single allele pair ($Pm_{16}pm_{16}$) is responsible; the allele (Pm_{16}) for *resistance* is dominant to the allele pm_{16} for *susceptibility* to powdery mildew. Only one chromosome can carry this gene. Therefore there can only be one critical line in the F_2. The F_2 progenies of monosomic lines 4A and 4B show high and very high levels of significance respectively. The significant difference in the chromosome 4B F_2 population can be attributed to chance. Moreover, the presence of 5 *susceptible* (nullisomics) plants among 45 is highly unlikely since such aneuploids are very rare among monosomic progeny (1-3%). The F_2 results for monosomic line 4A are not only significantly different from expected but there is a clear excess of *resistant* plants (39 of 40) in conformity with expectations. The single *susceptible* F_2 plant possesses a telocentric chromosome for one of the chromosome 4A arms. The most plausible explanation for its recessive phenotype is that the Pm gene is carried on the missing chromosome arm. Therefore chromosome 4A carries the gene for reaction to powdery mildew.

28. With the exception of two individuals (IV-5 and IV-7) whose genotype at the *Duffy* locus is Fy^aFy^a, all others (10) with a heteromorphic pair of number 1

chromosomes are heterozygous ($Fy^a Fy^b$) for the *Duffy* alleles; all the progeny with a pair of normal chromosomes 1's are $Fy^a Fy^a$ in all matings except the last one on the left in which case they are $Fy^a Fy^b$. Therefore the *Unl* and Fy^b alleles (and therefore the long chromosome 1 and Fy^b) segregate together in 10 of the 12 cases available for study. These results are significantly different ($\chi^2 = 14.8$, $P < 0.08$) from those expected with independent segregation and indicate that the two loci are linked. Since the *Unl* locus is on chromosome 1 the *Fy* locus must therefore also be on chromosome 1.

The matings producing these results can be diagrammed as follows:

All except last one on left — Last one on left

$$\frac{Unl \ Fy^b}{unl \ Fy^a} \times \frac{unl \ Fy^a}{unl \ Fy^a} \qquad \frac{Unl \ Fy^b}{Unl \ Fy^b} \times \frac{unl \ Fy^a}{unl \ Fy^a}$$

Since 2 (IV-5 and IV-7) of 22 progeny are recombinants, the percent recombination between *Unl* and *Fy* is 2/22 x 100 or 9%. Each of the recombinants occurred as a consequence of crossing-over between the *Duffy* and *Uncoiler* loci in their respective dihybrid parent in the third generation.

Chapter 16
Chromosome Aberrations

8. Chromosome duplications increase gene content by providing members of a species with one or more duplicates of a particular gene(s). Only one of each kind of gene is usually necessary to perform a given gene's function. One of the duplicate genes may mutate to a qualitatively different form (neomorph) which can then perform a different function from that of other alleles at the original and duplicate loci. If the new function is beneficial, the mutant allele will be selected and increased in frequency. Eventually it may replace all the other alleles at such a locus. Duplications increase the amount of genetic material and therefore provide the opportunity for the formation and addition of new kinds of genes to the genetic complement.

12. a)

   ```
          D E F
    A B C     G H I J
    ─────────────────
    a b       g h i j
   ```

   b)

Zygotes	Type	Ratio
ABCDEFGHIJ//ABCDEFGHIJ	normal	1
ABCDEFGHIJ//abghij	deletion heterozygote	2
abghij//abghij	deletion homozygote	1

   In plants, gametes with the deficient chromosome *abghij* usually will not be produced. The probable reason for this is that the gametophyte is a multicellular organism, the normal development of which requires the presence of a balanced haploid complement of chromosomes and genes. Exceptions are known, in which case it must be assumed that the genes of the deleted segment are not essential to the normal development of the gametophyte or gametes. Thus, zygotes of categories (2) and (3) will not usually be found.

   c) Deletion heterozygotes are more likely to be found in animals. This is because deletions are not usually lethal to their gametes. However, most heterozygotes will probably undergo zygotic or embryonic abortion.

13. The phenotype of individuals homozygous for the deletion of the segment carrying the dominant allele Y is the same as that of homozygous recessive (yy) individuals. This implies a complete lack of function for the y allele; y is an amorph.

14. True-breeding *normal* and *waltzer* strains are VV and vv respectively. All F_1's are expected to be Vv (*normal*). Thus the single F_1 *waltzer* ♀ is an exceptional individual.
    a) The first two explanations are eliminated.
    (i) Mutation of a dominant (V) to a recessive (v) allele in the normal parent is eliminated since such a change would result in the F_1 *waltzer* female being vv. All her progeny from matings with *normal* (VV) males would be *normal* and heterozygous (Vv). Crosses between these *normal* sibs (Vv x Vv) should produce both *normals* ($V\_$) and *waltzers* (vv) in approximately a 3 : 1 ratio. Since three females of this *normal* progeny, when crossed with two of their brothers produced 60 *normal* progeny, this indicates that the F_1 *waltzer* female could not have been vv; therefore mutation of $V \to v$ can be ruled out.
    (ii) Mutation of a suppressor of v (*waltzer*) also cannot account for this breeding data.
    Reasons: Since all F_1's would be Vv and the exceptional F_1 female expresses the recessive *waltzer* trait, regardless whether the *normal* strain was homozygous for a dominant or recessive suppressor, mutation of the suppressor would produce a *normal* not a *waltzer* F_1 female (or male).
    (1) If the *normal* strain was homozygous for a dominant allele S of a suppressor locus and the *waltzer* strain was homozygous for the recessive non-suppressor allele, mutation of S to s would yield a Vv ss F_1 with the *normal* gait.
    (2) If the *normal* strain was homozygous for the recessive allele s of the suppressor locus and the *waltzer* mice carried the non-suppressor dominant allele in the homozygous condition, mutation of s to S would produce a Vv SS F_1 which would also express the *wild-type* phenotype.
    (a) A suppressor is an allele at a separate locus which changes a mutant phenotype, determined by a non-allelic gene, to a *wild-type* one.
    (b) The *waltzer* parent of the F_1 cannot carry a suppressor since the trait is a mutant one. The *normal* parent of the F_1 could carry a suppressor (assume this is so in this case) which would be ineffective in the presence of allele V is responsible for the *normal* (*wild-type*) phenotype.

    Either of the explanations 3 or 4 can account for the unexpected female. In both cases the F_1 *waltzer* female would be hemizygous (vo) at the V locus. Non-disjunction of the chromosome carrying V in the *normal* parent would produce an n-1 gamete. Union of an n-1 (o) and an n (v) gamete from the *waltzer* male would produce a (2n-1) (vo) monosomic F_1 *waltzer* individual.
    A deletion in the *normal* parent, of a segment of the chromosome carrying V would produce an F_1 with a *normal* chromosome number containing a single heteromorphic chromosome pair. Specifically, one chromosome (with a deletion of the segment containing allele V) would be shorter that the normal homologue carrying allele v. Such an individual would be hemizygous at the V locus and express the *waltzer* phenotype.

    b) (i) The original *waltzer* female was due to a deletion of a portion of the chromosome that carried the dominant allele V. This accounts for the fact that, although all progeny of the F_1 *waltzer* female had 40 chromosomes like the *normals*, some of the *waltzer* progeny had a heteromorphic chromosome pair with one chromosome shorter than its homologue. Had the *waltzer* female been the consequence of non-disjunction she would have possessed 39 chromosomes as would her *normal* and *waltzer* progeny.

16.

```
        Long arm    short arm
                     Sh
    9 ─────────●────*──          9 ─────────●────*──
                                                  sh
    9 ─────────●────*──    ×     9 ─────────●────*──
                     sh                           sh

    Testcross progeny
                     Sh
    9 ─────────●────*──
                     Sh                     86% Sh sh (full)
    9 ─────────●────*──
                     sh                     14% sh sh (shrunken)
    9 ─────────●────*─X
                     sh
    9 ─────────●────*──          9 ─────────●────*──
                     sh                           sh

                                 a crossover chromosome
                                 of normal length
```

The recessive types appear because of crossing-over in the *Shsh* heterozygote between non-sister chromatids of the heteromorphic chromosome pair in the segment distal to the *shrunken* locus and the end of the short arm. This exchange replaces the missing segment of the *sh* homologue distal to the crossover with a normal region. It produces a chromosome of normal length with the *sh* (not the *Sh*) allele which is transmitted by the pollen like all other chromosome 9's of normal length. Since 14% of the kernels were *shrunken*, approximately 28% crossing-over occurred in the *sh*-end region of the short arm. All *shrunken* kernels are the consequence of an exchange in this region.

17. Segmental duplications, not necessarily in tandem, are the most likely cause of the differences in chromosome size and DNA content between the two *Allium* species. **Reasons:**
(1) Segmental duplications are changes that involve segments of chromosomes. They increase the DNA content by an amount less than that in a genome because each chromosome is likely to have only a few such changes and those in any one chromosome will not duplicate all the segments. Such changes should produce larger chromosomes than those in the ancestral species. Moreover, hybrids between ancestral species are expected to form observable heteromorphic bivalents, if the duplications are sufficiently large.
(2) The types of chromosomal changes that can be eliminated:
a) Deletions cannot account for these differences if *A. fistulosum* is the ancestral species because:
(i) *A. cepa* has significantly less DNA and smaller chromosomes.
(ii) With a significant decrease in DNA content (27%), many important genes would be missing. It is highly unlikely that the plants with such deletions would be viable. This would also apply if *A. fistulosum* was derived from *A. cepa*.

b) Neither inversions nor translocations can produce these results since such structural rearrangements do not normally increase or decrease the amount of DNA. Reciprocal translocations of the centric fusion and dissociation types effect a change in both DNA content and chromosome number. Moreover, when such chromosome aberrations are heterozygous they cause the formation of meiotic configurations different from those observed. Inversion heterozygotes form loops involving both homologues whereas reciprocal translocations form cross-shaped configurations and rings or chains of four.

c) Aneuploidy cannot account for the differences since it results in a different chromosome number in the derived type. Moreover, aneuploids are unstable reproductively and highly sterile, which neither of these species is.
Polyploidy (increase in the number of chromosome sets) will result in 50, 100 or 150% more DNA than in the ancestral species if the derived species is triploid, tetraploid and pentaploid respectively; the derived type will also possess a different (higher) chromosome number than the ancestral one.

18. a) Mechanisms that could have produced the double recessive F_1's:
(i) Inactivation of *D* and *Se* through position effect.
If the F_1's are actually *DSe*//*dse* and an inactivator, e.g., a heterochromatic region or a mutation occurs elsewhere on the chromosome carrying *D* and *Se*, it should be possible to separate the inactivator from *D* and *Se* by crossing over and thus restore the *wild-type* phenotype. The inactivator would cause pseudodominant expression of both *d* and *se*.
(ii) Deletion of the chromosome segment carrying *D* and *Se*.
This is the most plausible mechanism. Deletion of this small chromosome segment (0.16 m.u.) carrying these loci in *wild-type* mice would not only result in viable F_1 progeny but these animals would also express the recessive traits *dilute* and *short* ear because of pseudodominance; the F_1's would be hemizygous *d se*, allowing these alleles to express their potentials.

b) Mechanisms that could not have produced the double recessive F_1's:
(i) Simultaneous mutation of *D* to *d* and *Se* to *se*.
The occurrence of a mutation at any one locus is independent of mutation at an other. The chance of simultaneous mutations *D* to *d* and *Se* to *se* is given by the product of the chances of a mutation at each of these loci. This will be significantly lower than the occurrence of a mutation at each locus and therfore precludes this explanation.
(ii) Nondisjunction to produce a monosomic *d se*/*O*.
Highly unlikely since monosomy for an autosome is incompatible with viability. Fifteen double recessives among several hundred F_1's indicates a higher than normal rate of non-disjunction; although possible, highly unlikely.
(iii) Nondisjunction in both parents.
For the double recessive F_1's to be *dd sese* would not only require non-disjunction in both parents but specifically nondisjunction of the same (*d, se* carrying) chromosome pair during meiosis. Although this mechanism could be responsible for one or two *dilute, short* ear F_1's, it certainly could not account for the occurrence of all 15 such F_1's.

19. a) The results demonstrate that the *Vil Des* and *Akp-3* genes are present in the hemizygous state in the Sp^r/Sp^+ mouse and in double dose in the Sp^+/Sp^+ mate. Therefore the second copy of these genes is missing from chromosome 1 in the Sp^r/Sp^+ mouse. Since the mutation Sp^r involves a deletion of only the C4 band, these 3 genes must be located in the C4 band segment. Since the *Acrg* (=*A*) gene has the same copy number in both mice and therefore is present in double dose in both Sp^r/Sp^+ and Sp^+/Sp^+ mice, it must map outside the C4 band.

b) Obtain Sp^r/Sp^+ mice heterozygous for an amorphic (=null = non-leaky) allele (*a*) at the *A* (acetylcholine receptor) locus. The phase (coupling or repulsion) of the dihybrids must be known. Cross the dihybrids with $Sp^+A^+//Sp^+A^+$ individuals. Select Sp^r/Sp^+ mice and determine the percent that are recombinants. This will give the distance between the deletion and the gene (*A*). For example, assume the dihybrids are heterozygous in the coupling phase:

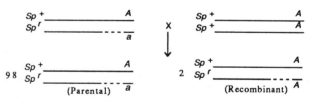

% Recombination: 2/100 x 100 = 2
A is 2 map units from the Sp^r deletion

20. a) The isochromosome is from the father. Female propositae, who do not have Xg^a antigen and therefore do not possess the *Xg* allele in some cells, have 2 X-chromosomes: a normal X (long and short arm) and an isochromosome (2 long arms). The reasoning used is the same for both families.
Since the father of the propositus is *affected*, he is expected to transmit his X or iso-X and therefore *Xg* allele to his daughter (propositus) who should be *affected*. Since she is *unaffected*, and a normal X chromosome cannot arise from an iso-X, the paternal X chromosome is not an isochromosome.
The female parent who is also *affected* could carry an iso-X with Xg^a (highly unlikely) and a normal X without the allele. She probably has 2 normal X's and is heterozygous, with the X without *Xg* being transmitted to the propositus. If

the mother has an iso-X it would not carry Xg because the daughter is *unaffected*. Since the propositus is *unaffected* with 2 X's and the one from the mother is normal, implies that the iso-X must be from the father. It is probably due to an aberration that occurred in either the father's meiocytes or gametes to produce a long-arm iso-X with no Xg.

b) The Xg locus is on the short arm. If it was on the long arm, the propositus would produce Xg^a and be *affected* since the iso-X is from the father who is *affected*.

26. All regions of chromosome 9 are represented by deletions and/or duplications in the 7 patients.
a) Patient 7 with the duplication q31→qter showed 43% increase in AK activity (50% greater than in normal (diploid) individuals). This indicates that the region is present 3 times and not twice per diploid somatic chromosome complement. Thus, the AK gene must be on chromosome 9.
Patient 7 with the q31→qter duplication showed significantly increased AK activity and patient 5 with the q11→q33 expressed normal AK activity. This indicates that the AK gene is located in the terminal band of the long (q) arm of chromosome 9. Specifically the gene is assigned to the 9q33→qter region at the distal end of its q arm.

b) Clone the gene. Use it as a probe in *in situ* hybridization and/or Southern blotting experiments involving genomic DNA of individuals with normal chromosomes and those heterozygous for deletions of different segments of the 9q33→qter region. In individuals heterozyous for deletions of the region carrying the AK gene, the hybridization signal should be less intense in the Southern blotting experiments.

28. a) In the progenies where the phenotypic ratio is 1:1:1:1 the results are due to independent segregation of the 2 allele pairs (on different chromosome pairs 2 and 3). In the remaining few progenies (ratio 1:1) the results are due to a reciprocal translocation involving the chromosomes 2 and 3.

b) Cross-shaped configuration at pachytene. A figure 8 or ring-of-four configuration at metaphase I.

c) Not necessarily: c/o occurs in females but not in males in *D. melanogaster*.
(i) If gene is close to breakage point - yes.
(ii) If gene is far away from breakage point - no. C/o would occur to produce *red, normal* as well as *red, miniature* as well as *purple, normal* translocation heterozygotes.

30. The calculated percent recombination from Sturtevant's data for each of the four regions *st-sr*, *sr-e*, *e-ro* and *ro-ca* of chromosome 3 is given below along with the expected (standard) percent recombination for each of these regions.

Percent recombination	Regions			
	st-sr	*sr-e*	*e-ro*	*ro-ca*
Sturtevant's data	9.70	0.17	0.02	0.02
Expected 18.00	8.70	20.40	9.60	

The simultaneous reduction in percent crossing-over in the four regions studied is most plausibly explained by the occurrence of a paracentric inversion in chromosome 3 in a line with a standard segmental arrangement. **Reasons:**

(i) Inversion homozygotes and heterozygotes possess a full complement of genes; they therefore are viable.
(ii) Inversions prevent crossing-over within the loop of inversion heterozygotes if the inversion is small and allow exchanges in the loop if it is large. However, in *D. melanogaster* females, chromosomes resulting from single exchanges in the loop of paracentric inversion heterozygotes are not included in eggs; those resulting from double crossovers, although few, can be included in the gametes.
(iii) Inversions reduce crossing-over outside the inversion loop. The reduction is greatest next to the inverted region and decreases as the distance from the loop increases. Since the *wild-type* females produced results consistent with the known effects of inversions in the heterozygous condition, they were heterozygous for an inversion of the appropriate segment of chromosome 3.

There were no single crossover types in the *e-ro* and *ro-ca* regions. This indicates that the inversion involved at least this chromosome segment. There was a significant reduction in percent single crossover phenotypes from 8.0 to 0.17 in the *sr-e* region, which is in juxtapostion to the *e-ro* region. This indicates that the inversion probably extended partly into this segment. Percent recombination in the *st-sr* region is still fairly high (9.7), although significantly reduced from expected (18.0 → 9.7). This indicates that the *st-sr* region is definitely outside the inversion. Only two unusual genotypes and phenotypes were obtained (the last two in the table); both of these can be attributed to double crossovers within the loop. Thus, one of the breaks producing the inversion probably occurred between *sr* and *e* and the other between *ca* and the region to its right (beyond *ca*).

35. The segmental arrangements in the Pike's Peak (PP) and Arrowhead (AR) strains differ from that in the standard (ST) line by a single inversion. AR and ST differ in that they possess a different sequence of the 5 6 7 8 9 genes; PP and ST have different sequences with respect to the genes 3 4 5 6 7. Each of these comparisons by itself does not indicate the direction of evolutionary change, e.g., AR →ST or ST → AR.
A comparison of the gene sequences in AR and PP indicates that the relationship between these two strains requires two inversions if one arragnement is to be derived directly from the other. For example, if PP is derived from AR, the first inversion in AR would involve the 9 8 7 6 5 region to produce the sequence 1 2 3 4 5 6 7 8 9 10 11 12 13 and the second one would involve the 3 4 5 6 7 segment to produce the PP gene sequence (1 2 7 6 5 4 3 8 9 10 11 12 13). If AR was derived from PP, the same two inversions would have to occur but in the reverse sequence.
The strains most closely related to each other and the ancestral one should all differ from each other by the fewest number of inversions, whereas those most distantly related by a larger number of inversions. Since AR and PP each differ from ST by a single inversion and from each other by two inversions, the relationship among the three arrangements is most likely AR-ST-PP rather than AR-PP-ST or PP-AR-ST. In this relationship, ST might have originated first and then produced AR and PP, or the initial arrangement might have been either AR or PP with the other arrangement evolving through ST as an intermediary. Thus, the three possible phylogenetic sequences are:
(i) AR←ST→PP (ii) AR→ST→PP (iii) AR←ST←PP
The decision as to the actual sequence will depend upon data from the analysis of many other arrangements.

40. If linkage between *Pl* and the breakage point was complete, all semisterile progeny would be *purple* and all fully fertile ones, *green*. Linkage is therefore incomplete. In the translocation heterozygote the normal chromosome carries the recessive allele *pl* and the translocated chromosome carries the dominant allele, *Pl*.

Progeny: 141 *semisterile, purple*   P
         137   *fully fertile, green*   P
          55   *semisterile, green*   R
          69   *fully fertile, purple*   R

Percent recombination = (124/402) x 100 ≈ 31.
Pl is approximately 31 map units from the translocation point. Therefore crossing over occurs in approximately 62% of the meiocytes in the *Pl*-breakage point segment. (See 1 in the pachytene configuration).

46. The chromosome constitutions of the two plants heterozygous for reciprocal translocations were different. Specifically, that of one plant was 1.2 1.4 3.2 3.4 and that of the other plant was 1.2 1.3 2.4 3.4 because the arms involved in the reciprocal translocations producing the two plants were different. Only if different arms of the same two nonhomologous chromosome are involved in reciprocal translocations to produce two translocation heterozygotes can a cross between the two heterozygotes produce progeny, 3/4 of which will exhibit a ring of four and one-quarter, two bivalents at metaphase I of meiosis. Had the same arms, e.g., 2 and 4, been involved in both interchanges, half the progeny of the cross between the two translocation heterozygotes would have formed a ring of four; the other half would have formed two bivalents at metaphase of the first division of meiosis.

49. b) Centric fusion is the mechanism responsible for karyotype evolution in *Mus minutoides*. **Reasons:**
(i) The basic karyotype possesses 36 acrocentric chromosomes each with a localized centromere.
(ii) Derived subspecies have fewer chromosomes than the ancestral species.
(iii) Each decrease in chromosome number is associated with a concomitant increase in the number of metacentrics.
(iv) The 2n=18 subspecies has half as many chromosomes as are present in the basic karyotype and all 18 are metacentric. These observations imply that each decrease in chromosome number has been associated with the replacement of two acrocentric chromosome pairs by a metacentric pair. In species with localized centromeres such changes can only come about by centric fusion (unequal reciprocal translocations between two nonhomologous acrocentrics with the subsequent loss of the very small metacentric). Neither deletions, inversions, duplications nor polyploidy can produce such results.

51. a) The most noticeable differences between the chromosome complement of each of the three species of apes and that of humans are with respect to relative numbers of the long acrocentric chromosomes: humans have 6, the chimpanzee 8, the gorilla 12 and the orangutan 16. This increase in the number of acrocentrics is accompanied by a concomitant decrease in the number of metacentric chromosomes; 34 in humans and the northern chimpanzee, 36 in the pigmy chimpanzee, 30 in the gorilla and 26 in the orangutan. These comparisons of the numbers of each of the kinds of chromosomes suggest that the chimpanzee and humans are closest relatives, followed by the gorilla and then the orangutan. Similar comparisons between the ape species indicate that the chromosomes of the chimpanzee and the gorilla show about the same degree of divergence from each other, and also from humans, as those of orangutan do from the remainder of the group.
If this karyotypic analysis is correct, the following evolutionary tree shows the relationships of the four primate species including humans:

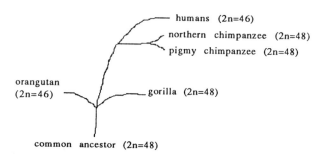

b) It is likely that at a critical stage in the evolution of humans there was a reduction in chromosome number from 2n=48 to 2n=46, initiated by an unequal reciprocal translocation (centric fusion) between two nonhomologous acrocentrics in a primitive human-ape population. This gave rise to a large and a small metacentric chromosome, with subsequent loss of the latter. The reduction in chromosome number was related to a reduction in the number of acrocentric chromosomes and an increase in the number of metacentric chromosomes in the complement. Thus individuals with four acrocentric (48 chromosomes) would have existed side by side with those with one metacentric and two acrocentrics (47 chromosomes) as well as those with two metacentrics (46 chromosomes). The derived monosomics (2n-1=47) would initially have mated with individuals with the normal human-ape karyotype (2n=48) and eventually with others like themselves as well as normals. The 2n-1=47 individuals formed n and n-1 gametes on both male and female sides. A mating of two such individuals produced normals, derived monosomics and derived nullisomics (2n-2=46) with the new (human) karyotype. Over a long period all three types co-existed in the same population; the translocated metacentric diverged genetically from the two normal acrocentrics. Eventually a barrier to reproduction was established between the human-ape population (2n=48) and primitive man (derived nullisomics, 2n-2=46). Subsequent inbreeding established a population of individuals with 46 chromosomes.

54. a) Since each parent forms haploid (n) gametes, that of the normal house mouse *Mus musculus* will contain 20 acrocentrics, one chromosome of each pair. The gamete of the normal tobacco mouse, *M. poschiavinus*, will contain 13 chromosomes of which six will be acrocentric and seven metacentric. Therefore, hybrids from crosses between the two species should possess 33 chromosomes of which 26 should be acrocentric and 7 metacentric: 20 of the acrocentrics will be from the house mouse and 6 acrocentrics plus 7 metacentrics will be from the tobacco mouse. This expectation was realized. Of the 26 acrocentrics, 6 occur in like pairs and 14 have no homologues, further confirming each parent's chromosome complement constitution.
A comparison of the chromosome complements of *M. musculus* and *M. poschiavinus* shows that since the latter species arose from the former one, 7 metacentric chromosome pairs have replaced 14 acrocentric pairs. From this comparison it appears that *M. poschiavinus* arose from *M. musculus* via a series of 7 centric fusions. In the course of each such reciprocal translocation, 2 acrocentrics were replaced by 2 metacentrics, a large one and a small one. Subsequent to the event the small genetically inert metacentric was lost. Thus, each centric fusion would not only reduce the somatic chromosome number by

one chromosome pair but 2 acrocentric pairs will be replaced by a metacentric which would contain all the euchromatin (essential genes) of the nonhomologous acrocentrics. **Reasons:**

(i) The karyotype of *M. poschiavinus* contains 7 metacentric and 6 acrocentric pairs. This is precisely what is expected if 7 centric fusions occurred beginning with a 2n=40 karyotype containing only acrocentric chromosomes.

(ii) *M. poschiavinus* has less DNA than *M. musculus;* this provides further support as the small metacentric and its DNA, produced by each centric fusion, would be lost.

(iii) The meiotic behaviour (type of chromosome pairing during diakinesis) in the hybrids reveals the presence of trivalents and 6 bivalents, one of which is the XY pair (in males). This is expected if Robertsonian fusions are responsible for the karyotypic differences. If each of the tobacco mouse metacentrics is derived from 2 nonhomologous acrocentrics in the house mouse complement, it should be partially homologous with 2 acrocentric chromosomes and these 3 chromosomes should form a trivalent. Since the tobacco mouse possesses 7 metacentrics, each derived from 2 different house mouse acrocentrics, 7 trivalents are expected and were observed, during diakinesis. Six bivalents including the XY one in males and the XX one in females are also expected. Six acrocentric pairs in the tobacco mouse would be unmodified relative to their homologues in the house mouse and those acrocentrics would pair with their complete homologues from the *M. musculus* complement to form bivalents.

b) Establish linkage groups in both species.

(i) In each species, the number of such associations should be the same as the haploid number of chromosomes. Specifically, there should be 20 linkage groups in the house mouse and 13 in the tobacco mouse.

(ii) If the centric fusion hypothesis is correct, 14 of the *M. musculus* linkage groups should no longer be present in *M. poschiavinus*. They should be replaced by 7 new linkage groups, each possessing the genes of 2 specific separate linkage groups in the house mouse. If linkage groups 1 and 2 are carried by acrocentrics 1 and 2 in *M. musculus*, the new linkage group should possess all the essential genes of the 2 previously independently inherited linkage associations because the metacentric is derived from acrocentrics 1 and 2 by an unequal reciprocal interchange.

Chapter 17
Chemistry, Structure and Replication of Genetic Material and Chromosomes

3. a) The main protein components of eukaryotic chromosomes are histones which in most species are of 5 types: H1, H2A, H2B, H3 and H4, coded for by different structural genes. Histones play a major role in maintaining chromosome structure. They also function in gene regulation. In this respect they are non-specific inhibitors of gene action.

b) Nucleosomes are the main subcomponents of the fundamental repeating structural unit of the basic chromosome (= chromatin) fibre known as the nucleofilament which is ≈100Å in diameter. The other component of these units is spacer (linker) DNA. A complex of 8 histone molecules and, depending on the species, ≈150-240 nucleotide pairs of the DNA duplex form a basic repeating unit. The nucleosome is composed of a core of 8 histone molecules, 2 each of H2A, H2B, H3 and H4, around which ≈145 nucleotide pairs of the DNA coil almost 2 complete turns. The remainder of the DNA duplex, anywhere from ≈5 to 100 nucleotide pairs long forms the spacer (linker) component of the repeat unit. In the nucleofilament the linker DNA is usually not associated with any other chemical components. This segment of the DNA duplex extends from one end of the chromosome to the other and becomes important in forming the solenoid (250-300Å diameter) - the secondary structure of chromosomes.

c) Since the H1 histone is not a component of nucleosomes and is not associated with linker DNA in nucleofilaments, it is not necessary for the structure of either. H1 however appears to be the primary, if not sole determiner of the second level of coiling, changing the nucleofilament (in which the overall length of the DNA duplex is condensed by a factor of ≈7) into a solenoid structure. Although the details are still unknown, H1 histone associates with linker DNA to effect the second level of coiling which arranges, on the average, 6 nucleosomes per turn in a helix of this diameter. This further level of coiling increases the total condensation of the fully extended DNA duplex by a factor of about 7 x 6 = 42.

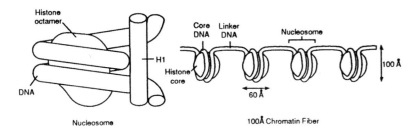

4. There are 2 main reasons why DNA in eukaryotes should be double-stranded:
(i) With few exceptions, all somatic cells in eukaryotes possess identical chromosome complements, genotypes and DNA base-pair contents and sequences. This can only be achieved if all chromosomes possess double-stranded and not single-stranded DNA molecules.

(ii) The base sequences in the complementary chains of each DNA duplex do not carry the same information, e.g., the codons AAA and TTT (=UUU) code for different amino acids.
If the DNA in eukaryotic chromosomes was single-stranded and the double helix existed only during replication, following mitosis the 2 daughter cells and their chromosome complements would contain different (alternative) sets of genetic information; this could be responsible for different, alternative functions.

5. A total of 9 errors are present.
(i) The leading and lagging strands have been incorrectly placed and identified. The leading strand is synthesized continuously along the top template strand 5'→3' not 3'→5'. The lagging strand is formed discontinuously along the bottom template strand, also in the 5'→3' direction.
(ii) Although the RNA primers are correctly placed at the 5' ends of the leading strand and lagging fragments, their synthesis is catalyzed by a RNA polymerase called RNA primase not by DNA polymerase I.
(iii) Unwinding proteins (helicases, rep proteins) not SSBPs denature the helix at the replication fork; SSBPs not unwinding proteins stabilize the denatured helix keeping the complementary strands separated.
(iv) DNA gyrase plays no role near or at the replicating fork. It relieves the tension created by unwinding farther down the duplex by causing breaks in both strands of the DNA duplex.
(v) The function of DNA polymerase III is also misplaced. This enzyme is involved in elongation, adding nucleotides complementary to those in the template strands, one at a time, to the 5' ends of primers and existing chains.
(vi) DNA ligase does not perform the function indicated (adding DNA nucleotides to RNA primers). Its function is to join Okazaki fragments and to seal other discontinuities between adjacent segments of the duplex.
The removal of RNA primers and their replacement by DNA nucleotides is carried out by DNA polymerase I.
Redraw the illustration for yourself and indicate the errors you have corrected.

6. a) It is from RNA because it contains uracil.

b) With few exceptions, no. **Reason:** except for some identical genes in the two species which would possess identical base pair contents and sequences and whose complementary single strands would hybridize, cattle and rats possess different genes with different base pair sequences; hybridization would not occur between single strands from the two sources. Identical AT/GC ratios do not imply identical base pair sequences.

13. It is unlikely that the Amphiuma contains 20 and 70 times as much DNA as the toads and birds respectively. A more likely explanation is that the amount of non-coding DNA (highly repetitive, middle repetitive) is increased to account for most of the differences in DNA content.

17. A difference in ploidy would account for the differences in the DNA contents of the 4 strains of yeast. Strain 1 could be haploid, strains 2 and 3 diploid and triploid respectively, while strain 4 is tetraploid.
One could test the above suggestion by crossing the various strains and determining the amount of DNA per cell in the resulting progeny. For example, if strain 2 was diploid and strain 4 was tetraploid, a hybrid should be triploid, and have an amount of DNA equivalent to that of strain 3 approximately 6.8 ± 0.54 pgms per cell. One could also determine the number of chromosomes in each strain as well as conduct genetic studies.
An alternative suggestion to ploidy would be polyteny (multistrandedness), where there could be a variation in the degree of strandedness of the

chromosomes. An increase in the number of strands per chromosome would result in an increaed amount of DNA per cell.

18. a) The information provided is insufficient to determine whether the DNA is double- or single-stranded. The other information necessary is the ratio of purines (A+G) to pyrimidines (C+T).

b) Length of the double helix = 52μm= 52 x 10^{-6} m
Distance between two base pairs = 3.4Å= 3.4 x 10^{-10} m
Number of nucleotide pairs = 5.2 x 10^{-5}/3.4x10^{-10}=152942
Therefore molecular weight of the molecule = 152942 x 660
$\approx 101 \times 10^6$

19. a) and b)

Cell type	Amount	N-value	C-value	Number of chromosomes	Number of chromatids
1st polar body	127±3	1	2	5	10
sperm	61±1	1	1	5	5
male pronucleus	133±7	1	2	5	10
prophase of first cleavage	263±10	2	4	10	20
telophase of first cleavage	124±3	2	2	10	10

20.

Species	Nucleic Acid	Strandedness	Reasons
(1)	DNA	double	because of absence of uracil; because A=T and C=G
(2)	RNA	double	uracil present; because A=U, G=C
(3)	DNA	double	same as species 1

(4) Insufficient data to conclude whether DNA or RNA. Single stranded because purines more common than pyrimidines. In double-stranded nucleic acids purines=pyrimidines.

| (5) | DNA | single | presence of T; because base ratios are unequal |
| (6) | RNA | single | uracil present; -purines not = to pyrimidines |

(7) Insufficient data to conclude whether DNA or RNA. The ratio A+G/C+T(U) is equal to 1.00. Since A+T(U)/G+C is 1.26 indicating A=T(U) and C=G the nucleic acid is double-stranded.
(8) Insufficient data to conclude whether DNA or RNA. Since A=T=G=C (and therefore purines equals pyrimidines) it is most likely double-stranded. Since base and base-pair sequences do not follow one another in a regular repeating sequence, e.g., any sequence can occur for any segment of single- or double-stranded DNA or RNA, it is highly unlikely that a single-stranded nucleic acid would possess the bases in equal proportions.

21. NA causes bidirectional (AT ↔ GC) single base pair mutations of the transition type.
Barring contamination, every clone (or plaque) arises from a single phage. The contrasting results for phage A and B can be explained by the assumption that phage B contains only one copy of the genetic information while phage A contains 2 copies. More than 2 copies can be ruled out since viruses are either single-stranded or double-stranded. The number of copies of genetic information can be directly related to the number of strands in the DNA molecule of a virus. In the single-stranded phage B, mutation of one base to another on the DNA e.g. C→U would result in the production of mutant DNA strands in all the subsequent generations. Therefore the plaque would contain *mutant* phages only. If the phage is double-stranded like A, a change of the base(s) in one strand would leave the other strand unaltered, (*wild-type*). Thus half of the progenies would be *normal* (*wild-type*) and the other half would be *mutants* since both strands replicate independently.

22. DNA can act as a template for replication only when it is single-stranded. The findings showed that calf thymus polymerase alone does not activate native DNA (double-stranded) in replication. After being heated to 99°C, DNA molecules would denature, i.e., they open up to form two primer strands, then calf thymus polymerase can activate DNA replication by linking the nucleotides together (by forming a bond between the 3 position of one carbon and the 5 position of the next).

24. a) Both vaccinia and M13 viruses possess DNA rather than RNA since their nucleic acids contain thymine rather than uracil. The differences in their base compositions are due to the fact that vaccinia DNA is double-stranded (amount of A=T and the quantity of G=C) whereas the DNA of M13 is single-stranded.

b) The explanation can be tested in different ways:
(i) One of the best ways is to label, denature, renature (reanneal) and then centrifuge the DNA of each virus as follows:
(a) Label one sample of the virus with heavy nitrogen (N^{15}) and another with light nitrogen (N^{14}).
(b) Extract the DNA from each of the two labelled samples.
(c) Next, mix the DNA from the two sources (in aqueous solution) and then heat the mixture to approximately 100°C.
(d) Cool the solution slowly after heating.
(e) Centrifuge the DNA in a CsCl solution to determine the number and position of DNA bands that are likely to be formed.

Expectations:
(i) Vaccinia should form 3 bands ($N^{15}N^{15}$, $N^{15}N^{14}$, $N^{14}N^{14}$) if its DNA is double-stranded since the complementary single-strands of heavy and light DNA will hybridize at random with each other.
(ii) M13 should form only 2 bands if it is single-stranded. Those from the labelled sample will be heavy (N^{15}) and form one band and those from the unlabelled sample will be light (N^{14}) and form the other band. These 2 bands should be at positions corresponding to the heavy and light bands formed by double-stranded vaccinia DNA.

27. a) Mutation is not a probable cause of the genetic change (RII → SIII) because RII cells would back-mutate to SII type only. A second mutation in the same cell would be required in order to change the cell to SIII type. Since the rate of one mutation is about 1 in 10 million, the likelihood of a double mutation would be about 1 in 100 million. This mutation rate is much lower than the observed frequency of true-breeding SIII cells. If a mutator gene is present it could increase the mutation rate significantly but only in one direction, e.g., RII→SII. Therefore a mutator gene is unlikely to be responsible for the genetic change (RII→SIII).
(ii) There are 4 known ways by which a bacterial cell can acquire new characters: mutation, transduction, conjugation and transformation. Mutation has already been eliminated by the reasons given in part (i). Transduction and

conjugation can also be eliminated because the former requires a mediator (phage) and a living host (SIII cells were heat-killed); the latter also requires living SIII and RII cells. Therefore transformation is responsible for the gentic change. Transformation requires free DNA material from SIII and competent RII cells - two conditions that were probably satisfied in the experiment.

b) DNA is the component responsible for the hereditary changes because all the other components of the cell were demonstrated to have little or no effect on the transforming activity; but if DNA alone was degraded (by DNase), the transforming activity was impaired. Therefore DNA is the substance that carries the genetic information from the donor (SIII) cells to the recipient (RII) cells.

When cells are heat-killed, their chromosomes (DNA) are undamaged and free chromosomal material liberated somehow from the heat-killed cells, can pass through the cell wall of the dead cells and subsequently be taken up by living recipient cells. Then part or all of the genetic material that is taken up would be integrated into the recipient's genome by replacing the corresponding part. Illustrate this for yourself.

30. The results support the semi-conservative mechanism of replication. Before *E. coli* are transferred to N^{14} medium, all the DNA should be labelled with N^{15}; consequently only one band (heavy) should be observed. After the first replication, in N^{14} medium, the density of DNA was found to be intermediate between N^{14} and N^{15}. One heavy (N^{15}) and one light (N^{14}) band are expected after the first replication if the mechanism of replication is conservative. The results after the second round of replication in N^{14} (one light and one heavy band of equal thickness) definitely rule out dispersive replication. Only one band, intermediate between N^{14} and N^{15}, is expected if dispersive replication prevails.

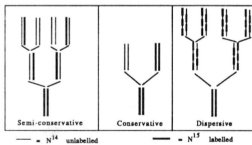

b) The results indicated that the intermediate band must have contained hybrid duplex DNA consisting of one heavy and one light chain. These results further confirmed that the dispersive mechnism can be ruled out. They support the semi-conservative mechanism but they cannot rule out the conservative type if *E. coli* chromosomes consist of two or more duplexes in parallel instead of one duplex in each chromosome.

c) It is necessary to show that the *E. coli* chromosome consists of one duplex only in order to prove that the semi-conservative mechanism is the correct method of replication. One can prove this by determining the length/mass ratio for DNA.

32. a) Chromosome is uninemic, i.e., one DNA duplex per chromosome before replication and two DNA duplexes per chromosome after replication.

The results suggest that DNA replicates semi-conservatively so that after one round of replication in the presence of tritium all chromosomes should be labelled; specifically both chromatids of a chromosome after replication should be labelled. After a 10-hour colchicine treatment the cells should contain the diploid number of 12 chromosomes, each being completely (both chromatids) and uniformly labelled. The entire situation is illustrated for 1 pair of homologues.

b) Not necessarily, as shown below. The distribution could be semiconservative and the replication conservative.

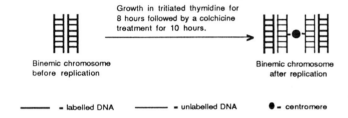

c) The few cells that have 48 chromosomes after 34 hours of colchicine treatment would be such that half of the chromosomes in the complement would be completely unlabelled and the other half would be labelled, specifically with one chromatid labelled and the other unlabelled.

33. a) The DNA of *D. pneumoniae* is double-stranded, which upon heating, denatures into its component single strands. Slow cooling permits renaturation into a double-stranded molecule. Subsequent centrifugation results in a specific band in the cesium chloride density gradient depending upon the density of DNA. Heating a mixture of heavy (N^{15}) and light (N^{14}) double-stranded DNA will result in denaturation of each of the 2 types of DNA molecules into their specific (N^{15} or N^{14}) single strands. Subsequent cooling will result in renaturation such that complementary strands of the original DNA molecules pair so as to restore the 2 types (N^{15} or N^{14}) of DNA. In addition, a heavy single strand can pair with a complementary light single strand thereby giving rise to a hybrid DNA molecule (N^{15}-N^{14}). Centrifugation of the cooled mixture will result in 3 bands in the density gradient - heavy, light and intermediate.

b) The two species probably have DNA which is sufficiently different so that when the separate strands are brought together upon cooling, strands of one species will not pair with strands of the other. Thus only N^{15} and N^{14} types of double-stranded DNA molecules will be present upon cooling and only 2 bands will be evident in the density gradient.

c) From the observations, it is evident that pairing of the strands will occur only if the sequences of nucleotides are complementary. Two species can have the same base ratio, but the base-pair sequences of their DNA molecules will be different.

36. a) Histones are the primary chemical components of chromosomes other than DNA, and the 2 together are responsible for chromosome structure. Once histones associate covalently with DNA they retain this stable relationship throughout most of the cell cycle; that is, histones are not rapidly metabolized like most other proteins. DNA replicates only during S and each daughter chromatid and DNA duplex has the same histone content (quantitatively and qualitatively) as the parent chromosome and DNA duplex. Therefore, it is essential that the synthesis of histones be synchronized with DNA replication so that each of the chromatids of a chromosome is identical to the parent chromosome with respect to DNA and histones. Moreover, since histones do not perform other functions and their synthesis during S phase fulfills their cellular role, their synthesis at other times during the cell cycle is suppressed.

b) We would expect the genes coding for the 5 types of histones to be present in multiple copies per genome. **Reasons:** Each chromosome likely possesses many (1000s) histones, all of which have to be synthesized during the S phase of the cell cycle. There would be insufficient time to produce all these proteins if the genes for each of these proteins were present once per genome.

c) This fact indicates that, with few exceptions, each segment of each histone type performs an important role in maintaining chromosome structure and repressing gene action. Therefore mutations resulting in altered amino acid sequences cannot be tolerated because they would either abolish or drastically alter the primary function of the protein.

d) The electronmicrographs show that one of the daughter DNA duplexes at the replication fork is beaded (contains histones), whereas the other is naked. Therefore old and newly synthesized histones are distributed conservatively during replication. Had their distribution been semiconservative histones would have been observed along both daughter duplexes. The electronmicrographs indicate that histones do not dissociate from DNA during replication. Old histones stay with the daughter duplex containing the leading strand and new histones assemble on the daughter DNA duplex containing the lagging strand. Therefore, the daughter strand of the replicating duplex that is deficient in histones is the lagging one. The most likely reason for the difference is that histones bind much more strongly to double-stranded than to single-stranded DNA. Old histones probably do not follow the duplex formed along the lagging strand because this strand contains single-stranded regions prior to the ligation of its Okazaki fragments.

38. a) When cellular or nuclear DNA is fragmented and centrifuged in a cesium chloride gradient all the fragments collect in a minor band or bands referred to as satellite DNA. All the DNA fragments in a particular band have an identical or very similar base pair content (%GC). If satellite DNA is nuclear in origin it usually consists of highly repeated short base-pair sequences, arranged in tandem in one or more regions of the different chromosomes of the genome. Within a species more than one kind of short base-pair sequence may be present. Satellite DNA is usually present in constitutive heterochromatic regions of a chromosome. Regardless whether satellite DNA is nuclear or cytoplasmic (mitochondrial or chloroplast) in origin, it may be heavier or lighter (contain a higher or lower %GC) than main band DNA derived from the majority of the regions of all chromosomes in the nuclear genome. The density of DNA is proportional to its GC content. The higher the %GC, the higher the buoyant density. Satellite DNA is almost always present in constitutive heterochromatin which in turn is often located on either or both sides of centromeres and in secondary constrictions. However, it can also occupy small or large segments of chromosome arms, and complete chromosome arms, particularly in many plant species.
Note: satellite DNA and chromosomal satellites are not synonymous although the latter may possess satellite DNA.

b) The presence of satellite DNA in somatic cells is determined by isolating the DNA from all of its organelles, fragmenting this DNA and centrifuging it in a cesium chloride gradient and finally determining (visualizing) the size and position of the bands by UV absorbance at 260 nm.

c) (i) The buoyant density of 1.692 represents main band DNA since it represents the largest amount of DNA fragments from the majority of the nuclear chromosomal regions. The 1.701 and 1.707 buoyant densities represent satellite bands with a higher %GC content than the main band DNA. The DNA in the fragments in these bands is derived from a minority of all the DNA sequences in the cells. Since whole cell DNA was fragmented and centrifuged it is highly likely that one of the two satellite bands contains DNA from the chloroplasts and the other contains DNA from mitochondria. This can be confirmed or refuted by isolating these organelles, extracting, fragmenting and centrifuging their DNA and determining its buoyant density. If it is the same as that in one of the minor bands, the source of the minor band has been identified. That one of the minor bands represents nuclear chromosomal satellite DNA cannot be excluded from the information provided.
(ii) %GC in DNA bands with buoyant densities of 1.692, 1.701 and 1.707 is ≈31, 40 and 46 respectively.

d) The essential features of *in situ* hybridization are:
(i) Standard cytological preparations are treated with HCl and ribonucleases to remove histones and RNA.
(ii) Denaturation of the DNA duplexes *in situ* using heat or alkali.
(iii) Exposure of the denatured DNA to radioactively labelled single-stranded DNA or RNA fragments from satellite DNA.
(iv) Allow the mixture to cool slowly so that duplex formation can occur. The satellite DNA or RNA fragments will hybridize (form DNA-DNA or DNA-RNA duplexes) with chromosomal regions which possess such DNA. After a suitable period the slides are washed so that only hybridized radioactive DNA or RNA remains associated with the cells. When cRNA is used, the slide is further exposed to ribonucleases to digest all RNAs that did not hybridize.
(v) Finally the slide is subjected to autoradiography and stained, e.g., with Giemsa. After a suitable period of time the slides are scrutinized visually to determine the locations of hybridization and therefore regions of chromosomes that contain highly repetitive base-pair sequences.

e) There are three satellite bands with buoyant densities of 1.692, 1.688 and 1.671. This indicates that the *D. melanogaster* genome contains at least 3 kinds of highly repetitive sequences which differ not only in base-pair (%GC) content but probably in length, extent of multiplicity and chromosomal location.

f) No. *E. coli* does not possess specific chromosomal regions whose %GC is significantly different from that of other chromosomal segments. For example when *E. coli* DNA is fragmented into segments ≈10^4 base-pairs long the buoyant density of these fragments is quite uniform and corresponds to a GC content of ≈51%.

40. a) The evidence presented here indicates that the *pol* alleles are in the structural gene for DNA polymerase I. The strongest support for this statement is provided by the analysis of the pure form of the enzyme in the *mutants* and the *wild-type* strain. The enzyme from *pol*-6 and *wild-type* strains show different temperature sensitivity with respect to enzyme activity. This indicates that the two forms of the enzyme are structurally different. If the mutation occurred in a regulatory gene that produced normal *pol* 1 but in a reduced quantity, the pure enzyme should show an identical sensitivity to temperature.

b) It cannot be unequivocally resolved from these data whether polymerase I is involved in DNA replication. *Pol* 1 may only be required in minute quantities and the amber mutation may be leaky. Moreover, if only a few molecules per cell are necessary and these are incorporated into a larger enzyme complex, their activity may not be detectable *in vitro*. However, since the *pol*-A1 mutants can replicate DNA rapidly in the presence of 1% of normal level of polymerase I, it is highly likely that there is at least one other enzyme in *E. coli* cells that is responsible for replicating DNA *in vivo*. The data of Gefter and Okazaki imply that one of these other enzymes could be DNA polymerase II.

c) Since *pol*-A1 mutants are sensitive to UV light (possess a defective repair mechanism), unlike *wild-type E. coli* and excise dimers, indicates that DNA polymerase I is involved in DNA repair. The data provide no clue as to when the enzyme would repair DNA defects. However, bacterial cells possess a multi-step dark repair system for repairing UV-induced DNA damage which involves (i) endonucleolytic nicking, (ii) excision of damaged lesion (iii) resynthesis of removed nucleotide sequence and (iv) sealing the gaps in the polynucleotide chains. It is highly likely that this enzyme would be involved in catalyzing steps (ii) or (iii) or both.

41. a)(i) The genetic proof would be demonstrating that all genes show a typical completely X-linked inheritance pattern and only one of the two alleles at each locus in all males in all generations is functional and is always maternally derived.
For discussion purposes let F (dominant, *fast*-growing) vs. f (recessive, *slow*-growing) represent all allele pairs. Expect the following results in reciprocal crosses between true-breeding *fast* and *slow* growing lines if the heterochromatic chromosome set is paternal in origin and genetically inert:

Cross 1 $♀E^FE^F$ (*fast*) x $♂ E^fH^f$ (*slow*)→F_1 $♀E^FE^f$ (*fast*); $♂E^FH^f$ (*fast*)→ F_2 1/2♀ (1/4 E^FE^F, 1/4 E^FE^f); 1/2♂ (1/4 E^FH^F, 1/4 E^FH^f)
3/4 *fast*-growing : 1/4 *slow*-growing (all males)

Cross 2 $♀E^fE^f$ (*slow*) x $♂ E^FH^F$ (*fast*)→F_1 $♀E^FE^f$ (*fast*); $♂E^fH^F$ (*slow*)→ F_2 1/2♀ (1/4 E^FE^f, E^fE^f); 1/2♂ (1/4 E^FH^f, E^fH^f)
Within both sexes - 1/2 *fast*- and 1/2 *slow*-growing.

E-euchromatic chromosome set, functional; H-heterochromatic set, genetically inert.
(ii) By completely dispensing with the heterochromatic genome as has happened in the Iceryine species of insects and in the hymenoptera (bees, wasps, ants, etc.).

b) The chromosome aberrations in the male progeny in both experiments occur in one chromosome set only (E or H). This indicates that all the chromosomes within a set (E or H) are derived from one parent only.

(i) The results of experiments 1 and 2 indicate that the H genome in males is paternal in origin (and that this chromosome set in males is the E genome in somatic cells in their paternal parent). Had the H set in males been female in origin then in experiment 1 all chromosome aberrations would have occurred in the H set and such breaks would have been non-existent in male offspring in experiment 2.

(ii) Irradiation of females (2 E genomes) (experiment 1) affects progeny of both sexes more or less to the same extent. With an increase of radiation dosage the number of survivors in both sexes decreases significantly, indicating that the E set of chromosomes is genetically active. Irradiation of males (experiment 2) does not affect the viability and number of female offspring. This indicates that since males receive the genome that becomes heterochromatic from their father, the H set is genetically inert. If it was genetically active, both sexes should have been affected and more or less to the same extent.

(2) H is not a permanent characteristic of either a chromosome segment, an entire chromosome or a genome in insects with the lecanoid system. This chromosome set behaves like the X chromosome in mammalian species and therefore possesses facultative not constitutive heterchromatin. Some of the evidence for this conclusion is as follows:
(a) Males transmit only their E genome which becomes heterochromatic in males (evidence discussed above).
(b) Genetic experiments demonstrated that although males are diploid, they nevertheless are functionally haploid and express only the alleles at all loci that were derived from their mothers. These same alleles and genome are genetically inactive among the male offspring but active in female progeny.

42. Diagram of a normal somatic cell in the *Drosophila* species with 12 chromosomes (6 pairs).

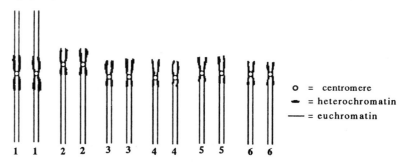

a) One chromosome pair must be either metacentric or submetacentric with constitutive heterochromatin located next to the centromere in both arms; the rest of the arms possess euchromatin. This chromosome pair can account for 2 arms in the configuration of the polytene chromosomes.

b) One arm of each of the homologues of each of the other 5 chromosome pairs must possess constitutive heterochromatin located next to its centromere; the rest of the chromosome arm must be euchromatic. The other arm of each homologue of each of these 5 chromosome pairs has to possess constitutive heterochromatin only. In theory these arms could be long or short. With rare exceptions when complete chromosome arms are heterochromatic, they are usually short; therefore such chromosomes are acrocentric as shown. An alternative is that 1, 2, 3, 4 or all of them are telocentric. The latter chromosome types are uncommon in nature. Each of these 5 pairs of homologous chromosomes accounts for 1 of the 5 remaining arms in the configuration of polytene chromosomes.

c) The reason why 2n=12 chromosomes in somatic cells form 7 arms in the polytene configurations of the cells of the salivary glands is (i) the chromosomes are of the morphologies indicated above and (ii) the heterochromatin of all of the chromosome pairs fuses to form the chromocenter.

43. A sperm is haploid containing a single chromosome set (=genome) with most of the DNA sequences therein being present only once. This implies that each chromosome before replication and each chromatid thereafter must possess only one continuous DNA duplex (or 2 or more in tandem joined by linkers, e.g., proteins for which there is no evidence). The presence of a sizeable fraction of unique sequences per genome precludes a polynemic chromosome structure.

45. a) No. The size of a DNA molecule does not depend on whether the chromosome is metacentric, acrocentric or any other type. Evidence - the largest chromosomes in *wild-type D. melanogaster* are metacentric and in the inversion homozygote strain of this species they are acrocentric; yet the molecular weights (MWs) (and therefore lengths) of the extracted DNA molecules from both are of the same length. Therefore the size of the largest (longest) DNA molecule depends on chromosome size and not on chromosome morphology.
Note: The MWs of the longest DNA molecules extracted using the viscoelastic technique and the MWs of the DNA in the chromatids of the largest chromosomes at metaphase are approximately the same. This indicates that the number of large DNA molecules per chromosome before replication and per chromatid must be small, 1 or 2.

b) The evidence is almost unequivocal. The DNA in each chromosome before replication and in each chromatid thereafter is highly likely to be continuous from one end (telomere) of the chromosome and chromatid through their centromere to the other. The possibility that each chromosome before replication and each chromatid possesses 2 DNA molecules in tandem, with one in each chromosome arm is excluded by the data presented for *D. melanogaster*:
(i) The same results (MWs and DNA contents) were obtained with DNA from both the *wild-type* and inversion strains. The inversion in the longest chromosome changed the arm ratio from 1:1 in the *wild-type* strain to ≈7:1 in the inversion strain without changing the total DNA content or size of the longest DNA molecule extracted from the chromosome. If each arm contained one DNA molecule, and therefore the DNA was discontinuous at the centromere, the inverted chromosome would contain DNA molecules ≈75% larger than those of the *wild-type* strain.
(ii) The MW of the longest extracted DNA molecule from the translocation strain, in which chromosome 3 was increased in length by ≈37%, was about 40% larger than the longest DNA molecule in the *wild-type* strain.

c) All the data, indicate that eukaryotic chromosomes are uninemic in structure. That is, each chromosome before replication and each chromatid in each chromosome contains only one continuous linear and unbranched DNA molecule (DNA duplex). Although the data do not provide absolute proof that there are no non-DNA linkers, e.g., proteins in chromosomes holding DNA duplexes in tandem, it also does not provide any evidence that such molecules exist.

47. a) No, there is no true relationship between the C value and the number of genes in eukaryotes. Evolutionary heirarchy and complexity do not relate to amount of DNA. This lack of correlation is referred to as the C-value paradox.

b) (i) No, it is not likely that *L. hirsutus* has three times as many different kinds of genes as *L. angulatus*. The difference is most probably due to *L. hirsutus* having many more repetitive DNA sequences, i.e., satellite DNA.
(ii) Most of the difference in total amount of DNA would be expected to be due to highly repetitive DNA and some middle repetitive DNA sequences.
(iii) These changes should not be localized. Instead all the chromosomes in the genome should be affected more or less equally. This could be due to the fact that the function of the repetitive DNA is nonspecific, required by all areas of genetic activity and that a mechanism responsible for the creation of repetitive sequences probably operates in all chromosomes. Bivalents should appear morphologically different (heteromorphic) among the species as they would not be qualitativiely different. They would differ only in size because of differences in amounts of repetitive DNA and some middle repetitive DNA.
(iv) Unequal crossing-over, for example, is a mechanism that could have given rise to differences in DNA content.

Chapter 18
Mutation and Repair

1. a) This indicates that a higher proportion of the ultraviolet induced mutations are genic.

12. a) and b) The *mutD* gene codes for the epsilon subunit of the polIII DNA polymerase holoenzyme which is responsible for the 3' - 5' exonuclease activity and the proofreading function of the polymerase. The high mutation rate is due to the lack of this function in the *mutD* mutants. Any mismatch that is incorporated in the DNA is not corrected at the replication fork. The polymerase most likely generates transitions, where the wrong pyrimidine is most often incorporated opposite a purine. Transversions will be the next most common, due to pyrimidine-pyrimidine and purine-purine pairings.
Frameshifts will likely be due to the decreased stability of the polymerase complex in the absence of the epsilon subunit.
The extremely high rate of mutation exhibited by *mutD* mutants is indicative of the vital importance of the proofreading function of the enzyme. In the absence of the 3' - 5' exonuclease activity, the only repair system available is the *mutHLS* methyl-directed repair process which is quickly overwhelmed.

18. a) The *short winged* F_1 ClB female (derived from a cross between a *normal* winged ClB female and an irradiated *normal* winged male) in crosses with *normal* winged F_1 males produces *normal* and *short winged* progeny in approximately equal proportions. This indicates that the hereditary factor for *short wings* is carried and transmitted by the *short winged* F_1 female. Therefore the *short winged* phenotype is not a phenocopy. If it was, all F_2's would have developed *normal* wings.

b) The *short winged* F_1 female is the consequence of a mutation in a gamete of the irradiated male parent. That the mutant phenotype is expressed in an individual of the immediate progeny of two phenotypically identical parents implies that the induced mutation is a dominant one. The F_2 results (testcross-like) confirm that the *short winged* F_1 female is heterozygous. Since the determiners for *normal* vs. *short wings* segregate independently of the ClB vs. non-ClB homologues and equal proportions of individuals with the two phenotypes are observed in both sexes, one can conclude that the induced dominant mutation for *short wings* ($s \rightarrow S$) is autosomal. The entire cross can be illustrated as follows:

Normal winged ClB ♀ x *normal winged* irradiated ♂
XX*aa   XYaa

F_1 XXaa, X*Xaa, *normal winged* ♀'s; 1 X*XAa *short winged* ClB ♀; XYaa, *normal winged* ♂; X*Yaa, inviable ♂

Short winged ClB F_1 ♀ x *normal winged* F_1 ♂
XX*Aa   XYaa

F_2 XXaa, X*Xaa, *normal winged* ♀'s; X*XAa, XXAa *short winged* ♀'s; XYaa, *normal winged, viable* ♂; XYAa, *short winged, viable* ♂; X*Yaa, *inviable* ♂, X*YAa, *inviable* ♂
Note: * - ClB X chromosome renders X*Y males *inviable*

c) (i) Deletion of all or part of the *a* locus.
(ii) Chemical change at the *a* locus ($a \rightarrow A$).
One can distinguish between these alternatives genetically by studying large numbers of progeny from matings between *aa* and *AA* flies. If reversion or a mutation to any other allele at the locus occurs, the initial mutation involved a chemical change and not a deletion. If reversion to the *normal* phenotype and genotype does not occur, the initial change probably was a deletion. Hybridization of the molecular probe of the *a* gene would not occur if the entire locus or most of it was deleted. If only the *a* locus was deleted, cytological analysis could not distinguish between the two kinds of changes.

d) The reverse mutation occurred at a suppressor locus ($s \rightarrow S$) since the revertant *normal* appears among the immediate progeny of a mating between individuals from the same true-breeding line.

True-breeding *short* ♀ (*AAss*) x True-breeding *short* ♂ (*AAss*) 1 in 50,000 is *AASs normal* winged
True-breeding revertant (*AASS*) x original *normal* (*aass*)
F_1 *AaSs* - *normal* → F_2 9 *A_S_* - *normal*; 3 *A_ss* - *short*; 3 *aaS_* - *normal*; 1 *aass* - *normal* i.e., 13 *normal* : 3 *short*

Had the reverse mutation occurred at the original locus (*a*), crosses between the true-breeding revertant and the original *normal* strain would have produced only *normal* progeny (*aass* x *aass* → *aass*).

19. a) In such populations the frequency of individuals showing the recessive phenotypes is not a direct measure of the mutation rate since one does not have estimates of the frequencies of recessive alleles already in the population.
Mutation rates can be determined from the appearance of dominant phenotypes among the offspring of parents homozygous for recessive alleles because the dominant alleles must be the direct result of a mutation in either of the parents.

b) For mutations to the dominant.
Frequency of mutation   =   No. of mutated alleles
                                Total no. of alleles

                          =   No. of *affected* individuals
                                2 x total no. of individuals

Frequency for *brachydactyly* = 11/2 x 735,000 = 1 in 134,000.

Frequency for *aniridia* = 3/2 x 735,000 = 1 in 490,000.

c) Incomplete penetrance at birth or death of the zygote or early embryo may cause errors in these estimates. Also, the absence of reporting or noting such defects in certain individuals, particularly the less serious defects may also affect the estimate.

20. *Duchenne muscular dystrophy* is a hereditary, X-linked condition. A characteristic feature of X-linked disorders is that males, barring meiotic mishaps, always receive their sole X chromosome from their mothers. Unless the son was an aneuploid, e.g., a Klinefelter (XXY) male, it is unlikely that he received an X chromosome and thus the causative gene from his father, an X-ray technician in the industrial plant. Instead the boy received the causative X-linked gene from his mother. Absence of the X-linked condition in the father's family for the past four generations has no bearing on the case. The sudden appearance of the affliction in the boy, a member of the fifth generation, must

be due to the occurrence of an X-linked mutation on the maternal side of the family. The mutation must have occurred in the primordial germ cells of either the boy's mother or his grandmother since the condition first makes its appearance in his generation, the fifth generation. Had the mutation occurred in the great-grandmother, then the condition would have made its appearance in members of the third generation. Both sexes would have been *affected* if the mutation were a dominant one while only male collateral relatives would have expressed the condition if the mutation were a recessive one. The precise nature of the mutation could be ascertained on the basis of whether or not the mother has other *affected* offspring and as to whether they are of both sexes. Only *affected* male offspring, would suggest the nature of the mutation to be recessive and sex-linked. Thus, an argument based on the mode of inheritance of sex-linked characters renders the technician's claim unjustifiable. The son's abnormality is not the consequence of an X-ray induced mutation in the father who is therefore not justified in suing the plant for improper protection against radiation.

b) The fact that a couple's ancestry does not reveal any cases of *amaurotic idiocy* does not render it impossible for them to have a child with the disorder. The sudden occurrence of the condition, if hereditary, could be due either to a mutation (induced by the radiation amounts that the couple were exposed to) or the consequence of the segregation of a rare recessive allele for which the couple would both have to be heterozygous. On the other hand, if the condition were not hereditary, the afflicted child could be a phenocopy, where, because of environmental modifications the child's genotype (*normal*) expresses an altered phenotype (*abnormal*).
That the disorder is hereditary and determined by an autosomal recessive allele, requires that both parents contribute the causative allele to the *affected* child's genotype. However, neither parent's ancestry reveals a history of the condition suggesting therefore that both of them lack the recessive allele (*a*) in their genotypes. For two such individuals to have an *affected* child, each would have to incur a recessive mutation at the specific locus. The radiation that the couple were exposed to would have to induce a mutation at the same locus and in the same direction ($A \to a$) in each parent. Since mutations are rare events and each locus is characterized by a specific mutation rate (μ), it is highly improbable that the radiation would induce a similar type of mutation in each parent at the locus in question. The probability of such an occurrence is extremely low; therefore one can argue that the appearance of the disorder in the couple's offspring is not the consequence of a radiation induced mutation in both parents.
Alternatively, despite an absence of the rare disorder in the couple's ancestry, it is possible for the rare recessive allele to be present in some members of both families. The couple in question could both be heterozygous for the recessive allele. The probability of two unrelated individuals being heterozygous for a specific recessive allele would depend upon the frequency of that allele in the population. Since allele frequencies are greater than mutation frequencies at that locus, it is less likely that the occurrence of the *affected* child is due to the effects of radiation. Instead, it is the consequence of the segregation of a rare recessive allele. The possibility that the child is a phenocopy is rendered invalid because of the hereditary nature of the condition.

21. a) The F_1 progeny should all be genotypically identical (*Bb*) and therefore phenotypically similar (*black*) if the autosomal allele *B* is completely penetrant and dominant. All F_1 progeny backcrossed to the *white* parent should therefore produce two types of offspring, *black* and *white*, in equal proportions. The four *white* F_1 progeny in backcrosses with the *white* strain can be accounted for as follows:

(i) Since one of the F_1 *whites* behaves predictably in the backcross, it must be heterozygous (*Bb*) but expresses a *white* phenotype because it is a phenocopy.
(ii) A second *white* F_1 when backcrossed produces offspring of two types, *black* and *white*, in approximately a 1:3 ratio because of a mutation at another autosomal (suppressor) locus in the germ cells of its *black* parent. Specifically the mutation at the suppressor locus is from recessive (*a*) to dominant (*A*) and is epistatic to the dominant allele for *black* at the *B* locus. Thus: F_1 *white* (*BbAa*) x *white* strain (*bbaa*) → 1 *BbAa*, *white*; 1 *bbAa*, *white*; 1 *bbaa*, *white*; 1 *Bbaa*, *black*.
(iii) The remaining two F_1 *whites* in crosses with the true-breeding *white* strain produce only *white* offspring suggesting that they must be homozygous for *b*. This would be possible only if a mutation $B \to b$ occurred in the germ cells of the *black* parent.

b) (i) Of the four exceptional F_1 *whites* two are due to recessive mutations at the *B* locus and one is the result of a dominant mutation at a suppressor locus. The overall mutation rate per gamete therefore is the sum of the mutation rates at each locus, i.e., 2/250,000 (mutation rate at *B* locus) + 1/250,000 x 0.5 (mutation rate at the suppressor locus) = 5/500,000 = 1/100,000
(ii) The mutation rate from *B* to *b* = 2/250,000 or 1/125,000.

22. a) The 12:3:1 phenotypic ratio of kernels observed in a self-fertilized ear of true-breeding deep Mexican sweet corn is an anomaly explicable only if one invokes the occurrence of a recessive mutation ($A \to a$) at the locus determining *aleurone colour*. The mutation renders the plant heterozygous (*Aa*) at the aleurone locus. The appearance of the third phenotype *dotted* and the proportions of the different phenotypes in the self-fertilized progeny necessitate a second interacting locus (*Dtdt*) for which the plant must also be heterozygous. Thus segregation at two heterozygous and interacting loci would result in the observed 12:3:1 phenotypic ratio provided the nature of the interaction between the two loci was as follows:
A - dominant allele for *coloured* (deep) aleurone; epistatic to *Dt*; *a* - recessive allele for *colourless* aleurone; *Dt* dominant mutator allele for *dotted*; hypostatic to *A* but causes *a* to mutate to *A* in some of the aleurone cells thereby resulting in *dotted* kernels on a *colourless* background; *dt* - recessive allele, amorphic.
The cross may be illustrated as follows:
$AaDtdt \times AaDtdt \to 9 A\_Dt\_ : 3 A\_dtdt$ (*deep*); 3 $aaDt\_$ (*dotted*); 1 $aadtdt$ (*colourless*)

b) *A* - *deep colour*; unaffected by alleles of the *Dt* locus.
A^b - *deep colour*; not affected by the presence of *Dt*.
a - *colourless* in absence of *Dt*; when the latter is present, it causes *a* to mutate to *A* on some of the aleurone cells and therefore produce dots.
Dt - causes *a* to mutate to *A*; *dt* - no mutational effect.
Given the above interactions figure out the results of the crosses for yourself.

c) The induction of mutation during a period towards the end of aleurone development would explain why only some of the aleurone cells are affected and appear as coloured spots. If the induced mutation occurred at an earlier stage, a part or all of the endosperm would be coloured. When homozygous, *a* causes the absence of anthocyanin pigment production which leads to a lack of *purple* colour in the endosperm. However, if the dominant allele *Dt* is also present, *a* can mutate to other alleles of the *A* locus that permit anthocyanin production in this newly formed heterozygote (*Aa*).

29. The formation of the UV lesion is such that the 3' pyrimidine is pulled out of base-pairing position, but the 5' base is not. As a result, during *umuCD/polIII* bypass replication, the 5' base is available to be read by the *polIII* enzyme but

the 3' is not. The *polII* enzyme incorporates an A opposite the 3' base, but correctly inserts the complementary base opposite the 5' base of the lesion. The difference in the mutability of the two sequences lies in the type of pyrimidine in the 3' position. Incorporation of an A is not mutagenic for a T, but is for a C.

33. a) The data suggest that 53 of the 99 UV induced mutations are transition type mutations because they are revertible by base analogues. Of these, 47 reverted by AP and BU but not by HA. The former two base analogues can induce transitions in both directions (A-T ↔ G-C) but HA can do so in only one direction (G-C → A-T). Among the 53 UV-induced transitions (i) 47 must have been in the G-C → A-T direction, i.e., thymine induced in the same direction as those produced by HA since none were revertible by HA and (ii) 6 were in the A-T → G-C direction since they were revertible by both HA and the base analogues AP and BU. The other 46 UV-induced mutations were of another type(s). On the basis of the above information one can conclude that in T4, UV light primarily induces transitions in the G-C → A-T direction.

b) Since most of the remaining mutations did not respond to base analogues they may not have been transitions. Since these mutations were reverted by proflavine, which is known to induce base-pair additions and deletions, UV light probably induced base-pair additions and deletions indirectly.

UV light induces dimer formation within each strand of DNA. During the process of repair and replication of structurally altered DNA, e.g., with dimers, single-strand breaks are present. Mispairing can occur in the regions near these breaks, resulting in extrahelical loops and gaps in one strand of the DNA. When the gap is filled in, the loop may still remain. During the next replication, one or more base pairs may be added or deleted. If the loop (buckle) forms in the broken strand, addition of one or more base pairs will occur; if it occurs in the intact strand, deletion of one or more base pairs will be effected. This slipped mispairing model for the origin of frameshift mutations was proposed by G. Streisinger in 1966 and is schematically illustrated below.

Frameshift mutations occur because of mispairing in regions of reduced stability of the DNA duplex, such as near single-stranded breaks. Although the mutagenic action of intercalating agents (e.g., proflavine) is not completely understood, evidence indicates that these agents bind to and stabilize the loops of DNA near single-stranded gaps.

34. a) Since phage S13 is a single-stranded DNA virus, single bases not base pairs are involved in each mutation at a given site. More complete information about the specificity of mutagens is possible from studies involving single-stranded rather than double-stranded DNA. The base change (substitution) involved in forward mutation in each mutant is as follows:

Mutation	Inferred base change
$h^+ \to h_1 1$	T → C
$h^+ \to h_1 2$	A → G
$h^+ \to h_1 1$	G → A
$h^+ \to h_1 2$	G → A
$h^+ \to h_1 63$	G → A
$h^+ UR48 \to h_1 UR48s$	C → T

b) NA is known to deaminate A, G and C, converting A to hypoxanthine, G to xanthine and C to uracil. Hypoxanthine assumes the bonding properties of guanine resulting in a AT → GC transition. Uracil pairs like thymine and causes a GC → AT transition. Xanthine has the same pairing properties as G. Therefore only deamination of A and G are expected to be mutagenic. However, in S13 all 4 bases were mutated by NA to varying degrees. The data indicate that NA induces GC → AT as effectively as it induces AT → GC transitions. C → T transitions were induced at a high frequency and T → C substitutions occurred at a low frequency. NA might exert a mutagenic effect on T and G indirectly by reacting with a neighbouring base.

c) Other than G, EMS also acts on C, T and A. The first and last mutational changes showed that C mutated as frequently as G. The first and fourth mutational changes showed that T and A mutated at low frequencies. Presumably the bases are more exposed to attack by the alkylating agent when DNA is single-stranded. The alkylated bases may then tautomerize and bring about transitions.

35. *lex*⁻ There are two possible modes of action.
(i) The *lexA* mutant results in a loss of *lexA* function. This means that the SOS system is fully induced at all times. Therefore, the error prone repair system is able to start repair of the benzopyrene lesions immediately, rather than only at higher levels of damage. Thus the mutation rate increases.
(ii) The *lexA* mutant results in a *lexA* protein which is not cleavable by activated *recA*. Therefore the SOS response is not turned on, and repair of the lesions fails to occur. The result is a drop in mutation frequency, but a severe drop in survival of the treated bacterial cells.

uvrB⁻ - The *uvr ABC* genes code for the proteins that perform the error-free component of the SOS response. *uvrB* mutant strains will have an inactive *uvr* repair system, so all repair is channelled into the error-prone pathway. Therefore the mutation rate will rise sharply.

his⁻ - The histidine synthesis pathway is not involved in the repair of lesions induced in DNA by benzopyrene. Thus, there will be no change in the overall mutation rate.

41. a)

Mutant	Amino acid change		RNA code word		DNA base change
	From	To	From	To	From To
A223	Ile	Ser	AUU AUC	AGU AGC	A → C
A78	Cys	Gly	UGU UGC	GGU GGC	A → C
A58	Asp	Ala	GAU GAC	GCU GCC	T → G
A23	Arg	Ser	AGA	AGC AGU	T → G
A46	Glu	Ala	GAA	GCA	T → G

It is clear from the DNA base changes induced in each of the 5 mutants to produce revertants that the same type of base-pair change, i.e. an AT → CG transversion, appears to be favoured by the mutator gene. The reverse substitutions, CG → AT, if it were caused by the mutator gene, should have given Arg → Ile changes (AGA → AUA), DNA change - C → A) at the A23 position, but these were not observed. Thus, the mutator gene (i) is specific in its effects (ii) unidirectional and (iii) induces transversions.

b) The mutator gene may (i) produce mutagenic base analogues which lead to replication or incorporation errors or (ii) produce a mutant polymerase which results in errors during DNA replication of defective proofreading function.

42. HA reacts with cytosine and 5-hydroxymethylcytosine (which is phage specific) to cause GC → AT transitions. Since 11 rII mutants have GC at the mutated site, HA can revert these mutants to *wild-type*, r$^+$, by inducing GC → AT transitions. The *wild-type* phenotypes will not appear immediately (shortly) after the HA treatment unless C is on the transcribed strand. Therefore, if the revertants are plated on the restrictive host, *E. coli* strain K, only those that have C on the transcribed strand in the mutant allele can grow because they can produce *wild-type* RNA from the mutated (C → T) DNA strand. Mutants that have C on the non-transcribed strands will not grow on the restrictive host because the transcribed strands are still mutant in nature. However, when grown on the permissive host, *E. coli* strain B, the HA-reacted C will lead to the incorporation of A instead of G in the transcribed strand during DNA replication, yielding newly synthesized strands identical to those of r$^+$. Therefore the resulting phage will be able to grow on the restrictive host.

Chapter 19
The Gene: Its Genetics and Interallelic Complementation

3. A mixture of blood from an A and a B hemophiliac has *normal* clotting capacity because of complementation. One hemophiliac produces a substance necessary for clotting which the other cannot. This is because each carries the dominant allele (for a given clotting factor) of a gene for which the other has a recessive allele (cannot produce this factor). That is the recessive mutant alleles in the two hemophiliacs are non-allelic (in different functional genes). Specifically, hemophilia A individuals cannot produce antihemophilic globulin (AHF) while those with hemophilia B are deficient for plasma thromboplastin (PTC). The two factors, probably produced by different biochemical pathways, are required for normal clotting. Each mutation causes a metabolic block as shown below:

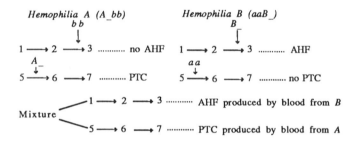

8. No. The hypothesis requires a qualification in view of the fact that discontinuous genes possess both exons and introns. Since only the exons possess the coding sequences of such genes there is still a perfect colinearity between the coding sequences in the exons and the amino acid sequence in the polypeptide chain specified by the gene.

10. In transduction an even number of crossovers is required to integrate a donor segment into the recipient genome. If the donor is *wild-type* or possesses a mutant gene non-allelic with the mutant gene in the recipient, the probability is relatively high that the *wild-type* allele will integrate into the recipient genome by crossing-over to produce *wild-type* recombinants. If the mutant gene in the donor is allelic with the mutant gene in the recipient, the mutations at corresponding (identical) sites (identical, homoalleles) or at non-corresponding (different) sites (non-identical, heteroalleles). If mutations occur at corresponding sites, *wild-type* recombinants can not arise by crossing-over since the process would merely replace one mutant site with another. If the mutant sites are at non-corresponding positions crossing-over can occur between them to produce *wild-type* alleles and recombinants. Very few such alleles and recombinants are expected since the mutant sites are very closely linked, even if at opposite ends of the functional region. Numbers of *wild-type* recombinants fall into 3 classes in the different crosses: Very many, e.g., 1,822 in *wild-type* x *trp*-10 transductions, a large number, e.g., 208, 240, 270, 280 and 602 in other crosses and none or very few (0, 4, 7, 12, 22) in others. From this we can conclude that those crosses producing large numbers of *wild-type* recombinants involve non-allelic genes (mutations in different functional regions). Crosses producing very few *wild-type* recombinants involve allelic genes (mutations at different sites in the same functional region) and those producing no such recombinants involve allelic genes with mutations at corresponding sites.

12. a) *Large* colonies originate by complete transduction, a process in which the donor fragment in merozygotes integrates by crossing-over (an even number of exchanges) into the recipient genome. The transductions that give rise to many *large* colonies involve (i) replacement of a mutant allele in the recipient genome by the *wild-type* allele on the donor fragment or (ii) nonallelic mutant genes. In both cases the fragment integrated into the genome is relatively large and it is much easier to integrate a large chromosome segment (crossovers far apart) than a short one.

Those transductions that produce a few *large* colonies involve mutant strains that carry allelic mutant genes that are heteroalleles. Since the mutant sites are close together the probability is low that two changes will ocur in the same gene (short region), one on either side of the mutant site in the recipient. Transductions that produce no *large* colonies involve mutant strains that possess identical mutant alleles (homoalleles) - mutations at corresponding sites. These mutants can give rise to *large* colonies only if mutation occurs from the mutant to the *wild-type* allele at the locus.

b) *Minute* colonies arise through abortive transduction. Since the donor fragment is not integrated into the recipient genome and one cell in the entire *minute* colony remains merozygotic (partially diploid). When *minute* colonies occur between two mutant strains unable to synthesize the same (or different) compounds, this indicates that the strains carry mutant genes that are functionally nonallelic. That is, they carry mutant alleles at different gene loci. If no *minute* colonies arise in transductions between two mutants strains this indicates their mutant genes are functionally allelic.

c) Mutants *ath*A-2, 5, 8 and 10 are functionally allelic (i) since no abortive transductants (*minute* colonies) are observed and (ii) frequency of *large* colonies, due to crossing-over, is low.

d) Mutants *ath* A-4, 5 and 10 are homoallelic because crossing-over does not occur between them and therefore *large* colonies are not observed.

e) The *wild-type* condition is dominant otherwise merozygotes would not live and grow and minute colonies would not be produced.

f) At least three genes are involved. The *athA* one at which all *athA* mutant alleles arose. *AthC-5* represents another gene. The mutations in the other mutants may constitute one or more gene.

13. a) Mutants 51 and 102 do not complement, therefore they are allelic. Similarly mutants 47, 101, 104, 106 are also allelic. Mutants 51 and 102 complement all the mutants 47, 101, 104 and 106; therefore they are non-allelic to the latter 4 mutants. These results indicate that the *rII* region consists of a minimum of 2 genes. If each of all other *rII* mutants fails to complements any of the mutants 51 and 102, 47, 101, 104 and 106 then the *rII* region consists of two genes only. However, if some of the other mutants complement all 6 of these mutants, then more than 2 genes occur in this region. Additional complementation studies would be necessary to determine the number of genes in the *rII* region.

b) Would expect a normal yield of phage and normal plaque formation because the *wild-type* (r^+) phage would provide the normal alleles at all loci in the *rII* region.

15. a) Since the F_1's in the first two crosses express the *wild-type* phenotype and those in the last cross are phenotypically mutant, only the mutant genes in strains 3 and 4 are allelic. The mutations in the four strains 1, 2 and 3(and 4) are therefore at three different (non-allelic) loci that determine flower colour.

    b) They are not linked since a 9:7 F_2 phenotypic ratio can only be obtained if the allele pairs assort independently. Whether these genes are syntenic or not cannot be determined from these data.

    c) Cannot determine whether they are functioning in the same or in different pathways. If they are acting in different pathways, both pathways must be complementary in the production of pigment.

16. a) (i) The mutant genes for *white* (w) and *coral* (w^{co}) are allelic. **Reason:** F_1 females heterozygous in the trans configuration are not *wild-type*. Therefore there is no complementation.
    P_1 ♀ $X^w X^w$ (*white*) x ♂ $X^{w^{co}}Y$ (*coral*)→F_1 ♀$X^{w^{co}}X^w$ (*light coral*); ♂ $X^w Y$ (*white*).

    w and w^{co} are recessive alleles of the same X-linked gene. The results do not permit determination of whether the mutations in the two recessive alleles are at the same or different sites in the functional region of the W locus for eye colour.

    (ii) The mutant genes for *white* (w) and *carmine* (c) are non-allelic because the F_1 females express the *wild-type* eye colour as a result of complementation.
    P_1 ♀ $X^w X^w CC$ (*white*) x ♂ $X^W Y cc$ (*carmine*)→F_1 ♀$X^W X^w Cc$ (*wild-type*); ♂ $X^W Y Cc$ (*white*).

    w and c are recessive alleles of different genes. The genes may either be on different chromosomes (w on X and C on one of the autosomes) as assumed above or on the same chromosome (X). The results do not permit an unequivocal location of the genes.

    b) (i) *Coral* ♀ ($X^{w^{co}}X^{w^{co}}CC$) x *carmine* ♂ ($X^W Y cc$)→F_1 ♀$X^W X^{w^{co}}Cc$ (*wild-type*); ♂ $X^{w^{co}}Y Cc$ (*coral*)
    (ii) *Carmine* ♀ ($X^W X^W CC$) x *coral* ♂ ($X^{w^{co}}Y CC$)→F_1 ♀$X^W X^{w^{co}}Cc$ (*wild-type*); ♂ $X^W Y Cc$ (*wild-type*).

17. a) Since (i) the mutation rate is low (ii) mutations arise at random with respect to position and direction within a locus and *Drosophila melanogaster* has thousands of genes (assume 10,000) it is highly unlikely that all three mutations should arise at the w locus and always from w and/or w^{co} to W (*red*).
    b) Crossing over between non-sister chromatids in the region of the functional alleles w and w^{co}.

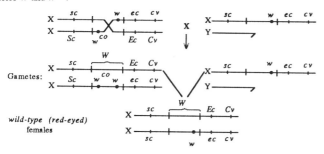

Reason - all 3 *wild-type* females possess recombinant phenotypes for the outside markers (genes).
Note: (i) w and w^{co} are allelic mutations at non-corresponding positions within the W locus.
(ii) W *wild-type* allele due to c/o - no defective sites
(iii) allele $w^{co}w$ would be a double mutant

c) w^{co} and w are heteroalleles of the W locus located between the loci sc and ec and in the sequence sc-w^{co}-w-ec. If w was closer to sc than w^{co} and *vice-versa* relative to ec the three *wild-type* female progeny would be Sc-W-ec-cv and not sc-W-Ec-Cv.

18. From fact one, none of the three *biotin-requiring* mutants are due to mutant alleles at the locus determining the last step in the biotin pathway.
    a) Since prototrophic recombinants occurred less frequently in bi_1 x bi_3 crosses than in bi_1 x bi_2 crosses, bi_2 is closer than bi_3 to bi_1. Since the results of the bi_2 x bi_3 are not given we cannot determine whether bi_2 and bi_3 are on the same or different sides of bi_1. If they were on opposite sides of bi_1 prototrophic recombinants would be expected with a frequency of 1/7,000 progeny. Otherwise recombinants would be expected with a frequency of 1/3,000 progeny.
    Since prototrophic recombinants always showed recombination for markers on either side of the biotin locus, this indicates that bi_1, bi_2 and bi_3 represent mutations at different sites in the same gene. That bi_1, bi_2 and bi_3 are different mutant alleles of the same gene is evidenced by the fact that bi_1-bi_2, bi_1-bi_3 and bi_2-bi_3 heterokaryons and diploid mycelia were *biotin-requiring*. This is only possible if the mutant genes are allelic and do not show complementation.

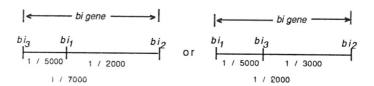

b) Roper's findings completely support Pontecorvo's hypothesis. The results show that the three mutations are at different points in a linear array within a chromosomal segment ($\approx$ 0.2 m.u.) that performs a single function.
Gene - a specific chromosomal segment which (i) performs a single specific function and (ii) consists of a linear sequence of potentially mutable sites, each of which can result in an alternative chemical form and between two of which crossing-over can occur.
Allele - One of two or more chemical forms of a gene. Each mutant allele of a gene differs from the *wild-type* and other alleles at the locus by at least one mutable site.

20. The method for determining the order of genes or sites within a gene is by means of 3-factor reciprocal crosses or, if more factors are involved in the cross, by the analyses of results using 3 factors at a time.
    In these reciprocal 3-factor crosses one of the strains is a single mutant and the other a double mutant. Such crosses, in which each parent is alternatively used as donor and recipient, are set up by growing the transducing phage on each

parent (or obtaining the DNA form each parent) and then using the phage (or DNA) to tranduce (transform) the other strain. *Wild-type* recombinant progeny are selected for. The proportions of these in reciprocal crosses, which are dependent on the number of exchanges required to produce them, indicate the correct gene or site sequence. The expectations for both sequences are compared with the actual results. For one of the sequences the proportions of *wild-type* recombinants should be the same or insignificantly different in reciprocal crosses because the same number of exchanges (2) is required to produce them regardless whether the single or double mutant is the donor or the recipient. For the other sequence, the proportions of *wild-type* recombinants should be significantly different in reciprocal crosses. If the single mutant is used as donor and the double mutant as recipient, 4 exchanges are required to produce *wild-type* recombinants, whereas if the double mutant is the donor and the single mutant the recipient, only 2 crossovers are required to produce such types.

a) The order of the mutant sites with respect to each other and *thr* and *leu* is *thr ara-2 ara-1 ara-3 leu*. In the cross between *ara-1* and *ara-2*, if *ara-2* is closer to *thr* than *ara-1* is, one can expect more ara^+leu^+ recombinants when *ara-2* is the donor than when it is the recipient. This is because ara^+leu^+ recombinants can arise as the result of 2 exchanges (double crossing-over). In the reciprocal cross ara^+leu^+ recombinants would require four exchanges.

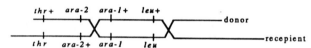

If *ara-2* is the receipient, $ara+ leu+$ recombinants are the consequence of 4 exchanges (quadruple crossing-over.)

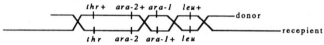

By similar reasoning there should be fewer ara^+thr^+ recombinants when $ara-2^-$ is the donor than when it is the recipient. These expectations are realized in the first set of reciprocal crosses. Applying the above reasoning to the other 2 sets of reciprocal crosses we find that in the set of reciprocals involving $ara-1^-$ and $ara-3^-$, *ara-1* is closer to *thr* than is *ara-3* and in the set of reciprocals involving *ara-2* and *ara-3*, *ara-2* is closer to *thr* than *ara-3*. Therefore, since *ara-2* is closer to *thr* than both *ara-1* and *ara-3*, and *ara-1* is closer to this marker than *ara-3*, the order of the mutant sites must be *thr ara-2 ara-1 ara-3 leu*.

b) The *large* colonies are due to integration, by an even number of exchanges, of a donor segment with a *wild-type* site into the recipient chromosome thus replacing the corresponding mutant site and verifying that the mutant sites 1, 2 and 3 are at different (non-corresponding) positions. The *large* colonies therefore, do not provide additional information.

The *minute* colonies are the result of abortive transduction, which can be used to determine whether 2 or more phenotypically identical or similar mutants are functionally allelic or non-allelic (in the same or different genes (cistrons). The absence of *minute* colonies in transduction of one mutant strain by phage grown on another indicates that the mutant sites in the two mutant strains are in the same gene. Therefore, the mutant genes in the two strains are allelic. This conclusion is drawn because the abortive transductants are diploid for the gene (or genes) concerned. Neither mutant can perform the function the other cannot carry out (there is no complementation - no *minute* colonies). Thus the mutant genes must be allelic. The presence of *minute* colonies in transduction of one mutant strain by another, indicates that the mutant genes in the two

strains are non-allelic (in different functional regions). This conclusion is made because complementation occurs to produce *minute* colonies. Complementation indicates that each mutant strain carries the *wild-type* allele of the mutant gene in the other strain. This, of course, is possible only if the mutant genes in the 2 strains are non-allelic. Since *minute* colonies form when mutant 3 is infected with phage grown on mutants 1 and 2 and when mutant 1 is infected with phage grown on mutant 2 and 3 the presence of such colonies indicates non-allelism; the 3 mutant sites must exist in 3 different genes. Therefore, the arabinose region can be schematically represented as follows:

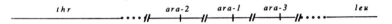

21. a) Complementation data suggest there are three genes involved. Mutants 210, 246, 300 belong to one gene. Mutants 222, 225, 261 belong to the second gene and mutants 244 and 279 are due to mutations in a third gene.

b) The two genes which include the mutations in mutants 210, 246, 300, 222, 225 and 261 appear to be contiguous because the deletion mutant 290 was deficient in both but not in the gene that included the mutations in mutants 244 and 279.

c) Reverting mutants are point mutants and nonreverting mutants are deletion (multi-site) mutants. The mutations in mutants 210, 222, 225, 246, 261 and 300 are point mutations. Those in mutants 238, 292. 282, 277, 290, 253 and 293 are multisite. The nature of mutants 244 and 279 cannot be determined from these data.

d)

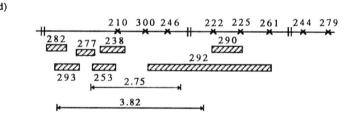

23. The difference in size between the ovalbumin gene and its protein is due to the fact that the gene is a discontinuous one. Only the exons of the genes (coding sequences) are spliced into the mRNA. If it was a continuous gene the ovalbumin protein should have been ≈2,830 amino acids (AA's) long since each codon is a triplet. The exons in the gene occupy a minority of its nucleotide pairs. Specifically, since the protein is composed of 386 AAs, the coding sequence must constitute only 386 x 3 = 1158 nucleotide pairs. Disregarding the chain-terminating codon, a large part of the gene (8,490-1158=7,332 nucleotide pairs) is of a non-coding nature (intron or introns).

24. **Rationale for the ordering of the mutational sites:** Consider two sites, 1 and 2, relative to each other and the outside marker, *anth* (=a). In a given cross the frequency of $anth^+$ among trp^+ recombinants will be high or low.

Donor (single mutant)   A1 $\dfrac{a^+\ 1^+\ 2^-}{a^-\ 1^-\ 2^+}$ HIGH   A2 $\dfrac{a^+\ 1^-\ 2^+}{a^-\ 1^+\ 2^-}$ LOW

Recipient (double mutant)

Donor (double mutant)   B1 $\dfrac{a^-\ 1^-\ 2^+}{a^+\ 1^+\ 2^-}$ LOW   B2 $\dfrac{a^-\ 1^+\ 2^-}{a^+\ 1^-\ 2^+}$ HIGH

Recipient (single mutant)

In crosses A1, B1 B2 the frequency of $anth^+$ among trp^+ recombinants will be high because only two recombination events (non-reciprocal crossover) are required to produce $anth^+ trp^+$ progeny. In cross A2, four recombination events are required to produce such progeny. The probability of four events occurring is significantly lower than the chance of occurrence of two such events. Therefore $anth^+ trp^+$ will occur with significantly lower frequency in the A2 cross than in the others.

a) By comparing the frequencies of $anth^+ trp^+$ recombinants in reciprocal crosses as illustrated above. The sequence of mutational sites can be readily ascertained. The sequence of the markers in each of the 5 sets of reciprocal crosses is as follows:

First set of reciprocal crosses anth-B6-B4
Second set of reciprocal crosses anth-B7-B4
Third set of reciprocal crosses anth-B6-B7
Fourth set of reciprocal crosses anth-B6-B14
Fifth set of reciprocal crosses anth-B14-B7

Therefore the order of the 4 *trp* sites is - anth-B6-B14-B7-B4

b) (i) By determining whether a mutation in a given mutant can revert to the *wild-type* or some other form (allele) of the gene. Point mutants can revert, multisite (deletion) mutants rarely if ever do.
(ii) Point mutants recombine with other point mutants at different sites and with many multisite mutants. On the other hand the latter mutants do not recombine with many point mutants (those in the region of the chromosome deleted from the multisite mutant) and deletion mutants that are missing segments of the chromosome that they do not possess.

25. a) This conclusion is based on the number of R-loops that are formed both between the denatured β gene and its primary transcript and its mRNA. Only one continuous R-loop is formed between the primary transcript and the gene. This indicates that the gene and its primary transcript are colinear (the DNA and RNA possess the same number of nucleotide pairs and nucleotides respectively and the sequence in the RNA is complementary to that in the DNA strand with which it hybridizes). Since two R-loops are formed when the denatured β gene is hybridized with the mRNA, the gene and mRNA must not be colinear; the β gene must be discontinuous. One sequence of consecutive nucleotides within the β gene is missing from its mRNA.

b) Since only 2 R-loops are formed and a segment of the DNA of the β gene between the 2 R-loops also loops out, the β gene must contain 2 exons and 1 intron.

c) Since only one continuous R-loop is formed between the β gene and its primary transcript and 2 such loops are produced when mRNA is hybridized with the β gene, the intron must be excised after the primary transcript is formed. If it was excised before formation of the primary transcript it would have formed two R-loops upon hybridization with the gene.

d)

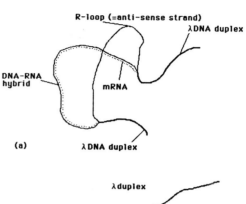

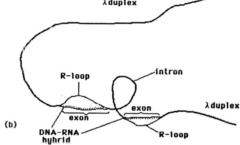

28. Since codons are three consecutive bases long and the φX174 genome consists of 5,375 nucleotides, the 9 genes if they do not possess introns, should code for a maximum of 5375/3=1791 amino acids (AAs). However, the proteins specified by these genes contain 2,050 AAs, i.e., 259 more than would be expected if all the genes were without introns. This incongruency can be attributed to the fact that at least some of the nine genes overlap. That is, part of one gene is contained within another. When this occurs the reading frames of the 2 genes are different (offset by 1 or 2 nucleotides). If they were not offset, nonsense codons in the region common to both genes would lead to the synthesis of incomplete polypeptides specified by the two genes.

29. a) Genes D and E must be overlapping. **Reasons:**
(i) Genes D and E are completely linked and (ii) *amber* (nonsense mutations) = nonsense codons *am3* and *am6* in gene E reside within restriction fragment Z7, which completely encompasses gene D.

b) Gene E starts at nucleotide 179 (A) and ends at nucleotide 454, the last base of nonsense codon TGA. The base sequence of the first three codons is ATGGTACGC. These should and do code for the first three amino acids (AAs) methionine, valine and arginine in the gene E protein. The last two codons (before the stop codon) start at nucleotide 446 and possess the sequence AAGGAG. These should and do code for the last two AAs, lysine and glutamic acid respectively at positions 89 and 90. The overlapping of the two genes and the partial AA sequences of these proteins are shown on the next page:

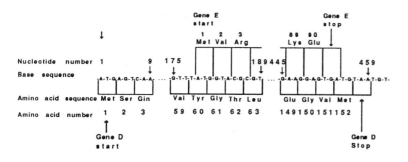

c) Because their reading frames are different, i.e., they are out of phase by one nucleotide as shown above.

31. Each mutant allele produces a polypeptide (pp) with defective primary, secondary and tertiary structures because a single site missense mutation results in a single amino acid substitution at the corresponding position in the pp chain. In homokaryons e.g. $ad^{F4}ad^{F4}$ and $ad^{F39}ad^{F39}$ the quaternary structure of adenylosuccinase would be defective with little catalytic activity (<1% of that of the *wild-type* form of the enzyme and therefore a mutant phenotype). In heterokaryons each mutant allele will specify the synthesis of its particular defective pp chain. the polypeptides specified by the two alleles, if produced in equal proportions, will unite at random to form three kinds of dimers (enzymes) in the proportions shown below:

Enzymes (dimers) in heterokaryons (F4,F39)

allele	polypeptide	types and proportions	quaternary structure	% activity relative to *wild-type*
ad^{F4}	•————	1/4 *mutant* 1	almost completely defective	< 1%
		1/4 *mutant* 2	almost completely defective	< 1%
ad^{F39}	————•	1/2 hybrid*	more normal than other 2 forms of enzyme	~25%

34. a)

```
        H88
   ———————————————————
   B37  C51   C33   347   B138
   ———  ———   ———   ———   ————
             184   782
             ———   ———
```

Insufficient information to unequivocally place mutants B37 and B138. All that is known is that both these mutants are outside the rII region within which mutants C51 and 347 reside, but both mutants overlap at least part of the H88 deletion. The information is also insufficient to determine whether the H88 deletion embraces all the other smaller deletions completely or partially, with the remainder of their missing segments extending beyond the bounds (ends) of the larger deletion mutant.

b) Conduct two complementation tests with each deletion mutant. One with a mutant possessing a mutation in the A gene and the other with a mutant possessing a mutation in the B gene. If complementation occurs in the first cross the deletion mutation occurs in the B gene and *vice-versa* if the other cross exhibits complementation. If complementation fails to occur in both crosses the deletion mutant must partially extend into both genes A and B.

35. a) Mutations 47, 101, 104 occur in the A gene because they fail to complement each other in mixed infections of *E. coli* strain K. As a result no progeny (*mutant* phenotype) are produced. Mutation 102 occurs in the B gene because complementation and a normal burst size (*wild-type*) phenotype) occurs in mixed infections of the host strain.

b) Mutant 104 is probably a deletion mutant.

Chapter 20
Biochemical Genetics

7. Since Tyr and His are each bound to two amino-acids, these must occur internally in the tetrapeptide. Histidine is flanked by phenylalanine and tyrosine whereas tyrosine is flanked by histidine and leucine. From the data it cannot be determined whether Phe and Leu are at the carboxyl and amino termini of the tetrapeptide or *vice versa*.
Therefore the amino acid sequence in this tetrapeptide is either NH_2 - Phe-His-Tyr-Leu-COOH or
NH_2 - Leu - Tyr- His - phe - COOH

b) NH_2-Met-Gln-Arg-Tyr-Glu-Ser-Leu-Phe-Ala-Gln-Leu-Lys-Glu-Arg-Lys-Glu-Gly-Ala-Phe-Val-Pro-Phe-Val-Thr-Leu-Gly-Asp-Pro-Gly-Ile-COOH

8. *Red* eyes (*wild-type*) may be the result of the presence of both *brown* and *scarlet* pigments, each produced by a separate biochemical pathway. For example:

... A → B → C → *brown* pigment
                              → *dull red* (*wild-type*)
... R → S → T → *scarlet* pigment

If we assume that the allele *Bw* of a gene produces an enzyme that converts S to T in the pathway producing *scarlet* pigment and that the allele *bw* does not, then *bwbw* individuals will have *brown* eyes. Furthermore, allele *St* of another gene produces an enzyme that converts A to B in the pathway producing *brown* pigment. If allele *st* does not function in this respect, *stst* flies will not synthesize *brown* pigment and therefore will have *scarlet* eyes. If flies are of genotypes *BwBwStSt*, *BwbwStSt*, *BwBwStst* or *BwbwStst*, both pigments will be produced and the flies will have *dull red* eyes. If flies are *bwbwstst* neither pigment will be produced and they will have *white* eyes.

9. a) (i) The 3:1 F_2 phenotypic ratio in each of the first three crosses indicates that *blue*, *lilac* and *deep-pink* are determined by the alleles R, r' and r respectively of a single gene. The allele R is completely dominant to r' and r whereas r' is dominant to r.
(ii) The F_1 phenotypes and the 9:3:3:1 F_2 phenotypic ratio in each of the last two crosses indicate that in addition to alleles at the R locus, alleles at another independently inherited locus M are involved. Genotypic explanation of the last two crosses:
Blue (RRMM) x *salmon-pink* (rrmm) → F_1 RrMm (*blue*) → F_2 9 R_M_ (*blue*) : 3 R_mm (*slaty-blue*) : 3 rrM_ (*deep-pink*) : 1 rrmm (*salmon-pink*).

b) General - the alleles at the R locus via their specific forms of the enzyme R determine the number and position of hydroxyl (OH) groups on the lateral benzene ring of anthocyanidin molecules: R results in OH's at 3', 4' and 5' positions, r' in hydroxyl groups at 3' and 4' and r only at the 4' location. Alleles at the M locus via their specific forms of enzyme M determine the number and position of sugars (R) attached to the anthocyanidin molecule: M (dominant) results in a sugar being attached at both the 3 and 5 carbon positions; m (recessive) results in one sugar molecule at the 3 carbon position.
Specific - The phenotypes in the first three crosses have the following biochemical basis: *blue* colour occurs because of hydroxyl groups at 3', 4' and 5' positions and the attachment of two R molecules (delphinidin molecules), *lilac* results from OH's at 3' and 4' and 2 Rs (cyanidin molecules) and *deep-pink* occurs because of an OH at 4' and two Rs (pelargonidin molecules). In the cross *blue* x *salmon-pink*, *salmon pink* occurs because of an OH at 4' and only one R; *slaty-blue* is the result of OHs at 3', 4' and 5' and only one R. The differences between *blue* and *slaty-blue* on one hand and *deep-pink* and *slaty-pink* on the other are due to the number of R molecules attached to the anthocyanidin molecules: 2Rs in *blue* and *deep-pink* and only one in *slaty-blue* and *salmon-pink*.
In the cross *lilac* x *salmon-pink* the biochemical basis for *slaty-lilac* colour is due to OHs at 3' and 4' and the attachment of only one R. The difference between *lilac* and *slaty-lilac* has the same molecular basis as the difference between *blue* and *slaty-blue*.

c) *Slaty-lilac* molecules possess hydroxyl groups at 3' and 4' positions on the lateral benzene ring and one sugar molecule at the 3 carbon location.

10. a) Four *wild-type* genes mutated to produce the six mutants.
(i) The mutant genes b_1 and b_2 for *brown* 1 and *brown* 2 respectively are allelic since there is no complementation. The F_1 of the cross between them and all F_2's are mutant (*brown*).
(ii) The mutant genes r_1 and r_2 for *red* 1 and *red* 2 are allelic. There is no or only partial complementation in the F_1 of the cross between them and the F_2 shows a monohybrid 1:2:1 ratio (the heterozygotes express a *reddish-blue* phenotype and the two homozygotes express the parental (*red*) phenotypes.
(iii) b_1 and b_2 are non-allelic to the mutant genes for *red* 1, 2, 3 and 4 since the F_1's express the *wild-type* (*purple*) phenotype and the F_2's of the first cross show the classical dihybrid ratio 9:3:3:1.
(iv) The mutant genes for *red* 3 and *red* 4 are non-allelic with r_1 or r_2 since in the third cross the F_1's are *wild-type* and the F_2 phenotypic ratio is a dihybrid (9:7) one.
(v) The mutant genes re and rd which determine *red* 3 and *red* 4 are non-allelic as evidenced by the fact that the F_2 of the last cross shows a dihybrid (5:4:7) ratio. The F_1 phenotype does not indicate whether the genes are allelic or not.

B gene	R gene	Re gene	Rd gene
b1	R	Re	Rd
b2	r	re	rd

b) The four genes are not linked. In all non-allelic (two gene) crosses the F_2 ratios (9:3:3:1, 9:7, 5:4:7) are those expected with independent segregation. The genes are either on different chromosomes or the same one 50 or more map units apart.

c) The B enzyme produced by the *brown* locus contains one type of polypeptide chain and is in all likelihood a monomeric enzyme. The R enzyme produced by one (*red* 1 and *red* 2) of the three *red* loci contains two types of polypeptide chains and is therefore at least dimeric in composition. The enzymes A, D, E and P are in all likelihood tetrameric structures as evidenced by the results of cross 5.

d) The four genes act in different biochemical pathways:

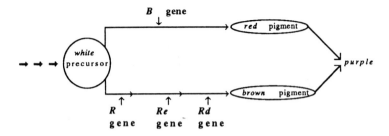

The *blue* phenotype is due to double heterozygosity at the *Re* and *Rd* loci (*RereRdrd*). The *purple* phenotype is due to heterozygosity at one or none of the *Re* and *Rd* loci (*ReReRd_*). The *white* phenotype would be the consequence of homozygosity for the recessive alleles at the *B* locus and one or more of the three loci for *red* pigment.

11. a) The primary structure refers to the amino acid sequence of a polypeptide (pp) chain. Therefore all proteins, regardless of the number of pps in their structure, will possess a primary structure.

b) The secondary structure is the regular configuration of each pp chain which in ribonuclease, insulin, alkaline phosphatase and human adult hemoglobin molecules is an α-helix. Therefore all proteins also possess a secondary structure.

c) Each pp and therefore all proteins possessing only one pp chain will possess a final tertiary 3-D structure which determines the function of the pp and protein. Of the four proteins discussed in this question, only the final structure of ribonuclease will be a tertiary one.

d) The three proteins, insulin, alkaline phosphatase and human adult hemoglobin, comprising two or more pps will also have a fourth level of structural organization, a quaternary structure. In these proteins, each component pp chain will have its own primary, secondary and tertiary structure. Superimposed on this is a specific association of the component pps to form the final complex configuration which constitutes the biologically active form of these proteins.

12. Since allele *C* produces the maximum amount of pigment in either the homozygous or heterozygous condition, it is considered to be a dominant allele. Its biochemical basis maybe that the allele *C* produces an enzyme of sufficient activity to convert all the available substrate to pigment. Thus *CC* produces no more pigment than *C_*.

The alleles c^k, c^d and c^r are hypomorphic since they contribute towards the dominant effect. C^k has greater effect than c^d which in turn has greater effect than c^r; $C^k c^k$ produces more pigment than $c^d c^d$ which in turn produces more pigment than $c^r c^r$.

The hypomorphic activity of the allels c^k, c^d and c^r may be the result of the production of an enzyme by each of these alleles which is less efficient than the *wild-type* enzyme, although it acts in the same reaction in producing melanin.

Since pigment is either not produced or not detected in $c^a c^a$ individuals, allele c^a either produces no enzyme, a partial enzyme or a full length but completely inactive enzyme. Absence of enzyme or its presence in incomplete length and defective structure is due to a nonsense (chain-terminating) codon. A full-length completely functionless enzyme is due to a missense mutation in a codon specifying a specific amino acid at or near the active site of the enzyme. Such an amino acid substitution has the effect of drastically altering the final structure of the enzyme and therefore causing the enzyme to be functionless. Regardless of the nature of the c^a enzyme, substrate will not be converted to pigment in $c^a c^a$ individuals.

15. a) The reason why the cross second *revertant* x *wild-type* (w.t.) produced *scarlet* segregants is that the mutation in *scarlet* flies did not occur at the *scarlet* locus from *st* to *St*. Instead mutation occurred at another (suppressor) locus from the recessive *su* (non-suppressor allele) to the dominant *Su* (suppressor) allele. Therefore original *scarlet* flies possess the genotype *ststsusu* and the homozygous *wild-type* revertants must be *ststSuSu*. Explanations of cross: second w.t. revertant (*ststSuSu*) x original w.t. (*StStsusu*) → F_1 *StstSusu* (w.t.) → F_2 9 *St_Su_* (w.t.) : 3 *St_susu* (w.t.) : 3 *ststSu_* (w.t.) : 1 *ststsusu* (*scarlet*) → 15 w.t. : 1 scarlet.

Note: First *revertant* x *original* w.t. → all w.t. because reversion to w.t. occurred at the *scarlet* locus, most likely at the same base-pair site of the original mutation *St* to *st*. For example *St* (CG at this base-pair site 5'TAC3'/3'ATG5' in codon 50 which specifies tyrosine) → *st* (GC at this site in codon 50 resulting in a TAG3'/3'ATC5' nonsense = chain terminating codon → *St* (substitution of GC by CG thus generating the original tyrosine specifying codon). Since the genetic code is degenerate the reverse mutation could also have occurred at the third bp site in codon 50 from CG to TA, creating another codon 5'TAT3' which would also specify the incorporation of tyrosine in the corresponding postion in the pp chain.

b) The reversion in the second cross may be due to the fact that the suppressor gene codes for a specific minor tRNA molecule. If so, the suppressor allele (in this case dominant *Su*) changes the way in which the mRNA specified by the mutant allele *st* is read. Suppressors exist for both missense and nonsense codons. Transfer RNAs with altered anticodons are the molecular basis of suppressor gene action in these cases. Assume that the mutant allele *st* possesses a nonsense codon 5'TAG3'/3'ATC5' at position 50 which arose from the tyrosine specifying codon 5'TAC3'/3'ATG5' and therefore the pp and protein in *ststsusu* flies is short and nonfunctional. These flies do not synthesize *brown* pigment because of a metabolic block in the biochemical pathway; their eyes are *scarlet*, due to *red* pigment synthesized via another pathway. In molecular terms, *ststsusu* flies would possess the normal allele at the gene locus coding for the minor tRNA which recognizes tyrosine codons in the mRNA produced by the alleles at the *St* and other structural gene loci. As a consequence, the *st* mRNA containing a nonsense codon would not be recognized as a sense codon with the result that a nonfunctional incomplete protein would be produced. However, if a mutation occurs in the *scarlet* (*ststsusu*) strain from *su* to *Su* (suppressor allele) in the region of the gene specifiying the anticodon segment of the minor tRNA carrying tyrosine, e.g., 5'GTA3'/3'CAT5' → 5'CTA3'/3'GAT5', thus changing the anticodon region in minor $tRNA^{tyr}$ from 3'AUG5' to 3'AUC5', the nonsense codons 5'UAG3' would be recognized by the mutant minor tRNA molecules as sense and since it like the *wild-type* minor tRNA carries tyrosine, this amino acid would be inserted into its original position. Full length fully functional original type protein would be produced. Therefore *brown* pigment would be synthesized and flies would express a *wild-type* eye colour.

16. a) In strain one the metabolic block is for the biochemical reaction cysteine → cystathionine.
In strain two the metabolic block is for the biochemical reaction cystathionine → homocysteine.

In strain three the metabolic block is for the biochemical reaction homocysteine → methionine.
Explanation: Nutritional requirement is satisfied by supplying a substance after but not before the metabolic block.

b) By determining whether or not supplying the precursor satisfies the nutritional requirement of the mutant strain.

17. a) The growth requirement of a particular biochemical mutant on MM is satisfied by adding a substance produced after the metabolic block. Since the growth requirement of all four mutants is satisfied by supplementing MM with thiamine, this compound must be the end product of the pathway (or the last product produced in the portion of the pathway affected in these four mutants).
When an attempt is made to order the precursors into a linear pathway the data argue against such a pathway: (i) if the sequence was pyrimidine → thiazole → cmpd B → thiamine, a pyrimidine mutant strain should grow on MM when supplemented with thiazole - this does not occur. (ii) If the sequence was thiazole → pyrimidine → cmpd B → thiamine, the thiazole mutant should grow on MM when the latter is supplemented with pyrimidine - this is also not observed. Therefore the pathway must be branched.

b) The sequence of biochemical steps is as shown:

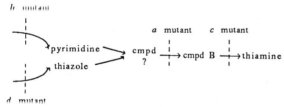

c) Mutant *a* blocks the conversion of cmpd B to thiamine. Mutant *b* prevents pyrimidine synthesis. Mutant *c* is blocked for the last step in the pathway. Mutant *d* prevents thiazole production.

18. Since the mutant allele $proα1^-$ causes severe multiple phenotypic effects, substitution of cysteine for glycine in a critical region of the α1 polypeptide (pp) chain (at position 988) must affect the secondary and tertiary structures of the α1 pp and the quaternary structure of the type I collagen. From the information provided one cannot determine the specific effect(s) of the amino acid (AA) substitution on the triple-helix other than it must significantly alter the quaternary structure and function of type I collagen. This may result in inefficient and reduced secretion of the protein, a lower melting temperature than that of the normal protein and instability in the cell, all of which must interfere with bone formation, blood vessel development and other effects mentioned in the question. Lethality occurs probably because of a number of reasons including significantly reduced amounts of secretion of type I collagen, a large portion (≈75%) of which is of the mutant type and is unstable.

b) Type I collagen is a component of many different tissues and organs: bones, blood vessels, ears, heart, eyes and others. When this protein is rendered defective it prevents all these tissues from developing normally.

c) Because the mutant allele determining the disease expresses its potential in a heterozygote whose parents were both *normal*. Had the mutant allele been recessive, the heterozygote would have been *unaffected*. One cannot determine from studying this heterozygote alone whether the allele is completely or incompletely dominant. This can only be done by comparing phentoypes of individuals that are homozygous and heterozygous for the mutant allele.

d) Since the affected individual was heterozygous at the $proα1$ locus ($proα1^+ proα1^-$) only and homozygous for the normal allele $proα2$ coding for the α2 pp, all type I proteins would have a normal α2 pp. Since (i) type I collagen is a heterodimer composed of two α1 pps and one α2 pp (ii) if the normal and mutant α1 pps are produced in equal proportions and any two of them can assemble with equal probability with an α2 pp to form type I collagen and (iii) if a type I collagen molecule with one or two mutant pps is defective, we can expect 75% of all proteins to be abnormal. This can be determined from the binomial expression $(a+b)^n$ where a = chance of any α1 pp being normal, b = chance of any α1 being mutant and n=2; the number of such pps per heterodimer is $2ab + b^2 = 1/2 + 1/4 = 3/4$.

24. Since (i) human adult hemoglobins are tetramers (ii) such molecules are never hybrid, e.g., $α^A β^A β^C$ and (iii) only four types of adult hemoglobins (A, C, G, X) are generated from the four different polypeptide chains $α^A$, $α^G$, $β^A$, and $β^C$ the following conclusions can be drawn:
a) The four polypeptide chains (pp) are synthesized separately and must be specified by different genes.
b) The two pps in each dimer must be identical, i.e., two $α^A$ pps, another two $α^G$ chains and still others two $β^A$ and two $β^C$ chains. Dimerization obviously follows pp synthesis. The mechanism preventing hybrid dimer formation is unknown.
c) Since adult hemoglobins are of four types only, such molecules cannot possess four identical pps; nor can they be composed of α chains of two kinds or β chains of two kinds. Assemblage of the dimers into hemoglobin molecules always involves (i) an α and a β dimer (ii) each of the two α dimers $α^A$ and $α^G$ can then be assembled with each of the β dimers $β^A$ and $β^C$

27. The mutants that do not revert or recombine with some of the TSase mutants indicate they involve deletions of segments of the *E. coli* genome. In these mutants since all three reactions are abolished this indicates that (i) the genes A and B which code for polypeptides A and B respectively in TSase are closely or completely linked and (ii) the deletions in these mutants are missing at least parts of both genes.
Indoleglycerol phosphate (InGP) is the immediate precursor of tryptophan and TSase catalyzes the conversion of InGP + serine to tryptophan.

28. The mutations in the two nonsense mutants 5972 and 9778 arose in different genes. **Evidence:**
(i) Their metabolic blocks are different at least in part. For example, mutant 5972 cannot convert chorismic acid to anthranilic acid because of a defective anthranilate synthetase. Mutant 9778 not only lacks anthranilate synthetase activity but also PR transferase activity required to convert anthranilic acid to PRA.
(ii) When extracts (including polypeptides), from the two mutants were mixed, the enzyme anthranilate synthetase possessed normal (*wild-type*) activity. Thus there was complementation between the products (polypeptides) produced by the two mutants. Since this is so, two genes must be involved. If one gene locus was involved, only one biochemical reaction would have been affected by both mutants, e.g., conversion of chorismic acid to anthranilic acid. Moreover, complementation would have been interallelic involving a multimeric enzyme with polypeptides specified by different mutant alleles (heteroalleles) of the same gene.

The enzyme anthranilate synthetase is composed of two different kinds of polypeptide chains, each present at least once. Therefore this enzyme is at least a dimer. The evidence for this comes from the following observations:
(i) Although both mutants produce a defective anthranilate synthetase, mutant 9778 also gives rise to a defective PR transferase. This is impossible if both mutants were due to different alleles of the same gene since different alleles of the gene would determine different forms of one kind of polypeptide only which may be a component of one or more enzymes. Therefore, either all mutant alleles produce defective forms of one specific enzyme only or all of them produce defective forms of the same two or more different enzymes. This does not occur in this case.
(ii) Extracts from the two mutants show normal anthranilate synthetase activity when mixed.

The enzyme PR transferase is in all probability composed of only one kind of polypeptide chain, present either only once or more times in this enzyme.
Evidence:
(i) Mutant 9778 alone produces defective PR transferase.
(ii) Very rarely is the polypeptide specified by a given structural gene a component of two enzymes. Since the polypeptide specified by the mutant gene in 9778 occurs in a subunit of anthranilate synthetase and is also present in PR transferase, it is most likely that PR transferase is a monomeric enzyme (composed of only one polypeptide chain). It could be a multimeric one with all subunits coded for by the same gene.

Summary

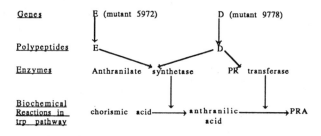

29. The results presented contain two significant facts that have a bearing on the one-gene one-enzyme hypothesis: (i) Hb-S hemoglobin differs from normal (Hb-A) hemoglobin by a single amino acid (AA) substitution in the β chain; HO-2 hemoglobin differs from normal by a single AA change in the α chain (ii) The mating between a normal (Hb-A = no Hb-S, no HO-2) parent and a double mutant (Hb-S, HO-2) one, produce offspring of four phenotypes: Two of these, normal (Hb-A) and double mutant (Hb-S, HO-2) were parental (P) types and the other two, Hb-S, no HO-2 and no HO-2 were recombinant (R) types.
Conclusions:
a) Since both P and R types occur among the offspring two pairs of alleles (Hh and Ss) are involved. If one gene specified the synthesis of the two kinds of chains only parental types of progeny (Hb-A and Hb-S,HO-2) would have been possible.
b) HO-2 hemoglobin is abnormal in its α chain and hemoglobin Hb-S is abnormal in its β chain; mutant alleles H and S of two different genes produce HO-2 and Hb-S hemoglobin. This indicates that different kinds of pp chains are coded for by different (non-allelic) genes. These data were the first proof of this fact.

Subsequent to these findings, it has been shown that it is universal that different kinds of pp chains are specified by different genes.
The results presented require a modification from the one-gene one-enzyme to a one-gene one-pp hypothesis. The structural and functional specificity of each kind of pp chain is totally determined by one particular structural gene. This is the primary and sole biochemical function of each structural gene. If a protein contains one kind of pp chain present one or more times, the one-gene one-enzyme hypothesis is equivalent to the one-gene one-pp hypothesis. Not only does the gene specify one protein, but the total specificity of such a protein is determined by that one gene. If a protein is composed of two or more kinds of pp chains, e.g., α and β chains as in human adult hemoglobin, each gene determines the specificity of one pp chain but the total structural and functional specificity of the protein is determined by the two or more genes involved.

33. **Mode of specification of esterase:** Since each of the three possible crosses N x F, N x S and F x S involving the true-breeding lines a with slow (S), b with normal (N) and c with fast (F) electrophoretic forms of this enzyme produces a 1:2:1 F_2 phenotypic ratio, each of the three different forms of the enzyme is specified by a different allele of the same gene: E^N (for normal electrophoretic migration), E^F (for fast mobility) and E^S (for slow migration).
Number of polypeptides: Since the F_1's in each cross produce three electrophoretic bands, two in the same positions as those formed by the two parents and the third in an intermediate position (i) the enzyme is a dimer and (ii) the intermediate bands represent a hybrid enzyme. For example, in the F_1 from the cross N ($E^N E^N$) x F ($E^F E^F$):

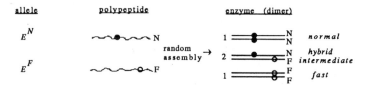

Had the enzyme been a monomer only two bands would have been produced by the F_1 heterozygote. If esterase was a trimer, F_1 heterozygotes would have produced four bands.

34. a) The polypeptides (pp) A and B from LDH-1 and LDH-5 respectively, when mixed and then reassembled, form five different isozymes. This indicates that LDH is a tetramer - each isozyme is composed of four pps. If the enzyme was a monomer it could exist in only two isozymic forms; if it were a dimer or trimer it would occur in three and four molecular forms respectively. The five isozymic types of LDH-1 through LDH-5 are formed in the proportions 1:4:6:4:1. This indicates that the A and B chains assemble at random to form the tetramers. Expansion of the binomial $(a+b)^n$ will indicate the expected types and proportions of the five isozyme types derived from a mixture of an equal number of A and B chains. Note: a represents the A pp chain's probability of incorporation into any isozyme = 1/2; b represents the B chain's probability of incorporation into any tetramer also 1/2; n=4 indicates the number of pps per LDH molecule.

Term	a^4	$4a^3b$	$6a^2b^2$	$4ab^3$	b^4
Probability	1/16	4/16	6/16	4/16	1/16
LDH isomzyme	4A*	3A+1B*	2A+2B	1A+3B	4B

*A and B refer to the kinds of polypeptides in each isozyme

b) (i) Only one allele pair is responsible for the phenotypic differences as evidenced by the results of the second cross. Assume that the allelic differences occur at the B locus with B^a and B^b determining the *normal* and *mutant* forms of the enzyme. In crosses between homozygous *normal* and heterozygotes, males and females were distributed approximately equally within the two phenotypic classes. This indicates that the B locus is an autosomal one. Although the number of progeny from the third cross is too small to observe the expected 1:2:1 phenotypic ratio, there is sufficient evidence to conclude that these results are determined by the $B^a B^b$ allele pair. Since the A and B chains of LDH are specified by genes A and B respectively, all the animals studied must be homozygous for A.

(ii) Expect only one pp to exist in alternative forms since only one gene (B) in these mice exists in different allelic froms B^a and B^b. Also in heterozygotes, as in *normal* and *mutant* homozygotes, band 5 contains only one kind of pp, specifically A. Moreover, in heterozygotes, 5, 4, 3 and 2 subbands occur in bands 1, 2, 3 and 4. This suggests that these subbands contain different numbers and combinations of *normal* and *mutant* forms of pps of one kind of chain only. If both A and B polypeptides existed in alternative forms many more subbands would have occurred including subbands of band 5.

(iii) Polypeptide B exists in alternative forms. Bases for the conclusion: (a) Only the B gene in these animals exists in different allelic forms B^a and B^b which determine two different electrophoretic (isozymic) forms of the pp. (b) There are no subbands in band 5 which contain pp A indicating that the A chain exists in only one electrophoretic form.

(iv) Genotypes: homozygous normal - $AAB^a B^a$, homozygous mutant - $AAB^b B^b$, heterozygous - $AAB^a B^b$.

The polypeptide constitution of each band and subband in individuals with the three genotypes and the relative proportions of the subunits in each isozyme in heterozygotes are as follows:

LDH band	Homozygous normal ($AAB^a B^a$)	Homozygous mutant ($AAB^b B^b$)	Heterozygotes* ($AAB^a B^b$)	
1	$B^a B^a B^a B^a$		1/16	$B^a B^a B^a B^a$
			4/16	$B^a B^a B^a B^b$
			6/16	$B^a B^a B^b B^b$
			4/16	$B^a B^b B^b B^b$
		$B^b B^b B^b B^b$	1/16	$B^b B^b B^b B^b$
2	$AB^a B^a B^a$		1/8	$AB^a B^a B^a$
			3/8	$AB^a B^a B^b$
			3/8	$AB^a B^b B^b$
		$AB^b B^b B^b$	1/8	$AB^b B^b B^b$
3	$AAB^a B^a$		1/4	$AAB^a B^a$
			1/2	$AAB^a B^b$
		$AAB^b B^b$	1/4	$AAB^b B^b$
4	$AAAB^a$		1/2	$AAAB^a$
		$AAAB^b$	1/2	$AAAB^b$
5	$AAAA$	$AAAA$		$AAAA$

*Subunit pps are A, B^a and B^b. The fractions preceding the heterozygous tetramers indicate the relative proportions of each subband within the major band.

35. a) α chain - 29,500;   β chain - 49,500;   tryptophan synthetase - 159,000
Since each amino acid has a molecular weight of approximately 110, the number of amino acids/chain can be calculated as follows:
α chain 29,500/110 = 268;   β chain 49,500/110 = 450

Similarly, the enzyme tryptophan synthetase contains 159,000/110 = 1,445 amino acids. If this enzyme was composed of one of each of the two kinds of chains (monomeric for α and β chains), it would possess 718 amino acids (268 + 450).

b) Tryptophan synthetase must contain two kinds of polypeptides - α and β chains each present twice. When the number of amino acids in the two chains are multiplied by two and their sums added, the total corresponds closely to the number of amino acids in the enzyme 1,445 amino acids, i.e., [(2 x 268) + (2 x 450) = 1436].

The discrepancy between the two estimates of the number of amino acids in tryptophan synthetase is due at least in part to the fact that the weight of amino acids is approximated.

Chapter 21
Protein Synthesis: Transcription and Translation

2. b) (i) **Ribosomal RNA (rRNA)**
(a) Located in ribosomes in the cytoplasm (and also in mitochondria and chloroplasts)
(b) 1) In eukaryotes - the 18S rRNA in the small subunit and the 28S rRNA in the large subunit of ribosomes are specified by the nuclear DNA of the repeated 18S and 28S genes in the nucleolus-organizing region. The 5S rRNA in the larger subunit is coded for by the DNA in repeated 5S genes at specific but different locations in the genome in different species.
2) In prokaryotes (bacteria) - the 5S, 16S and 23S rRNAs are transcribed from 5S, 16S and 23S genes which are present as a repeating unit at various places in the genome, e.g., in *E. coli* this unit is believed to be present at 6 different positions in the *E. coli* genome.
(c) The rRNAs are single stranded. They probably play a structural role and bind tRNA and mRNA to ribosomes.

(ii) **Messenger RNA (mRNA)**
(a) Located in the cytoplasm (and also in mitochondria and chloroplasts).
(b) Is highly heterogenous; transcribed from the DNA sense strands of structural genes.
(c) Since it is transcribed from the sense strand by RNA polymerase, which functions like the DNA polymerases, the mRNA is single-stranded with a base sequence complementary to the template strand. Since different structural genes code for different polypeptide and proteins and these vary in size (length) and amino acid composition, their mRNAs will also vary in size and nucleotide sequence. Some mRNAs are larger than others because they are transcribed from operons and are polygenic messengers coding for two or more kinds of polypeptides. mRNAs are generally short-lived in bacteria but not necessarily so in eukaryotes. They associate with ribosomes during translation but are not incorporated into ribosome structure. In eukaryotes primary transcripts are processed to form mRNAs - capping the 5'-end, adding poly-A tails to the 3' ends of the mRNA and excising the introns of discontinuous genes.

(iii) **Transfer RNA (tRNA)**
(a) Located in the cytoplasm and in mitochondria and chloroplasts.
(b) Approximately 30-40 types of tRNAs are specified by non-redundant and redundant tRNA genes in the genomes of bacteria and eukaryotes respectively.
(c) tRNAs are also single-stranded. These molecules are 77-87 nucleotides long, uniform in size (MW=25,000) and possess a uniform sedimentation constant (4S). Each tRNA possesses a clover-leaf like structure, transports a specific amino acid to the mRNA on the ribosomes and recognizes a codon on the mRNA. tRNAs read a gene's message and ensure that amino acids are assembled in a sequence in the polypeptide chain as specified by the structural gene. tRNAs are also processed from pre-tRNA transcripts.

c) Viruses can only code for tRNAs and mRNAs. Since they parasitize host cells which produce ribosomes and provide the raw materials for the synthesis of all viral proteins, viruses do not need to code for and synthesize rRNA.

f) They are not coupled because each message is read (translated) separately. The mRNA specified by each structural gene begins at a chain initiating codon (AUG) and terminates with at least one chain-terminating (stop = nonsense) codon. Since such a codon (UGA, UAA, UAG) does not code for amino acid incorporation into the growing polypeptide chain, the latter is released from the ribosomes, preventing further addition of amino acids.

6.

7. The initiation complex is an aggregate of various molecules at the 5' end of a mRNA that initiates the translation of the mRNA codons into the polypeptide chain. It comprises mRNA, the smaller ribosomal subunit, initiator tRNA (specific in prokaryotes), GTP, Mg^{++} and at least three protein initiation factors. The initiation complex is formed as follows in *E. coli*: (i) The small ribosomal subunit (30S) binds to the initiation protein F3. This complex recognizes the Shine-Delgarno sequence of up to six ribonucleotides. This sequence precedes the chain-initiation AUG codon and binds to mRNA. The F3 protein binds the 30S subunit to the mRNA. (ii) Subsequently, the 30S subunit moves along the mRNA until its "P" (peptide binding) site is opposite the AUG codon. (iii) Several other initiation proteins including F1, F2 and also Mg^{++} and GTP then facilitate the binding of charged formylmethionyl-tRNA (Met-tRNA$_F$) to the P site in the 30S-mRNA complex.

The following events occur to begin polypeptide synthesis:
(i) A large (50S) subunit is added to the initiation complex to form a 70S ribosome. The F1 protein which probably helps bind the 50S subunit to the complex is released once the binding occurs. (ii) A second tRNA with its AA specified by the second codon becomes bound to the complex at the "A" site with the help of elongation factors EF-Tu and Ts. (iii) The first peptide bond is formed by peptidyl transferase (an integral part of the 50S subunit) between the COOH end of the first AA and the NH_2 end of the AA specified by the second codon. As a result the growing pp chain is attached to the "A" site. Simultaneously, the tRNA is released from the "P" site. The pp chain therefore grows from the NH_2 to the COOH end. The binding of the first two AAs is the first step in pp elongation.
(iv) The entire mRNA-tRNA-AA_1-AA_2 complex then moves a distance of one codon along the ribosome in the direction of the "P" site. This movement of the complex from the "A" site to the "P" site requires several proteins such as EF-Ts and EF-G as well as GTP. The second codon in the mRNA will now be at the "P" site to which the growing pp chain is attached. The third codon in the mRNA is now at the "A" site, it is free to hydrogen bond with the correct AA-tRNA complex.
The sequence of elongation is repeated over and over until the complete pp is synthesized. A single AA is added to the growing pp chain each time the mRNA

advances through the ribosome. The termination of protein synthesis is signaled by the presence of one of three codons (UAG, UAA or UGA) in the mRNA. These codons neither specify an AA nor do they direct tRNA into the "A" site. The completed pp is initially attached to the terminal tRNA at the "P" site. A chain-terminating codon signals the action of GTP-dependent release factors (RF1, RF2 RF3) which cleave the pp chain from the terminal tRNA. After cleavage, the tRNA is released and the ribosome dissociates into its subunits.

Translation in eukaryotes: The general aspects of translation in eukaryotes are similar to those in prokaryotes. The differences are: (i) Translation occurs on larger ribosomes in eukaryotes whose rRNAs and proteins are more complex than those in prokaryotes. (ii) Although protein factors that guide initiation, elongation and termination of translation in eukaryotes are similar to those in prokaryotes, more of them appear to be required at each step in eukaryotes. (iii) Initiation of eukaryotic translation does not require the AA formylmethionine. Nevertheless, the AUG codon is essential for the formation of the initiation complex and a unique $tRNA_i^{met}$ is used to initiate translation. (iv) Unlike prokaryotes, eukaryotic RNA transcripts are extensively modified to form mRNAs.

8. a) The UAA, UGA and UAG triplets are nonsense codons which signal the termination of translation.

b) (i) A mutation changes a sense codon, which specifies the incorporation of a specific amino acid in the polypeptide (pp) chain, to a nonsense codon that terminates translation, resulting in the formation of incomplete pp chains. The closer the nonsense mutation is to the front end of the gene the shorter the pp chain will be. In most cases mutant alleles with nonsense codons are amorphic; they produce a nonfunctional protein.
(ii) A missense mutation changes a codon which codes for one particular amino acid (AA) into one that specifies a different AA at the same position in the pp chain.
(iii) A frame-shift mutation alters all the codons from the point of addition or deletion of a nucleotide pair. This results in a pp with an AA sequence that is also changed from the point of the mutation. The AA sequence is usually significantly different from that of the *wild-type* pp, that it renders the pp functionless or almost so.

13. The growing pp chain must be bound to the ribosomes and not to mRNA. If this were not so, degradation of mRNA by ribonuclease should release the pp chain from the ribosomes.

15. a) (i) The thin fiber that is destroyed by DNase is a portion of the *E. coli* chromosome (DNA). (ii) The granules on the strings which RNase removes from the fiber (DNA) are the ribosomes. (iii) The mRNAs are the strings attached to the DNA fiber, to which the granules (ribosomes) in turn are attached.

b) Transcription and translation are coupled. The strings (mRNAs) are attached to the DNA; the granules (ribosomes) are associated with the mRNAs on the chromosome. If the latter was not true, ribosomes would not be attached to the mRNAs; the mRNAs would separate from the DNA and be translated elsewhere in the cell.

c) In all likelihood, the granules on the fiber (DNA), one of which has an arrow pointing to it, are RNA polymerase molecules which associate with DNA, select the sense strand and transcribe it.

20. It was already known at that time that genes (which are transcribed) are present on chromosomes in the nucleus and proteins (the products of translation) are produced in the cytoplasm.

22. The six nucleotide long sequence 5'-TATAAT-3' of bases at position -10 in prokaryotic promoters is called the Pribnow box, after its discoverer. It occurs 10 nucleotides upstream from the start of transcription. Although variations in this sequence occur in different promoters, e.g., GATAAT, TATGTT, the sequence represented by each nucleotide present most frequently in all promoters is referred to as the consensus sequence. The sequences are those of the coding strand of DNA.
The second 6 nucleotide long consensus sequence, 5'-TTGACA-3' is found about 35 nucleotides upstream (-35) from the first base transcribed and has no special name.
Mutation studies have revealed the following roles for these two sequences:
(i) Both affect the efficiency of initiation of transcription.
(ii) The -10 sequence is necessary for the initial melting of the DNA duplex to expose the template strand. It is also the region where the core enzyme of the holoenzyme makes contact with the DNA.
(iii) The -35 sequence is necessary for recognition by the sigma factor.

Both strong and weak promoters are known such that initiation of transcription varies from about once every 1 to 2 seconds to once every 10 to 20 minutes.

24. a) Protein synthesis continues to occur after the addition of actinomycin-D because the mRNAs already present and unaffected by the antibiotic, continue to be translated into proteins.

b) The average lifetime of an *E. coli* mRNA is about 20 minutes. This conclusion is based on the fact that once DNA dependent RNA synthesis is terminated, translation of mRNAs continues for approximately 20 minutes.

29. a) An RNA (intron) with enzymatic properties is a ribozyme. Self-splicing refers to the removal of the intron(s) in an RNA molecule followed by the ligation of the exons without the aid of enzymes. The intron behaves catalytically to ligate the exons.

b) **Group I self-splicing introns** require a guanine-containing nucleotide (GMP, GDP or GTP) to be present. The following diagram shows how self-splicing occurs in rRNA of *Tetrahymena thermophila*:

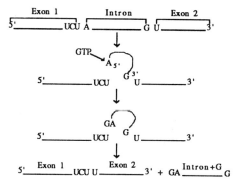

During self-splicing the U-A bond at the left (5') side of the intron is transferred to the GTP. The unbonded U displaces the G at the right (3') side of the intron, religating the RNA with a U-U connection and releasing the intron. All bonds are transfer ones rather than new ones; therefore no external energy source is required.

Group II self-splicing introns - in mitochondrial genes of yeast. These introns do not require an external nucleotide. Instead the first bond is transferred to an adenine nucleotide within the intron, resulting in a lariat structure. A second bond transfer releases the intron. For the lariat to form, the ribose sugar of the adenine nucleotide must make three phosphodiester bonds. The following diagram shows how self-splicing occurs in this group:

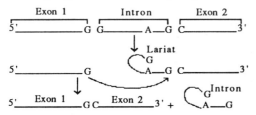

33. The observations permit an unequivocal answer to the question. The mRNA code, specifically its codons, is recognized by the tRNA molecules and not the amino acids (AAs). If the mRNA code was read by the AAs, the alanine that was attached to the cysteine tRNA molecule (the cysteine was changed to alanine by Raney nickel) could not have been incorporated into the peptide. Further support is derived from the fact that when alanine is bound to its own tRNA it was not incorporated into peptide 13.

34. a) The new RNA (present in *E. coli* cells after they are infected by T2 phage) is highly likely mRNA transcribed from the T2 genes. It is responsible for the intracellular production of T2 phage proteins as suggested by the following evidence:
(i) The new RNA and mRNAs share a common property - both have a high turnover rate.
(ii) The A+U/G+C ratio of 1.5 for the new RNA is similar to the A+T/G+C ratio of 1.7 for T2 phage DNA but significantly different from the A+T/G+C ratio of 1.0 for *E. coli* DNA. Thus, it is highly likely that the new RNA is transcribed from T2 DNA.

b) The T2 DNA is used as template for the synthesis of the new RNA. This conclusion is supported by the fact that the new RNA and T2 DNA have similar base compositions. In the DNA-RNA hybridization experiments of Hall, B.D. and S. Spiegelman, PNAS 47: 137, 1961, the heat-denatured T2 DNA hybridized only with the new RNA but not other RNAs. This would indicate that the new RNA and the T2 DNA had complementary base sequences, thereby confirming the above hypothesis.

35. a) (1) **The data of Elsdale et al. (1958) suggest that:**
(i) The nucleolus-organizing (NO) region of DNA resides in a specific secondary constriction region of a particular chromosome.
(ii) Both chromosomes of a pair of normal homologues have NO regions in corresponding secondary constriction positions.
(iii) Deletion of the DNA in an NO region of a chromosome abolishes nucleolus formation on that homologue.
(iv) Toads containing one normal chromosome with the NO region and one mutant homologue without the NO region are heterozygous (*Nunu*) and *uninucleolar*.

(v) The heterozygous *uninucleolar* toads express a normal phenotype and are viable because one NO and nucleolus is sufficient to carry out the essential functions determined by this segment of the DNA.
(vi) The DNA in the NO region must possess genes for essential organismal functions since homozygous recessive (*nunu*) *anucleolar* toads die as tadpoles.

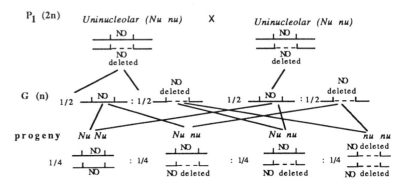

(2) **The data of Brown and Gurdon (1964) indicate that:**
(i) The genes for both 18S and 28S rRNA are located in the same segment of DNA of the NO region. This can account for their co-ordinate synthesis.
(ii) 18S and 28S RNAs are components of ribosomes, the workbenches (organelles) on which proteins are synthesized. If these RNAs are not transcribed (because the 18S and 28S genes have been deleted) ribosomes will not be formed and therefore proteins will not be synthesized. The nucleolus is the site of ribosome synthesis.
(iii) *Anucleolar* (*nunu*) toads survive to the tadpole stage by producing proteins on ribosomes synthesized in the heterozygous (*Nunu*) mother's oocytes.

In summary, deletion of the NO-region DNA and therefore 18S and 28S rRNA genes prevents nucleolus and ribosome formation, both deemed essential for viability.

(3) **Possible genetic explanations for the phenotypic effects:**
(i) Deletion of the NO region (18S and 28S rRNA genes).
(ii) An alternative explanation is that the anucleolate condition might be a consequence of a defect in nucleolus formation which secondarily results in the absence of ribosomal RNA synthesis.
(iii) The alternative hypothesis to that in (ii) would be to have the primary genetic defect prevent ribosomal RNA synthesis. This suggests that the nucleolus marks the location of ribosomal RNA and ribosome synthesis, and the presence of the nucleolus is secondary to these synthetic processes.

b) Wallace and Bernstiel's (1966) results support the conclusion that the NO region possesses the 18S and 28S rRNA genes; deletion of this entire region prevents ribosome formation. **Reasons:**
(i) Failure of rRNA from *wild-type* toads with normal NO regions to hybridize with DNA from the *anucleolar mutants* indicates that these *mutants* did not possess the DNA in the NO (18S and 28S genes) region.
(ii) Intermediate levels of DNA-RNA hybrids in *uninucleolar mutants* are expected if one chromosome possesses an NO region and its homologue does not.
(iii) Some *mutants* showed a linear reduction in NO DNA-rRNA hybridization proportional to dosage of the mutation, suggesting that the larger the segment of

36. a) (i) The molecular weight must be divided by two to estimate the number of genes because only one of the two strands of DNA in each region is transcribed.
(ii) There are approximately 160 18S and 28S rRNA genes in a diploid cell. **Basis for answer:** The MW of DNA in a genome of *D. melanogaster* is 1.2×10^{11}. The MW of 18S + 28S rRNA which is represented by 0.27 of the *wild-type* genome, is 2.1×10^6. Since only one of the DNA strands is transcribed, the amount of DNA that will hybridize to rRNA genes is $(1.2 \times 10^{11}/2 \times 0.27\%) = 1.6 \times 10^8$. Therefore the number of stretches of 18S and 28S rRNA per genome is $1.6 \times 10^8 / 2.1 \times 10^6$ which is ≈ 80.

b) (i) (a) Number of 18S rRNA genes = $2.0 \times 10^8 / 2 \times 0.7 \times 10^6 \approx 140$
(b) Number of 26S rRNA genes = $4 \times 10^8 / 2 \times 1.4 \times 10^6 \approx 140$
(c) Number of genes for all tRNA genes - upper limit $2 \times 10^7 / 2 \times 2.5 \times 10^4 \approx 400$
lower limit - $1.6 \times 10^7 / 2 \times 2.5 \times 10^4 \approx 320$

(ii) Average number of genes for each species of tRNA = 320/60 = 5 to 400/60 $\approx 7 = 5$ to 7.

37. a) The data indicate that tRNA molecules recognize mRNA codons because: (i) the mRNA was recognized by tRNAs and not by amino acids (AAs) and (ii) tRNAs were capable of binding to specific AAs.

b) Poly-UG mRNA will contain the codons for phenylalanine and cysteine. If the codons in the mRNA were recognized by AAs, only phenylalanine would be incorporated into the poly-UG mRNA after the reduction of cysteine to alanine.

c) The data suggest that the attachment of tRNA to the AA was catalyzed by an enzyme, capable of recognizing both the structure of the tRNA and the AA side chain. Once the attachment was accomplished, alteration of the AA side chain had no effect on the recognition of mRNA by tRNA implying that (i) the AA binding site of the tRNA was different from the mRNA recognition site and (ii) tRNA did not bind to the AA side chain.

38. a) In *wild-type* and all the *mutants* tested, the amount of DNA that hybridized with 28S rRNA was twice the amount that hybridized with 18S rRNA. Since the MW of 28S rRNA (1.4×10^6 daltons) is also twice that of 18S rRNA (7.0×10^6 daltons), the number of genes for the two kinds of rRNA must be approximately equal.

b) Since the ratio of 2 over 3 (2/3) for the six genotypes is 2.02, 2.09, 2.09, 2.07, 2.05 and 2.05 respectively, the ratio between the amount of DNA complementary to 28S rRNA and that complementary to 18S rRNA is $\approx$ two.

c) Regardless of the size of the deletion of the NO region, the ratio between DNA complementary to 28S rRNA and that complementary to 18S rRNA was always two. This implies that the two kinds of rRNA genes are interspersed in the NO region. If the genes for 28S rRNA were clustered in one block and those for 18S rRNA in another, the above ratio would have varied depending on the extent and the positions of the deletions.

39. The Dintzis data unequivocally support the fourth mode of polypeptide (pp) synthesis, i.e., pp chains are synthesized sequentially by the stepwise addition of single amino acid (AAs) from the amino end to the carboxyl end. **Reasoning:** In the experiment where reticulocytes are exposed to labelled AAs for 30 seconds, only the pps that were partly produced will have had time to be completely synthesized. All the labelled pps will show a gradient of radioactivity with more label present at the COOH end. Moreover, there would not be any complete chains that were labelled at the NH_2 end; all complete chains would be labelled at the COOH end as observed. In the experiment when the cells are exposed to labelled AAs for three minutes, entire new chains will have had time to be synthesized in the presence of labelled AAs. Some pps will now be labelled at the NH_2 end and not at the COOH end. The reverse however was true, again supporting the sequential synthesis of pps from the NH_2 end to the COOH end. If peptides were synthesized separately and then combined (assembled) to form the pp chain, radioactive label should have been found in all or most peptides after any length of exposure - labelled segments of pps should have been interspersed with unlabelled regions in both the experiments. If the chains were formed sequentially starting at either end, the pattern of labelling observed and one opposite to it should have been obtained. If chain formation began internally and proceeded towards each end, pps that completed their synthesis in the 30 second experiment should have been unlabelled in the internal region where synthesis was initiated. They should have been labelled continuously from this region through to and including both ends. In the 3-minute experiment the above results should have been observed in addition to pps being completely labelled. If the chains were formed sequentially, beginning at the carboxyl end, the results should have been the opposite of those obtained.

40. a) The A+T/G+C ratio of DNA is $\approx$ equal to the A+U/G+C ratio of RNA in all organisms studied. This indicates that DNA and RNA not only have similar base compositions and sequences but that DNA serves as a template for RNA synthesis. This deduction is supported by the results obtained with synthetic DNA polymers: when poly-T was used as a template the synthetic RNA contained adenine, whereas that produced with poly-AT as template contained uracil and adenine in alternating sequences.
Since DNA serves as a template for RNA synthesis, all four ribonucleoside 5'-triphosphates are required for RNA production. The RNA synthesized from DNA templates has a base sequence complementary to the template strand as indicated by the studies using synthetic DNA polymers. RNA polymerase transcribes the sense strand of DNA into a single-stranded RNA molecule whose base sequence is complementary to the template. It probably does so by catalyzing the polymerization of ribonucleoside-5' triphosphate into RNA. The fundamental mechanism for RNA synthesis is probably similar to DNA replication. The sense strand is transcribed in the 3' to 5' direction and therefore RNA, like the new DNA strands is synthesized in the 5' to 3' direction by the sequential addition of nucleotides specified by the template strand.

b) The additional information does not help to elucidate the role of RNA polymerase. It does, however, confirm the functions of this enzyme as outlined in (a).

c) **RNA is single-stranded because:**
(i) Neither the amounts of A and U nor those of G and C are equal. If RNA molecules were double-stranded the amounts of A and U should be equal as should be those of the complementary bases G and C.
(ii) The base composition of RNA transcribed from single-stranded ϕX174 DNA is complementary to its template. ϕX174 DNA is single-stranded; the RNA transcribed from it is probably single-stranded because the base composition of RNA is complementary to its base composition. If RNA was double-stranded, the % of A and U should equal the % A and T of ϕX174 DNA.
Synthesis of RNA chains proceeds in the 5' to 3' direction. This is because the attachment of 3'-deoxyadenosine triphosphate to the 3' end of RNA molecules

inhibits further RNA synthesis. If it occurred in the reverse direction, the blocking of the 3' terminal should have no effect on RNA synthesis since the 5' position would still be available for further addition of ribonucleotides.

d) (i) (a) The core enzyme is required for the polymerization of ribonucleotides during RNA synthesis. It has no strand specificity because it can initiate synthesis of RNA chains anywhere along either strand of a gene. As during DNA replication the bases in the RNA chain being synthesized are complementary to those in the DNA template.
(b) Sigma (σ) plays a regulatory function and is involved in recognition of the promoters (start signals) along the DNA template, where RNA transcription is initiated. Since a promoter sequence is found on one of the two DNA strands only, σ must select the correct (sense) strand for transcription.
(c) The rho (ρ) factor appears to bind to the stop signal of a gene to terminate RNA synthesis since chain elongation continues beyond the stop signal in the absence of ρ.

(2) Illustrate the process for yourself.

41. The data suggest that only the lighter strand is transcribed into mRNA because the phenotype is altered almost immediately when the lighter strand was used in transformation. On the other hand, when the heavier strand is used for transformation, phenotypic expression is delayed for 45 minutes, a period only slightly longer than the time required for one-generation (40 minutes). This indicates that the heavier strand must have to replicate before transcription can occur. Thus, the results indicate that mRNA is copied from only one of the two strands of DNA although both can transform a cell's genotype.

44. (i) The β^- allele affects transcription. This conclusion is based on the fact that the AT→GC base-pair substitution at position (-29) occurs in the highly conserved proximal promoter element, the "TATA" box which affects the efficiency of transcription. In the $\beta^-\beta^-$ individual transcription is reduced by 75%. The result is significantly fewer β-polypeptides and hemoglobin molecules with expression of *mild thalassemia*.
The β^o allele affects RNA processing. **Reasoning**: The AT→GC base-pair substitution in the second base-pair position of IVS-2 is in a critical region of the gene. It occurs within the acceptor site of the second intervening sequence. This mutation completely abolishes normal RNA splicing and the formation of normal RNA, thereby resulting in *severe thalassemia*. In $\beta^o\beta^+$ individuals, the IVS-2 is only partially excised such that the mRNA cannot code for a normal β pp.
(ii) Normal hemoglobin molecules can be extracted in the $\beta^-\beta^-$ individual since the mutation is external to the coding regions (genes) for the β and other hemoglobin chains.

45. a) Only the sense strand is transcribed *in vivo*. In SP8 the H strand is the sense strand and therefore the mRNA formed *in vivo*, hybridizes with this strand only.
b) Both DNA strands must be transcribed *in vitro* since both can hybridize with RNA synthesized *in vitro*.
In vivo synthesis of RNA is very specific; it involves the recognition of the promoter site by the sigma factor and the selection of the sense strand before the core enzyme of RNA polymerase can transcribe the sense strand into RNA. Since the promoter sites are found on one of the two strands only, transcription is asymmetrical. *In vitro* transcription is nonspecific and therefore both DNA strands can be transcribed. The circular DNA molecules are fragmented during isolation and so lose their specificity. Symmetrical transcription may also be due to the lack of sigma factors in the *in vitro* system.

46. Assume that the cytoplasmic genes in this case are located only in chloroplasts.
a) Since *wild-type Euglena* contains both nuclear and chloroplast DNA whereas the W$_3$BUL mutant contains only nuclear DNA, one should compare synthesis vs. non-synthesis of tRNAs and synthetases in the two strains. Synthesis of a particular tRNA or synthetase by both strains indicates that the molecule is coded by nuclear genes. If it is synthesized only by *wild-type* cells it must be specified by cytoplasmic (chloroplast) genes.

Molecule	Presence (+) vs. absence (-) in *wild-type* cells grown in		Presence (+) vs. absence (-) in W$_3$BUL cells grown in		Location of gene(s) for molecule
	light	dark	light	dark	
Species I of tRNA for ile	+	-	-	-	cytoplasm
Species II of tRNA for ile	+	-	-	-	cytoplasm
Species I of tRNA for phe	+	-	-	-	cytoplasm
Species II of tRNA for phe	+	-	-	-	cytoplasm

If the tRNA or enzyme is inducible, comparisons should be made between strains grown in the light. If they are constitutive, compare presence vs. absence in *wild-type* and mutant strains under light- and/or dark-grown conditions.

Molecule	Presence (+) vs. absence (-) in *wild-type* cells	Presence (+) vs. absence (-) in W$_3$BUL cells	Location of gene(s) for molecule
ile tRNA synthetase I	+	+	nuclear
ile tRNA synthetase II	+	-	cytoplasm
phe tRNA synthetase II	+	+	nuclear
phe tRNA synthetase II	+	+	nuclear

c) Since phenylalanyl-tRNA synthetase I is not only found in isolated chloroplasts but also in W$_3$BUL cells which contain neither chloroplasts nor chloroplast DNA, indicates that this sythetase is synthesized in the cytoplasm. If it were synthesized in the chloroplasts, W$_3$BUL cells would not have produced any of this enzyme since they lack this organelle.

d) The data only permit a tentative conclusion:

(i) Genes for both species I and II of tRNA for each of isoleucine and phenylalanine and for isoleucyl tRNA synthetase II are located in the chloroplasts.
(ii) Isoleucyl-tRNA synthetase II and phenylalanyl-tRNA synthetase I are found in isolated chloroplasts.

48. a) Yes. The reasons are as follows:
(i) The base compostions of chloroplast DNA (satellite band) and RNA are both higher than those of nuclear DNA and cytoplasmic RNA respectively. This is expected if the chloroplast DNA is template for chloroplast RNA.
(ii) Actinomycin D inhibits DNA-dependent RNA synthesis and DNase degrades DNA. Treatment of isolated chloroplasts with both inhibits RNA synthesis. These observations indicate that chloroplast DNA codes for chloroplast RNA. If it did not act as template for RNA synthesis, such treatment of chloroplasts should not inhibit RNA formation.

b) One way to provide support for the hypothesis is to separately isolate chloroplast DNA and RNA, denature the former and perform DNA-RNA hybridization experiments.

c) Extensive DNA-RNA hybridization and hybrid duplex formation can be expected to occur if the chloroplast DNA acts as template for chloroplast RNA. Otherwise few or no double helices would be formed.

49. This information indicates that:
(i) Chloroplasts contain DNA as well as RNA.
(ii) Some of the RNA (at least part of which is mRNA) is specified by chloroplast DNA and,
(iii) Protein synthesis, which uses mRNA as template, also occurs in these cytoplasmic organelles.

Bases for conclusions:
a) Actinomycin D is known to bind to DNA and to inhibit DNA-dependent RNA synthesis. Treatment of isolated chloroplasts with this drug inhibits protein synthesis.
b) Amino acids are incorporated into proteins when these organelles are isolated.
c) Treatment of isolated chloroplasts with ribonuclease, known to degrade RNA, inhibits assembly of amino acids into proteins.

50. (i) Individuals with Hb-CS have α chains that are 172 amino acids (AAs) long and have an AA sequence that is identical to that of normal α chains up to AA in position 142. This abnormality is most plausibly explained by the following hypothesis: There is at least a 96 bp sequence between the chain terminating codon of the α determining gene and the promoter region of the next gene. A single base-pair (bp) mutation in the chain terminating codon, e.g., ATT/TAA (UAA in the mRNA) of the normal α gene changes the nonsense codon into a sense one. If the first bp in the nonsense codon was changed from AT to GC this would generate a GTT/CAA codon (CAA in the mRNA) which would code for glutamine, the first AA in the series of 31 additional AAs in Hb-CS. The elimination of the normal α gene's terminating codon in this manner would permit translation, codon by codon of the next 90 bases in the mRNA until a new chain terminating codon is reached.
(ii) The Hb-W1 α chain is 146 AAs long and differs from the normal α pp by 8 AAs beginning with AA in position 139. The most plausible explanation for this mutant α chain is a frameshift mutation in the α gene involving a single bp. A deletion of any one of the bps in codon 139 causes a shift in the reading frame and results in a new sequence of eight codons and therefore AAs until a new chain terminating codon is reached in the mRNA base sequence. The following illustration involving codons 139 to 146 and beyond in *normal* and *mutant* α alleles serves as a possible explanation:

α allele and α pp	139	140	141	142	143	144	145	146	147	148
Normal mRNA	AAA	CCC	UCA	UAA*	GUA	CUU	GGA	ACG	UUA	GCA
Normal pp	LYS	PRO	SER	-	-	-	-	-	-	-
A ↓ deletion										
Mutant mRNA	AAC	CCU	CAU	AAG	UAC	UUB	GAA	CGU	UAG*	CAC
Mutant Hb-W1 pp	ASN	PRO	HIS	LYS	TYR	LEU	GLU	ARG	-	-

*UAA and UAG are chain terminating codons.

53. The results imply that peptidyl transferase activity is associated with the 50S subunit alone. If this was not the case peptide bond formation could not occur in the absence of the template strand and 30S subunits.

54. They must have known the number of amino acids (AAs) that were present in plastocyanin. Specifically, 168 AAs. Since codons are three base-pairs long, 507 base pairs (168 x 3 = 504 + 3 base-pairs in the chain terminating codon) would be found in the gene. The gene therefore cannot contain any introns.

56. a) R-loop mapping is a procedure that permits visualization of DNA-RNA hybrids in the electronmicroscope. The technique involves partial denaturation of duplex DNA and incubation of denatured DNA in the presence of complementary RNA at temperatures below the melting temperature to allow hybridization of the single-stranded RNA with the complementary strand of denatured DNA. This hybridization results in the displacement of the DNA strand with the same nucleotide sequence as that in the RNA molecule. The remainder of the DNA duplex remains intact. Observation of such DNA under the electron microscope permits visualization of the displaced single-strand of DNA which forms an "R-loop". The number of "R-loops" observed in DNA-RNA hybrids depends on the lengths of both the DNA and RNA strands and whether a segment or segments of nucleotide pairs in one strand is missing in the other.

b) (i) Because more than one "R-loop" is formed this indicates that there are nucleotide sequences in the ovalbumin gene that are missing in its mRNA.
(ii) There are seven introns and eight exons in the gene. **Basis for answer:** There are seven "R-loops" in the hybrid molecule formed between the ovalbumin gene DNA and ovalbumin mRNA. Each "R-loop" indicates that the corresponding region (intron) in the mRNA is missing. Since there are seven "R-loops" the gene contains seven introns. All non "R-loop" regions of the hybrid molecule are exons, this includes the segments to the left and right of the A and G loops and the six hybrid regions between R-loops A and G.

E = exon; I = intron

(iii)

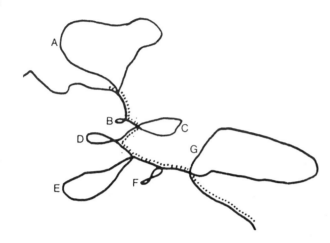

c) One R-loop in the hybrid between the denatured DNA of the gene and the primary transcript indicates that the base sequence in the primary transcript (RNA) is completely complementary to the entire sequence of the gene (DNA) The transcript is therefore not missing any nucleotides. Since the molecule derived from hybridizing the denatured DNA of the gene and its mRNA has seven R-loops, this indicates that the mRNA is deficient in seven segments present in the primary transcript. Since the mRNA is derived from the primary transcript the introns are excised after the RNA is formed.

57. a) Three nucleotides constitute a codon which codes for a single amino acid (AA). The distance between nucleotides in the mRNA is 3.4Å. Therefore the length of the mRNA that codes for the β chain of human hemoglobin is 146 x 3 = 438 nucleotides x 3.4Å ≈ 1500Å.

b) Number of AAs in an enzyme whose MW is 300,000 is 300,000/100 = 3,000. Since a codon consists of three nucleotides, the gene determining this enzyme must consist of 3,000 AAs x 3 = 9,000 nucleotides long, plus three nucleotides, for a stop codon = 9,003 nucleotides.

c) The number of AAs that could be coded for by φX174 DNA which is 4,500 nucleotides long is 4,500/3 nucleotides per AA = 1,500 AAs. MW of 1,500 AAs is 1,500 x 100 daltons per AA = 150,000 daltons. The number of proteins of MW 30,000 that the φX174 genome could code for is 150,000/30,000 = 5.

d) *Aspergillus* possesses 4×10^7 nucleotide pairs/1,500 nucleotide pairs per gene, which is $\approx 2.7 \times 10^4$ genes.

58. The base sequence in the synthetic mRNA is 5'-AAA. . . AAC[(A)$_n$C]-3', with AAA and AAC coding for lysine and asparagine respectively. The pp that this mRNA specifies has the amino acid sequence NH$_2$-Lys-Lys-. . . Lys-Asn-COOH. Therefore:
(i) The mRNA must be read sequentially from the 5' to the 3' end and (ii) the COOH end of the pp corresponds to the 3' end of the mRNA. If mRNA translation was 3'-5', the first triplet CAA would code for glutamine at the NH$_2$ end, with lysine coded for by AAA, being located at the COOH end of the pp.

59. AAA and AAG are the codons for lysine. The codons - UUU and UUC code for phenylalanine. Since the message is 5'-AAAUUU-3' and the product of its translation is NH$_2$-Lys-Phe-COOH, the direction of translation of mRNA must be 5' to 3'.

61. a) The formylated Met-tRNA (Met-tRNA$_F$) is capable of incorporating methionine into polypeptide (pp) chains only in the start position. This implies that Met-tRNA$_F$ is involved in pp chain initiation only. The nonformylatable Met-tRNA (Met-tRNA$_M$) incorporates methionine into pps in the internal positions only, indicating that it is involved in chain elongation only.

b) The formyl group in F-Met may prevent peptid bond formation at the NH$_2$ end. This would impose a direction on protein synthesis so that it proceeds from the NH$_2$ end to the COOH terminus.

c) Both AUG and GUG are chain initiating codons since Met-tRNA$_F$ can utilize either to incorporate methionine in the start positions (NH$_2$ end) of pp chains. AUG codes for methionine and GUG for valine, if these codons are located within genes.

63. The two Ser-tRNA species respond to codons with unrelated nucleotide sequences: one responds to AGU and AGC and the other responds to codons UCA and UCG. This implies that the anticodons of the two Ser-tRNA species also probably have different nucleotide sequences. Since there is only one Ser-tRNA synthetase, specific for two tRNA species with different anticodons, the latter are probably not the recognition sites for the synthetase.

64. a) As indicated in the table, either R_1 or R_2 factors can facilitate the release of pp chains from ribosomes. The S factor alone cannot perform this function, indicating that the S factor is not required for the release of pps.

b) The data do not provide any information that has a direct bearing on the question posed.

Chapter 22
Coding, Colinearity and Suppressors

2. a) A degenerate code is one in which an amino acid (AA) is specified by more than one codon. For example leucine is specified by six mRNA codons UUA, UUU, CUU, CUC, CUA and CUG.
Degeneracy in translation is facilitated by wobble - the ability of the third base in the tRNA anticodon 5' end to hydrogen bond with any one of two or three different bases at the 3' end of the codon is the mRNA - e.g., leucine tRNAs with a 3'AAU5' anticodon can recognize two leucine codons whereas leucine tRNAs with a 3'GAI5' anticodon can recognize three leucine codons. The codon 5'CUG3' can only be recognized by a leucine tRNA with the anticodon sequence 3'GAC5'.

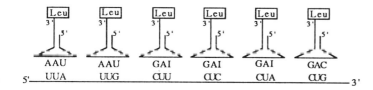

b) Almost all of the degeneracy is confined to the third base of four different codons ACU, ACC, ACA and ACG, all of which are identical in the first two positions but each possesses a different base in the third position.

c) Since tyrosine is coded for by the 5'UAU3' and 5'UAC3' codons, the tRNA molecules that can recognize these codons cannot have inosine at the 5' end of the anticodon. This is because of the wobble at the 5' end of the anticodon. The two histidine codons can be recognized by a tRNA carrying histidine with G at the 5' position in the anticodon. Similarly the two tyrosine codons can be recognized by a tRNA carrying tyrosine with G at its 5' location in the anticodon.

d) This species of alanine-tRNA can recognize the three alanine codons because (i) of the wobble at the third position of the anticodon and (ii) it possesses inosine at the 5' end of the anticodon enabling the tRNA with the anticodon sequence 5'IGC3' to hydrogen bond to any of the three alanine codons 5'GGU3', 5'GCA3' and 5'GCC3'.

4. Ribosomes can be located at various distances from the initial and final codons. If we assume that the first ribosome was positioned over the first two codons and the last one over the second and third last codons of the mRNA with the last triplet in the mRNA being a chain-terminating codons, then: (i) The nine interribosomal regions will encompass 150 x 9 = 1350Å of the mRNA. If each codon occupies 3.4 x 3 = 10.2Å, the nine interribosomal regions will code for ≈1350/10.2=132.3AAs.
(ii) Each ribosome has a P and an A site and therefore extends slightly over two codons length or 10.2 x 2 = 20.4Å. Nine of the ribosomes will occupy two sense codons each. Only the last ribosome will occupy one sense codon. The 9 x 2 + 1 = 19 codons associated with ribosomes will code for 19 AAs and occupy 19 x 10.2 = 193.8Å of the mRNA. This mRNA (1350Å + 193.8Å) = 1543.8Å long will therefore code for 1543.8/10.2 = 151.3 = 151 AAs.

11. The genetic code cannot be universal because TAA and TAG in the non-coding strand of the *wild-type* G surface antigen gene of *P. primaurelia* are not nonsense codons (=UAA and UAG in mRNAs) as they are in most genes in prokaryotes and eukaryotes. Their presence within this gene should result in the termination of translation and the synthesis of an incomplete polypeptide chain of the G surface antigen. Since all G surface antigens are of *normal (wild-type)* length, these codons must be sense and not nonsense codons in this gene in *P. primaurelia*. This is also probably true of the other genes in this species. Additional information is necessary to determine which amino acids are specified by these two codons.

12. It is important to note that all the mutations were induced by 2-AP which causes single base-pair (bp) transitions (AT ↔ GC or their equivalent). The mutation substituting isoleucine for valine indicates that a codon (181) has at least one bp and the codon can exist in at least two forms. The second 2-AP induced mutation in codon 181 substitutes methionine for isoleucine. This mutation could not be at the same bp site that produced the original mutant as the second mutation would have changed the substituted bp back to the original one. This would have given the same codon that was there originally and therefore should have coded for valine not methionine at position 181. Thus, a codon has at least 2 bps and codon 181 of the second mutant differs from the original one for valine by 2 bps. If the methionine codon differs from the original codon by two bps, 2-AP treatment of this mutant could not produce a revertant with valine at position 181 by effecting a transition at the second mutational site (bp). Therefore the revertant was either induced at the site of the original bp change or at yet another (third) bp site.
Conclusions (i) A codon and therefore gene can exist in at least four different forms. (ii) The genetic code is degenerate since the bp content of the revertant codon is different from the original one coding for valine. (iii) Codons consist of at least two bps

13. The results have a bearing only on (c). They clearly indicate that the code is not overlapping. Since NA causes single base-pair substitutions, e.g., AT→GC, if the code was overlapping, a missense mutation in one bp should simultaneously result in the substitution of three adjacent amino acids unless the substitution was in the first bp of the first codon and the last bp in the last codon in which case only one AA would be substituted or the second bp of the first codon and the second last bp of the last codon in which case two adjacent AAs would be substituted.
The results do not have a bearing on codon size. At least three mutants with AA substitutions at the same position and recombination data from crosses between these mutants would be necessary to show that each codon consists of three consecutive bps. The information provided also does not indicate whether the code is degenerate. This would require data from experiments using synthetic messengers with known base compositions and sequences and trinucleotide binding studies which permit assignment of functions to each of the codons. The data also do not have any bearing on how the code is read (translated). One requires genetic information involving single, double and triple frameshift mutations in the same gene and/or the results of studies with deletion mutants which are missing segments of two adjacent genes e.g.

The effects of frameshift mutations in gene A on the activity (phenotype) of gene B would indicate whether the mRNA is read from a fixed starting point in a polarized manner.

15. a) These results show that the role of tRNAs is that of an adaptor. It allows for the placement of amino acids (AAs) in the growing chain as specified by the codons in the mRNA by recognizing the mRNA codons. This is evidenced by the fact that different tRNAs carrying the same AA (leucine) show mRNA specificity, i.e., recognize different codons in different messengers for the AA they carry.

b) Yes, coding degeneracy is involved because two different codons in two different messengers (poly-UG and poly-UC) specify the incorporation of the same AA (leucine) into pps. This is evidenced by the following facts: (i) The AA that is incorporated to the greatest extent in the pps specified by poly-UG (3:1) and poly-UC (3:1) messengers is phenylalanine; this indicates that the most common codon (UUU) in these mRNAs codes for the AA phenylalanine.
(ii) UUU is the only common codon in poly-UC and poly-UG messengers, both of which also code for the incorporation of leucine. Since the other codons in the two mRNAs will be different, the code must be degenerate. At least two different codons specify the incorporation of leucine into pps. The degeneracy is due to the fact that different tRNAs that carry the same AA (e.g., leucine), have different anticodon sequences which must therefore recognize different codons in the mRNA.

16. a) In 1961 it was established that (i) a codon in mRNAs was three bases long (ii) the synthetic mRNA with U only coded for a polypeptide (pp) with phenylalanine and (iii) the pp was not synthesized if the mRNA with U was omitted from the *in vitro* experiment or paired with a polynucleotide chain containing A. Therefore it was concluded that (i) at least one codon for phenylalanine is UUU and (ii) the mRNA must be single-stranded to be translated.

b) These results imply that only one of the two strands of DNA in each gene is transcribed. If both were transcribed, complementary sequences should code for the same AA, not different ones as is the case in these observations.

18. a) (i) Each of the three mRNAs is composed of two bases in different proportions - the first contains A and C in a 5:1 ratio, the second comprises A and C in a 1:5 ratio and the third has U and C in a 5:1 ratio.
(ii) Eight different codons are possible for each of the mRNAs and are given by the binomial expansion $(a+b)^n$ where $n=3$. In experiment 1, $a=A=5/6$, $b=C=1/6$; in experiment 2, $a=A=1/6$, $b=C=5/6$ and in 3, $a=U=5/6$, $b=C=1/6$.

Codons (5'-3')	Experiment 1 A5/6:C1/6	Experiment 2 A1/6 : C5/6	Codons (5'-3')	Experiment 3 U5/6 : C1/6
AAA	125/216	1/216	UUU	125/216
AAC	25/216	5/216	UUC	25/216
ACA	25/216	5/216	UCU	25/216
CAA	25/216	5/216	CUU	25/216
ACC	5/216	25/216	UCC	5/216
CAC	5/216	25/216	CUC	5/216
CCA	5/216	25/216	CCU	5/216
CCC	1/216	125/216	CCC	1/216

Codon Assignments
Experiment 1 (a) Since AAA is the most commonly occurring codon, it should code for the most common amino acid (AA) in the polypeptide (pp)s - therefore AAA codes for lysine.

b) The second most commonly occurring codon will be those with 2As and 1C. The ratio of the most common codon AAA to each of these codons is 125/25=5:1. Amino acids specified by these codons should be present in a ratio of 1:5 (lysine). Since the % incorporation of asparagine, glutamine and threonine is 24.2, 23.7 and 26.5 respectively, each must be specified by one of the three codons AAC, ACA or CAA. These data do not permit unequivocal identification of the specific codons for these three amino acids.

c) The next most frequently occurring codons will be those with 1A and 2Cs. However, they will be 25 times less frequent than the codon AAA and therefore any AA specified by one of these codons should be present in a ratio of ≈1:25 (lysine). Since the % incorporation of histidine and proline is 6.5 and 7.2 respectively, these AAs must be coded for by two of the three codons ACC, CAC or CCA. A third AA should be specified by the remaining codon with 1 A and 2 Cs. The coding dictionary indicates that threonine, present with a 26.5% frequency in this experiment, is specified by not only one of the codons with 2As and 1C (actually ACA) but also by one of the three codons with 1A and 2Cs (ACC). Its % incorporation of 26.5 is higher than that of asparagine and glutamine because it is specified by two codons whose combined % incorporation (26.5) is only minutely different from expected. The codon CCC is expected to occur with a frequency of 0.8%. The AA it specifies should therefore be incorporated with ≈ the same low frequency. Such small amounts of incorporation of AAs may not be detected or as in this case the CCC codon specifies the AA proline which is also specified by a codon with 1A and 2Cs (actually CCA). Therefore a small % (0.8) of the 7.2 value may be attributed to proline specification by CCC.

Experiment 2
Using the same reasoning as for experiment 1 we can conclude that CCC codes for proline, while histidine and threonine are coded for by two of the three codons ACC, CAC, CCA. The third codon in this group also codes for proline as has been determined by other experiments. Asparagine and glutamine are coded by two of the three codons with 2As and 1C. The third codon in this group specifies a third AA (threonine) which is obscured in this experiment by the fact that a codon with 1A and 2Cs, actually ACC (expected frequency ≈20%) also specifies the incorporation of this AA. Lysine is coded for by the least commonly occurring codon AAA as verified in experiment 1.

Experiment 3
With similar reasoning we can conclude that UUU codes for phenylalanine; leucine and serine are coded by two of the three codons with 2Us and 1C and proline is specified by one of the three codons with 1U and 2Cs.

b) The collective data from the experiments, indicate that the code is degenerate for threonine and proline.

19. a) (i) Since *wild-type* recombinants fail to occur in the cross between mutants A46 and A47, the mutations in the two strains msut have occurred at the same mutable site in codon 210. These results show that a mutable site can occur in at least three forms - one in the *wild-type* allele, e.g., AT, another in the A46 mutant allele, e.g., GC and still another in the A47 mutant alelle, e.g., CG.
(ii) At least two adjacent mutable sites must specify each amino acid. This is evidenced by the fact that *wild-type* recombinants occur in the cross between mutant strains A46 and A23. The very low frequency of recombination

(0.002%) suggests that the mutable sites are very close together. If the mutations were at the same site, *wild-type* recombinants could not be produced by intragenic recombination (crossing-over). The maximum number of mutable sites per codon can be determined by crossing the mutant strains with AA substitutions at the same position, e.g., 210, in all possible combinations and ascertaining whether or not *wild-type* recombinants occur.

b) Since either glycine or serine at position 210 yields a fully active enzyme whereas threonine at this position results in a partially active one, the activity of an enzyme does not require a unique AA sequence.

20. a) The level of binding of ^{14}C-Leu-tRNA to ribosomes in the presence of the trinucleotide UUG is significantly greater than binding of this labelled AA-tRNA to ribosomes in the absence of any mRNA. Therefore the base sequence of the codon for leucine must be UUG.

b) Leu-tRNA anticodon is 5'-CAA-3'.

21. a) The mRNA sequence is:
5'-GUA UGC GUG GCU AAG UUA GUA - 3'. The sequence of AAs coded for by this mRNA is: NH$_2$ - Val Cys Glu Ala Lys Leu Val -COOH.

b) The resulting message is:
5'- GUA UGC GAG GUC UUA GUU AGU A-3'.
                        ↑ stop codon
The resulting AA sequence is NH$_2$- Val Cys Glu Val.

22. A comparison of the codons in the human mitochondrial DNA and the amino acids (AAs) they specify with the codons and the AAs that they specify in nuclear genes leads to two conclusions: (a) except for UGA, all the other codons in both the mitochondrial and nuclear genes code for the same AAs, (b) UGA is used as a tryptophan codon in the human mitochondrial genome and not as a chain terminating codon as it is in the nuclear structural genes.

24. a) Since the level of binding of ^{14}C-aminoacyl-tRNA to ribosomes in the presence of doublets is no greater than the level of binding in the absence of template RNA, doublets cannot code for amino acids.

b) Only the tested triplet GUU and both poly-U and poly-UG mRNAs code for AAs. GUU codes for valine as evidenced by the significantly higher level of its binding with ^{14}C-val-tRNA than in the absence of any mRNA. Moreover, tRNAs carrying phenylalanine and leucine bind at the control level. Since UUG and UGU do not code for valine and poly-UG mRNA codes for valine at a high level, this is due to the presence of the GUU codon is this messenger. Since poly-U mRNA codes for phenylalanine as evidenced by the large amount of ^{14}C-Phe-tRNA binding to the nitrocellulose membrane, at least one codon for this AA is UUU. Since ^{14}C-Phe-tRNA also binds to the membrane in the presence of poly-UG mRNA at a significantly higher level than in the control, this is also probably due to the frequent presence of UUU in this mRNA.

26. (i) These results are excellent proof that translation of mRNA is sequential from the 5' to the 3' end. Consider the mRNA with the repeating tetrabase (UAUC)$_n$. It will contain the four codons-5'-UAU-CUA-UCU-AUC-3' repeated over and over. The four codons code for the AAs Tyr, Leu, Ser and Ile respectively from the NH$_2$ to the COOH end. If the direction of translation was reversed the repeating tetrapeptide sequence would have had the opposite orientation -

NH$_2$-Ile-Ser-Leu-Tyr-COOH. The results with the 5'-(UUAC)$_n$-3' mRNA verify this conclusion.

(ii) The results unequivocally verify that codons are three bases long. If each codon was one base long, since each repeating sequence contains three different bases, they should code for three and not four AAs. If each codon was two bases long each repeat sequence of four bases would possess only different two codons specifying only two and not four different AAs. Codons with four or more bases cannot produce a repeating four AA sequence. Therefore triplet codons account completely for the repeating four AA sequence specified by each of the mRNAs. The sequences in both these mRNAs will give a repeated four AA sequence with both a non-overlapping and an overlapping type of code. If the mRNA is (UAUC)$_n$:

non-overlapping:

5' UAU CUA UCU AUC 3'

NH$_2$ Tyr Leu Ser Ile ---- COOH

overlapping:

If we also take into consideration the AAs specified by each codon then the sequence of the four AAs will be different with overlapping and nonoverlapping codes. It will be Tyr-Leu-Ser-Ile with the latter and Tyr-Ile-Ser-Leu with an overlapping code.

28. a) One mutation occurred to produce the original *mutant lysozymeless* strain; a subsequent single mutation occurred to produce the revertant strain.

b) The changes were additions and deletions which involved single base pairs. Since more than one amino acid was altered the mutations could not have been transitions or transversions. Results can also be accounted for on the basis of an initial deletion, followed by an addition.

wild-type
mRNA 5'-AUG-ACU-AAG-AGU-CCA-UCA-CUU-AAU-GCU-UAG-3'
      NH$_2$-Thr- Lys - Ser -Pro- Leu -Asn - Ala-COOH

mRNA 5'-AUG-ACU-AAG-A*GU-CCA-UCA-CUU-AAU-GCU-UAG-3'
                   delete A
original   ↓
mutant 5' AUG-ACU-AAG-GUC-CAU-CAC-UUA-AUG-CUU-AG-3'

original *mutant*
protein NH$_2$-Thr-Lys-Val-His-His-Leu-Met-Leu-. . .

original
mutant 5' AUG-ACU-AAG-GUC-CAU-CAC-UUA-AUG*-CUU-AG-3'
                                       add G
                                         ↓
double
mutant 5' AUG-ACU-AAG-GUC-CAU-CAC-UUA-AUG-GCU-UAG-3'

double *mutant*
protein NH$_2$-Thr-Lys-Val-His-His-Leu-Met-Ala-COOH

29. Each *pseudo wild-type* x *wild-type* cross produces recombinants that are mutants of two types, e.g., *pseudo wild-type* x *wild-type* produces eJ17 and eJ44 recombinants. This indicates that the *pseudo wild-types* were double mutants. If *pseudo wild-types* were single mutants such crosses would produce parental types (*pseudo wild-type* and *wild-type*) only except in very rare cases when mutation would occur.

b) Proflavine induced frameshift mutations; specifically base-pair (bp) additions (+s) and/or deletions (-s) as deduced from a comparison of the amino acid (AA) sequences in peptide A coded for by gene e^+ and *pseudo wild-types* 1 and 2. Such mutations (+s and -s) change the bp sequence as well as the codons and therefore the sequence of AAs from the point of mutation to the end of the gene. Single bp substitutions cause single AA changes. If two frameshift mutations of oppostie sign (+ and -) occur in the same gene, the bp sequence will be incorrect only in the region between the two mutations; therefore only the corresponding region of the mutant pp specified by the double (+- or -+) mutant allele will differ from the pp specified by the *wild-type* allele. For discussion purposes, compare the AA sequences of peptide A produced by e^+ and *pseudo wild-type* 2. The two pps have an identical AA sequence except for 5 AAs between lysine and alanine in peptide A. Therefore the two mutations can only be of the frameshift type. The comparison shown below of the base sequences of the mRNAs and AAs specified by e^+ and the double mutant indicate that the explanation is correct.

	Lys	Ser	Pro	Ser	Leu	Asn	Ala	
e^+	AA^A_G	AGU	CCA	UCA	CUU	AAU deletion of A	GC addition of G or A	
eJ42 eJ44	AA^A_G	GUC	CAU	CAC	UUA	AU^A_G	GC	
	Lys	Val	His	His	Leu	Met	Ala	

33.   Genetic map of gene A
                  Position 48   Position 174

DNA Sense strand	3' - CCT	ATA - 5'
DNA Antisense strand	5' AGG	TAT - 3'
mRNA	5' - GGA	UAU - 3'
A protein	NH$_2$ - Gly	Tyr - COOH

(i) All new polynucleotide strands are synthesized 5' to 3'. The base sequence is complementary to that of the template strand which is transcribed in the 3' to 5' direction. Therefore, the mRNA will have a base sequence that is complementary to the sense (template) strand except that U will replace T.
(ii) The mRNA is translated 5' to 3'. Amino acids in the polypeptides are incorporated sequentially from the NH$_2$ to the COOH end. Therefore, the pp would have the AA sequence NH$_2$...Gly......Tyr....COOH.
                48   174

34. a) Since all the mutants on the non-suppressor strain *E. coli* B produced incomplete head proteins, the base analogues must have induced nonsense mutations in all 10 of them. Since each mutant produced different incomplete polypeptides (pp) of unique length, this indicates that the nonsense mutations and codons were induced at different locations in the gene coding for the head protein.

b) The *mutant* strains grew on *E. coli* CR63 but not on *E. coli* B because the former has a suppressor allele Su at a suppresor locus and the latter does not. The suppressor allele permits the formation of complete pps (proteins) and therefore allows growth. Although suppressors function in different molecular ways, the suppressor allele Su in strain CR63 may be a mutant allele at a locus that codes for a minor tRNA, the altered (mutant) anticodon sequence of which reads a nonsense codon as sense, thus allowing the synthesis of a complete pp. For example if the nonsense codon is 5'-TAG3'/3'-ATC-5' (transcribed), the non-suppressor (su) strain would possess the *wild-type* (su) allele for minor tyrosine-tRNA 5'GTA-3'/3'-CAT-5' (transcribed). A normal anticodon sequence 3'-AUG-5' in this tRNA would recognize the tyrosine codons 5'-UAC-3' and 5'-UAU-3' but not the nonsense codon 5'-UAG-3'. Therefore, only incomplete pps will be produced. The mutant allele Su at the minor tyrosine tRNA locus in strain CR63 would be 5'-GTA-3'/3'-CAT-5' (transcribed) (3'-AUC-5' in the mRNA) at the anticodon specifying region. This tRNA has a base sequence complementary to that in the nonsense codon. It can H bond with the nonsense codon. Tyrosine will be inserted into its original position producing a functional full length pp (protein). The nonsense codon is therefore suppressed.

c) Since all the mutants produced incomplete chains of the same pp (protein) and each pp is coded for by one specific structural gene, the 10 mutations must have occurred in one (the same) gene.

d) In all species 2-factor crosses permit an estimation of the genetic distance between two points while 3-factor crosses permit determination of the sequence of the three mutations. Since the gene occupies a short segment of the genome this indicates that probably only one crossing over event occurs to produce *wild-type* recombinants in the first cross but two such events are required to produce the recombinants in the second cross. The probability of two such events is significantly lower than that of one such event occurring. A significantly lower %R in the 3-factor cross than in the 2-factor ones indicates that the mutational site in the single mutant must be located between the mutational sites in the double mutant. Therefore the sequence of the mutational sites in the 3-factor cross B17 x H11+B278 is H11-B17-B278.
Analyzing all the data gives the following genetic map:

```
  B17    B272  H32        B278 C137   A489
  |-------|-----|-----------|----|------|
    1.15   0.5     1.82     0.33  1.26
```

In some cases the recombination frequencies represent the average of several crosses.

37. a) (i) Both AUG and GUG are pp chain initiating codons.
Evidence that AUG is a start codon: (i) Poly-(AUG)$_n$ mRNA codes for a pp containing methionine only which requires both Met-tRNA$_F$ and Met-tRNA$_M$.
(ii) Using poly-UAG mRNA, which contains, among others, AUG codons and Met-tRNA$_F$, methionine was incorporated into polypeptides only in the start position.

Evidence that GUG is a start codon: (i) GUG binds Met-tRNA$_F$ and therefore can code for F-Met as well as valine (ii) The poly-(UG)$_n$ messenger codes for a pp with the AA sequence F-Met-(Cys-Val)$_n$. This indicates that GUG codes for methionine only in the start position (and valine internally) and therefore is a chain-initiating codon.

(2) Function of tRNA$_F$ - recognizes only the chain-initiating codons AUG and GUG for methionine. Evidence: (i) Using Met-tRNA$_F$, methionine was incorporated into pps only in the start position regardless whether the mRNA had GUG or AUG in the start position (ii) Poly-(AUG)$_n$ codes for a pp with methionine only and it requires both Met-tRNA$_F$ and Met-tRNA$_M$. Met-tRNA$_M$ incorporates methionine internally as evidenced by studies with poly-UAG.

Function of tRNA$_M$ - recognizing only the AUG codon, since GUG in non-start positions in the mRNA codes for valine. Moreover it recognizes the AUG codons internally only as evidenced by (i) poly-(UG)n codes for methionine in the start position which requires Met-tRNA$_F$. (ii) Poly-(AUG)$_n$ requires both forms of Met-tRNA for pp synthesis and (iii) using Met-tRNA$_M$ and poly-UAG, methionine is incorporated only in the internal positions in the pps.

(3) The formyl group probably increased the rate of translation. Moreover, since the NH$_2$ group is replaced by CHO it ensures that the pp cannot join any others.

b) Polypeptide translation is described in the answer to question 7 in Chapter 21.

39. If a 2-AP induced transition can convert an ochre into an amber codon, while HA cannot, ochre mutants must have an A-T base pair (bp) which corresponds to a G-C pair in the amber codon because HA does not affect A-T bps. If neither amber nor ochre *mutants* revert to *wild-type* with HA treatment, it must be because both codons have an A-T pair that corresponds to a G-C in the *wild-type*.

b) Without any knowledge of the genetic code it is impossible to determine which of the three bps are in the sense or antisense strands. The only inference from the mutation data is that in a 2-AP or an HA induced mutation a purine replaces purine and a pyrimidine replaces pyrimidine in the same strand.

c) NH$_2$OH induces mutations by reacting with cytosine, forming a T analogue, and causing a C-G→T-A transition. To induce r mutations in the first cycle of growth NH$_2$OH must react with cytosine in the sense strand because the r mutation is expressed before DNA replication. If NH$_2$OH reacted with a C in the antisense strand the mutation would not be expressed until after replication (the second cycle of growth). The amber codon contains a G-C, pair which corresponds to an A-T pair in the ochre codon, which ever position that may be in that codon. Treatment with NH$_2$OH cannot cause a mutation at that position in the *wild-type* codon. G-C is present in the amber codon at that site, and a reaction with NH$_2$OH would produce an ochre codon. Thus a mutation from *wild-type* to amber and ochre codons must occur at one of the other two bp positions. If mutations are detected in both the first and second cycle of growth, it can be deduced that NH$_2$OH reacts with cytosine in both the sense and the antisense strands to form an amber or ochre codon. Since an amber or ochre codon can be detected in either the first or second cycle of growth, there must be different *wild-type* codons that mutate to the amber codon that have a C in either the sense or the antisense strand. A C-G pair in the *wild-type* codon mutates to A-T in the amber and ochre codons and if there are forms of the *wild-type* codon that have C in the sense and antisense, the mutant (amber or ochre codons) must have two T-A pairs in opposite orientation to each other, in the two positions that do not differ is between amber and ochre codons. From question (a) amber and ochre codons have at least one bp in common and one that differs. Because the data show that the two T-A base pairs must be in opposite orientation, amber and ochre codons must have two base pairs in common and one that is different.

Since the exact orientation of the two common bases is not known the following sequences are possible for the two types of nonsense codons in the mRNA:

Amber: UAG   UAC   AUG   AUC
Ochre: UAU   UAA   AUU   AUA

d) Amber : UAG;   Ochre : UAA
The sequence of bases for the amber codon can be determined as follows: The first two bases in the codon are U and A (without knowing the order). A single base transition in CAG or UGG can result in the same mutant codon only if the resultant mutant codon is UAG. To determine the ochre codon, you know that ochre has an A-T pair in the codon position that the amber codon has a G-C pair. You also know that at least one A-T pair is common to both, so the T-A must be common as well to enable the ochre mutation to be expressed in either the first or second cycle of growth. Therefore the ochre mRNA codon is UAA.

41. The base sequences of the codons for tryptophan TGG/ACC (sense) and glutamine CAG/GTC (sense) are different. Each differs from the nonsense codon TAG/ATC (sense) by a single base-pair, but the position at which they differ is not the same. A single base-pair change in each of these codons can generate the nonsense codon:

Glutamine   CAG/GTC (sense) → CG→TA in 1st bp→ TAG/ATC (sense) → mRNA (nonsense codon)→ 5'UAG3'.

Tryptophan TGG/ACC (sense) → GC→AT in 2nd bp → TAG/ATC (sense) → mRNA (nonsense codon) → 5'UAG3'

The *Su1* suppressor codes for a mutant serine tRNA with an altered anticodon sequence 3' - AUC - 5' that recognizes the nonsense codon UAG, regardless which of the original codons gave rise to the nonsense codon

42. a) Mapping of mutational sites A46 and PR8
Since T is to the left of the A gene two sequences are possible for the two mutational sites: (i) T - A46 - PR8 and (ii) T - PR8 - A46. Since T is relatively close to the A gene, crossing-over will be rare in the T-A region. The majority of the selected $A46^+PR8^+$ and $A46^-PR8^-$ recombinants should possess parental (P) genotypes at the T locus because of the absence of crossing-over; a minority should have a recombinant genotype at the locus because of rare crossing-over. Therefore if the sequence is T - PR8 - A46, among the $A46^+PR8^+$ recombinants from the cross $T^+PR8^-A46^+ \times T^-PR8^+A46^-$ the majority should be T^- and a minority should be T^+. Among the $A46^-PR8^-$ recombinants, the majority should be T^+ and a minority T^-. The results conform to these expectations. The sequence of T and the two mutational sites is T-PR8-A46. If the order was T-A46-PR8, the opposite results to those obtained would have occurred among the recombinants.

b) Significance of the fact that double mutant $PR8^-$-$A46^-$ is *pseudo wild-type*.

The single mutants $A46^-$ (with an amino acid (AA) substitution (Gly→Glu) at position 210) and $PR8^-$ (with an AA change Tyr→Cys at position 174) do not grow on minimal medium. This indicates that each of these AA substitutions is in a critical region of the A pp and each results in a defective 3-D configuration and an inactive enzyme. Since the double mutant $A46^- PR8^-$ grows (slowly) on MM this indicates that its secondary and 3-D configuration is more normal than that of the single mutants. It is a partially active enzyme resulting in a *pseudo wild-type* phenotype. Thus, a combination of two incorrect AAs produces an enzyme with an active 3-D configuration.

Chapter 23
Development and Regulation

3. a) The *tortoise-shell* phenotype is a variegated one (patches of black hair and patches of orange hair) which is determined by an X-linked pair of alleles, B (for *black*) and *Bl* (for *orange*). The reasons why only some females (XX) and XXY males express the *tortoise-shell* phenotypes are:
(i) These individuals are heterozygous $X^B X^{B1}$.
(ii) Because of dosage compensation in individuals with 2 or more X chromosomes by random inactivation (heterochromatization) of all Xs except one in each somatic cell early in embryogenesis. Once the decision is made it is imprinted and transmitted to all the descendent cells. In theory, 1/2 the randomly located early embryonic cells will have the X^B chromosome inactivated. The descendent clones from each of these cells will have a functional euchromatic X^{B1} chromosome and express an *orange* color. The other 1/2 of the embryonic cells will have the X^{B1} chromosome inactivated. The clones derived from these cells will have a functional X^B chromosome and express a *black* color.

b) No. Because these genes are on the homologous portions of the X and Y which, like autosomal genes, are present in the same double dose in both sexes.

c) Some of the reasons for this deduction are as follows:
(i) Heterochromatic (= condensed) Y chromosomes like that in *D. melanogaster* carry few if any genes.
(ii) In normal body cells, genes express their potential in interphase nuclei, when chromosomes are relatively uncondensed, but not in mitotic and meiotic cells when chromosomes are condensed.
(iii) Many B chromosomes are genetically inert. These chromosomes are heterochromatic.
(iv) In species like the mealy bug, with the lecanoid meiotic system, one chromosome set in males is heterochromatic and genetic studies indicate that all the chromosomes and genes in this genome are nonfunctional.

d) The question refers to heterozygotes in mammals. Dosage compensation occurs by random inactivation of X chromosomes (see 3a)(ii)). Individuals that are heterozygous for X-linked recessive alleles will have different proportions of the cells and their clones express the potentials of the 2 alleles. In humans *E* (*normal*) vs. *e* (*anhidrotic ectodermal dysplasia* = absence of sweat glands, teeth and nails) are an X-linked pair of alleles; *e* is recessive. In different heterozygous ($X^E X^e$) females different proportions of the 2 Xs may be inactivated. In one such female 40% of X^Es may be inactivated randomly; in another 40% of X^es may be inactivated. Both will show a mosaic phenotype of skin patches with and without sweat glands, variation in proportions and distribution of missing teeth and variation in the development of nails. Such variability in expression does not occur among individuals heterozygous for autosomal recessive alleles because inactivation of one of the homologues does not occur.

5. The difference exists to equalize the phenotypic effects of the genes on the differential portion of the X chromosome in both sexes. The X chromosome carries many genes and dosage compensation is by random inactivation of Xs in somatic cells. This mechanism results in females being phenotypically equivalent to males for characters determined by X-linked genes; only 1 X remains active in each somatic cell in females. In males, the alleles at all loci on the differential segment of the X express their potential in all cells in which the gene functions producing a non-variegated phenotype. In females, only the alleles on the euchromatic X express their potential (are transcribed); those on the heterochromatic X are not. If a female is heterozygous for an X-linked pair of alleles - some cells will express the coded information determined by one allele; other cells will express the potential determined by the other allele. Such females therefore will express a variegated phenotype. For example, in humans the X-linked alleles G^A and G^B code for different forms of the G-6PD enzyme; in $G^A G^B$ females some cells produce the A form of the enzyme and others the B form.

7. Histones probably suppress gene action (transcription) in animal and plant chromosomes by condensing (coiling) chromosomes and thus making the DNA inaccessible to RNA polymerase. Although chromatin at the first level of coiling (nucleofilament = nucleosomes (DNA + H2A, H2B, H3 and H4 histones) plus linker DNA) can be transcribed, this is not so for chromosomes at the second (solenoidal) and further levels of coiling. The condensation of nucleofilaments into solenoidal structures is carried out primarily, if not solely, by the H1 histone. This nucleoprotein may also be partly responsible for the additional levels of condensation.

11. a) The organizing tissue, by its activity evokes and determines the differentiation (direction and extent of development) of other groups of cells (target tissue). The organizing tissue is in the gonads (testes, ovaries and the hormones they produce) and the target tissue is the skin within which the feathers develop. The bases for this conclusion are: (i) Fowl of genotype *hh* develop *cock-* and *hen-feathering* in males and females respectively; (ii) *Hh* birds whose gonads have been removed, develop *cock-feathering* (*CF*); transplantation of testes and/or ovaries into these birds reverses feather development.

b) Sexual differences in feathering are determined by 2 factors: (i) *Genotype*. Whereas all $H\_$ fowl are *hen-feathered* (*HF*) *hh* males are *cock-feathered* and *hh* females are *HF* because *h* permits the skin to respond differently to the male and female hormones. (ii) *Hormones*. Plumage differences between the sexes in *hh* breeds depend on the hormonal differences between males and females, which in turn are dependent on the sex determining mechanism of the gonads. In females (AAZW) the ovaries produce female hormones. In males (AAZZ) the testes produce male hormones. **Bases for conclusion:** (i) Skin transplantation studies. If skin from an *hh* male (potentially CF) is transplanted onto a female, regardless of her plumage genotype, the transplant develops HF. If the skin from an *hh* female is transplanted onto a $H\_$ or *hh* male, the transplant develops CF. In both cases the transplant acquires the expression of the host. (ii) Females whose ovaries have been removed, developed CF. If a testes is transplanted into such a female she continues to express CF. If ovaries are transplanted into the gonadless females, they develop HF.

Note: In *HH* breeds, both sexes are HF because *H* suppresses differential response of the skin to male and female hormones. *H* inhibits CF in the presence of both hormones. This conclusion is based on the facts presented in i), ii), and iii).

c) The initial (and subsequent) effects of the *H* and *h* alleles are solely in the target tissue.

14. An answer to the question requires recognition of 2 facts: (i) The lz^{cl} alleles at the lz^+ locus affects females only (it has a sex-limited expression). (ii) The *tra* allele at the transformer locus transforms XX flies into phenotypic males. $lz^{cl}\ lz^{cl}$ *tra tra* flies are phenotypic males without any of the abnormalities observed in females. This indicates that sex must be determined before the *lz* locus can produce its effects. Therefore the *tra* locus must function in development before the *lz* gene performs its functions since *tra* transforms females into males without any of the mutant effects. If the *lz* gene acted before the *tra* gene performed its functions the $lz^{cl}\ lz^{cl}$ *tra tra* flies should have been **affected,** and not **unaffected,** males.

20. a) The *variegated* phenotype in *Ww* translocation heterozygotes is not due to mutation of *W* to some other allele. This allele, produces a *mottled* phenotype when it is on the translocated chromosome next to heterochromatin. When it is relocated to its normal position on the X (and therefore far from heterochromatin) it expresses its full potential in all cells of the eyes of *yy Ww Spl Spl* translocation heterozygotes - it produces a *non-variegated* phenotype - eyes completely *dull red*. When w^a is used in place of *W* (the "mottled" allele) and placed on the translocated chromosome, in the X-chromosome 4 translocation heterozygotes next to heterochromatin, it also produces a variegated phenotypic effect in $w^a w$ flies.

The *variegated* phenotype is due to position effect. It occurs when heterochromatin inactivates genes in adjacent euchromatic regions in some somatic cells. Position effect is probably due to heterochromatinization of euchromatic segments and their genes. In *Yy Ww Spl spl* translocation heterozygotes, the *W* and *Spl* alleles are on the translocated chromosome next to heterochromatin which inactivates them in some somatic cells and gives use to a *variegated* phenotype. If *w* and *spl* were on the translocated chromosome phenotypic variegation would not occur because (i) these alleles are amorphs and (ii) *W* and *Spl* are in the normal locations and therefore not inactivated in any cells.

b) This is because the *spreading effect* of inactivation moves linearly outward from the heterochromatin through the sequence of genes in the adjacent euchromatin. The spreading effect and heterochromatinization decreases as the distance of the genes from the heterochromatic region increases.

22. a) If the pedigree is typical of others for *Duchenne muscular dystrophy* (DMD), the reasons the condition occurs almost exclusively in boys because: (i) It is determined by a recessive allele, *m*, of an X-linked gene and (ii) *m* is rare in the population because of its lethal effects. Since *affected* males die before they reproduce, females can only be heterozygous, *M m*. Barring mutation they always receive their X chromosome with *m* from their mother. Each X chromosome in females is transmitted to the children, such that half the male offspring of heterozygous mothers are expected to be *affected* but all the daughters *unaffected* only (or *subclinically affected*) because their father is unaffected ($X^M Y$).

b) First male second generation - $X^M Y$ first female second generation - $X^M X^M$; second female second generation - $X^M X^m$; second male third generation $X^m Y$.

c) Some females are *clinically affected* (less than 50% of their muscles are normal (functional)) and others are *clinically unaffected* (more than 50% of these muscles are normal with dystrophic muscle fibers ranging from few to near 50%). All these females must be $X^M X^m$. The observations are in complete accord with the Lyon hypothesis which states that in females, 1 of the 2 Xs in each somatic cell is genetically inactivated randomly, early in embryogenesis. Once inactivation occurs, the state is transmitted to all the descendent cells in the clone. This random inactivation of the X accounts for the range of *affected* phenotypes in heterozygous females from *clinically affected* (more than 50% of the muscle cells and fibers have the X^M inactivated) to various degrees of *subclinical phenotypes* (more than 50% of the muscle fibers have the X^m inactivated).

25. In view of the following facts: (i) The G6PD locus is X-linked; (ii) the late-replicating X chromosome in females is paternal in origin and becomes heterochromatic. (iii) Females express the potential of the allele at the G6PD locus that they receive from their mother; the allele they obtain from the male parent is inactivated. Therefore, we can conclude: (a) that dosage compensation occurs in kangaroos, and (b) its mode is different from that in eutherian mammals where X inactivation is random.

Model of evolution of eutherian from marsupial X inactivation. Assume that the marsupial system of dosage compensation is ancestral. During marsupial spermatogenesis, an element similar to controlling elements (transposons) in maize, is excised from the Y chromosome and inserted into the X. In eutherians, this system is modified so that in early embryogenesis in females, the element is excised from the paternal X and reinserted at random into one of the 2 Xs. Assume there is only one controlling element per X chromosome and that the site of insertion is always the same. Since the element is genetic there should be mutants in the system which hinder the excision of the controlling element and/or its subsequent reinsertion. Non-random paternal X-chromosome inactivation in marsupials may occur because the excision of the controlling element is impaired.

28. a) At the regulator locus, i^+ is dominant to i^-. This is because functional repressor is made by it which can bind to the operator o^+. i^- cannot make repressor protein. The i^s allele is dominant to the first two alleles i^+ and i^- because once it is present the repressor protein is irreversibly bound to the operator. The i^d allele is dominant to all other alleles with the exception of i^q. This is so because the i^d product aggregates i^+ repressor protein and inactivates it. The i^q allele is dominant to all other alleles except i^s because, despite overproduction of functional repressor by i^q, it cannot circumvent binding by inducer, which i^s can.

In the case of the operator locus, the o^c allele is dominant to o^+ (assuming all other loci in the operon are *wild type*).

b) The $i^-o^+z^+y^+$ genotype verifies that the i^+ locus codes for a repressor which inhibits synthesis of enzymes by binding to the operator; the i^- allele in genotype 4 must not recognize the operator, because there is no enzyme repression at all under both induced and non-induced conditions. It is possible from genotype 5 to conclude that i^s codes for a super-repressor that does not respond to inducer, as having an inducer present does not alleviate its repression of the operon. The $i^+o^cz^+y^+$ genotype confirms that the o^c mutant operator is not recognized by the repressor as there is enzyme production whether there is inducer present or not.

c) i) The alleles z^+ and y^+ are both dominant to their mutant z^- and y^- counterparts as shown in diploid 1 which is heterozygous for the 2 loci. Both the enzymes are produced in the presence of the inducer.
ii) The allelic relationships of the repressor and the operator as outlined in a) are not contradicted by the diploid data.
iii) Diploids 2 and 3 show that the repressor proteins will function in cis or trans with an operon that is $o^+z^+y^+$. Diploids 9, 10 and 11 show that the different alleles of the operator affect only the genes that are on the same strand of DNA as the mutant operator (it only acts in the cis position); constitutive production of the enzymes provided the alleles at the 2 loci are *wild type*, z^+ and y^+.

29. a) The C locus must be a regulator because it induces the production of all the enzymes in the presence of arabinose as shown in the first merodiploid. It must also play some positive role because its absence fails to cause induction in the third haploid.
b) C^+ must interact with arabinose to stimulate or activate enzyme production, because when both arabinose and the C^+ regulator are present maximum enzyme production is observed. The C^c allele allows the production (although in lower quantities than the induced *wild-type* operon) of enzyme even in the absence of arabinose (as seen in diploid 2), and in *wild-type* quantities in the presence of arabinose (as seen in haploid 2). It can be concluded that the C^c allele stimulates enzyme production to some extent without arabinose induction, but requires the presence of arabinose for maximum enzyme production. This allele should be dominant to *wild-type* because its product will stimulate enzyme production without induction. The C^- allele fails to result in enzyme production, probably due to an indifference to arabinose or the promoter (where messenger production would be stimulated by the C-arabinose complex).
c) 1) The diffusible protein coded by C^+ would stimulate the transcription of both enzymes when arabinose is present, but neither when arabinose is absent.
2) C^- will not produce the diffusible protein necessary to stimulate the operon, regardless of the presence or absence of arabinose, enzymes would not be produced in either case.
3) In the absence of arabinose, the C^c protein will stimulate greater enzyme production than the *wild-type* non-induced level, but less than the *wild-type* non-induced level, but less than the *wild-type* induced level. In the presence of arabinose, enzyme production will be at the same rate as in the *wild-type* induced state.
4) The phenotype will be the same as that in (3), as the C^c product will act on both operons of the diploid in the absence of arabinose: both the C^c and C^+ products will stimulate enzyme production in the presence of arabinose.

d) Region X is probably the location of the promoter; despite C^+ and A^+ being present in the diploid, isomerase was not produced at the expected induced levels. Because the copy of the operon with the A^+ allele is missing the X region, the deletion is acting in cis fashion; this would indicate a promoter deletion in a positive-control system.

30. In the presence of inducer the first 2 genotypes produce normal enzyme and the third does not; it can be concluded that gene b is the structural gene. In the fifth genotype the c product is acting in trans (i.e., the c^+ product from the second operon in the diploid is able to regulate the enzyme from the first set) so it must be the regulator. The last 4 genotypes show that the a^- mutation acts in a cis manner; gene a must be the operator locus.

31. a) The operator must be at the E end of the operon, and mRNA synthesis must begin there as well. For all enzymes to be affected by deletions at the E end of the operon, the genes specifying these enzymes must all be translated off the same transcript; any genes that are regulated by the operator must be between the operator and the end of the region of DNA coding for the transcript.

    b) i) If a nonsense mutation in a structural gene is proximal to the operator, rather than distal, there is less likelihood that functional product will be made because the stop codon would occur early in the transcript; translation would cease, producing only a small polypeptide with little or no *wild-type* function. If the nonsense mutation is distal to the operator, the stop codon would occur further down the transcript, permitting the translation of most, if not all of the genes in the operon. The nonsense mutation 2 occurs in the E gene, near enough to the end of (closer to gene D) the gene; there might be enough of the messenger translated to produce normal E protein (although the rest of the genes D, C, B and A would not be translated).
    ii) Mutation 1 would have the most drastic effect because it occurs in gene E which is closer to the operator end of the tryptophan operon. The closer a nonsense mutation is to the origin of translation, the shorter the polypeptide will be, and the less likely that the polypeptide will be functional. Mutation 2 might retain normal E function depending on where in gene E it occurs; if there is no essential region past the nonsense mutation, it will probably retain *wild-type* activity. Mutation 3 would be the least drastic because gene B is distal to the operon A transcript with mutation 3 would allow the translation of genes E, D, C, and maybe B, depending on where in gene B the stop codon occurs.

32. Suppressors are mutant tRNA molecules that can recognize nonsense codons as sense and insert amino acids in the positions of the stop codons. The mutations in the regulator genes R_1 and R_2 must be nonsense mutations which are recognized as sense by an external suppressor. The repressor must be a protein. If the repressor were RNA it could not suppress the constitutive mutation, as the RNA is synthesized straight off the DNA, without an intermediate.

33. The first type of mutation is a missense mutation which substitutes one amino acid with another, and in turn alters the activity of the protein enough to cause a *mutant* phenotype. As there is no change in the reading frame and a nonsense codon is not produced, the mutation affects one codon only. As for the latter type of mutation, if a mutation were a frameshift, the messenger would code for a protein with no function. If it were a nonsense mutation, the reading of the messenger would stop prematurely. In either case, there would be no *wild-type* production of the enzymes distal to these two types of mutations.

34. a) Class 1 mutations are missense mutations, and class 2 are nonsense mutations. Suppressors would only be able to alleviate a nonsense mutation because the suppressor product is a tRNA molecule that places an amino acid in a nonsense codon and allows translation and protein synthesis to continue. There are no polarity effects from the class 1 mutations, and the fact that class 1 mutants are CRM^+ indicate that most of the protein is identical to that of *wild-type*, but the active site is non-functional. This is evidence of a missense mutation.
    b) The difference is the location of the mutation on the DNA strand. In class 1 mutations it is the gene that is mutated that does not produce active protein. In nonsense mutations any genes proximal to the gene containing the nonsense mutation will produce normal protein, while those distal to it will produce proteins in minimal quantities. As for the gene that actually contains the nonsense mutation, the location of the nonsense mutation within the gene will determine the severity of the mutation. If the stop codon appears early in the gene it is likely that the active portion of the enzyme is coded for further on in the gene, and no activity will be seen. If, however, the stop codon appears near the end of the gene, it is likely that the active portion of the enzyme will be translated before the ribosome reaches the nonsense codon, and the activity of that enzyme will be near normal.
    c) Some activity (although greatly reduced) is seen in genes after the nonsense mutations (in class 2), with there being more activity in the genes further away from the mutation (for a given mutant). This could be explained if each gene had its own start codon. Maximum production occurs when the ribosome attaches at the proper initiation site and translates the entire messenger without falling off (the stop codon at the end of a protein does not cause the ribosome to fall off if there is a new start codon adjacent to it). If there is a nonsense mutation prior to a specific gene, say the PRTase gene, that gene will still be translated to some extent by some ribosomes attaching at its start codon. Genes distal to the PRTase gene, will show even more activity because the ribosome will be able to attach to the start codon at the PRTase gene, the distal gene and any genes in between.

36. The deletion strains and amber mutants have smaller polyribosomes than *wild-type* and revertant strains. This clearly indicates that there is only one mRNA per operon. If a separate messenger was transcribed from a single gene, the

deletion mutants deficient in *a* and *y* genes should have had fewer β-galactosidose polyribosomes, but were of the same size as those in the *wild-type* strain. Only the amber mutants for the 2 genes should have produced smaller polyribosomes if each gene produced a separate mRNA.

38. a) *Fertility* vs. *sterility* (*grandchildless*): is determined by a single autosomal pair of alleles; *Gs* for *fertility* is dominant to *gs* for *sterility*. The recessive allele causes *sterility* in both sexes of the progeny of a female homozygous for it. In males, the *gsgs* genotype has no such effect; unless the female mates of such males are also *gsgs*, the progeny (males and females) will be *fertile*. Therefore *sterility* (and also *fertility*) of an individual depends on its mother's genotype and not on its own genetic constitution. The maternal genotype (*gsgs* or *Gs*_) has its effect (*sterility* or *fertility*) delayed one generation. This phenomenon is referred to as maternal influence or inheritance.
Evidence that a single autosomal pair of alleles is involved and that Gs is dominant to gs comes from cross A x C.

All 97 females in the progeny were *fertile*. Twenty-four of these females produced sterile offspring (1 ♂ : 1 ♀) only; 73 produced fertile offspring 1 ♂ : 1 ♀) only (a 1:3 ratio of *sterile* : *fertile* progenies). This indicates that a single pair of alleles is involved with *Gs* for *fertility* dominant to *gs* for *sterility*. The 97 fertile females must have possessed the 3 possible genotypes *GsGs*, *Gsgs* and *gsgs* in a 1:2:1 ratio which is only possible if the locus is autosomal. Had it been X-linked, *fertile* females would have possessed two genotypes in a 1:1 ratio. The results of the D x E cross support the conclusion that a single pair of alleles is involved.

Evidence for maternal influence:
gs, causes *sterility* in both male and female progeny of *gsgs* females and not in such females themselves. This is indicated by the results of crosses A x C and D x E. In the cross Z x C, the 97 females possess the three genotypes *GsGs*, *Gsgs* and *gsgs* in a 1:2:1 ratio, yet all of them are *fertile*. The progeny of 24 (1/4) of them (the *gsgs* ones) are *sterile*. The results obtained with 29 of 61 *fertile* females in the D x E cross also support the conclusion that the maternal genotypic effect occurs among her progeny.

That *gsgs* males are *fertile* is indicated by the results of the D x E cross. Since the female parent D is *Gsgs* the male parent E must be *gsgs* since 1/2 (29) of the 61 *fertile* female progeny produced only *sterile* offspring and the other 1/2 (32) of them produced only *fertile* males and females. Since *Gs*_ females are *fertile* and the cross D x E produces viable progeny, the *gsgs* males must also be *fertile*. That *gsgs* does not have the same effect in males as in females is indicated by the breeding behavior of the progeny of D x E which are *Gsgs* and *gsgs* in a 1:1 ratio among both sexes. Although 1/2 the females produce only *sterile* progeny, all

males produce *fertile* individuals only. Since both *gsgs* males and females produce progeny, *gs* has no effect on its homozygous carriers; *gsgs* flies are indistinguishable from *Gs*_ ones phenotypically. All are *fertile* except those that are progeny of *gsgs* females.

Gene action: One possibility is that the *gs* gene (when homozygous) causes a biochemical block in females resulting in the accumulation of precursor materials in the egg cytoplasm. This prevents the normal development of the gonads during early development and has a permanent effect on their morphogenesis. Another possibility is that the *wild-type* allele, *Gs*, may be responsible, in the female, for the production of some substance which is contributed to the cytoplasm of eggs and is essential for the proper development of the gonads during embryogenesis.

b) One should note that pole cells are germ cell precursors and polar plasma is germplasm. In *wild-type Gs*_ flies, the nuclear multiplication stage cleavage nuclei divide normally and synchronously and distribute uniformly to all parts of the embryo. After several nuclei reach the posterior poles, pole cells bud off the embryo. By the final stage of the blastoderm, poles cells increase in number. Subsequent development produces a normal polar plasma with normal polar cells.

Since mutant (*grandchildless*) females either lack polar plasma and polar cells or these are defective, the *gs* gene may retard division and migration of cleavage nuclei to the posterior pole of the embryo. This would lead to the failure of pole cell formation and various degrees of malformation of blastoderm cells and yolk nuclei in the posterior region of the embryo.

40. a) (i) The mutant alleles at the *Pc* locus are recessive to the *wild-type* allele Pc^+ for the following reasons: (1) The phenotype of viable and fertile adults heterozygous for any mutant allele, e.g., Pc^+Pc^3 is more similar to that of Pc^+Pc^+ flies than it is to the phenotype of mutant individuals, e.g., Pc^3Pc^3; (2) Only homozygotes and hemizygotes for the mutant alleles express the mutant phenotype and die as embryos. (ii) The cross should produce 1/4 Pc^+Pc^+ (*wild-type* - no transformation of thoracic and abdominal segments): 1/2 Pc^+Pc (*pseudo-wild-type* - less extreme transformations of thoracic and abdominal segments than in mutants): 1/4 *PcPc* (complete transformation of all thoracic and abdominal segments to A8) because of incomplete dominance of Pc^+ and recessiveness of *Pc* and other mutant alleles.

b) The explanation must account for the following facts: (i) There is an anterior-posterior gradient of gene activity in the BX-C region; (ii) All thoracic and abdominal segments of mutants homozygous for deletions of the *Pc* locus resemble segment A8. This suggests that all genes in the BX-C region are turned on in all segments; (iii) The *Pc* locus interacts with the genes of the BX-C genes in an inactive state. This repressor is more concentrated in anterior segments, where fewer genes are turned on, than in the posterior segments. The BX-C region contains regulatory sites that interact with the repressor to keep the BX-C genes turned off. The regulatory regions of the most distal genes would have a high affinity for the regulatory product. The findings of Zink and Paro indicate that a DNA-binding protein performs this regulatory function. The *wild-type allele Pc$^+$* would produce a normal protein which functions by interacting with specific sites in the DNA of the BX-C complex. In mutant homozygotes and homozygotes for deletions of the *Pc* locus, the protein would either not be produced or produced in defective form. This prevents binding at any sites with the result that all genes function in all segments.

41. **Experiments involving *wild-type* and single *mutants***

    The transfer of cytoplasm and its cell from donors to recipients in the first 6 transplants reveals the following:

    1. $Wt \rightarrow bcd$. The maternal information from anterior segments that is defective in eggs from *bcd* females is a cytoplasmic component located at the anterior pole of the egg. It can function at quite a distance from the anterior region since it changes the abdomen in *bcd* flies to thorax in the middle of the embryo.

    2. $Wt \rightarrow bcd$. Partial rescue when cytoplasm from the middle portion of the donor embryo is used instead of cytoplasm from the anterior region suggests that the maternal information is distributed in an anterior → posterior gradient.

    3. $Wt \rightarrow osk$. Rescue of *oskar* with cytoplasm from the posterior pole of the *wild-type* embryo indicates a localized, maternally determined *Osk* dependent activity necessary for determination (specification) of posterior segments.

    4. $Wt \rightarrow Wt$. The *bcd* phenocopy indicates that a cytoplasmic factor at the posterior end of the embryo inhibits the activity specified by the *Bcd* locus at the anterior pole.

    5. $osk \rightarrow Wt$. From the results in 4 we cannot determine the cause of the inhibition. This experiment shows that the inhibition depends on the product of the *Osk* gene or something regulated by it.

    6. $bic \rightarrow osk$. On the basis of the results of the previous experiments the *bic* phenotype could be explained by ectopic expression of the *Osk* activity at the anterior pole. This hypothesis is supported by the findings in this experiment which indicate that there is *osk*-rescue activity at the anterior pole of *bic* embryos.

    Experiments with double mutants

    1. *bic; osk*. In this double mutant *osk* is epistatic. This is expected if *bic* acts by relocalization of *Osk* function.

    2. *bcd; swa*. The single mutants *bcd bcd* and *swa swa* have similar functions. If *swa* was required only for anterior localization of the *bcd* gene product, the same phenotype would be expected in *bcd bcd swa swa* and *bcd bcd Swa Swa* flies, since *Swa* would not have any product to localize. The *bic*-like phenotype of *bcd bcd, swa swa* suggests that *swa* may play an important role in posterior localization of *Osk* function.

    3. *osk; exu*. Additional information is required to explain the reverse of *bic* phenotype.

42. a) Yes. By definition homeotic mutants are those that cause a tissue normally determined to form a specific organ or body part to alter its differentiation and form another structure - these mutants can convert one part of a body segment into another part of the same or different segment. All 4 mutations perform such functions and are therefore homeotic.

    b) The term *ground state* refers to an evolutionarily primitive development pathway from which other developmental pathways arise because of the function of additional genes. Of the genes that determine the sequential development of segments T2, T3, A1, A2, A3, A4, A5, A6, A7, A8 from the anterior to the posterior region of *Drosophila*, the gene that is responsible for development of T2 (primitive or original) state is the *ground state* gene and therefore the T2 segment is the developmental *ground state*. The expression of the first gene in addition to other specific ones, is also required for the development of other segments. For example T3 formation requires the activity of the *wild-type* allele at the first locus and the *wild-type* allele at a specific second locus. The differentiation of A1 requires the action of *wild-type* alleles of 3 genes: T2 and T3 and an additional one. With each successive segment the activity of 1 more additional gene is required for differentiation.

    That the T2 segment is the developmental *ground state* is evidenced by the fact that in flies homozygous for deletion of the entire BX-C region, T3 and each of the A segments resemble T2.

    c) 10 genes are expected to be present in the BX-C region. One (A) for T2, 2 (A and B) for T3, e (A, B and C) for A1 ...... 10 for A8 (A, B, C, D, E, F, G, H, I, J).

    d) The genes in this complex are expressed successively from front to back in the embryo. Each additional gene in the complex allows the next segment from T2 through A8 to differentiate. This conclusion is based on facts i), ii) and iii) which indicate that each recessive mutant gene, e.g., *bx* (when homozygous) causes a particular segment, e.g., T3, to resemble the segment, e.g., T2, formed before it.

    e) The *wild-type* genotype is AA, BB, CC, DD, EE, FF, GG, HH, II, JJ. Genotype of flies that converting (i) A1 to T3 is AA, BB, cc, DD, EE, FF, GG, HH, II, JJ; (II) A2 to A1 is AA, BB, CC, dd, EE, FF, GG, HH, II, JJ; and (iii) A3 to A2 is AA, BB, CC, DD, ee, FF, GG, HH, II, JJ.

43. a) Yes, dosage compensation (DC) occurs in *D. melanogaster* as evidenced by the fact that the w^a allele at the X-linked *white* (w) locus produces the same amount

of pigment in both males ($AAXY$) and females ($AAXX$). If dosage compensation did not occur females would produce twice as much pigment as males.

b) Females heterozygous for *mutant* and *wild-type* alleles at X-linked loci are phenotypically uniform at all levels. This indicates that DC is not achieved by X-chromosome inactivation. If it was, females heterozygous for X-linked allele pairs would express mosaicism at the biochemical, cellular, serological and other phenotypic levels.

c) i) The cells of $Pgd^a Pgd^b$ females produce 3 different types of 6-PGD enzyme because: (a) both alleles and therefore both X's are active; (b) each allele produces many polypeptides (pps) of its type, and (c) these pps unite at random to produce 3 molecular forms of the enzyme.

Allele	Polypeptide	6-PGD dimer	Band
Pgd^a	•—————	1/4 •—•	Fast (F)
	Random assembly	1/2 •—•	Intermediate (SF)
Pgd^b	—————•	1/4 •—•	Slow (S)

ii) These results confirm the conclusion that dosage compensation does not occur by X-chromosome inactivation; they imply hyperactivity for the X-chromosome genes in males or hypoactivity for such genes in females.

d) The results in (iv) indicate that dosage compensation acts at the transcriptional level. Specifically the X in males is hypertranscribed so as to be equivalent to 2 Xs in females. This explanation accounts for: (i) equivalent transcription levels in the 2 sexes ; (ii) the fact that the polytene X in males has half as much DNA as the polytene Xs in females, and (iii) the width differences of the polytene Xs in the 2 sexes.

45. a) i) All the data indicate that *bladder carcinoma* is inherited and not a phenocopy. If the cancer was a phenocopy its cause would not be transmitted. Cells of the *normal* mouse that were transformed with DNA from human *bladder carcinoma* cells formed minitumors *in vitro*. Inoculation of the minitumor cells into *normal* mice caused tumor formation. Transformation of *normal* cells inoculation of *normal* mice with DNA from the minitumors also caused tumor formation. Similar experiments with *normal* human and *normal* mouse cells did not induce cancer development.

ii) Transformation involves transfer and integration of a very short segment of DNA (1 or very few genes) for the donor to the recipient cells. This indicates that only 1 or very few genes in humans cause bladder carcinoma. ii) The gene(s) produce the same phenotypic effects in mice as in humans. iii) Minitumors occurred in some of the first generation cells of *normal* mice transformed by DNA from human *bladder carcinoma* cells. This indicates that the alleles of the gene(s) for this form of cancer are dominant. The normal condition is due to the recessive alleles of the gene(s). The serial transformations (transfections) also confirm that a single short segment of human DNA, causing the cancerous phenotype, is inherited in a dominantly fashion.

b) That the transforming DNA is entirely within the EcoRI fragment of DNA from the cancerous mouse cells proves that the fragment is human in origin. Since: (i) the EcoRI fragment of DNA from the cancerous mouse cells, transformed by DNA from the human *bladder carcinoma* line, can transform normal cells into malignant ones, and (ii) normal mouse cells do not carry the *bladder carcinoma* gene(s). This fragment must be human in origin and contain the dominant alleles of the gene(s) for *bladder carcinoma*. These data do not prove that there is only 1 gene for this cancer. Such proof must come from other experiments which involve: (a) the cloning and sequencing of both the proto-oncogene (*wild-type* = normal allele) and the oncogene (*mutant* allele), and (b) identifying the mutational site(s) and changes that converted the proto-oncogene into the oncogene. If the gene is a structural one and the protein specified by the gene has been identified, one could compare the amino acid sequences of the normal and mutant forms of the protein.

c) The p21 protein coded by ras^+ may phosphorylate other proteins. The mutant form of p21 produced by ras may cause errors in phosphorylation. This could affect the activity of many proteins; one or more of these could be critical in regulating cell cycle events. The protein may act in cell nuclei where it may bind with DNA, specifically with specific target sequences to regulate (enhance) expression of a gene(s) involved in determining an event(s) in the cell cycle and cell growth.

48. a) i) Gene mutation is ruled out as a cause of BL because the protein produced by the *c-myc* gene in BL cells is identical to that synthesized in normal B cells. If BL was due to a mutant allele (oncogene) at the *c-myc* locus the *c-myc* protein would have had a different amino acid composition and sequence. ii) Position effects are due to chromosomal rearrangements which juxtapose euchromatic segments and their genes next to heterochromatic regions. The translocations in BL involve only euchromatic segments of chromosomes 8 and 14. Therefore position effects cannot be the cause of BL.

b) The *c-myc* gene in its normal location is transcribed and translated at a normal rate. The normal amount of *c-myc* protein results in (regulates) normal cell division and cell growth. In Bl, the *c-myc* oncogene becomes deregulated as a result of its proximity to genes that code for antibodies. These antibodies are known to be produced at high rates in B cells because of high transcription rates. Specifically, the chromosome translocation may place a *c-myc* oncogene in juxtaposition to DNA enhancer sequences that increase levels of transcription of genes over a considerable distance. The *c-myc* gene would then be expressed (transcribed and translated) at the same high level in B cells with the translocation as immunoglobulin genes are expressed in normal B cells - at a very high rate. The consequence is much more *c-myc* gene product than synthesized normally. This results in increased cell division and cell growth accompanied by other changes and eventually cancer.

c) Disregarding technical details, one should measure and compare the level of *c-myc* mRNA in B cells containing 2 normal homologues of each of the chromosomes 8 and 14 with that in BL cells which possess a normal 8 and a normal chromosome 14 and the 2 translocated chromosomes. If the explanation in b) is correct the normal B cells and BL cells should have low and high levels of *c-myc* mRNA respectively.

50. a) The normal (Rb^+) allele is dominant at the cellular level. It acts to regulate growth; specifically it is a growth suppressor. This is evidenced by the fact that expression of *retinoblastoma* at the cellular level requires homozygosity or hemizygosity for: (i) the mutant allele Rb^-, or (ii) deletion of the *Rb* locus. Rb^+Rb^- individuals do not suffer from *retinoblastoma*; Rb^+ is a tumor suppressor = regulator of normal cell growth.

b) The data unequivocally confirm the explanation in a). The *retinoblastoma* cell line (Rb^-Rb^-) forms tumors *in vitro*. The transfer of Rb^+ to this line results in Rb^+Rb^- or $Rb^+Rb^-Rb^-$ cells with suppressed neoplastic phenotype (reversion of uncontrolled cell proliferation to normal cell growth). Therefore Rb^+ is a dominant allele that suppresses both tumor and cell growth.

52. a) Reciprocal-shift experiments are those conducted to determine the dependent relationships between gene-determined events during the *cell division cycle* (*cdc*).

b) To determine whether any 2 genes mediated steps in the cell cycle are dependent, interdependent or independent 2 reciprocal-shift experiments are performed for each pair of *cdc* mutants. The event determined by one gene is restricted and that determined by the other is permitted in the first incubation with the conditions being reversed in the second incubation. The order of restriction and permission in the reciprocal experiment is the opposite of the order in the first experiment. The expectations are different and easily distinguishable in these experiments as shown below for the 2 cell cycle events *A* and *B* determined by genes *A* and *B* respectively.

		Completion of cell cycle events	
Relationship between the two events		First incubation: restrict A, permit B Second incubation: permit A, restrict B	restrict B, permit A permit B, restrict A
Dependent	A B	−	+
Dependent	B A	+	−
Independent	A B	+ +	+ +
Interdependent	AB	−	−

c (i) The cell cycle events determined by the *cdc* 2 and *cdc* 18 genes are independent since the restriction of either of these steps does not prevent the other from being completed. ii) The sequence of the steps determined by the 3 genes is $cdc\,8 \rightarrow cdc\,14 \rightarrow cdc\,4$ based on the rationale in b) and the data in the Table.

53. a) *Tumor*. A mass of cells that multiplies when it should not. A *benign tumor*, e.g., warts, remains localized (does not spread) in the place of origin and therefore is not cancerous. Such tumors are slow growing and compact. They may become quite large and cause damage indirectly by pressing against a vital organ. The karyotypes of benign tumor cells are usually normal.

Malignant tumor. A mass of cells that multiplies when it should not, and becomes cancerous. Its characteristics are: (i) The tumor does not respond to the mechanisms that normally restrict the number of cells to those required for growth and replacement. (ii) Cancer cells invade adjacent tissues causing damage to them. They may also leave the site of their origin and circulate through the blood or lymph to other places in the body, where they may initiate secondary cancer. (iii) Cancer cells usually have non-normal karyotypes which include chromosome aberrations of various kinds, including aneuploidy and polyploidy. A given type of cancer may possess a specific chromosome aberration as in *Burkitt's lymphoma*.

b) Dosage compensation by inactivation of one or the other X chromosome will cause each cell to produce the A or B form of G6PD. Since (i) the millions of cells in each uterine tumor produce only 1 form of the enzyme and (ii) each separated tumor produces either the A or B form of G6PD, strongly supports the single-cell origin for cancers. Otherwise each tumor would have had some cells producing the A form and other cells producing the B form of the enzyme.

55. i) A tyrosine specific protein kinase is one that phosphorylates only tyrosine. The *Src* specified tyrosine protein kinase phosphorylates many different proteins.

ii) One of the proteins that is phosphorylated by this kinase is a compound vinculin, involved in the formation of adhesion plaques that help bind cells to each other. A mutant allele (*src*) at the *Src* locus could possibly cause increased phosphorylation of vinculin in cancer cells; this could be responsible for the rounded shape of cancer cells and their decreased binding of cells.

Chapter 24
Population Genetics, Inbreeding, Outbreeding and Evolution

1. a) That inbreeding *per se* does not favour an increase in the frequency of recessive alleles is best demonstrated if one considers the consequences of the most extreme form of inbreeding - self-fertilization.
   Consider a randomly mating population in which the frequencies of the alleles A and a are p and q respectively. The proportions of genotypes in the randomly mating population will be: $p^2(AA) + 2pq(Aa) + q^2(aa)$. If all individuals of all genotypes are self-fertilized the proportions of genotypes in the ensuing generation will be:
   1st self-fertilization $\{(p^2+2pq/4)AA + (2pq/2)Aa + (q^2+2pq/4)aa\}$.
   Reasons: The proportion of heterozygotes for every generation of self-fertilization is reduced by one-half and the two homozygous types are each correspondingly increased by one-quarter. If all the individuals are again self-fertilized, the proportions of genotypes will be:
   2nd self-fertilization $\{(p^2+2pq/4+2pq/8)AA + (2pq/4)Aa + (q^2+2pq/4+2pq/8)aa\}$.
   After n generations of self-fertilization, the proportions of the three genotypes will be as follows:
   n^{th} self-fertilization $\{[p^2+\Sigma n(2pq)/2(2^n)] + (2pq)/2^n] + [q^2+\Sigma n(2pq)/2(2^n)]\}$
   It should be apparent from the above that the proportions of the genotypes AA, Aa and aa are altered in each successive generation of self-fertilization. The frequencies of the two alleles $A(=p)$ and $a(=q)$ however, remain unchanged, generation after generation of self-fertilization as illustrated below:
   Consider the 2nd self-fertilized generation
   Genotypes contributing to the frequency of the A allele are AA and Aa.
   Their proportions are: $(p^2+2pq/4+2pq/8)$ and $(2pq/4)$.
   Therefore, frequency of the A allele: $\{(p^2+2pq/4+2pq/8) + 1/2 \text{ of } (2pq/4)\} = \{p^2 + pq/2 + pq/4 + pq/4\} = \{p^2+pq\} = \{p(p+q)\} = p$, since $p+q=1$
   Similarly it can be shown that successive generations of self-ferilization will not result in an alteration of the frequency (q) of the recessive allele a. Inbreeding therefore affects only the distributions of alleles among different genotypes. It does not alter allele frequencies from one generation to the next and therefore cannot favour an increase in the frequency of recessive alleles in the population.

3. a. Self-fertilization:
   Aa - 100% heterozygous → self-fertilized $AA + 2Aa + aa$; rate of reduction of heterozygosity = 1/2 or 1/2 heterozygotes and 1/2 homozygotes

   b. Brother-sister mating (both parents common)

   c. Half-brother, half-sister mating (only one common parent)

   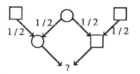

   d. Cousin matings

   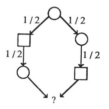

4. The reduction in the percentage of heterozygotes after each generation of self-fertilization of a single heterozygous (Tt) plant = 50%. To attain approximately 94% homozygosity, one would have to reduce the percentage of heterozygotes in the population to 6%, i.e. ≈1/16.
   Number of generations of self-fertilization that will be required to attain this is:
   $(1/2)^n = 1/16$, where n = number of generations, i.e., n=4.
   Therefore four generations of self-ferilization of a single heterozygous plant will attain 94% homozygosity.

5. The percentage of the progeny population that will be heterozygous for *starchy* endosperm ($Wxwx$), after one generation of self-fertilization of three plants of which only one is heterozygous ($Wxwx$) = 100 x $(1/2)^1$ x 1/3 ≈ 16%
   Percentage of heterozygotes in the population after four generations of self-fertilization of the three plants, only one of which is heterozygous = 100 x $(1/2)^4$ x 1/3 ≈ 2%

16. By inbreeding is implied the mating of individuals that are related to each other by ancestry. A coefficient of inbreeding of an individual, is a measure of the degree of relatedness of the individuals parents, i.e., the probability that the alleles carried by the parental gametes that produce the individual, are identical by descent.
    a) Coefficient of inbreeding for the progeny of an uncle-niece marriage:

    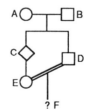 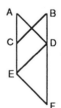

    F has two common ancestors in A and B. The number of individuals in the path from F through parent E to ancestor A onto parent D and back to individual F = 4. Similarly the number of individuals in the path from F through parent D to ancestor B onto C and back to individual F through parent E = 4. Therefore the inbreeding coefficient of individual $F = (1/2)^4 + (1/2)^4 = 1/8$.

b) For the progeny of a first cousin marriage:

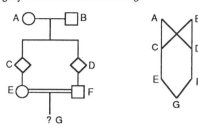

Tracing the paths of individual G back to each of the two common ancestors A and B, the inbreeding coefficient of G can be calculated as: $(1/2)^5 + (1/2)^5 = 2/32 = 1/16$.

c) For second cousin marriages:

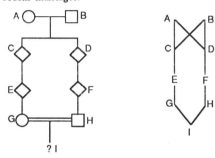

Coefficient of inbreeding for individual I = $(1/2)^7 + (1/2)^7 = 2/128 = 1/64$.

d) For half first cousins:

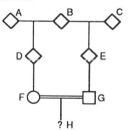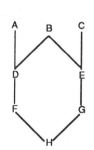

Individual H has only one common ancestor, B. Coefficient of inbreeding for individual H = $(1/2)^5 = 1/32$

e) Half-brother and half-sister:

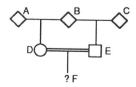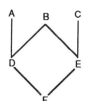

Individual F has only one common ancestor, F. Coefficient of inbreeding = $(1/2)^3 = 1/8$.

17. *Susceptibility* to 'pathenogenic disease' is shown in many instances to be genetically determined, as are 'hereditary diseases', characterized by Garrod as "inborn errors of metabolism". Original artificial selection as practised by dog fanciers ignored these facts, so that certain hereditary disorders (recessive) were rendered homozygous by inbreeding (e.g., *blindness* among Irish Setters, etc.). Undoubtedly, *susceptibility* to pathogenic disease is due to the same cause. Mongrels between breeds differing at many loci are highly heterozygous; their increased vigor results partly from the masking of deleterious recessive alleles by their dominant counterparts, and partly from heterotic effects.

19. A single generation of self-fertilization of heterozygotes reduces the proportion of heterozygotes in the offspring by 50%. Therefore two generations of self-fertilization should reduce the proportion of heterozygotes to 25%. However, since only 40% of the initial population of plants were heterozygous, after two generations of self-fertilization, one should expect 10% (25/100 x 40/100 = 1/10) of the offspring to be heterozygous. Therefore the proportion of plants that are likely to be homozygous for a given gene locus = 100 - 10 = 90%.

20. The parents of Domino - Publican and Donna Anna 22nd - have two common ancestors in Beau Brummel and Lamplighter. Therefore there are two contributing paths to the inbreeding coefficient of Domino. The contribution to the coefficient with Beau Brummel as an ancestor = $(1/2)^5$ since there are five individuals in the path (Publican → Pretty Lady → Beau Brummel → Paladin → Donna Anna 22nd). Similarly, the contribution to the coefficient with Lamplighter as an ancestor = $(1/2)^5 (1+1/8)$ since there are five individuals in the path (Publican → Paladin → Lamplighter → Donna Anna 14th → Donna Anna 22nd) and Lamplighter itself has an inbreeding coefficient of 1/8. Therefore the inbreeding coefficient of Domino = $(1/2)^5 + (1/2)^5(1+1/8) = 1/32 + 1/32 + 1/256 = 17/256$.

27. a) The total number of alleles at the *haptoglobin* locus in the given population of 219 Egyptians = 2 x 219 = 438. The allelic relationship at the autosomal locus specifies the following frequencies of the two alleles in the small population: frequency of $Hp-1$ = # of *haptoglobin-1* alleles/total # of alleles = (2x9)+(135)/438 = 153/438 = 0.30
Each *Haptoglobin-1* individual will contain two $Hp-1$ alleles while each heterozygote will contain one $Hp-1$ allele.
Similar reasoning will enable the calculation of the frequency of the *haptoglobin-3* allele as 285/438 ≈ 0.70

b) No, the distribution of the three phenotypes among the population of 219 individuals is not characteristic of a large and randomly mating one. The expected frequencies of the three haptoglobin types should have been:
$(Hp-1, Hp-1) = (0.30)^2$ x 219 ≈ 19
$(Hp-1, Hp-3) = 2(0.70)(0.30)$ x 219 ≈ 93
$(Hp-3, Hp-3) = (0.70)^2$ x 219 ≈ 107

32. a) *Nontaster* individuals are *tt* and constitute 30% of the given Caucasian population in Hardy-Weinberg equilibrium, i.e., frequency of genotype *tt* = 30/100 = 0.3.
Therefore frequency of allele *t* = q = square root of 0.3 = 0.55 and of allele *T* = (1- q) = p = (1- 0.55) = 0.45

Frequency of genotype $TT = p^2 = 0.205$
Frequency of genotype $Tt = 2pq = 0.495$
Frequency of genotype $tt = q^2 = 0.3$

b) A mating between two *taster* individuals can be of the following types: (i) TT x TT; (ii) Tt x TT; (iii) TT x Tt; (iv) Tt x Tt. Only mating (iv) can give rise to a *non-taster* child, barring mutation at the T locus.
Probability of a *taster* in the given population being heterozygous is ≈49.5%
Probability of a mating between two *tasters* that are heterozygous = 0.495 x 0.495. Probability that a child from such a mating will be a *non-taster* = 0.25. Therefore overall probability of a *non-taster* child = 0.495 x 0.495 x 0.25 = ≈4%.

c) Calculate the probability for yourself using the same type of reasoning as in (b).

34. Let the frequencies of the three alleles I^A, I^B and i which determine the A, B and O antigens, be represented by p, q and r respectively. Therefore p = 0.6, q = 0.3 and r = 0.1; p+q+r=1.
The frequencies of the four different blood groups can be determined as follows:
Frequency of the A blood group = freq. of $I^A I^A$ + freq. of $I^A i$ genotypes, since I^A is dominant to i.
= $p^2 + 2pr = (0.6)^2$ x $2(0.6)(0.1) = 0.48$
Similarly, the frequency of the B blood group = freq. of $I^B I^B$ + freq. of $I^B i$ genotypes, since I^B is dominant to i.
= $q^2 + 2qr = (0.3)^2$ x $2(0.3)(0.1) = 0.15$
Frequency of the AB blood group = freq. of $I^A I^B$ since I^A and I^B are co-dominant alleles = $2pq = 2(0.6)(0.3) = 0.36$
Frequency of the O blood group = freq. of ii genotype since i is the recessive allele in the series = $r^2 = (0.1)^2 = 0.01$
Thus $p^2 + 2pq + q^2 + 2pr + 2qr + r^2 = 1.0$

37. a) *Colour blindness* is determined by a recessive allele of an X-linked gene; therefore the frequency of *affected* males will reflect the frequency of the allele, as males are hemizygous for X-linked genes. In the population in question, since *affected* males are 20 times more frequent than *affected* females, the relationship can be represented as follows: 20q = 1 or q = 1/20 = 0.05, where q is the frequency of the recessive allele.
(i) Thus the frequency of the allele for *colour blindness* = 0.05.
(ii) If q = 0.05, then p = 1-q is given by 1 - 0.05 = 0.95, i.e., the frequency of the allele for *normal* colour vision = 0.95. The frequency of heterozygous females in the population = 2pq = 2 x 0.95 x 0.05 = 0.0950.

b) If 9% of males in a randomly mating population at Hardy-Weinberg equilibrium are *colour-blind*, then the frequency of the causative allele = 9/100 = 0.09., i.e., q = X^c (X-linked recessive allele for *colour blindness*) = 0.09.
Since *affected* females will be homozygous for the causative recessive allele, the frequency of *colour-blind* females in the population will be given by $q^2 = (0.09)^2$ = 0.0081.

80. a) Because no males result from normal crosses between species, and yet males result when XXY females are used for the crosses it appears that either (i) the Y chromosome of each species, if carried by the sperm, forms an inviable zygotic combination with an X-bearing egg of the other species, or (ii) that the Y-bearing sperm cannot ferilize eggs of the other species. (It is assumed that the normal female parents of the crosses produced X gametes only, and that the male parents produced no "O" gametes).

b) For both hypotheses, the assumptions listed should be tested by cytological studies; females should be XX, males XY.
(i) Male hybrids with Y and egg (cytoplasm) from *melanogaster* and X plus X-linked traits from *simulans*, should produce male and female offspring in crosses with *melanogaster* females, but only female offspring in crosses with *simulans* females. The same kind of situation should hold for male hybrids from the reciprocal cross.
(ii) This hypothesis would be difficult to distinguish from (i); one would have to study the fertilization process itself.

81. The mechanism of speciation was by introgression. Amphidiploidy is ruled out since an amphidiploid should have a 2n number equal to the sum of the chromosome numbers of the other two species; *gypsophilum* is obviously the derived species, but has 2n=16, and not (2n, 16 + 2n, 16 = 32). Recombinant isolates (species) are unlikely to be intermediates between other isolates (species) as in *gypsophilum*. In contrast the experimental behaviour of the *hesperium-recurvatum* hybrids and *gypsophilum* obeys the expectations for an introgressively derived species. In particular, the interfertilitiy of the group comprising *gypsophilum*, F_1 hybrids and progeny from crosses within the group, as contrasted with their consistent higher sterility in crosses with *hesperium* and *recurvatum* would provide a mechanism for the reproductive isolation of *gypsophilum* from the other species.

88. a) A measurement of the extent to which two or more species are genetically different is referred to as the "genetic distance" between them. The more distantly related two species are, the greater the value of this parameter.

b) Techniques that are currently used as part of the molecular approach to phylogenetic analysis are those that facilitate: (i) a comparison of the amino acid sequences of homologous proteins (gel electrophoresis, chromatography); (ii) a measurement of the similarities and differences in nucleotide sequences as detected by interspecific hybridization of DNA molecules; (iii) measurement of the cross reactivities of antisera to purified proteins.

c) Despite a small "genetic distance" between them, the major biological differences between humans and chimpanzees in terms of their anatomy, way of life, etc. are due to regulatory mutations. These are changes or alterations to the mechanisms that control or regulate the expression of their genes.

d) It is not likely to provide the necessary insights into understanding the basis for evolution until developmental and regulatory mechanisms are better understood. In the interim, information from a molecular approach to phylogenetic analysis will serve as an adjunct to results obtained from fossil records and other comparative studies.

89. a) The various types of proteins perform different functions within an organism. The more vital a function of a particualr protein, the less likely that it will be able to tolerate a change to its amino acid sequence. Therefore the rate of change in such protein molecules will be slow.

b) Class c protein is a type that is least amenable to change. It probably performs a function that is essential for life. Faster rates of change are indicative of molecules that can withstand change and still carry out their biological functions. Thus: class c corresponds to cytochrome oxidase, class b corresponds to hemoglobin and class a corresponds to fibrinogen.

c) Rate of change in histones - nucleoproteins - would be extremely low. They are tolerant of a few minor changes. Any major alterations to their amino acid sequence (primary structure) could render them incapable of binding to the DNA molecule, thus affecting the packaging and condensation of DNA in the chromosomes. Their ability to regulate gene expression may also be affected by alterations to their amino acid sequences.

. The more distantly related two species are, the greater is the "genetic distance" between them as detected in terms of amino acid substitutions within a given protein, e.g., cytochrome c. On the basis of the number of amino acid substitutions detected in cytochrome c of the different organisms it can be concluded that chipanzees and the rhesus monkey are both closely related to humans and to one another. The horse and the donkey are closely related to one another but distantly related to humans and the other two primates. Yeast is an organism that is distantly related to the equine species and most distantly related to the primates.

Chapter 25
Current Approaches to Genetic Analysis: Somatic Cell Hybrids, RFLPs and Recombinant DNA

3. These cells are autotetraploid with four sets of autosomes, 2X chromosomes and 2Y chromosomes. They have the same 2:1 ratio of sets of autosomes to # of X chromosomes that the cells of diploid males have. Therefore, all X chromosomes will remain active. The 4n=92 cells will synthesize both G6PD and HGPRT because they will not have any heterochromatic X chromosomes.

4. a) The genes for IDH and MDH are completely concordant in their expression as are the genes LDH-B and PEP-B; therefore the former two genes are syntenic (belong to the same linkage group) and the latter form another syntenic group. All the other genes are discordant or only partly concordant in expression with each other and the four syntenic genes; therefore they are asyntenic.

b) It is impossible to determine if syntenic genes are linked or not because of the lack of recombination data. Such data cannot be obtained from studies of somatic-cell hybrids and their clones; these studies only permit a decision as to whether two or more genes are syntenic or asyntenic.

c) Yes, chromosome 9 carries the CEL gene since it is the only chromosome present in all the hybrids produced that have CEL activity.

d) The β-galactosidase gene is present on chromosome 22 since it is the only gene that is always expressed when chromosome 22 is present. Determining whether there is complete concordance between the presence of a given gene and a particular chromosome permits the identification of the chromosomes that carry the other six genes. This leads to the following conclusions: (i) chromosome 1 carries the ENOL gene; chromosome 2 carries genes for IDH and MDH, chromosome 12 carries genes coding for LDH-B and PEP-B and chromosome 15 carries the gene specifying HEX-A.

e) (i) The AGA gene may not be present on any of these chromosomes.
(ii) The gene may be present in an amorphic (non-leaky) mutant form so that is does not produce AGA.

f) Since the hybrid clones possessed all the chromosomes from a mouse cell line that did not produce HGPRT which is required for survival of the hybrid clones on HAT medium, the gene for HGPRT must be present in all clones and the same is true for the chromosome carrying it. Only the X chromosome is present in all six clones; it must be the carrier of the HGPRT gene.

g) No; because of a lack of perfect concordance between the presence of the Y chromosome and the expression of any of the nine genes. Also in (c) and (d) we had assigned eight of the nine genes to chromosomes and in (e) indicated that AGA is probably not on any of the chromosomes present in these clones.

5. a) Whenever chromosome 8 is present, the c-myc gene is also present. No such correlation between any of the other genes and c-myc. This proves that the c-myc locus is on chromosome 8. The 8q⁻ chromosome is a product of reciprocal translocation which carries a short terminal region of the q arm of chromosome 14 and the 14q⁺ chromosome is a product of the same interchange and it carries a short terminal segment of the q arm of chromosome 8. Since the clone with 14q⁺ carries the c-myc gene but clone 8q⁻ does not, the c-myc gene must be located in the terminal region of the q arm of chromosome 8.

b) The reason why B cells in *BL* individuals become malignant may be as follows: The *c-myc* in its normal position functions normally because of normal regulation of its activity to control cell division and cell growth. In B cells of *BL* patients the *c-myc* gene is placed close to or next to the gene coding for the heavy protein chains of antibodies which is expressed at a high rate in the B cells. The translocation enables the *c-myc* gene to evade the mechanism(s) that normally control its expression. It may be that the *c-myc* gene in the translocated chromosomes in B cells is juxtaposed to enhancers which can activate transcription over considerable distances. Thus the *c-myc* gene is expressed, in *BL* individuals, like the immunoglobin genes in normal B cells, at a very high rate and this may lead to *BL*.

6. a) The mutant chromosomes are the result of reciprocal translocations between chromosomes 14 and X. This is evidenced by a comparison of the chromosomes morphologies of these chromosomes and the normal X and 14 chromosomes.

b) Yes. The NP gene is at a non-terminal location on the long arm of chromosome 14. This is indicated in B by the perfect concordance between the presence of the large translocated chromosome (missing the telomeric region of chromosome 14) and the NP gene.

c) (i) The perfect concordance in hybrid clones with the large translocated chromosome, which has almost all of the long arm of the X chromosome, and the expression of the HGPRT, G6PD and PGK genes indicates that these genes are on the long arm of the X.
(ii) Since the second clone in C with a translocated chromosome containing only the distal region of the X expresses G6PD activity but not the potential of the other two genes, only gene G6PD is located in the distal segment of the q arm of the X. Since the fourth clone in C, with a translocated chromosome possessing ≈1/2 the q arm of the X expresses the information specified by genes G6PD and HGPRT but not that coded by PGK, the HGPRT gene must be somewhere in this region of the q arm but more proximal to the centromere than the G6PD locus.
(iii) The third clone in C, with most of the q arm of the X, expresses the activity specified by all three genes. Reasoning as before, it can be concluded that the PGK gene is the nearest of the three to the centromere. Therefore, the sequence of these three genes on the q arm of the X chromosome is: centromere - PGK - HGPRT - G6PD - telomere

7. a)

	Segment of chromosome 11 deleted		Chromosome regions (6)
Deletion	Breakpoint location(s)	Arm terminus	Cytogenetic markers
J1-23	p13 →	pter	1. pter → p13
J1-10	p1208 →	pter	2. p13 → p1208
J1-9	p1205 →	pter	3. p1208 → p1205
J1-7	p11 →	pter	4. p1205 → p11
J1-11	q13 →	qter	5. p11 → q13
			6. q13 → qter

Note: these five breakpoints divide chromosome 11 into six regions, each of which is characterized by specific cytological (cytogenetic) markers.

b) (i) Since the J1 strain, which has the entire chromosome 11, gives a positive reaction with all six genes, all of them must be on this chromosome.

(ii) (a) Since deletion strain J1-11 carries the SA11-1 and SA11-3 genes, they cannot be in the q13→qter region of the chromosome. Since individuals with the other four deletions do not carry these genes and the J1-23 is the shortest of the aberrations and common to all others, these two genes must be in region 1 (qter→p13).

(b) The LDH-A gene is present in strains J1-23 and J1-11 but absent from strains J1-10, J1-9 and J1-7. The J1-9 and J1-7 deletions have breakpoints at p1208 and p1205 respectively. The J1-9 deletion is the shorter of the two, therefore the LDH-A resides in the region between p1208 and p13.

(c) The only deletion strain that does not carry gene ACP-2 is J1-7. Since ACP-2 is present in J1-9 (breakpoint at p1205) but absent from J1-7 (breakpoint at p11), the gene is in the p1205→p11 region.

(d) Using the same type of reasoning gene SA11-2 can be assigned to the q13→qter region.

Note: Regions 3 (p1208→p1205) and 5 (p11→q13) do not carry any of these genes.

(e) The Hb β gene hybridized to DNA from clones J1-23, J1-11 and J1-10, indicating the presence of the human β globin gene in a segment of chromosome 11 in these clones. Therefore this gene is not present on the portion of the short arm deleted in J1-10. Hybridization did not occur between the Hb β gene and DNA from clones J1-9 and J1-7. These results indicate that the β globin gene is lost when the breakpoint of the deletion is at p1205 but it is present when the deletion in the p arm has its breakpoint at p1208. Therefore the β-globin gene must be located on the short arm of chromosome 11, in the region p1208→p1205.

c) The answer requires a lengthy discussion. Suffice it to say that an extension of these studies will permit a thorough if not complete mapping of every chromosome in humans. The benefit of this information with respect to diagnosis and detection of genes determining human disease, alleviation of the undesirable effects produced by these genes, prognosis and counselling of individuals and families with such genes is obvious.

a) Clones CF57-14 and CF57-1, which have the same translocation (9 pter→9q24::Xq12→Xqter), had the same restriction bands as total human genomic. Clone CF11-4, which possessed the 9pter→9q34::Xq13→qter translocation, did not have these bands. Therefore the segment of chromosome 9 that is translocated in clones CF57-14 and CF57-1 minus the region translocated in clone CF11-4 indicates the segment of chromosome 9 within which the CEL gene resides which is in the q arm between q24 and q34.

b) The Lamp 92 probe hybridized to DNA from clone CF11-4 and CF57-14. This indicates that this marker is proximal to the breakpoint of CF57-14 (9q24). The other three markers hybridized to DNA from clone CF57-14 containing 9p24→9qter but not to DNA from clone CF11-4 containing 9pter→9q34. Therefore these three markers are located between the two translocation breakpoints 9q24 and 9q34 leaving the Lamp 92 marker closest to the centromere.

We cannot determine the sequence of the three markers distal to Lamp 92 because we do not have other translocations with different breakpoints beyond 9q24.

(i) 3 bands - 5, 12 and 17 kb in length
(ii) 2 bands - 14 and 20 kb in length.
(iii) 4 bands - 5, 8, 9 and 12 kb in length.

11.

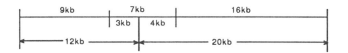

The 3 kb and 4 kb fragments are products of double digestion. They are produced by digesting the 7 kb *EcoRI* fragment with *BamHI*. The 9 kb and 16 kb EcoRI fragments are unaltered by *BamHI* because they do not possess *BamHI* sites.

12. The overlay is not correct because the double digest gel lacks a 25 bp fragment. Moreover, depending on the location of the *HindI* site in the *EcoRI* fragment, either the 150 bp or 125 bp fragment in the double-digest gel should not have been present.

13.

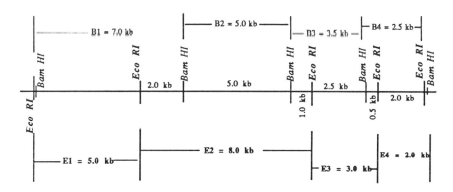

b) Fragments E_1 and E_3 would show hybridization to fragment B3 by Southern blotting. (There is an *EcoRI* site in B3 that is common to fragments E_1 and E_3).

14. a) Since the 5.6 and 3.2 kb fragments obtained by digestion of the *EcoRI* 23.7 fragment with *HpaI*, are also found within the single digest by *HpaI*, these fragments must be between the *EcoRI* sites of the 23.7 fragment. The 10 and 5.4 kb fragments are overlaps in the terminal positions. The 8.3 kb fragment does not have a *HpaI* restriction site within it because it is found both in the *EcoRI* and *HpaI* double digest and the *EcoRI* single digest. The 10 kb and 3.5 kb fragments are terminal overlaps. The 10 kb fragment is common to both the *EcoRI* (23.7 kb), *HpaI* and *HpaI* (20.6 kb) + *EcoRI* double digests. Therefore the restriction map of the restriction enzyme sites is as follows:

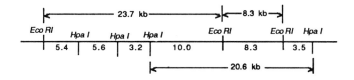

ambiguity - The sequence of the 3.2 and 5.6 kb fragments cannot be resolved without further information.
b) Yes it does. The order of the *HpaI* fragments is 5.4, 5.6, 3.2 and 10 kb. If 3.2 and 5.6 kb fragments were in the reverse sequence, the digestion of the 9.6 kb fragment with *HpaI* should have produced 5.4, 3.2 and 1.0 kb fragments.

15. (i)

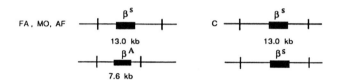

(ii) Assuming no crossing-over, the genotype of the fetus is $\beta^A \beta^S$ and it will have the *sickle-cell* trait.
(iii) Linkage must be close to reduce the probability of crossing-over between the DNA marker and the disease gene. The mating must be informative, that is, the parents should be heterozygous for the RFLP. Linkage phase in a family must be first established. This requires analysis of the parents and at least one child, or the grandparents or lateral relatives of the parents. Thus, this approach is rarely valuable for prenatal diagnosis in a first pregnancy unless a particularly large and informative pedigree is available. For disorders determined by X-linked genes, if there is only one *affected* male in a family, this may represent a new mutation. In this situation linkage with a RFLP may not be helpful.

17. In mutant A, a single base-pair change has occurred in the original 400 kb fragment region to produce a *XhoI* restriction site which is 275 bp to the right of the 75 bp end segment and 75 bp to the left of the 50 bp segment. Upon cleavage 125 and 175 bp fragments are produced instead of the 400 bp fragment. Mutant B has the *XhoI* site eliminated by a point mutation between the 75 bp and 400 bp segments resulting in a 475 bp fragment instead of two separate ones 75 bp and 400 bp in length. Since the DNA in both mutants A and B is normal in length the mutations in these mutants were single base-pair substitutions. The DNA in mutant C is 25 bp longer than in the normal strain and the duplication resides in the 400 bp segment. Unequal crossing-over resulting in a tandem duplication of a 25 bp segment is the most plausible explanation of this mutation.

18. Complete digests of the DNA would break the DNA duplex at all four restriction sites and produce 5 bands - 50 bp, 75 bp*, 175 bp 200 bp and 250 bp*. Partial digests may occur at any one site, any two sites or any three sites. The number and size of bands expected in each case is as follows:
At any one site - 1-75bp*, 675 bp* 2-250bp* and 500 bp* 3-325 bp*, 425 bp* 4-250 bp* and 500 bp*.

At any two sites - 1 and 2 - 75* 200, 475*; 1 and 3 - 75*, 250, 425*; 1 and 4- 75*, 425, 250*; 2 and 3-275*, 50, 425*; 2 and 4-275*, 225, 250*; 3 and 4- 325*, 175, 250*.

At any three sites - 1, 2 and 3 - 75*, 200, 50, 425*; 1, 2 and 4 - 75*, 200, 225, 250; 1,3 and 4 - 75*, 250, 275, 250*; 2,3 and 4- 275*, 50, 175, 250*.

Total number of bands (13) - 50, 75*, 175*, 200, 225, 250, 275*, 325*, 425*, 475* 500* 675*, 750.

20. Initially establish a genomic library of *wild-type N. crassa* DNA in a plasmid that is able to replicate in both *N. crassa* and *E. coli*. The plasmid library would consist of a population of plasmids each with a specific segment of the *N. crassa* genome (each with specific gene(s)) and collectively all segments (genes) from all chromosomes. Each plasmid would be present in a given *E. coli* cell; a large number of host cells should carry the entire plasmid library. Next, transform cells of the *tryptophanless* (*tryp⁻*) strain with plasmids of the library, plating the transformed cells on a medium lacking tryptophan. The cells that grow on this medium must possess $tryp^+$. That is cells with the *wild-type* allele are easily selected. Next, isolate the plasmids from a large population of transformed cells or their derivatives. The plasmids in this pure clone of *E. coli* cells should contain the *wild-type* allele at the *tryp* locus. The $tryp^+$ allele could then be isolated from the plasmids using standard accepted procedures.

22. Use reverse transcriptase which copies the mRNA base sequence of the structural gene, e.g., β-globin gene in humans, into DNA (cDNA), which can then be cloned. This is the procedure used for developing clones of many transcriptionally active genes in many different types of cells. Since the mRNA and not the primary transcript is cloned, such cDNA from discontinuous genes will not have any introns.

23. First make a cDNA library derived from mRNAs from sea urchin embryos. Use *E. coli* as the host for multiplication of the bacterial or phage vectors and their cDNAs. Then synthesize an antibody to the histone and use it as a probe in searching the cDNA library. The probe should be radioactively labelled and permit detection of bacterial or phage plaques and therefore vectors that produce the histones in the bacterial cells. Presence of histones will indicate that the cDNA genes coding for histones are present.

25. a) The viral DNA is linear. If it were circular with 4 *BamHI* sites, only four fragments would have been produced.
b) Cleave the aster virus DNA with *BamHI*. Separate the fragments by electrophoresis and then Southern blot them. Hybridize the blot with a radioactively labelled probe derived from the cloned cDNA of the *Cp* gene mRNA. Then obtain autoradiographs in the usual manner. THe DNA fragment that carries the *Cp* cDNA of the *Cp* gene will be easily detected by autoradiography. It is unlikely that the other fragments, will "light up".

27. a) **Rationale behind the dideoxy method of DNA sequencing:**
Nucleotides that are missing two hydroxyl groups on the sugar (in the sugars that are important in sequencing, the hydroxyl groups are missing at the 2' and 3' positions) are called 2',3' dideoxy nucleotides. Four such nucleotides can be generated since there are four kinds of bases A, C, G and T.
The procedure relies on the incorporation of these nucleotides into growing strands of DNA which is normal. However, the missing hydroxyl at the 3' carbon position, where the incoming (next) nucleotide should bind to the growing strand cannot do so because dideoxy nucleotides cannot form phosphodiester bonds with the nucleotide (deoxynucleoside-5'triphosphate). As a consequence the growth of the DNA chain is terminated at that nucleotide. For example, if a strand of DNA is being synthesized in the presence of guanine dideoxynucleotide, this nucleotide can be added normally to the growing strand but it creates a dead end. The strand elongation is terminated at guanine. Since

there are four kinds of 2'3'-dideoxynucleotides with A, C, G and T they can terminate chain growth at A, C, G and T respectively.

Protocol
Four reaction mixtures are established. Each contains DNA polymerase, all four normal deoxynucleotides (one or more of which are radioactively labelled) and a copy of the purified single-stranded DNA whose complementary copy, produced by DNA polymerase, is to be sequenced. Four tubes labelled A, C, G and T are set up, each containing the reaction mixture plus one of the nucleotides in dideoxy form in low concentration. For example, a tube labelled A will have normal amounts of all four deoxynucleotides plus the adenine dideoxynucleotide. During replication the deoxy form of will normally be incorporated in the new strand opposite T because there is more of it. Occasionally, the dideoxy form of the A nucleotide will be incorporated and will terminate replication. Therefore, in the A tube, dead ends on different new DNA strands after any and all adenine dideoxynucleotides incorporated into the growing strands will be produced. For example, assume the DNA template strand has the sequence 3'-ATGCTACT-5'. Let A, C, G and T represent the normal deoxynucleotides and A* represent the dideoxy one. During replication in the A tube, 3 molecules shown below will be formed:

3'-ATGCTACT-5'   3'-ATGCTACT-5'   3'-ATGCTACT-5'
5'-TA*-3'   5'-TACGA*-3'   5'-TACGATGA*-3'

When the DNA in tube A is denatured, in addition to the full length template strand, three other single-stranded DNA segments wil be present, all of the latter ending in adenine.

Tubes labelled C, G and T contain C, G and T dideoxynucleotides respectively and they will form populations of DNA segments ending, respectively in all the Cs, in all the Gs and all the Ts. Considered together, the four tubes possess all possible lengths of DNA: one nucleotide, two nucleotides and so on up to the total length of the DNA template. All the DNA segments can be separated and identified by gel electrophoresis.

In each lane the shortest DNA fragments will migrate the fastest and the farthest so that the fragment at the bottom of the gel is the first nucleotide added to the growing strand 5'→3'. The second shortest fragment will have the second fastest migration rate and will be next to the one at the bottom of the gel and so on. The fastest migrant is called position 1, the second fastest position 2, and so on. The sequence of nucleotides complementary to the template strand is read directly off an autoradiograph, an X-ray film which shows the positions of the radioactive bands in the gel.

b)(i) 5'-ACGAGCCGGAAGCATAAAGTGTAAGCCTGGGTGCGT
ATGAGTAGCTA-3'
(ii) Since each chain is formed 5'→3', the smallest fragment will be at the bottom of the gel as it will migrate the farthest, indicating the 5' end of the new strand.

28. a) Since FH is a rare disease and is not observed to skip generations, it must be determined by a dominant mutant allele at the LDL receptor locus. The mutant allele is incompletely dominant since heterozygotes are not as *severely affected* as the propositus. The locus is on one of the autosomes. If the gene was on the differential segment of the X all *affected* males should have expressed the extreme form of the disease.

b) The partial sequence of bases in exon four is given in the diagram.

c) A comparison of the base sequence in the *normal* and *mutant* alleles indicates they have identical codons except at position 167. The *normal* allele has the base sequence TAC at position 167 and the *mutant* allele possesses the sequence TAG; therefore the mutation must have involved a base substitution C→G or specifically a substitution of the CG bp by a GC one. The mutation changes a sense codon for tyrosine to a nonsense codon which leads to termination of polypeptide synthesis. As a consequence, incomplete and probably completely functionless proteins are produced.

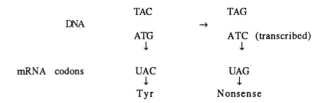

d) Because the propositus is homozygous for the *mutant* allele *LDL* receptors will not be produced. In the *mildly affected* heterozygous individual, the one *normal* allele produces sufficient receptor to keep *LDL-C* levels moderately low. Individuals homozygous for the *normal* alleles will produce *normal* kinds and amounts of the receptor, resulting in normal regulation of plasma cholesterol and therefore an absence of *FH* symptoms.

30. The atypical (mutant-affected) HaeII mtDNA cleavage pattern among the progeny of all matings is always identical to that of their mothers. Moreover, all the children of *affected* females are *affected* but none of the offspring of *affected* males express the mutant cleavage pattern. These facts indicate that the inheritance of mtDNA is maternal, via the cytoplasm of the egg and not the sperm. This is a case of cytoplasmic (=extranuclear), inheritance.

32. a) *Retinoblastoma* has an autosomal dominant mode of inheritance. **Reasons:** (i) it is rare in humans (ii) in the testcross-like pedigree it occurs equally frequently in both sexes and (iii) it does not skip generations. The most plausible explanation is that a single gene *Rb* is involved. Therefore in the pedigree presented (and probably in most if not all other cases), *affected* individuals are heterozygous (*Rbrb*) for the determining allele; *unaffected* individuals are homozygous (*rbrb*) for the recessive allele.

b) With respect to phenotypic expression of *retionblastoma* the data in (b) are at variance with the mode of inheritance presented in (a) because all the results in (b) indicate that eye-cancer occurs only when (i) there are two *mutant* alleles present at the *Rb* locus in an *affected* individual or (ii) one mutant allele is present on one chromosome 13 and the locus is deleted from the homologue of 13 or (iii) the locus is missing from both the homologues of chromosome 13. This incongruency can be explained as follows:

Only one Rb^- allele or deletion of the locus (-) is inherited by an *affected* individual from a carrier (heterozygous) parent. However, there is a very high probability that the normal allele Rb^+ derived from the other parent either mutates to Rb^- or the locus is deleted in primordial cells of at least one eye (*unilateral retinoblastoma*) but usually of both eyes (*bilateral retinoblastoma*), resulting in a Rb^-Rb^-, $Rb^-/-$ or $-/-$ individual with eye cancer. Thus, a recessive form of cancer shows dominant inheritance: an individual inherits a causative allele Rb^- or a chromosome 13 with a deletion of the Rb locus from the heterozygous parent and a normal dominant Rb^+ allele from the other parent; the likelihood of the latter mutating to Rb^- or being lost in primordial eye cells is so very high that the vast majority of individuals with one inherited recessive allele Rb^- develop eye cancer.

c) The data in b)v) indicate that the alleles at the *Rb* locus are involved in regulating cell division and cell growth. The Rb^+ allele is a growth suppressor; via its product it is able to limit the amount of cell division and cell growth that normally occurs. Since Rb^-Rb^-, *Rb*-/- and -/- individuals develop eye cancer; this indicates that the Rb^- allele has lost the ability to carry out the normal function of the *Rb* gene. It either produces no or little of the Rb^+ product or the product is highly defective. As a consequence there is no limitation on cell division. Cells in the eye likely divide and grow much more frequently than in normal individuals, leading to the formation of a malignant tumor.

33. The KEL locus is on chromosome 7, closely linked to the PIP locus on the q arm. **Evidence:** (i) The grandpaternal KEL2 allele is always transmitted with the PIP allele and the grandmaternal KEL1 allele is always traansmitted with her P1P2 allele. Therefore KEL2 is closely linked to P1P1 and KEL1 is closely linked to P1P2. The KEL locus must therefore be located on chromosome 7, and because of close linkage with P1P, it is most likely on the q arm. The family size is not large enough to determine whether linkage is very close or complete. The two genes could be on opposite sides of the centromere and still be completely linked, although this is highly unlikely.

Genotypes of individuals: ○□ - KEL2, PIP1/KEL2, PIP1; ◑◨ - KEL1, PIP2 / KEL2, PIP1.

All genotypes are parental.

38. The results indicate that topoisomerase II plays an important role in chromosome condensation. The mechanism by which it may do so is not indicated by the data provided and may not be currently known.

40. a) i) The *top2* gene and its enzyme are required for both chromosome condensation and separation of chromatids of mitotic chromosomes. This is evidenced by the results presented in b), d), e)(i).
(ii) The *nda3* gene and its enzyme are required for the synthesis of β-tubulin, a subunit of spindle fibers which are necessary for chromosome orientation at metaphase and separation of chromatids at anaphase. The evidence for this conclusion is presented in c), d), b)(i), (ii).

b) To maintain a culture of *ts top2-cs nda3* double mutant cells, one must use temperature shifts during the cell cycle. Specifically, grow the cells at 20 °C for the period of the cell cycle from the beginning of interphase to the end of prophase. The temperature is then shifted to 36 °C, the permissive temperature for *cs-nda3*, for a few minutes to allow spindle fibers to form. The temperature of the culture is then reduced to 20 °C to allow *ts top-2* to assist in the completion of mitosis.